Ergebnisse der Mathematik und ihrer Grenzgebiete

Band 60

Herausgegeben von

P. R. Halmos · P. J. Hilton · R. Remmert · B. Szőkefalvi-Nagy

Unter Mitwirkung von

L. V. Ahlfors · R. Baer · F. L. Bauer · R. Courant
A. Dold · J. L. Doob · S. Eilenberg · M. Kneser · G. H. Müller
M. M. Postnikov · B. Segre · E. Sperner

Geschäftsführender Herausgeber: P. J. Hilton

Shôichirô Sakai

C^*-Algebras and W^*-Algebras

Springer-Verlag New York Heidelberg Berlin
1971

Shôichirô Sakai
Professor of Mathematics, University of Pennsylvania

AMS Subject Classifications (1970): Primary 46 L 05, 46 L 10; Secondary 81 A 17

ISBN 0-387-05347-6 Springer-Verlag New York Heidelberg Berlin
ISBN 3-540-05347-6 Springer-Verlag Berlin Heidelberg New York

 Library of Congress Catalog Card Number 75-149121. Printed in Germany. Printing: Zechnersche Buchdruckerei, Speyer. Binding: Konrad Tritsch, Graphischer Betrieb, Würzburg.

To Masato and Kiyoshi

Preface

The theory of operator algebras in a Hilbert space was initiated by von Neumann [126] in 1929. In the introduction to the paper [119] in 1936, Murray and von Neumann stated that the theory seems to be important for the formal calculus with operator-rings, the unitary representation theory of groups, a quantum mechanical formalism and abstract ring theory. These predictions have been completely verified. Furthermore the theory of operator algebras is now becoming a common tool in a number of fields of mathematics and theoretical physics beyond those mentioned by Murray and von Neumann. This is perhaps to be expected, since an operator algebra is an especially well-behaved infinite-dimensional generalization of a matrix algebra. Therefore one can confidently predict that the active involvement of operator algebra theory in various fields of mathematics and theoretical physics will continue for a long time.

Another application of the theory is to the study of a single operator in a Hilbert space (see, for example, [99]). Nowadays we may add further the theory of singular integral operators and K-theory as other applications. Such diversifications of the theory of operator algebras have already made a unified text book concerning the theory virtually impossible. Therefore I have no intention of giving a complete coverage of the subject. I will rather take a somewhat personal stand on the selection of material—i.e., the selection is concentrated heavily on the topics with which I have been more or less concerned (needless to say, there are many other very important contributors to those topics. The reader will find the names of authors who have made remarkable contributions to those discussed in the concluding remarks of each section). Consequently parts of the book tend to be somewhat monographic in character.

Let me explain briefly about the contents. There are essentially two different ways of studying the operator *-algebras in Hilbert spaces. The first alternative is to assume that the algebra is weakly closed (called a W^*-algebra). These algebras are also called Rings of operators, and more recently, von Neumann algebras.

The earliest study along this line is due to von Neumann in 1929. In a series of five memoirs beginning with [119], Murray and von Neumann laid the foundation for the theory of W^*-algebras. Virtually all of the later work on these algebras is based directly or indirectly on their pioneering work.

We call a W^*-algebra a factor if its center is just the complex numbers. Murray and von Neumann concentrated most of their attention on factors. However, von Neumann [132] obtained a reduction theory by which the study of a general W^*-algebra may be to a large extent reduced to the case of a factor. At the same time a number of authors have pushed through the major portions of a global theory for general W^*-algebras (the reader may find a long list of papers by many authors in the bibliography in [37]).

The second alternative is to assume only that the algebra is uniformly closed (called a C^*-algebra). The earliest study along this line is due to Gelfand and Naimark [55] in 1943. A notable advantage of the C^*-algebra is the existence of an elegant system of intrinsic postulates, formulated by Gelfand and Naimark, which gives an abstract characterization of these algebras.

Using this approach, Segal [180] in 1947 initiated a study of C^*-algebras. The subsequent development which contains many beautiful results (cf. Chapters 1, 3, & 4) has been carried out by a number of authors.

The theory of C^*-algebras fits naturally into the theory of Banach algebras, and in certain respects they are among the best behaved examples of infinite dimensional Banach algebras.

In chapter 1, the characterization of W^*-algebras obtained in [149] is used to define W^*-algebras as abstract Banach algebras, like C^*-algebras, and to develop the abstract treatments of both W^*- and C^*-algebras.

Chapter 2 is concerned mainly with the classification and representation theory of W^*-algebras which are developed along classical standard lines. Some proofs may be new.

In chapter 3, the reduction theory is discussed. Here, a modern method, developed recently in [146], [159], [218] (i. e. the decomposition theory of states) is used. Also discussed are some recent results obtained by theoretical physicists.

Chapter 4 consists of some special topics from the theories of W^*-algebras and C^*-algebras. This chapter is the most personal in the book. All the topics covered are ones with which I have been more or less concerned. They are: Derivations and automorphisms on operator algebras; examples of factors; examples of non-trivial global W^*-algebras; type I C^*-algebras and a Stone-Weierstrass theorem for C^*-algebras.

I express sincere thanks to C. E. Rickart, S. Kakutani and I. E. Segal who made it possible for me to do research at Yale and M.I.T. during the various stages of the preparation of this book.

I express thanks to B. Sz. Nagy who invited me to write this book for the Ergebnisse series. Also the encouragement given to me by P. Hilton for the completion of the manuscript is much appreciated.

I thank R. Kallman and one of my students, Mrs. D. Laison, for their reading carefully the final manuscript.

My deepest thanks go to Miss P. Fay who with great patience and skill typed most of the manuscript.

I also wish to acknowledge the financial support which was given at various stages of the writing by the National Science Foundation (NSF–G–19041, NSFGP–5638, NSFGP–19845), John Simon Guggenheim Memorial Foundation and the University of Pennsylvania.

Finally I wish to express my appreciation to Springer-Verlag (especially, Dr. Klaus Peters) for their most efficient and understanding role in bringing this book to its completion.

Philadelphia, Pa. U.S.A., January 1971 S. Sakai

Contents

1. General Theory

1.1. Definitions of C^*-Algebras and W^*-Algebras 1
1.2. Commutative C^*-Algebras 3
1.3. Stonean Spaces 6
1.4. Positive Elements of a C^*-Algebra 7
1.5. Positive Linear Functionals on a C^*-Algebra 9
1.6. Extreme Points in the Unit Sphere of a C^*-Algebra 10
1.7. The Weak Topology on a W^*-Algebra 14
1.8. Various Topologies on a W^*-Algebra 19
1.9. Kaplansky's Density Theorem 22
1.10. Ideals in a W^*-Algebra 24
1.11. Spectral Resolution of Self-Adjoint Elements in a W^*-Algebra 26
1.12. The Polar Decomposition of Elements of a W^*-Algebra . . 27
1.13. Linear Functionals on a W^*-Algebra 28
1.14. Polar Decomposition of Linear Functionals on a W^*-Algebra 31
1.15. Concrete C^*-Algebras and W^*-Algebras 33
1.16. The Representation Theorems for C^*-Algebras and W^*-Algebras 40
1.17. The Second Dual of a C^*-Algebra 42
1.18. Commutative W^*-Algebras 45
1.19. The C^*-Algebra $C(\mathscr{H})$ of all Compact Linear Operators on a Hilbert Space $\mathscr{H}$ 46
1.20. The Commutation Theorem of von Neumann 48
1.21. *-Representations of C^*-Algebras, 1 50
1.22. Tensor Products of C^*-Algebras and W^*-Algebras 58
1.23. The Inductive Limit and Infinite Tensor Product of C^*-Algebras 70
1.24. Radon-Nikodym Theorems in W^*-Algebras 75

2. Classification of W^*-Algebras

2.1. Equivalence of Projections and the Comparability Theorem . 79
2.2. Classification of W^*-Algebras 83

2.3. Type I W^*-Algebras 87
2.4. Finite W^*-Algebras 89
2.5. Traces and Criterions of Types 95
2.6. Types of Tensor Products of W^*-Algebras 98
2.7. *-Representations of C^*-Algebras and W^*-Algebras, 2 . . . 102
2.8. The Commutation Theorem of Tensor Products 108
2.9. Spatial Isomorphisms of W^*-Algebras 111

3. Decomposition Theory

3.1. Decompositions of States (Non-Separable Cases) 121
3.2. Reduction Theory (Space-Free) 131
3.3. Direct Integral of Hilbert Spaces 137
3.4. Decomposition of States (Separable Cases) 140
3.5. Central Decomposition of States (Separable Cases) 146

4. Special Topics

4.1. Derivations and Automorphisms of C^*-Algebras and W^*-Algebras 153
4.2. Examples of Factors, 1 (General Construction) 171
4.3. Examples of Factors, 2 (Uncountable Families of Types II_1, II_∞ and III 183
4.4. Examples of Factors, 3 (Other Results and Problems) . . . 202
4.5. Global W^*-Algebras (Non-Factors) 216
4.6. Type I C^*-Algebras 219
4.7. On a Stone-Weierstrass Theorem for C^*-Algebras 236

Bibliography 243

Subject Index 251

List of Symbols 255

1. General Theory

1.1. Definitions of C^*-Algebras and W^*-Algebras

Let $\mathscr{A}$ be a linear associative algebra over the complex numbers. The algebra $\mathscr{A}$ is called a normed algebra if there is associated to each element x a real number $\|x\|$, called the norm of x, with the properties:

I $\|x\| \geq 0$ and $\|x\| = 0$ if and only if $x = 0$;
II $\|x+y\| \leq \|x\| + \|y\|$;
III $\|\lambda x\| = |\lambda| \, \|x\|$, λ a complex number;
IV $\|xy\| \leq \|x\| \, \|y\|$.

If $\mathscr{A}$ is complete with respect to the norm (i.e. if $\mathscr{A}$ is also a Banach space), then it is called a Banach algebra. A mapping $x \to x^*$ of $\mathscr{A}$ into itself is called an involution if the following conditions are satisfied

I $(x^*)^* = x$;
II $(x+y)^* = x^* + y^*$;
III $(xy)^* = y^* x^*$;
IV $(\lambda x)^* = \bar{\lambda} x^*$, λ a complex number.

An algebra with an involution * is called a *-algebra.

1.1.1. Definition. *A Banach *-algebra $\mathscr{A}$ is called a C^*-algebra if it satisfies $\|x^* x\| = \|x\|^2$ for $x \in \mathscr{A}$.*

1.1.2. Definition. *A C^*-algebra $\mathscr{M}$ is called a W^*-algebra if it is a dual space as a Banach space (i.e., if there exists a Banach space $\mathscr{M}_*$ such that $(\mathscr{M}_*)^* = \mathscr{M}$, where $(\mathscr{M}_*)^*$ is the dual Banach space of $\mathscr{M}_*$). We shall call such a Banach space $\mathscr{M}_*$ the predual of $\mathscr{M}$.*

Remark. It is not true in general that a dual Banach space is the dual space of a unique Banach space. For example, c^* and c_0^* are isometrically isomorphic to l^1, but c is not isometrically isomorphic to c_0, where c is the Banach space of all convergent sequences, c_0 the Banach space of all sequences convergent to zero, and l^1 is the Banach space of all summable sequences. However, we shall show later that a W^*-algebra is the dual space of a unique Banach space.

1.1.3. Definition. *The topology defined by the norm* $\|\ \|$ *on a* C^**-algebra* $\mathscr{A}$ *is called the uniform topology. The weak* $*$*-topology* $\sigma(\mathscr{M}, \mathscr{M}_*)$ *on a* W^**-algebra* $\mathscr{M}$ *is called the weak topology or the* σ*-topology on* $\mathscr{M}$.

1.1.4. Definition. *A subset* V *of a* C^**-algebra* $\mathscr{A}$ *is called self-adjoint if* $x \in V$ *implies* $x^* \in V$. *A self-adjoint, uniformly closed subalgebra of* $\mathscr{A}$ *is also a* C^**-algebra. It is called a* C^**-subalgebra of* $\mathscr{A}$. *A self-adjoint,* σ*-closed subalgebra* $\mathscr{N}$ *of a* W^**-algebra* $\mathscr{M}$ *is also a* W^**-algebra, because* $(\mathscr{M}_*/\mathscr{N}^0)^* = \mathscr{N}$, *where* $\mathscr{N}^0$ *is the polar of* $\mathscr{N}$ *in* $\mathscr{M}_*$. $\mathscr{N}$ *is called a* W^**-subalgebra of* $\mathscr{M}$.

1.1.5. Definition. *Let* $\{\mathscr{A}_\alpha\}_{\alpha \in \mathbb{I}}$ *be a family of* C^**-algebras. The direct sum* $\sum_{\alpha \in \mathbb{I}} \oplus \mathscr{A}_\alpha$ *of* $\{\mathscr{A}_\alpha\}_{\alpha \in \mathbb{I}}$ *is defined as follows: elements of* $\sum_{\alpha \in \mathbb{I}} \oplus \mathscr{A}_\alpha$ *are composed of all families* $(a_\alpha)_{\alpha \in \mathbb{I}}$ *such that* $a_\alpha \in \mathscr{A}_\alpha$ *and* $\sup_\alpha \|a_\alpha\| < +\infty$, *and the operations:* $\lambda(a_\alpha) = (\lambda a_\alpha)$ *(*λ *a complex number),* $(a_\alpha) + (b_\alpha) = (a_\alpha + b_\alpha)$, $(a_\alpha)(b_\alpha) = (a_\alpha b_\alpha)$, $(a_\alpha)^* = (a_\alpha^*)$ *and* $\|(a_\alpha)\| = \sup_\alpha \|a_\alpha\|$. $\sum_{\alpha \in \mathbb{I}} \oplus \mathscr{A}_\alpha$ *is again a* C^**-algebra.*

Let $\{\mathscr{M}_\alpha\}_{\alpha \in \mathbb{I}}$ be a family of W^*-algebras. The direct sum $\sum_{\alpha \in \mathbb{I}} \oplus \mathscr{M}_\alpha$ is a dual space since $\sum_{\alpha \in \mathbb{I}} \oplus \mathscr{M}_\alpha = \left(\sum_{\alpha \in \mathbb{I}} \oplus \mathscr{M}_{\alpha*}\right)^*_{l_1}$, *where the norm of an* element $(f_\alpha)(f_\alpha \in \mathscr{M}_{\alpha*})$ in $\left(\sum_{\alpha \in \mathbb{I}} \oplus \mathscr{M}_{\alpha*}\right)_{l^1}$ is defined by $\|(f_\alpha)\| = \sum_{\alpha \in \mathbb{I}} \|f_\alpha\|$; hence it is a W^*-algebra.

1.1.6. Lemma. *Let* $\mathscr{A}$ *be a* C^**-algebra; then* $\|x\| = \|x^*\|$ *for* $x \in \mathscr{A}$.

Proof. $\|x\|^2 = \|x^* x\| \leq \|x^*\|\,\|x\|$; hence $\|x\| \leq \|x^*\|$ and analogously $\|x^*\| \leq \|x^{**}\| = \|x\|$. q.e.d.

1.1.7. Proposition. *Let* $\mathscr{A}$ *be a* C^**-algebra without identity, and let* $\mathscr{A}_1$ *be the algebra obtained from* $\mathscr{A}$ *by adjoining the identity* 1. *For* $x \in \mathscr{A}$, λ *a complex number, define* $\|\lambda 1 + x\| = \sup_{\|y\| \neq 0} \dfrac{\|\lambda y + xy\|}{\|y\|}$. *Then* $\mathscr{A}_1$ *is a* C^**-algebra with norm* $\|\ \|$.

Proof. It is easy to see that $\mathscr{A}_1$ is a Banach $*$-algebra. Suppose μ is an arbitrary positive number less than 1. Then there exists a $y \in \mathscr{A}$ such that $\|y\| = 1$ and $\mu\|\lambda 1 + x\| < \|\lambda y + xy\|$.

Then,

$$\begin{aligned} \mu^2 \|\lambda 1 + x\|^2 < \|\lambda y + xy\|^2 &= \|(\lambda y + xy)^*(\lambda y + xy)\| \\ &= \|y^*(\lambda 1 + x)^*(\lambda 1 + x)y\| \leq \|(\lambda 1 + x)^*(\lambda 1 + x)\|. \end{aligned}$$

Hence $\mu^2\|\lambda 1+x\|^2 \leq \|(\lambda 1+x)^*(\lambda 1+x)\|$ and so

$$\|\lambda 1+x\|^2 \leq \|(\lambda 1+x)^*(\lambda 1+x)\|;$$

this implies $\|\lambda 1+x\|^2 \leq \|(\lambda 1+x)^*\|\,\|\lambda 1+x\|$ and so

$$\|\lambda 1+x\| \leq \|(\lambda 1+x)^*\|;$$

therefore $\|(\lambda 1+x)^*\| \leq \|(\lambda 1+x)^{**}\| = \|\lambda 1+x\|$; finally

$$\|\lambda 1+x\|^2 = \|(\lambda 1+x)^*(\lambda 1+x)\|. \qquad \text{q.e.d.}$$

Let $\mathscr{A}$ be a C^*-algebra with identity 1. The spectrum $\mathrm{Sp}(a)$ of an element a in $\mathscr{A}$ is the set of all complex numbers λ such that $x-\lambda 1$ is not invertible.

If $\mathscr{A}$ is a C^*-algebra without identity, the spectrum $\mathrm{Sp}(a)$ of an element a in $\mathscr{A}$ is the spectrum of a as an element of the C^*-algebra $\mathscr{A}_1$ obtained from $\mathscr{A}$ by adjoining the identity 1. In general, $\mathrm{Sp}(a)$ is a closed subset in the complex field.

An element a in $\mathscr{A}$ is called normal (resp. self-adjoint, unitary), if $a^*a=aa^*$ (resp. $a^*=a$, $a^*a=aa^*=1$).

A self-adjoint element e in $\mathscr{A}$ is called a projection if $e^2=e$. An element v is called a partial isometry if v^*v is a projection.

1.1.8. Proposition. $\mathrm{Sp}(ab)\cup(0)=\mathrm{Sp}(ba)\cup(0)$ *and* $\mathrm{Sp}(a^*)=\overline{\mathrm{Sp}(a)}$.

Proof. It is clear that $\mathrm{Sp}(a^*)=\overline{\mathrm{Sp}(a)}$. Suppose that $\lambda\neq 0$ and $ab-\lambda 1$ has the inverse u; then

$$\begin{aligned}(ba-\lambda 1)(bua-1) &= babua-ba-\lambda bua+\lambda 1\\ &= b(ab-\lambda 1)ua+\lambda bua-ba-\lambda bua+\lambda 1\\ &= ba-ba+\lambda 1=\lambda 1.\end{aligned}$$

Analogously, $(bua-1)(ba-\lambda 1)=\lambda 1$. Hence, $(ba-\lambda 1)$ is invertible and so $\mathrm{Sp}(ab)\cup(0)\supset \mathrm{Sp}(ba)\cup(0)$; proof of the reverse inclusion is analogous. q.e.d.

1.1.9. Corollary. *Let v be a partial isometry. Then vv^* is also a projection. v^*v is called the initial projection of v and vv^* is called the final projection* of v.

Remark. A C^*-algebra $\mathscr{A}$ can be defined by the following weaker condition: $\|x^*x\|=\|x^*\|\,\|x\|$ *for* $x\in\mathscr{A}$ (cf. [61], [136]).

References. [8], [17], [26], [28], [55], [124], [141], [149].

1.2. Commutative C^*-Algebras

Let $\mathscr{A}$, $\mathscr{B}$ be C^*-algebras, and let Φ be a homomorphism of $\mathscr{A}$ into $\mathscr{B}$ (i.e., $\Phi(\lambda x)=\lambda\Phi(x)$, $\Phi(x+y)=\Phi(x)+\Phi(y)$, $\Phi(xy)=\Phi(x)\Phi(y)$ for a complex number λ and $x, y \in \mathscr{A}$). Φ is called a *-homomorphism if $\Phi(x^*) = \Phi(x)^*$; Φ is called an isomorphism (resp. a *-isomorphism) if it is a one-to-one homomorphism (resp. *-homomorphism). We say that a C^*-algebra $\mathscr{A}$ is *-isomorphic to a C^*-algebra $\mathscr{B}$ if there exists a *-isomorphism of $\mathscr{A}$ onto $\mathscr{B}$.

Let K be a compact Hausdorff space, and let $C(K)$ be the algebra, under pointwise multiplication, of all complex valued, continuous functions on K. Define $\|a\| = \sup_{t \in K} |a(t)|$, and $a^*(t) = \overline{a(t)}$ for $a \in C(K)$. Then $C(K)$ is a commutative C^*-algebra.

1.2.1. Theorem. *Let $\mathscr{A}$ be a commutative C^*-algebra with identity. Then $\mathscr{A}$ is *-isomorphic to the C^*-algebra $C(K)$ of all complex valued, continuous functions on K, where K is the compact Hausdorff space of all maximal ideals of $\mathscr{A}$ (called the spectrum space of $\mathscr{A}$); therefore we can identify $\mathscr{A}$ with $C(K)$.*

Proof. By the commutative normed algebra theory of Gelfand (cf. [54]), there exists a homomorphism Φ of $\mathscr{A}$ into $C(K)$.

$$\begin{aligned}\|x^{2^n}\|^2 &= \|(x^{2^n})^* (x^{2^n})\| = \|(x^* x)^{2^n}\| \\ &= \|(x^* x)^{2^{n-1}}\|^2 = \cdots \\ &= \|x^* x\|^{2^n} = \|x\|^{2^{n+1}}.\end{aligned}$$

Hence $\|x\| = \lim_n \sqrt[2^n]{\|x^{2^n}\|} = \sup_{\lambda \in \mathrm{Sp}(x)} |\lambda| = \|\Phi(x)\|$. Therefore, Φ is isometric; hence it is one-to-one. Let h be a self-adjoint element in $\mathscr{A}$. Then $(\exp ih)^* = \exp(-ih)$ and so $(\exp ih)(\exp ih)^* = 1$. Hence

$$\|(\exp ih)^*\| \; \|\exp ih\| = 1;$$

also $\mathrm{Sp}((\exp ih)^*) = \overline{\mathrm{Sp}(\exp ih)}$; therefore $\lambda \in \mathrm{Sp}(\exp ih)$ implies $|\lambda| = 1$. Hence $\mathrm{Sp}(h)$ is a subset of the real numbers. Let $a = h + ik$ (h, k self-adjoint); then $\Phi(a^*) = \Phi(h - ik) = \Phi(h) - i\Phi(k)$ and so $\overline{\Phi(a)} = \Phi(a^*)$. q.e.d.

As a corollary.

1.2.2. Corollary. *Let $\mathscr{A}$ be a commutative C^*-algebra without identity. Then $\mathscr{A}$ is *-isomorphic to the C^*-algebra $C_0(\Omega)$ of all complex valued continuous functions which vanish at infinity on a locally compact Hausdorff space Ω (Ω is called the spectrum space of $\mathscr{A}$).*

1.2.3. Corollary. *Let $\mathscr{A}$ be a commutative C^*-algebra generated by a single self-adjoint element a, and let Ω be the spectrum space of $\mathscr{A}$. Then $\Omega \cup (\infty)$ is homeomorphic to $\mathrm{Sp}(a) \cup (0)$. Moreover, $\mathscr{A} = C_0(\mathrm{Sp}(a) \cup (0))$, where $C_0(\mathrm{Sp}(a) \cup (0))$ is the C^*-algebra of all complex valued continuous functions which vanish at 0 on the compact space $\mathrm{Sp}(a) \cup (0)$.*

Proof. Let Ω be the spectrum space of $\mathscr{A}$, and let $t \in \Omega$. Consider the mapping ξ of $\Omega \cup (\infty)$ onto $\mathrm{Sp}(a) \cup (0)$ defined by $\xi(t) = a(t)$ and $\xi(\infty) = 0$. ξ is one-to-one, since $\mathscr{A}$ is generated by a. ξ is continuous, and so it is a homeomorphism. q.e.d.

1.2.4. Theorem. *Let $\mathscr{A}$ be a commutative C^*-algebra, $\| \ \|_1$ another norm on $\mathscr{A}$ under which $\mathscr{A}$ is a normed algebra. Then $\|a\| \leq \|a\|_1$ for $a \in \mathscr{A}$.*

Proof. It suffices to prove the assertion in case $\mathscr{A}$ has an identity. Let $\mathscr{A} = C(K)$. Each point $t \in K$ will give a character of $\mathscr{A}$. Let K_1 be the set of all $\| \ \|_1$-continuous characters on $\mathscr{A}$; then $K_1 \subset K$. Let $\overline{K_1}$ be the closure of K_1 in K. Suppose that $\overline{K_1} \subsetneqq K$ and take an open set G such that $\overline{G} \subset \overline{K}_1^c$, where $\overline{K}_1^c$ is the complement of $\overline{K_1}$. Then there exists a continuous function b on K such that $b(t) = 1$ on $\overline{K_1}$ and $b(t) = 0$ on $\overline{G}$. Let $\tilde{\mathscr{A}}$ be the commutative Banach algebra obtained by the $\| \ \|_1$-completion of $\mathscr{A}$.

Suppose b is not invertible in $\tilde{\mathscr{A}}$; then there exists a $\| \ \|_1$-continuous character x_0 of $\tilde{\mathscr{A}}$ such that $b(x_0) = 0$. The restriction of x_0 on $\mathscr{A}$ will define a character t_0 belonging to K_1. Hence, $b(t_0) = 0$, a contradiction. Hence b is invertible in $\tilde{\mathscr{A}}$.

On the other hand, there exists an element c in $\mathscr{A}$ such that $c \neq 0$ and the support of $c \subset \overline{G}$; $bc = 0$ and so $c = 0$. Hence $\overline{K_1} = K$. Therefore $\|a\|_1 \geq \lim_n \sqrt[n]{\|a^n\|_1} = \sup_{t \in K_1} |a(t)| = \|a\|$. q.e.d.

1.2.5. Corollary. *Along with the assumptions of* 1.2.4, *suppose that $\|a^* a\|_1 = \|a\|_1^2$. Then $\|a\| = \|a\|_1$ for all $a \in \mathscr{A}$.*

Proof. $\|a\|_1 = \lim_n \sqrt[2^n]{\|a^{2^n}\|_1} = \sup_{t \in K_1} |a(t)| = \|a\|$. q.e.d.

1.2.6. Corollary. *Let $\mathscr{A}$, $\mathscr{B}$ be C^*-algebras and let Φ be a *-homomorphism of $\mathscr{A}$ into $\mathscr{B}$. Then $\|\Phi(x)\| \leq \|x\|$ for $x \in \mathscr{A}$. If Φ is a *-isomorphism, Φ is an isometry.*

Proof. Consider the direct sum $\mathscr{A} \oplus \mathscr{B}$ and the *-isomorphism Ψ of $\mathscr{A}$ into $\mathscr{A} \oplus \mathscr{B}$ defined as follows; $\Psi(a) = (a, \Phi(a))$ for $a \in \mathscr{A}$. Then

$$\|\Psi(a)\|^2 = \|\Psi(a)^* \Psi(a)\| = \|\Psi(a^* a)\| \leq \|a^* a\|$$

(by 1.2.5, since $\|b\|_1 = \max(\|b\|, \|\Psi(b)\|)$ for $b \in \mathscr{A}$ satisfies

$$\|b^* b\|_1 = \|b\|_1^2). \quad \text{q.e.d.}$$

Concluding remarks on 1.2

Theorem 1.2.1 is due to Gelfand [54]; Theorem 1.2.4 is due to Kaplansky [92].

References. [54], [92].

1.3. Stonean Spaces

Let K be a compact Hausdorff space. K is called a Stonean space if the closure of every open set is open.

1.3.1. Proposition. *Let K be a Stonean space. Then every element a in $C(K)$ can be uniformly approximated by finite linear combinations of projections.*

Proof. Let $a \in C(K)$. It suffices to assume that a is a positive function on K. For $\varepsilon > 0$, consider the finite division Δ of the interval

$$[0, \|a\| + 1]: 0 = \lambda_0 < \lambda_1 < \lambda_2 < \cdots < \lambda_n = \|a\| + 1$$

and $\lambda_{i+1} - \lambda_i < \varepsilon$ $(i = 1, 2, \ldots, n-1)$. By induction, we shall define a finite family $\{G_i\}_{i=1,2,\ldots,n}$ of open and closed sets.

$$G_1 = \text{the closure of } \{t \mid a(t) < \lambda_1, t \in K\},$$

$$G_i = \text{the closure of} \left\{ t \mid a(t) < \lambda_i,\ t \in K - \bigcup_{j=1}^{i-1} G_j \right\}$$

$(i = 2, 3, \ldots, n)$.

Then $G_i \cap G_j = (\emptyset)$ if $i \neq j$. Let x_j be the characteristic function of G_j; then it is continuous and a projection.

Moreover

$$\left| a(t) - \sum_{i=1}^{n} \lambda_i x_i(t) \right| < \varepsilon \quad \text{for } t \in K;$$

hence $\left\| a - \sum_{i=1}^{n} \lambda_i x_i \right\| < \varepsilon$. q.e.d.

1.3.2. Proposition. *Let K be a compact Hausdorff space. Suppose every bounded increasing directed set of real valued, non-negative functions $\{f_\alpha\}$ in $C(K)$ (i.e. $\alpha \leq \beta$ implies $f_\beta \leq f_\beta$, and $\|f_\alpha\| \leq k$ for all α with a fixed k) has a least upper bound in $C(K)$. Then K is Stonean.*

Remark. The converse is also true (cf. [192]).

Proof. Let G be an open set in K, and let (f_α) be the set of all continuous functions on K such that $0 \leq f_\alpha \leq 1$ and the support of $f_\alpha \subset G$. Then $\{f_\alpha\}$ is a bounded increasing directed set under the order $\leq$.

Put $f = \operatorname{l.u.b.}_\alpha f_\alpha$ in $C(K)$ and $g(t) = \operatorname{l.u.b.}_\alpha f_\alpha(t)$ for $t \in K$. Since $f_\alpha \leq f$, $g(t) \leq f(t)$ for all $t \in K$.

For $t \in G$, there exists an f_α such that $f_\alpha(t) = 1$; hence $g(t) = 1$ and so $f(t) = 1$ on G, since $f \wedge 1 \geq f_\alpha$ for all α, where $(f \wedge 1)(t) = \min(f(t), 1)$. Therefore $f(t) = 1$ on $\bar{G}$, where $\bar{G}$ is the closure of G. Suppose that there exists a $t_0 \varepsilon K - \bar{G}$ such that $f(t_0) > 0$. Take a positive h in $C(K)$ such that $h = 1$ on $\bar{G}$ and $h(t_0) = 0$. Then $f_\alpha \leq f \wedge h < f$, a contradiction. q.e.d.

References. [31], [192].

1.4. Positive Elements of a C^*-Algebra

Let h be a self-adjoint element of the C^*-algebra $\mathscr{A}$. h is said to be positive if $\mathrm{Sp}(h)$ is contained in the non-negative reals. We shall denote this by $h \geq 0$ or $0 \leq h$. Let $C_0(\Omega)$ be the commutative C^*-subalgebra of $\mathscr{A}$ generated by h. Then by 1.2.3, $\Omega \cup (\infty) = \mathrm{Sp}(h) \cup (0) \subset [0, \|h\|]$. Let $C_0([0, \|h\|])$ be the C^*-algebra of all complex valued continuous functions vanishing at 0 on the closed interval $[0, \|h\|]$. For $f \in C_0([0, \|h\|])$, let $f|\mathscr{A}$ be the restriction of f to $\mathrm{Sp}(h) \cup (0)$. Then $f|\mathscr{A} \in \mathscr{A}$. We shall denote $f|\mathscr{A}$ by $f(h)$. Clearly the mapping $\Phi: f \to f(h)$ of $C_0([0, \|h\|])$ into $\mathscr{A}$ is a $*$-homomorphism and $\Phi(C_0([0, \|h\|])) = C_0(\Omega)$, since an arbitrary complex valued continuous function g on the closed subset $\mathrm{Sp}(h) \cup (0)$ of $[0, \|h\|]$, which vanishes at 0, can be extended to an element of $C_0([0, \|h\|])$.

$$f(h) = h \quad \text{if } f(\lambda) = \lambda \quad \text{for } \lambda \in [0, \|h\|].$$

1.4.1. Proposition. *Let h be a positive element in $\mathscr{A}$. Then for an arbitrary positive integer n, there exists a unique positive element k in $\mathscr{A}$ such that $k^n = h$. This unique k is denoted by $\sqrt[n]{h}$ or $h^{1/n}$.*

Proof. In the above considerations, we shall consider a continuous function $f(\lambda) = \sqrt[n]{\lambda}$ on $[0, \|h\|]$. Then $f(h)^n = (f^n)(h) = h$. Now suppose that $k_1^n = h$ for some positive k_1 in $\mathscr{A}$. By the Stone-Weierstrass theorem, the function $\sqrt[n]{\lambda}$ can be uniformly approximated by polynomials p of λ such that $p(0) = 0$; hence $f(h)$ is a uniform limit of polynomials $p(h)$.

$$\text{Since } k_1 h = h k_1, \quad k_1 f(h) = f(h) k_1 .$$

Now let $C_0(\Omega)$ be the commutative C^*-subalgebra of $\mathscr{A}$ generated by k_1 and $f(h)$. Then $k_1(t), f(h)(t) \geq 0$ and $k_1(t)^n = f(h)(t)^n$ for $t \in \Omega$. Hence $k_1(t) = f(h)(t)$ for $t \in \Omega$. q.e.d.

1.4.2. Theorem. *Let P be the set of all positive elements in $\mathscr{A}$. Then P is a convex cone.*

Proof. We may assume that $\mathscr{A}$ has an identity. Let k be a self-adjoint element of $\mathscr{A}$ with $\|k\| \leq 1$.

By representing the commutative C^*-subalgebra of $\mathscr{A}$ generated by 1 and k as $C(K)$, we easily see that k is positive if and only if $\|1 - k\| \leq 1$. Clearly $k \in P$ implies $\lambda k \in P$ $(\lambda \geq 0)$. Let $k_1, k_2 \in P$; then

$$\left\| 1 - \frac{k_1 + k_2}{\|k_1\| + \|k_2\|} \right\| \leq \frac{\|k_1\| \, \|1 - k_1/\|k_1\|\| + \|k_2\| \, \|1 - k_2/\|k_2\|\|}{\|k_1\| + \|k_2\|} \leq 1 .$$

Hence $k_1 + k_2 \in P$. q.e.d.

Let $\mathscr{A}^s$ be the set of all self-adjoint elements in $\mathscr{A}$. By using P we can define a partial ordering $\leq$ in $\mathscr{A}^s$ as follows: $a \leq b$ (or $b \geq a$) if $b - a \in P$.

If $a \in \mathscr{A}^s$, we can easily see, by representing the C^*-subalgebra generated by a as a $C_0(\Omega)$, that a can be written as follows: $a = a_1 - a_2$ with $a_1 \cdot a_2 = 0$ and $a_1, a_2 \geq 0$. Moreover such a decomposition is unique—in fact, let $a = a_1' - a_2'$ where $a_1' \cdot a_2' = 0$ and $a_1', a_2' \geq 0$; then $(a^2)^{\frac{1}{2}} = a_1' + a_2' = a_1 + a_2$. Hence $\dfrac{a + (a^2)^{\frac{1}{2}}}{2} = a_1 = a_1'$ and so $a_2' = a_2$.

1.4.3. Definition. *The above decomposition $a = a_1 - a_2$ with $a_1 \cdot a_2 = 0$ and $a_1, a_2 \geq 0$ is called the orthogonal decomposition of a. a_1 (resp. $-a_2$) is called the positive (resp. negative) part of a, and a_1 (resp. a_2) is denoted by a^+ (resp. a^-). Moreover $a_1 + a_2$ is called the absolute value of a, denoted by $|a|$.*

1.4.4. Theorem. *The following conditions are equivalent:*

1. $h \geq 0$
2. $h = x^* x$ for some $x \in \mathscr{A}$.

Proof. 1. ⇒ 2. is clear.

2. ⇒ 1. We can assume that $\mathscr{A}$ has an identity.

Suppose $x^* x$ is not positive. Let $C(K)$ be the C^*-subalgebra of $\mathscr{A}$ generated by $x^* x$ and 1. There exists a point t_0 in K such that $x^* x(t_0) < 0$; hence there is a positive element a in $C(K)$ such that $a x^* x a < 0$. Let $x a = h_1 + i h_2$ $(h_1, h_2 \in \mathscr{A}^s)$. Then

$$(x a)^* (x a) = h_1^2 + h_2^2 + i h_1 h_2 - i h_2 h_1 \quad \text{and}$$
$$(x a)(x a)^* = h_1^2 + h_2^2 - i h_1 h_2 + i h_2 h_1 .$$

Hence $(x a)^* (x a) + (x a)(x a)^* = 2(h_1^2 + h_2^2) > 0$, and so

$$(x a)(x a)^* = -(x a)^* (x a) + 2(h_1^2 + h_2^2) > 0.$$

On the other hand, by 1.1.8, $\mathrm{Sp}((x a)^*(x a)) \cup (0) = \mathrm{Sp}((x a)(x a)^*) \cup (0)$; hence $(x a)(x a)^* \leq 0$, a contradiction. q.e.d.

1.4.5. Proposition. *Let $\mathscr{A}$ be a C^*-algebra with identity. Then every element of $\mathscr{A}$ is a finite linear combination of unitary elements.*

Proof. Let h be a positive element in $\mathscr{A}$; then $\dfrac{h}{\|h\|} \pm i\sqrt{1 - h^2/\|h\|^2}$ are unitary elements. Hence h is a finite linear combination of unitary elements. q.e.d.

Concluding remarks on 1.4.

Theorem 1.4.2 is due to Fukamiya [52] and Kelley and Vaught [101]. Theorem 1.4.4 is due to Kaplansky [175]. These results are the key lemmas which gave the positive answer to the non-commutative case of Gelfand-Naimark's query after a decade of mystery (cf. Concluding remarks in 1.16).

References. [52], [101].

1.5. Positive Linear Functionals on a C^*-Algebra

Let f be a linear functional on a C^*-algebra $\mathscr{A}$. Define $f^*(x)=\overline{f(x^*)}$ for $x\in\mathscr{A}$. Then f^* is also a linear functional on $\mathscr{A}$. f^* is called the adjoint of f. If $f^*=f$, we call f self-adjoint.

Every linear functional on $\mathscr{A}$ can be represented in the form $f=f_1+if_2$, where f_1, f_2 are self-adjoint. In fact, $f_1=1/2(f+f^*)$, $f_2=1/2i(f-f^*)$.

A linear functional φ on $\mathscr{A}$ is called positive if $\varphi(x^*x)\geq 0$ for $x\in\mathscr{A}$. φ is automatically self-adjoint.

We shall denote the positivity of φ by $\varphi\geq 0$ (or $0\leq\varphi$). For two self-adjoint linear functionals f_1, f_2, we shall denote $f_2-f_1\geq 0$ by $f_2\geq f_1$ (or $f_1\leq f_2$).

A positive linear functional φ satisfies the Schwartz inequality: $|\varphi(y^*x)|^2\leq\varphi(x^*x)\varphi(y^*y)$ for $x,y\in\mathscr{A}$.

1.5.1. Proposition. *Let $\mathscr{A}$ be a C^*-algebra with identity. Then a positive linear functional φ is bounded and $\|\varphi\|=\varphi(1)$.*

Proof. $|\varphi(x)|\leq\varphi(1)^{\frac{1}{2}}\varphi(x^*x)^{\frac{1}{2}}$, and $\|x^*x\|1-x^*x\geq 0$; hence $\varphi(x^*x)\leq\|x^*x\|\varphi(1)$ and so $|\varphi(x)|\leq\varphi(1)\|x\|$. q.e.d.

1.5.2. Proposition. *Let $\mathscr{A}$ be a C^*-algebra, and let f be a bounded linear functional on $\mathscr{A}$ such that $f(h)=\|f\|\,\|h\|$ for some positive $h(>0)\in\mathscr{A}$. Then f is a positive linear functional.*

Proof. By the Hahn-Banach theorem we can assume that $\mathscr{A}$ has an identity. Put $f_0=\dfrac{f}{\|f\|}$ and $h_0=\dfrac{h}{\|h\|}$. Then $f_0(h_0)=1$. First of all we shall show $f_0(1)=$ real. Suppose $f_0(1)=\alpha+i\beta$ (α,β reals, $\beta\neq 0$). Then for an arbitrary real λ, we have

$$|f_0(1+\lambda i h_0)|=|\alpha+i(\lambda+\beta)|\geq|\lambda+\beta|.$$

On the other hand,

$$|f_0(1+\lambda i h_0)|\leq\|1+\lambda i h_0\|\leq(1+\lambda^2)^{\frac{1}{2}}.$$

Hence

$$|\lambda+\beta|\leq(1+\lambda^2)^{\frac{1}{2}}\quad\text{for all real }\lambda,$$

and so

$$\lambda^2+2\lambda\beta+\beta^2\leq 1+\lambda^2\quad\text{for all real }\lambda.$$

This is a contradiction. Hence $f_0(1)$ is real.

Suppose $f_0(1)<1$; then

$$|f_0(1-2h_0)|=|f_0(1)-2|>1$$

and $|f_0(1-2h_0)|\le\|1-2h_0\|=\|(1-h_0)-h_0\|\le 1$, a contradiction. Hence $f_0(1)\ge 1$ and so $f_0(1)=1$.

Now, let a be a self-adjoint element of $\mathscr{A}$ and suppose $f(a)=\alpha+i\beta$ (α,β reals, $\beta\ne 0$). For an arbitrary real λ, we have

$$|f_0(a+i\lambda 1)|\ge|\beta+\lambda|$$

and $|f_0(a+i\lambda 1)|\le\|a+i\lambda 1\|=(\|a\|^2+\lambda^2)^{\frac{1}{2}}$. This is a contradiction, and so $f(a)$ is real. Let $h\ge 0$, $\|h\|\le 1$ and suppose $f(h)<0$; then $f(1-h)=1-f(h)>1$. On the other hand, $|f(1-h)|\le\|1-h\|\le 1$, a contradiction. q.e.d.

1.5.3. Definition. *A positive linear functional φ on a C^*-algebra is called a state if* $\|\varphi\|=1$.

Let $\mathscr{S}$ be the set of all states on a C^*-algebra $\mathscr{A}$. We can define a topology $\sigma(\mathscr{A}^*,\mathscr{A})$ on $\mathscr{S}$, where $\mathscr{A}^*$ is the dual Banach space of $\mathscr{A}$.

If $\mathscr{A}$ has an identity, one can easily see that $\mathscr{S}$ is a compact space.

1.5.4. Proposition. *Let h be a self-adjoint element in a C^*-algebra $\mathscr{A}$. Then* $\|h\|=\sup_{\varphi\in\mathscr{S}}|\varphi(h)|$.

Proof. Put $h=h^+-h^-$; then $\|h\|=\max(\|h^+\|,\|h^-\|)$. It is enough to assume that $\|h\|=\|h^+\|$ (otherwise, consider $-h$). By the Hahn-Banach theorem, there exists a bounded linear functional f on $\mathscr{A}$ such that $f(h^+)=\|h^+\|$ and $\|f\|=1$. By 1.5.2, $f\ge 0$ and

$$f(h)=f(h^+)-f(h^-)=f(h^+)=\|h\|,$$

since $\|h^++h^-\|=\|h^+\|$ and $f(h^++h^-)=f(h^+)+f(h^-)\le\|h^+\|$. q.e.d.

1.5.5. Corollary. *Suppose $h,k\ge 0$. Then* $\|h+k\|\ge\max(\|h\|,\|k\|)$.

Reference. [12].

1.6. Extreme Points in the Unit Sphere of a C^*-Algebra

1.6.1. Proposition. *Let $\mathscr{A}$ be a C^*-algebra and let S be its unit sphere. Then S has an extreme point if and only if $\mathscr{A}$ has an identity.*

Proof. Suppose $\mathscr{A}$ has an identity. We shall show that 1 is an extreme point in S. If $1=(a+b)/2$ $(a,b\in S)$, put $c=(a^*+a)/2$ and $d=(b+b^*)/2$. Then $1=(c+d)/2$ $(c,d\in S)$. Since $d=21-c$, d commutes with c.

Representing the C^*-algebra generated by $1, c, d$, we can easily see that $d = c = 1$. Hence $a^* = 21 - a$, so that a is normal. Hence $a = a^* = 1$, and 1 is an extreme point.

Conversely, suppose x is an extreme point in S. Let $C_0(\Omega)$ be the C^*-subalgebra generated by x^*x. Then, we can take a sequence $\{y_n\}$ of positive elements in $C_0(\Omega)$ such that $\|y_n\| \le 1$ for all n, $\|(x^*x)y_n - (x^*x)\| \to 0$ $(n \to \infty)$ and $\|(x^*x)y_n^2 - (x^*x)\| \to 0$ $(n \to \infty)$.

Suppose that at some point t of Ω, x^*x takes a non-zero value less than one. Then we can take a positive element c of $C_0(\Omega)$, non-zero at t such that $\gamma_n = y_n + c$, $s_n = y_n - c$, $\|(x^*x)\gamma_n^2\| \le 1$ and $\|(x^*x)s_n^2\| \le 1$. Hence $x\gamma_n$ and xs_n are in S.

On the other hand,

$$\|(xy_n - x)^*(xy_n - x)\| = \|x^*xy_n^2 - x^*xy_n - x^*xy_n + x^*x\| \to 0 \quad (n \to \infty).$$

Hence $xy_n \to x$, so that $x\gamma_n \to x + xc$ and $xs_n \to x - xc$.

Since $x + xc, x - xc \in S$ and $x = \dfrac{(x+xc)+(x-xc)}{2}$, $x = x + xc = x - xc$.

Hence $xc = 0$ and so $\|cx^*xc\| = \|x^*xc^2\| = 0$. This is a contradiction, because $x^*x(t)c^2(t) \neq 0$.

Therefore, x^*x has no non-zero value less than one on Ω. In other words, x^*x is a projection.

Put $x^*x + xx^* = h$, and let B be a maximal commutative C^*-subalgebra of $\mathscr{A}$ containing h. Suppose h is not invertible in B. Then there exists a sequence $\{z_n\}$ of positive elements belonging to B which satisfies $\|z_n^2\| = 1$, $\|hz_n^2\| \to 0$ $(n \to \infty)$. Hence,

$$\|xz_n\| = \|z_nx^*\| = \|z_nx^*xz_n\|^{\frac{1}{2}} \le \|z_nhz_n\|^{\frac{1}{2}} \to 0 \ (n \to \infty),$$

and analogously $\|z_nx\| = \|x^*z_n\| \to 0$ $(n \to \infty)$. Therefore,

$$\|z_n - xx^*z_n - z_nx^*x + xx^*z_nx^*x\| \to 1 \ (n \to \infty).$$

Now we use the symbolic notation: $y(1-x) = y - yx$, $(1-x)y = y - xy$.

We shall show that $(1 - xx^*)\mathscr{A}(1 - x^*x) = (0)$. Suppose

$$a \in (1 - xx^*)\mathscr{A}(1 - x^*x)$$

and $\|a\| \le 1$. Then

$$\|x \pm a\| = \|(x^* \pm a^*)(x \pm a)\|^{\frac{1}{2}} = \|x^*x \pm (x^*a + a^*x) + a^*a\|^{\frac{1}{2}}.$$

Since $a^*xx^*a = 0$, $x^*a = a^*x = 0$ and $x^*xa^*a = x^*x(1 - x^*x)a^*a = 0$. Hence $\|x \pm a\| = \max(\|x^*x\|^{\frac{1}{2}}, \|a^*a\|^{\frac{1}{2}}) \le 1$, so that by the extremity of x, $a = 0$.

On the other hand,

$$z_n - xx^* z_n - z_n x^* x + xx^* z_n x^* x \in (1 - xx^*)\mathscr{A}(1 - x^* x);$$

hence it is zero, a contradiction.

Therefore h is invertible in B. $h^{-1}h$ is the identity of B, and so it is a projection in $\mathscr{A}$ and the identity of $h^{-1}h\mathscr{A}h^{-1}h$.

Suppose $\mathscr{A}(1-h^{-1}h) \neq (0)$. Then there exists an element a $(\neq 0)$ in $\mathscr{A}(1-h^{-1}h)$. Since $a^*ah^{-1}h=0$, a^*a commutes with $h^{-1}h\mathscr{A}h^{-1}h \supset B$. This contradicts the maximality of B. Hence $h^{-1}h$ is the identity of $\mathscr{A}$. q.e.d.

1.6.2. Proposition. *Let P be the set of all positive elements of a C^*-algebra $\mathscr{A}$. The extreme points of $P \cap S$ are all projections of $\mathscr{A}$.*

Proof. Let e be a projection, and let $e=(a+b)/2$ $(a,b \in P \cap S)$. Then $a=2e-b$; since $0 \leq a \leq 2e$, $a \in e\mathscr{A}e$ and therefore a commutes with b.

By representing the commutative C^*-algebra generated by a and b as a $C(K)$, we can easily see $a=b=e$.

Conversely, suppose h is extreme in $P \cap S$. By using the function space representation of the C^*-algebra generated by h, we can easily see that h is a projection. q.e.d.

1.6.3. Proposition. *Let $\mathscr{A}^s$ be the self-adjoint portion of $\mathscr{A}$. The extreme points of $\mathscr{A}^s \cap S$ are all self-adjoint, unitary elements of $\mathscr{A}$.*

Proof. Let u be a self-adjoint unitary element in $\mathscr{A}$. Then clearly it is extreme in S since the mapping $x \to ux$ $(x \in \mathscr{A})$ is a linear isometry and 1 is extreme.

Suppose $a=a^+ - a^-$ is an extreme point in $\mathscr{A}^s \cap S$. Then a^+ and a^- must be extreme in $P \cap S$. In fact, let $a^+ = \dfrac{h+k}{2}$ $(h, k \in P \cap S)$; then $h-a^-, k-a^- \in \mathscr{A}^s \cap S$ and $a=\dfrac{h-a^- + k-a^-}{2}$. Hence $a=h-a^- = k-a^-$ and so $h=k=a^+$; analogously we can show that a^- is extreme in $P \cap S$. Hence, a^+ and a^- are projections by 1.6.2. Therefore, we can easily see that $a^+ + a^- = 1$. q.e.d.

1.6.4. Theorem. *An element x in S is extreme if and only if*

$$(1-xx^*)\mathscr{A}(1-x^*x)=(0).$$

Proof. We have proved already that $(1-xx^*)\mathscr{A}(1-x^*x)=(0)$ if x is extreme (cf. the proof of 1.6.1).

Conversely, suppose that $(1-xx^*)\mathscr{A}(1-x^*x)=(0)$ and $x \in S$. Let $x=(a+b)/2$ $(a,b \in S)$. $x^*(1-xx^*)x(1-x^*x)=x^*x(1-x^*x)^2=0$. Hence $\mathrm{Sp}(x^*x) \subset \{0,1\}$ and so x^*x is a projection. $\mathrm{Sp}(xx^*) \subset \{0,1\}$ by 1.1.8 and xx^* is also a projection.

Put $e=x^*x$ and $f=xx^*$. Then

$$\begin{aligned}(ex^*-x^*)(ex^*-x^*)^*&=(ex^*-x^*)(xe-x)\\&=ex^*xe-ex^*x-x^*xe+x^*x\\&=e-e-e+e=0\,.\end{aligned}$$

Hence $ex^*=x^*xx^*=x^*$, $xx^*x=x$ and $xe=x$.

$$e=x^*x=\frac{(x^*a+x^*b)}{2}=\frac{(x^*ae+x^*be)}{2}.$$

Since $x^*ae, x^*be\in eSe$, $e=x^*ae=x^*be$. Hence

$$x=xx^*x=xe=xx^*ae=fae=fbe.$$

On the other hand, $ae=fae+(1-f)ae$. Therefore

$$\begin{aligned}1&\geq\|ea^*ae\|=\|(fae+(1-f)ae)^*(fae+(1-f)ae)\|\\&=\|ea^*fae+ea^*(1-f)ae\|=\|e+ea^*(1-f)ae\|\,.\end{aligned}$$

Hence $(1-f)ae=0$ and so $ae=fae$. Analogously $be=fbe$ and so $x=ae=be$.

Now consider $x^*=(a^*+b^*)/2$. By symmetry, $a^*f=b^*f=x^*$ or $fa=fb=x$. Since $(1-f)\mathscr{A}(1-e)=(0)$,

$$a=fa(1-e)+ae=fb(1-e)+be=b. \qquad \text{q.e.d.}$$

1.6.5. Proposition. *Let p, q be projections in a C^*-algebra $\mathscr{A}$. An element x in the unit sphere of $q\mathscr{A}p$ is extreme if and only if*

$$(q-xx^*)\mathscr{A}(p-x^*x)=(0).$$

If x is extreme in the unit sphere of $q\mathscr{A}p$, then x is a partial isometry.

Proof. Suppose x is extreme in $S\cap q\mathscr{A}p$ and x^*x is not a projection. Then there exists a positive element c in $p\mathscr{A}p$ commuting with x^*x such that $x^*xc\neq 0$ and $\|x^*x(p\pm c)^2\|\leq 1$. This contradicts the extremity of x. Hence x^*x is a partial isometry. Put $x^*x=e$ and $xx^*=f$. We shall show that $(q-f)\mathscr{A}(p-e)=(0)$.

Suppose $a\in(q-f)\mathscr{A}(p-e)$ with $\|a\|=1$. Then

$$\begin{aligned}\|x\pm a\|&=\|(x^*\pm a^*)(x\pm a)\|^{\frac{1}{2}}=\|x^*x+a^*a\|^{\frac{1}{2}}\\&=\max\{\|x^*x\|^{\frac{1}{2}},\|a^*a\|^{\frac{1}{2}}\}\leq 1\end{aligned}$$

and $x=\dfrac{\{(x+a)+(x-a)\}}{2}$, a contradiction.

The proof of the converse is quite similar to the proof of 1.6.4.

1.6.6. Proposition. *Let E be a Banach algebra with identity* 1. *Then* 1 *is an extreme point in its unit sphere.*

Proof. Since E is isometrically representable as a subalgabra of the algebra $B(F)$ of all bounded linear operators on a Banach space F, it suffices to prove that the identity operator on F is extreme in the unit sphere of $B(F)$.

Let F^* be the dual of F; then $\|1 \pm a\| \leq 1$ implies $\|1^* \pm a^*\| \leq 1$. For any $f \in F^*$, put $f_1 = (1^* + a^*)f$ and $f_2 = (1^* - a^*)f$. Then $2f = f_1 + f_2$ and $\|f_1\| \leq \|f\|$, $\|f_2\| \leq \|f\|$. Therefore if f is an extreme point of the unit sphere of F^*, $f = f_1 = f_2$. Hence $a^* f = 0$ and so $a^* = 0$ and $a = 0$.

Now let $1 = \dfrac{b+c}{2}$ ($\|b\|, \|c\| \leq 1$) and put $a = 1 - b$. Then $1 - a = b$ and $1 + a = 21 - b = c$. Hence by the above considerations, $b = 1$ and so $c = 1$. q.e.d.

Remark 1. The converse of 1.6.6 is negative. For example, let $L^2(G)$ be the Hilbert space of all square integrable complex valued functions, with respect to the Haar measure, on an infinite compact group. Then $L^2(G)$ is a Banach algebra under the convolution multiplication and its unit sphere has many extreme points. But it does not have an identity.

Remark 2. 1.6.6 is due to Kakutani.

Concluding remark on 1.6.
Theorem 1.6.4 is due to Kadison [81] in case that $\mathscr{A}$ has an identity.
References. [81], [154].

1.7. The Weak Topology on a W^*-Algebra

Let $\mathscr{M}$ be a W^*-algebra—viz. a C^*-algebra which is the dual Banach space of a Banach space $\mathscr{M}_*$. Let S be the unit sphere of $\mathscr{M}$, $\mathscr{M}^s$ the self-adjoint portion of $\mathscr{M}$, P the positive portion of $\mathscr{M}^s$. In this section, we shall always use the $\sigma(\mathscr{M}, \mathscr{M}_*)$-topology on $\mathscr{M}$, the weak topology or the σ-topology on $\mathscr{M}$. It is well known that S is σ-compact. By the Krein-Milman theorem, there exist extreme points in S. Hence $\mathscr{A}$ has an identity.

1.7.1. Lemma. *$\mathscr{M}^s$ and P are σ-closed.*

Proof. We shall show first that $\mathscr{M}^s \cap S$ is closed. If $\mathscr{M}^s \cap S$ is not closed, there exists a directed set $\{x_\alpha\}$ in $\mathscr{M}^s \cap S$ which converges to an element $a + ib$ $(b \neq 0)$ $(a, b \in \mathscr{M}^s)$.

Suppose that there exists a positive number $\lambda > 0$ in the spectrum of b (otherwise, consider $\{-x_\alpha\}$). Then

$$\|x_\alpha + in1\| \leq (1+n^2)^{\frac{1}{2}} < \lambda + n \leq \|b + n1\| \leq \|a + ib + in1\|$$

for some large positive number n.

Since $\{x_\alpha + in1\}$ converges to $a+ib+in1$ and belongs to $(1+n^2)^{\frac{1}{2}} S$, the compactness of $(1+n^2)^{\frac{1}{2}} S$ implies $a+ib+in1 \in (1+n^2)^{\frac{1}{2}} S$. This contradicts the above inequality. Hence $\mathscr{M}^s \cap S$ is closed, so that by the Banach-Smulian theorem ([17]), $\mathscr{M}^s$ is closed.

Since $P \cap S \subset \mathscr{M}^s \cap S + 1 \subset P$, we have

$$P \cap S = \{\mathscr{M}^s \cap S\} \cap \{\mathscr{M}^s \cap S + 1\}.$$

Hence $P \cap S$ is closed, so that by the Banach-Smulian theorem, P is closed. q.e.d.

1.7.2. Lemma. *Let T be the set of all σ-continuous positive linear functionals on $\mathscr{M}$. Then for any self-adjoint element $a \notin P$, there exists an element φ in T such that $\varphi(a) < 0$. In particular, if $b \in \mathscr{M}$ such that $\psi(b) = 0$ for all $\psi \in T$, then $b = 0$.*

Proof. P is a σ-closed convex cone in the real locally convex space $\mathscr{M}^s$. Hence by the separation theorem, there exists a σ-continuous real linear functional g on $\mathscr{M}^s$ such that $\inf_{h \in P} g(h) > g(a)$.

Since P is conic, $\inf_{h \in P} g(h) = 0$. Hence $g(a) < 0$ and $g(h) \geq 0$ for $h \in P$.

Define $\varphi(a+ib) = g(a) + ig(b)$ for $a, b \in \mathscr{M}^s$. Then φ is a linear functional on $\mathscr{M}$. Moreover, since $\mathscr{M}^s$ is closed, the *-operation is σ-continuous. Hence φ is a σ-continuous positive linear functional such that $\varphi(a) < 0$. The remainder of the proof is clear. q.e.d.

1.7.3. Definition. *We call a directed set $\{x_\alpha\}$ in $\mathscr{M}^s$ increasing, if $x_\alpha \geq x_\beta$ whenever $\alpha \geq \beta$.*

1.7.4. Lemma. *Every uniformly bounded, increasing directed set converges to its least upper bound. Further, if $x = \text{l.u.b.}_\alpha\, x_\alpha$, then $a^* x a = \text{l.u.b.}_\alpha\, a^* x_\alpha a$.*

Proof. Let E be the set of all finite linear combinations of elements in T. It is clear that the topology $\sigma(\mathscr{M}, E)$ is weaker than $\sigma(\mathscr{M}, \mathscr{M}_*)$, and $\sigma(\mathscr{M}, E)$ is a Hausdorff topology. Since S is $\sigma(\mathscr{M}, \mathscr{M}_*)$-compact, $\sigma(\mathscr{M}, E)$ is equivalent to $\sigma(\mathscr{M}, \mathscr{M}_*)$ on bounded spheres. Therefore, to show that a uniformly bounded directed set $\{x_\alpha\}$ is a Cauchy directed set in $\sigma(\mathscr{M}, \mathscr{M}_*)$, it is enough to show that for any $\varphi \in T$ and positive $\varepsilon > 0$, there exists an index α_0 such that $|\varphi(x_\alpha - x_\beta)| \leq \varepsilon$ for $\alpha, \beta \geq \alpha_0$.

If $\{x_\alpha\}$ is a uniformly bounded increasing directed set of self-adjoint elements of $\mathscr{M}$, then for every $\varphi \in T$, $\{\varphi(x_\alpha)\}$ is a uniformly bounded increasing directed set of real numbers. Hence $\{x_\alpha\}$ is $\sigma(\mathscr{M}, \mathscr{M}_*)$-Cauchy so that by compactness of S, it converges to some element x. Moreover by 1.7.2, $x = \text{l.u.b.}_\alpha \{x_\alpha\}$. If u is an invertible element in $\mathscr{M}$, then clearly $\text{l.u.b.}_\alpha \{u^* x_\alpha u\} = u^* \{\text{l.u.b.}_\alpha\, x_\alpha\} u = u^* x u$.

If a is an arbitrary element of $\mathcal{M}$, there exists a suitable number $\lambda>0$ such that $\lambda 1+a$ is invertible. Then

$$\varphi((\lambda 1+a)^* x_\alpha(\lambda 1+a))=\lambda^2 \varphi(x_\alpha)+\lambda\varphi(a^* x_\alpha)$$
$$+\lambda\varphi(x_\alpha a)+\varphi(a^* x_\alpha a)\to\varphi((\lambda 1+a)^* x(\lambda 1+a))$$

for $\varphi\in T$.

On the other hand,

$$|\varphi(a^*(x_\alpha-x_\beta))|=|\varphi(a^*(x_\alpha-x_\beta)^{\frac{1}{2}}(x_\alpha-x_\beta)^{\frac{1}{2}})|$$
$$\leq\varphi(a^*(x_\alpha-x_\beta)a)^{\frac{1}{2}}\varphi(x_\alpha-x_\beta)^{\frac{1}{2}} \quad \text{for } \alpha\geq\beta$$

and $|\varphi((x_\alpha-x_\beta)a)|=|\varphi(a^*(x_\alpha-x_\beta))|$. Hence,

$$\lambda^2\varphi(x_\alpha)+\lambda\varphi(a^* x_\alpha)+\lambda\varphi(x_\alpha a)\to\lambda^2\varphi(x)+\lambda\varphi(a^* x)+\lambda\varphi(x a).$$

Therefore, $\varphi(a^* x_\alpha a)\to\varphi(a^* x a)$ and so $\operatorname{l.u.b.}_\alpha a^* x_\alpha a=a^* x a$. q.e.d.

1.7.5. Lemma. *If C is any maximal commutative C^*-subalgebra of $\mathcal{M}$, then its spectrum space is Stonean.*

Proof. Let $\{a_\alpha\}$ be a uniformly bounded, increasing directed set of positive elements in C and let $a_0=\operatorname{l.u.b.}_\alpha a_\alpha$ in $\mathcal{M}^s$. If u is any unitary element in C, then $\operatorname{l.u.b.}_\alpha u^* a_\alpha u=u^*\left(\operatorname{l.u.b.}_\alpha a_\alpha\right)u$ and

$$\operatorname{l.u.b.}_\alpha u^* a_\alpha u=\operatorname{l.u.b.}_\alpha a_\alpha=a_0.$$

Hence $u^* a_0 u=a_0$.

Since every element of C is a finite linear combination of unitary elements in C, a_0 commutes with every element of C. Hence $a_0\in C$ and the spectrum space is Stonean by 1.3.2. q.e.d.

1.7.6. Lemma. *Let e be any projection of $\mathcal{M}$. Then the subalgebra $e\mathcal{M}e$ is σ-closed and the mapping $x\to exe$ is σ-continuous.*

Proof. $e(P\cap S)e$ consists of those elements of $P\cap S$ which are $\leq e$. If $\{x_\alpha\}$ is a directed set in $e(P\cap S)e$ which converges to an element $x_0\geq 0$, then $e-x_\alpha\geq 0$, so that $e-x_0\geq 0$. Hence $e(P\cap S)e$ is closed.

Since $e(\mathcal{M}^s\cap S)e=e(P\cap S)e-e(P\cap S)e$, the compactness of $e(P\cap S)e$ implies that $e(\mathcal{M}^s\cap S)e$ is closed. Hence $e\mathcal{M}e$ is closed.

We shall next show the continuity of the mapping $x\to exe$. For this it is enough to show that the kernel $(1-e)\mathcal{M}+\mathcal{M}(1-e)$ of the mapping is closed, since $\mathcal{M}$ is the algebraic direct sum of $e\mathcal{M}e$ and

$$(1-e)\mathcal{M}+\mathcal{M}(1-e).$$

First, we shall show that if $\{e a_\alpha(1-e)\}$ $(a_\alpha\in\mathcal{M}^s\cap S)$ converges to a, $eae=(1-e)a(1-e)=0$.

For any integer n and complex number c ($|c|=1$),

$$\|ea_\alpha(1-e)+cne\| = \|\{ea_\alpha(1-e)+cne\}\{(1-e)a_\alpha e+\bar{c}ne\}\|^{\frac{1}{2}}$$
$$= \|ea_\alpha(1-e)a_\alpha e+n^2 e\|^{\frac{1}{2}} \leq (1+n^2)^{\frac{1}{2}}.$$

Now suppose that $eae \neq 0$ and that there exists a positive number $\lambda > 0$ in the spectrum of $\frac{eae+ea^*e}{2}$ (otherwise, consider $\{-a_\alpha\}$). Then

$$\|eae+ne+ea(1-e)+(1-e)ae+(1-e)a(1-e)\|$$
$$\geq \|e(a+n1)e\| \geq \left\|\frac{eae+ea^*e}{2}+ne\right\| \geq \lambda+n.$$

Therefore, $\|a+ne\| > (1+n^2)^{\frac{1}{2}}$ for a large positive number n. This is a contradiction. Hence $\frac{eae+ea^*e}{2}=0$.

Analogously, suppose there exists a positive number $\lambda>0$ in the spectrum of $\frac{iea^*e-ieae}{2}$ (otherwise, consider $\{-a_\alpha\}$). Then

$$\|a+nie\| \geq \|eae+nie\| \geq \left\|\frac{iea^*e-ieae}{2}+ine\right\|$$
$$\geq n+\lambda, \text{ a contradiction.}$$

Hence $\frac{iea^*e-ieae}{2}=0$ and so $eae=0$.

Similarly, suppose that $(1-e)a(1-e)\neq 0$; then

$$\|ea_\alpha(1-e)+cn(1-e)\| = \|\{(1-e)a_\alpha e+\bar{c}n(1-e)\}\{ea_\alpha(1-e)+cn(1-e)\}\|^{\frac{1}{2}}$$
$$= \|(1-e)a_\alpha ea_\alpha(1-e)+n^2(1-e)\|^{\frac{1}{2}} \leq (1+n^2)^{\frac{1}{2}}.$$

We obtain a contradiction, and so $a=ea(1-e)+(1-e)ae$. Therefore the closure of $(1-e)Se$ is contained in $e\mathcal{M}(1-e)+(1-e)\mathcal{M}e$. By symmetry, the closure of $eS(1-e)$ is also contained in $e\mathcal{M}(1-e)+(1-e)\mathcal{M}e$.

From the above remarks and the compactness of S, we can easily conclude that $eS(1-e)+(1-e)Se$ is closed, so that

$$e\mathcal{M}(1-e)+(1-e)\mathcal{M}e$$

is closed. Hence

$$(1-e)\mathcal{M}+\mathcal{M}(1-e)=(1-e)\mathcal{M}e+e\mathcal{M}(1-e)+(1-e)\mathcal{M}(1-e)$$

is closed. q.e.d.

1.7.7. Lemma. *If e is any projection of $\mathcal{M}$, then the mappings $x\to ex$ and $x\to xe$ are σ-continuous.*

Proof. Suppose that $\{ea_\alpha(1-e)\}$ $(a_\alpha \in S)$ converges to a and

$$(1-e)ae \neq 0.$$

By the proof of 1.7.6, $a = ea(1-e) + (1-e)ae$.

$$\begin{aligned}\|a+n(1-e)ae\| &= \|ea(1-e)+(n+1)(1-e)ae\| \\ &= \max\{\|ea(1-e)\|, (n+1)\|(1-e)ae\|\}\end{aligned}$$

for any positive number n. Hence, $\|a+n(1-e)ae\| = (n+1)\|(1-e)ae\|$ for a large positive number n. On the other hand,

$$\|ea_\alpha(1-e)+n(1-e)ae\| \leq \max\{1, n\|(1-e)ae\|\} = n\|(1-e)ae\|$$

for a large positive number n. This contradicts the above equality. Hence $e\mathcal{M}(1-e)$ is closed. Therefore, the mappings $x \to ex(1-e)$, $(1-e)xe$, ex, and xe are σ-continuous. q.e.d.

1.7.8. Theorem. *The mappings $x \to x^*$, ax, and xa are σ-continuous for x, $a \in \mathcal{M}$.*

Proof. $\mathcal{M} = \mathcal{M}^s + i\mathcal{M}^s$, $\mathcal{M}^s \cap i\mathcal{M}^s = (0)$, and by 1.7.1, $\mathcal{M}^s$ is closed; hence the mapping $x \to x^*$ is σ-continuous. Let C be a maximal commutative C^*-algebra of $\mathcal{M}$ containing a self-adjoint element h. Then the spectrum space of C is Stonean. Hence, by 1.3.1, for any positive number $\varepsilon\,(>0)$, there exists a finite family $\{e_i\}$ of mutually orthogonal projections belonging to C such that $\left\| h - \sum_{i=1}^{n} \lambda_i e_i \right\| < \varepsilon$, where $\{\lambda_i\}$ is a family of real numbers.

Let $\{x_\alpha\}$ $(\|x_\alpha\| \leq 1)$ be a directed set converging to 0; then for any σ-continuous linear functional g,

$$\begin{aligned}|g(hx_\alpha)| &= \left| g\left(\left(h - \sum_{i=1}^{n} \lambda_i e_i\right) x_\alpha\right)\right| + \left| g\left(\left(\sum_{i=1}^{n} \lambda_i e_i\right) x_\alpha\right)\right| \\ &\leq \|g\|\varepsilon + \sum_{i=1}^{n} |\lambda_i|\,|g(e_i x_\alpha)|\end{aligned}$$

Hence $\overline{\lim_\alpha}\, |g(hx_\alpha)| \leq \|g\|\varepsilon$. Therefore $\lim_\alpha g(hx_\alpha) = 0$. Hence a linear functional $l(x) = g(hx)$ is σ-continuous on S, so that by the Banach-Smulian theorem, $l(x)$ is σ-continuous on $\mathcal{M}$; so the mapping $x \to hx$ and therefore also $x \to ax$, is σ-continuous. Finally the mapping

$$x \to (a^* x)^* = xa$$

is continuous. q.e.d.

1.7.9. Corollary. *Let H be a subset of $\mathcal{M}$, $\mathcal{A}$ be the C^*-subalgebra of $\mathcal{M}$ generated by H, and let $\bar{\mathcal{A}}$ be the $\sigma(\mathcal{M}, \mathcal{M}_*)$-closure of $\mathcal{A}$. Then $\bar{\mathcal{A}}$ is a W^*-subalgebra of $\mathcal{M}$. $\bar{\mathcal{A}}$ is called a W^*-subalgebra of $\mathcal{M}$ generated by H. Moreover, $\bar{\mathcal{A}}$ is commutative if $\mathcal{A}$ is commutative.*

References. [149], [154].

1.8. Various Topologies on a W^*-Algebra

From the theory of locally convex spaces, we can identify the Banach space $\mathcal{M}_*$ with the Banach space of all $\sigma(\mathcal{M}, \mathcal{M}_*)$-continuous linear functionals. Now, let $\tau(\mathcal{M}, \mathcal{M}_*)$ be the Mackey topology on $\mathcal{M}$, that is, the topology of uniform convergence on all relatively $\sigma(\mathcal{M}_*, \mathcal{M})$-compact convex subsets in $\mathcal{M}_*$.

1.8.1. Notation. *Let $\mathcal{A}$ be a C^*-algebra, $\mathcal{A}^*$ the dual Banach space of $\mathcal{A}$. For $f \in \mathcal{A}^*$, denote $f(x) = \langle x, f\rangle$ for $x \in \mathcal{A}$, $\langle x, L_a f\rangle = \langle ax, f\rangle$, and $\langle x, R_a f\rangle = \langle xa, f\rangle$ for $x, a \in \mathcal{A}$. $L_a f$, $R_a f$* are in $\mathcal{A}^*$.

1.8.2. Definition. *Let V be a linear subspace of $\mathcal{A}^*$. V is said to be invariant if $f \in V$ implies $R_a f, L_a f \in V$ for $a \in \mathcal{A}$. V is said to be self-adjoint if $f \in V$ implies $f^* \in V$.*

Let $\mathcal{M}$ be a W^*-algebra, $\mathcal{M}^*$ the dual Banach space of $\mathcal{M}$, $\mathcal{M}_*$ the predual of $\mathcal{M}$. Then $\mathcal{M}_*$ is a closed linear subspace of $\mathcal{M}^*$. By 1.7.8, $\mathcal{M}_*$ is self-adjoint and invariant.

1.8.3. Proposition. *The mappings $f \to f^*$, $L_a f$, $R_a f$ in $\mathcal{M}_*$ are $\sigma(\mathcal{M}_*, \mathcal{M})$-continuous.*

Proof. Let $\{f_\alpha\}$ be a directed set of elements in $\mathcal{M}_*$ converging to 0. Then,

$$\begin{aligned} \langle x, f_\alpha^*\rangle &= \overline{\langle x^*, f_\alpha\rangle} \to 0\,, \\ \langle x, L_a f_\alpha\rangle &= \langle ax, f_\alpha\rangle \to 0\,, \quad \text{and} \\ \langle x, R_a f\rangle &= \langle xa, f_\alpha\rangle \to 0 \quad \text{for } x \in \mathcal{M}\,. \end{aligned}$$

q.e.d.

1.8.4. Proposition. *The mappings $a \to L_a f$, $R_a f$ of $\mathcal{M}$ with $\sigma(\mathcal{M}, \mathcal{M}_*)$ into $\mathcal{M}_*$ with $\sigma(\mathcal{M}_*, \mathcal{M})$ is continuous for each $f \in \mathcal{M}_*$.*

Proof. Suppose $\{a_\alpha\}$ converges to zero in the $\sigma(\mathcal{M}, \mathcal{M}_*)$-topology; then $\langle x, L_{a_\alpha} f\rangle = \langle a_\alpha x, f\rangle \to 0$. Hence $a \to L_a f$ and analogously $a \to R_a f$ are continuous. q.e.d.

1.8.5. Proposition. *The mappings $x \to x^*$, ax, xa are $\tau(\mathcal{M}, \mathcal{M}_*)$-continuous.*

Proof. Let $\{x_\alpha\}$ be a directed set in $\mathcal{M}$ converging to 0 in the $\tau(\mathcal{M}, \mathcal{M}_*)$ topology, and let G be a relatively $\sigma(\mathcal{M}_*, \mathcal{M})$-compact convex subset in $\mathcal{M}_*$. Then G^*, $L_a G$ and $R_a G$ are also relatively $\sigma(\mathcal{M}_*, \mathcal{M})$-compact. Hence

$$\begin{aligned} \langle x_\alpha^*, G\rangle &= \overline{\langle x_\alpha, G^*\rangle} \to 0 \qquad \text{(uniformly)}, \\ \langle a x_\alpha, G\rangle &= \langle x_\alpha, L_a G\rangle \to 0 \qquad \text{(uniformly)}, \\ \text{and} \quad & \langle x_\alpha a, G\rangle \to 0 \qquad \text{(uniformly)}. \end{aligned}$$

q.e.d.

1.8.6. Definition. *Let T be the set of all $\sigma(\mathscr{M},\mathscr{M}_*)$-continuous positive* linear functionals on $\mathscr{M}$. *For $\varphi\in T$, define $\alpha_\varphi(x)=\varphi(x^*x)^{\frac{1}{2}}$ for $x\in\mathscr{M}$. Then α_φ will define a semi-norm on $\mathscr{M}$. The family of semi-norms $\{\alpha_\varphi|\ all\ \varphi\in T\}$ defines a locally convex topology. This topology is called the strong topology or the s-topology on $\mathscr{M}$, and is denoted by $s(\mathscr{M},\mathscr{M}_*)$ or s.*

1.8.7. Definition. *For $\varphi\in T$, define $\alpha_\varphi^*(x)=\varphi(xx^*)^{\frac{1}{2}}$ for $x\in\mathscr{M}$. Then α_φ^* is a semi-norm $\mathscr{M}$. The locally convex topology on $\mathscr{M}$ defined by a family of semi-norms $\{\alpha_\varphi, \alpha_\varphi^*|\ all\ \varphi\in T\}$ is called the strong *-topology or the s*-topology on $\mathscr{M}$, and is denoted by $s^*(\mathscr{M},\mathscr{M}_*)$ or s^*.*

1.8.8. Notation. *Let Q be a set. Let t_1, t_2 be two topologies on Q. We shall write $t_1\leq t_2$ if t_2 is stronger than t_1.*

1.8.9. Theorem. $\sigma(\mathscr{M},\mathscr{M}_*)\leq s(\mathscr{M},\mathscr{M}_*)\leq s^*(\mathscr{M},\mathscr{M}_*)\leq \tau(\mathscr{M},\mathscr{M}_*)$.

Proof. Let f be a linear functional on $\mathscr{M}$ which is $\tau(\mathscr{M},\mathscr{M}_*)$-continuous on S, V_f the null space of f. Then $V_f\cap S$ is $\tau(\mathscr{M},\mathscr{M}_*)$-closed. Since $V_f\cap S$ is convex, $V_f\cap S$ is also $\sigma(\mathscr{M},\mathscr{M}_*)$-closed by Mackey's theorem. Hence V_f is $\sigma(\mathscr{M},\mathscr{M}_*)$-closed and so f is $\sigma(\mathscr{M},\mathscr{M}_*)$-continuous. Therefore f is $\tau(\mathscr{M},\mathscr{M}_*)$-continuous. Let $\{x_\alpha\}$ ($\subset S$) be a directed set converging to 0 in the $\tau(\mathscr{M},\mathscr{M}_*)$ topology. Then

$$\alpha_\varphi(x_\alpha)^2=\varphi(x_\alpha^* x_\alpha)=\langle x_\alpha^* x_\alpha, \varphi\rangle=\langle x_\alpha, L_{x_\alpha^*}\varphi\rangle .$$

Since $L_{x_\alpha^*}\varphi\subset L_S\varphi$ and $L_S\varphi$ is $\sigma(\mathscr{M},\mathscr{M}_*)$-compact (cf. 1.8.4), $\alpha_\varphi(x_\alpha)\to 0$. Hence $\{x_\alpha\}$ converges to 0 in the $s(\mathscr{M},\mathscr{M}_*)$ topology. Therefore, any $s(\mathscr{M},\mathscr{M}_*)$-continuous linear functional on S is also $\tau(\mathscr{M},\mathscr{M}_*)$-continuous on S. Hence it is also $\sigma(\mathscr{M},\mathscr{M}_*)$-continuous.

Conversely, we shall show that any $f\in\mathscr{M}_*$ is $s(\mathscr{M},\mathscr{M}_*)$-continuous. It suffices to assume that $f^*=f$. By the σ-compactness of S, there exists a self-adjoint a_0 in S such that $\|f\|=f(a_0)$. Let

$$\mathscr{E}=\{a|f(a)=\|f\|,\ a\in S\cap\mathscr{M}^s\};$$

then $\mathscr{E}$ is a σ-compact convex set. Let u be an extreme point in $\mathscr{E}$; then it is extreme in $\mathscr{M}^s\cap S$. In fact, let $u=(a+b)/2$ (a, $b\in\mathscr{M}^s\cap S$). Then $f(u)=\dfrac{(f(a)+f(b))}{2}=\|f\|$. Hence $f(a)=f(b)=\|f\|$. Hence $a, b\in\mathscr{E}$. By 1.6.3, u is self-adjoint and unitary. Also $(L_u f)(1)=f(u)=\|f\|$ and so $L_u f\in T$.

Conversely, $|f(x)|=|f(u^2x)|=|(L_u f)(ux)|\leq (L_u f)(u^2)^{\frac{1}{2}}(L_u f)(x^*x)^{\frac{1}{2}}$. Hence f is $s(\mathscr{M},\mathscr{M}_*)$-continuous. Therefore, the dual space of $\mathscr{M}$, with the $s(\mathscr{M},\mathscr{M}_*)$ topology, is $\mathscr{M}_*$. Hence by Mackey's theorem,

$$\sigma(\mathscr{M},\mathscr{M}_*)\leq s(\mathscr{M},\mathscr{M}_*)\leq \tau(\mathscr{M},\mathscr{M}_*).$$

Clearly $s(\mathscr{M},\mathscr{M}_*)\leq s^*(\mathscr{M},\mathscr{M}_*)$. The *-operation is τ-continuous. Hence $s^*(\mathscr{M},\mathscr{M}_*)\leq \tau(\mathscr{M},\mathscr{M}_*)$. q.e.d.

Remark. It is known that $s^*(\mathcal{M},\mathcal{M}_*)$ is equivalent to $\tau(\mathcal{M},\mathcal{M}_*)$ on S (cf. [1], [157]). Therefore, if a linear mapping Φ of a W^*-algebra into another W^*-algebra is σ-continuous, then it is s^*-continuous on bounded spheres, since Φ is τ-continuous.

1.8.10. Corollary. *Let f be a linear functional on a W^*-algebra $\mathcal{M}$. The following eight conditions are equivalent.*
1. *f is continuous in the $\sigma(\mathcal{M},\mathcal{M}_*)$ topology (resp.* 2. *$s(\mathcal{M},\mathcal{M}_*)$,* 3. *$s^*(\mathcal{M},\mathcal{M}_*)$ and* 4. *$\tau(\mathcal{M},\mathcal{M}_*)$);* 5. *$f$ is continuous on S in the $\sigma(\mathcal{M},\mathcal{M}_*)$ topology (resp.* 6. *$s(\mathcal{M},\mathcal{M}_*)$,* 7. *$s^*(\mathcal{M},\mathcal{M}_*)$ and* 8. *$\tau(\mathcal{M},\mathcal{M}_*)$)*

1.8.11. Corollary. *Let C be a convex set in $\mathcal{M}$. Then the following eight conditions are equivalent.*
1. *C is closed in the $\sigma(\mathcal{M},\mathcal{M}_*)$ topology (resp.* 2. *$s(\mathcal{M},\mathcal{M}_*)$,* 3. *$s^*(\mathcal{M},\mathcal{M}_*)$ and* 4. *$\tau(\mathcal{M},\mathcal{M}_*)$);* 5. *for every $\lambda>0$, $C\cap\lambda S$* is closed in the *$\sigma(\mathcal{M},\mathcal{M}_*)$ topology (resp.* 6. *$s(\mathcal{M},\mathcal{M}_*)$,* 7. *$s^*(\mathcal{M},\mathcal{M}_*)$ and* 8. *$\tau(\mathcal{M},\mathcal{M}_*)$).*

1.8.12. Proposition. *The mappings $x\to ax$, xa are $s(\mathcal{M},\mathcal{M}_*)$-continuous, and moreover the mapping $(x,y)\to xy$ is jointly $s(\mathcal{M},\mathcal{M}_*)$-continuous on $\lambda S\times\mathcal{M}$ $(\lambda>0)$ $(a, x, y\in\mathcal{M})$.*

Proof. Suppose $\{x_\alpha\}$ converges to 0 in $s(\mathcal{M},\mathcal{M}_*)$. Then, for $\varphi\in T$,

$$\alpha_\varphi(a x_\alpha)^2=\varphi(x_\alpha^* a^* a x_\alpha)\le\|a^* a\|\,\varphi(x_\alpha^* x_\alpha)=\|a^* a\|\,\alpha_\varphi(x_\alpha)^2\to 0$$

and

$$\alpha_\varphi(x_\alpha a)^2=\varphi(a^* x_\alpha^* x_\alpha a)=\alpha\, L_{a^*} R_a\varphi(x_\alpha)^2\to 0.$$

Moreover, suppose $y_\beta\to y$ in $s(\mathcal{M},\mathcal{M}_*)$ and $y_\beta\in S$. Then, for $\varphi\in T$,

$$\alpha_\varphi(y_\beta x_\alpha-y x)\le\alpha_\varphi(y_\beta(x_\alpha-x))+\alpha_\varphi((y_\beta-y)x)\to 0.$$

q.e.d.

Remark. In general, the $*$-operation is not $s(\mathcal{M},\mathcal{M}_*)$-continuous on bounded spheres (cf. 2.5).

Concluding remarks on 1.8.

1. $\mathcal{M}_*$ is $\sigma(\mathcal{M}_*,\mathcal{M})$-sequentially complete (cf. [1], [151]).

2. Let $\mathcal{M}$ be a type I-factor (cf. chapter 2) and let $(\varphi_n)\subset T$ be a $\sigma(\mathcal{M}_*,\mathcal{M})$-Cauchy sequence. Then (φ_n) is a Cauchy sequence in the norm topology of $\mathcal{M}_*$. (Cf. [1], [27]). (In this theorem, we can not replace T by $\mathcal{M}_*$ (cf. [151]).)

Problem. Is the converse of the theorem true?

3. There are many C^*-algebras, W^*-algebras and their preduals (cf. [155]) which are not topologically contained in the classes of the so-called classical Banach spaces $((M),(m),(C),(c),(C^{(p)})\,p\ge 1,(L^p)\,1\le p\le+\infty)$ mentioned by Banach (cf. [8]); therefore it is very meaningful to examine whether or not many unsolved problems concerning Banach space are positive in these examples.

Reference. [154].

1.9. Kaplansky's Density Theorem

Let $\mathcal{M}$ be a W^*-algebra and let $\mathcal{M}_*$ be the predual of $\mathcal{M}$. Let V be a self-adjoint, invariant linear subspace of $\mathcal{M}_*$ (i.e. $f \in V$ implies f^*, $L_a f$, $R_a f \in V$ for $a \in \mathcal{M}$) which is norm-dense in $\mathcal{M}_*$. Then we can consider the Hausdorff locally convex topology $\sigma(\mathcal{M}, V)$ on $\mathcal{M}$. Clearly $\sigma(\mathcal{M}, V) \leq \sigma(\mathcal{M}, \mathcal{M}_*)$. Using the methods of 1.8, we can easily show that the mappings $x \to ax$, xa and x^* on $\mathcal{M}$ are $\tau(\mathcal{M}, V)$-continuous.

1.9.1. Theorem. *Let $\mathcal{A}$ be a self-adjoint subalgebra of a W^*-algebra $\mathcal{M}$ which is $\sigma(\mathcal{M}, V)$-dense in $\mathcal{M}$. Then $\mathcal{A} \cap S$ is $\tau(\mathcal{M}, \mathcal{M}_*)$-dense in S, where S is the unit sphere of $\mathcal{M}$.*

Proof. It is enough to prove that $\mathcal{A} \cap S$ is $\sigma(\mathcal{M}, \mathcal{M}_*)$-dense in S, since the $\tau(\mathcal{M}, \mathcal{M}_*)$-closure of $\mathcal{A} \cap S$ is $\sigma(\mathcal{M}, \mathcal{M}_*)$-closed. Since V is norm-dense in $\mathcal{M}_*$ and $\mathcal{A} \cap S$ is uniformly bounded, it suffices to show that $\mathcal{A} \cap S$ is $\sigma(\mathcal{M}, V)$-dense in S. We can assume that $\mathcal{A}$ is uniformly closed. Let $a \in \mathcal{M}$. Since $\mathcal{A}$ is a convex set and $\mathcal{A}$ is $\sigma(\mathcal{M}, V)$-dense in $\mathcal{M}$, $\mathcal{A}$ is $\tau(\mathcal{M}, V)$-dense in $\mathcal{M}$. Hence there exists a directed set $\{a_\alpha\}$ in $\mathcal{A}$ such that the $\tau(\mathcal{M}, V)$-limit of the a_α is a.

Since $\|(1+a_\alpha^* a_\alpha)^{-1}\| \leq 1$, $\{R_{(1+a_\alpha^* a_\alpha)^{-1}} f\} \subset R_S f$ for $f \in V$ is relatively $\sigma(\mathcal{M}_*, \mathcal{M})$-compact and $R_S f \subset V$. Hence

$$|\langle a, R_{(1+a_\alpha^* a_\alpha)^{-1}} f\rangle - \langle a_\alpha, R_{(1+a_\alpha^* a_\alpha)^{-1}} f\rangle| < \varepsilon$$

for $\alpha \geq \alpha_0$ and so $|\langle a(1+a_\alpha^* a_\alpha)^{-1}, f\rangle - \langle a_\alpha(1+a_\alpha^* a_\alpha)^{-1}, f\rangle| < \varepsilon$ for $\alpha \geq \alpha_0$, for some α_0.

$$\begin{aligned} &a(1+a^* a)^{-1} - a(1+a_\alpha^* a_\alpha)^{-1} \\ &= a[(1+a^* a)^{-1}\{(1+a_\alpha^* a_\alpha)-(1+a^* a)\}(1+a_\alpha^* a_\alpha)^{-1}] \\ &= a(1+a^* a)^{-1}(a_\alpha^* a_\alpha - a^* a)(1+a_\alpha^* a_\alpha)^{-1} \\ &= a(1+a^* a)^{-1} a_\alpha^* a_\alpha(1+a_\alpha^* a_\alpha)^{-1} - a(1+a^* a)^{-1} a^* a(1+a_\alpha^* a_\alpha)^{-1}. \end{aligned}$$

Since $\|a_\alpha(1+a_\alpha^* a_\alpha)^{-1}\| \leq \frac{1}{2}$, $a_\alpha^* \to a^*$ and $a(1+a^* a)^{-1} a_\alpha^* \to a(1+a^* a)^{-1} a^*$ in the $\tau(\mathcal{M}, V)$-topology, we have

$$\begin{aligned} &|\langle a(1+a^* a)^{-1} a_\alpha^* a_\alpha(1+a_\alpha^* a_\alpha)^{-1}, f\rangle \\ &\quad - \langle a(1+a^* a)^{-1} a^* a_\alpha(1+a_\alpha^* a_\alpha)^{-1}, f\rangle| < \varepsilon \end{aligned}$$

for $\alpha \geq \alpha_1$, for some α_1.

On the other hand,

$$\begin{aligned} &a(1+a^* a)^{-1} a^* a_\alpha(1+a_\alpha^* a_\alpha)^{-1} - a(1+a^* a)^{-1} a^* a(1+a_\alpha^* a_\alpha)^{-1} \\ &\quad = a(1+a^* a)^{-1} a^*(a_\alpha - a)(1+a_\alpha^* a_\alpha)^{-1}. \end{aligned}$$

Hence

$$|\langle a(1+a^*a)^{-1}a^*(a_\alpha-a)(1+a_\alpha^*a_\alpha)^{-1},f\rangle|<\varepsilon$$

for $\alpha\geq\alpha_2$, for some α_2. Therefore,

$$\begin{aligned}
&|\langle a(1+a^*a)^{-1}-a_\alpha(1+a_\alpha^*a_\alpha)^{-1},f\rangle|\\
&\leq|\langle a(1+a^*a)^{-1}-a(1+a_\alpha^*a_\alpha)^{-1},f\rangle|\\
&\qquad+|\langle a(1+a_\alpha^*a_\alpha)^{-1}-a_\alpha(1+a_\alpha^*a_\alpha)^{-1},f\rangle|\\
&\leq|\langle a(1+a^*a)^{-1}a_\alpha^*a_\alpha(1+a_\alpha^*a_\alpha)^{-1}\\
&\qquad-a(1+a^*a)^{-1}a^*a_\alpha(1+a_\alpha^*a_\alpha)^{-1},f\rangle|\\
&\qquad+|\langle a(1+a_\alpha^*a_\alpha)^{-1}-a_\alpha(1+a_\alpha^*a_\alpha)^{-1},f\rangle|\\
&\qquad+|\langle a(1+a^*a)^{-1}a^*a_\alpha(1+a_\alpha^*a_\alpha)^{-1}\\
&\qquad-a(1+a^*a)^{-1}a^*a(1+a_\alpha^*a_\alpha)^{-1},f\rangle|<3\varepsilon
\end{aligned}$$

for $\alpha\geq\alpha_3\geq\alpha_1,\alpha_2,\alpha_0$, for some α_3.

Hence, the limit of $2(a_\alpha(1+a_\alpha^*a_\alpha)^{-1})$ in the $\sigma(\mathscr{M},V)$-topology is $2a(1+a^*a)^{-1}$.

Since $\|2a_\alpha(1+a_\alpha^*a_\alpha)^{-1}\|\leq 1$ and $2a_\alpha(1+a_\alpha^*a_\alpha)^{-1}\in\mathscr{A}$ (even if $\mathscr{A}$ does not contain 1), the $\sigma(\mathscr{M},\mathscr{M}_*)$-closure of $\mathscr{A}\cap S$ contains

$$\{2a(1+a^*a)^{-1}|a\in\mathscr{M}\}.$$

Let u be an extreme point of S. Then

$$2u(1+u^*u)^{-1}=2u(1+p)^{-1}=2u(1/2p+(1-p)),$$

where $p=u^*u$ is a projection (cf. the proof of 1.6.4). Hence

$$2u(1+p)^{-1}=up=u,$$

so that the $\sigma(\mathscr{M},\mathscr{M}_*)$-closure of $\mathscr{A}\cap S$ contains all extreme points of S, and thus coincides with S. q.e.d.

Concluding remarks on 1.9.

The density theorem of Kaplansky is one of the most useful theorems in the theory of operator algebras. The problem of extending this theorem to more general algebras is quite important.

1. It is easily seen that we can not replace the algebra by a linear subspace. For example, let $L^\infty(\Omega,\mu)$ be the commutative W^*-algebra of all essentially bounded measurable functions on a measure space Ω with a measure μ (cf. 1.18). Let $\mathscr{I}$ be a maximal ideal of $L^\infty(\Omega,\mu)$, and let E be a $\sigma(L^\infty,L^1)$-closed linear subspace of $L^\infty(\Omega,\mu)$ with deficiency one and $1\notin E$; then $L^\infty(\Omega,\mu)=\mathscr{I}+\mathbb{C}1=E+\mathbb{C}1$, where $\mathbb{C}$ is the field of complex numbers. Therefore we can construct a linear isomorphism Φ of $\mathscr{I}$ onto E. Let E_* be the predual of E, and let E^* (resp. $\mathscr{I}^*$) be the dual of E (resp. $\mathscr{I}$). Since E_* is $\sigma(E^*,E)$-dense in E^*, $\Phi^*(E_*)$ is $\sigma(\mathscr{I}^*,\mathscr{I})$-dense in $\mathscr{I}^*$, where Φ^* is the dual of Φ. If the unit sphere of $\Phi^*(E_*)$ is

$\sigma(\mathscr{I}^*, \mathscr{I})$-dense in the unit sphere of $\mathscr{I}^*$, then the unit sphere of $\mathscr{I}$ is $\sigma(\mathscr{I}, \Phi^*(E_*))$-compact and so $\mathscr{I}$ is the dual of $\Phi^*(E_*)$; hence $\mathscr{I}$ is a W^*-algebra.

On the other hand, if Ω has no atomic part, $\mathscr{I}$ can not have an identity. Therefore, in this case, $\Phi^*(E_*) \cap S$ is not $\sigma(\mathscr{I}^*, \mathscr{I})$-dense in S, where S is the unit sphere of $\mathscr{I}^*$.

2. Let E be a Banach space, and let $B(E)$ be the Banach algebra of all bounded linear operators on E. Let $\mathfrak{F}(E)$ be the algebra of all finite-rank linear operators on E.

Problem. Is the unit sphere of $\mathfrak{F}(E)$ dense in the unit sphere of $B(E)$ in the strong operator topology?

This problem is equivalent to the problem whether or not any compact linear operator on E can be uniformly approximated by finite-rank linear operators (cf. [65]).

References. [95].

1.10. Ideals in a W^*-Algebra

Let e be a projection of a W^*-algebra $\mathscr{M}$. Since the mappings $x \to ax$, xa are $\sigma(\mathscr{M}, \mathscr{M}_*)$-continuous, $\mathscr{M}e$ (resp. $e\mathscr{M}$) is a $\sigma(\mathscr{M}, \mathscr{M}_*)$-closed left (resp. right) ideal of $\mathscr{M}$.

1.10.1. Proposition. *Let $\mathscr{L}$ (resp. $\mathscr{R}$) be a $\sigma(\mathscr{M}, \mathscr{M}_*)$-closed left (resp. right) ideal of a W^*-algebra $\mathscr{M}$. Then there exists a unique projection p (resp. q) in $\mathscr{M}$ such that $\mathscr{L} = \mathscr{M}p$ (resp. $\mathscr{R} = q\mathscr{M}$).*

Proof. Let $\mathscr{L} \cap \mathscr{L}^* = \mathscr{N}$, where $\mathscr{L}^* = \{x^* | x \in \mathscr{L}\}$. Then $\mathscr{N}$ is a C^*-algebra and, since the *-operation is $\sigma(\mathscr{M}, \mathscr{M}_*)$-continuous, $\mathscr{N}$ is $\sigma(\mathscr{M}, \mathscr{M}_*)$-closed; hence $\mathscr{N}$ is a W^*-subalgebra of $\mathscr{M}$. Let p be the identity of $\mathscr{N}$. Then p is a projection in $\mathscr{M}$. Clearly $\mathscr{M}p \subset \mathscr{L}$. Conversely, if $a \in \mathscr{L}$, then $a^* a \in \mathscr{N}$. $pa^* ap = pa^* a = a^* ap = a^* a$; hence $(1-p)a^* a(1-p) = 0$ and so $a(1-p) = 0$. Hence $\mathscr{L} = \mathscr{M}p$.

Next, suppose $\mathscr{L} = \mathscr{M}p_1$, for some projection p_1; then $p_1 = ap$ and so $p_1 = p_1^* p_1 = pa^* ap$. Hence $p_1 \leq p$ and analogously $p \leq p_1$ and so $p = p_1$. q.e.d.

1.10.2. Proposition. *Let $\mathscr{M}^p$ be the set of all projections in $\mathscr{M}$. Then $\mathscr{M}^p$ is a complete lattice under the order $\leq$.*

Proof. Let $\{e_\alpha\}_{\alpha \in \mathbb{I}}$ be a set of projections, and let $\mathscr{L}_1$ be the $\sigma(\mathscr{M}, \mathscr{M}_*)$-closed left ideal generated by $\{\mathscr{M}e_\alpha | \alpha \in \mathbb{I}\}$ and $\mathscr{L}_2 = \bigcap_{\alpha \in \mathbb{I}} \mathscr{M}e_\alpha$. Then $\mathscr{L}_1 = \mathscr{M}e_1$ and $\mathscr{L}_2 = \mathscr{M}e_2$ for some projections e_1, e_2 in $\mathscr{M}$. Clearly,

$e_\alpha \leq e_1$ and $e_\alpha \geq e_2$ for all $\alpha \in \mathbb{I}$. Moreover, if p is a projection in $\mathcal{M}$ such that $p \geq e_\alpha$ for all $\alpha \in \mathbb{I}$, then $\mathcal{M}p \supset \mathcal{M}e_\alpha$ for all $\alpha \in \mathbb{I}$. Hence $p \geq e_1$ and so $\bigvee_\alpha e_\alpha = e_1$. Analogously $\bigwedge_\alpha e_\alpha = e_2$. q.e.d.

1.10.3. Definition. *Let a be an element in a W^*-algebra $\mathcal{M}$. Let $\mathcal{L} = \{x | xa = 0, x \in \mathcal{M}\}$. Then $\mathcal{L}$ is a $\sigma(\mathcal{M}, \mathcal{M}_*)$-closed left ideal. Hence $\mathcal{L} = \mathcal{M}e$ for a unique projection e in $\mathcal{M}$. Then $1-e$ is the least projection of all the projections q in $\mathcal{M}$ such that $qa = a$. $1-e$ is called the left support of a and is denoted by $l(a)$.*

Similarly, let $\mathcal{R} = \{y | ay = 0, y \in \mathcal{M}\}$. Then $\mathcal{R}$ is a $\sigma(\mathcal{M}, \mathcal{M}_*)$-closed right ideal. Hence $\mathcal{R} = f\mathcal{M}$ for a unique projection f in $\mathcal{M}$.

$1-f$ is the least projection of all the projections p in $\mathcal{M}$ such that $ap = a$. $1-f$ is called the right support of a and is denoted by $r(a)$. If a is self-adjoint, $l(a) = r(a)$ is called the support of a and is denoted by $s(a)$.

1.10.4. Proposition. *Let h be a self-adjoint element of a W^*-algebra $\mathcal{M}$, and let C be the W^*-subalgebra generated by h. Then the support $s(h)$ belongs to C.*

Proof. Let p be the identity of C. Then $ph = h$ and therefore $s(h) \leq p$. $(p - s(h))h = 0$; hence $(p - s(h))\overline{\{p(h) | p(h), \text{ all polynomials of } h\}} = 0$, where $\overline{\{\quad\}}$ is the $\sigma(\mathcal{M}, \mathcal{M}_*)$-closure of $\{\quad\}$, so that $(p - s(h))p = 0$. Hence $p = s(h)$. q.e.d.

1.10.5. Proposition. *Every $\sigma(\mathcal{M}, \mathcal{M}_*)$-closed two-sided ideal is of the form $\mathcal{M}z$ for some central projection z in $\mathcal{M}$.*

Proof. Let $\mathcal{I}$ be a $\sigma(\mathcal{M}, \mathcal{M}_*)$-closed two-sided ideal of $\mathcal{M}$. Then $\mathcal{I} = \mathcal{M}z_1 = z_2\mathcal{M}$ for some projections z_1, z_2 in $\mathcal{M}$, where z_1 (resp. z_2) is the identity of $\mathcal{I} \cap \mathcal{I}^*$. Hence $z_1 = z_2 = z$ and so $\mathcal{I} = \mathcal{I}^*$. Therefore $\mathcal{I}$ is a W^*-algebra with the identity z.

$xz = z(xz) = zxz$, and $zx = (zx)z = zxz$ for $x \in \mathcal{M}$. Hence, z is a central projection of $\mathcal{M}$. q.e.d.

1.10.6. Definition. *Let $\mathcal{Z}$ be the center of a W^*-algebra $\mathcal{M}$. Then $\mathcal{Z}$ is $\sigma(\mathcal{M}, \mathcal{M}_*)$-closed. Let p be a projection in $\mathcal{M}$. There exists a least central projection in $\mathcal{M}$ containing p. We call this central projection the central support (or envelope) of p, and is denoted by $c(p)$.*

1.10.7. Proposition. *Let p, q be two projections of a W^*-algebra $\mathcal{M}$ such that $p\mathcal{M}q = (0)$. Then $c(p)c(q) = 0$.*

Proof. Let $\mathcal{I}_1 = \{x | p\mathcal{M}x = (0), x \in \mathcal{M}\}$. Then $\mathcal{I}_1$ is a $\sigma(\mathcal{M}, \mathcal{M}_*)$-closed, two-sided ideal, and q is contained in $\mathcal{I}_1$. Hence, $c(q) \in \mathcal{I}_1$.

Similarly, if $\mathcal{I}_2 = \{y | y\mathcal{M}c(q) = (0), y \in \mathcal{M}\}$, then by the same reasoning we have $c(p)c(q) = 0$. q.e.d.

1.11. Spectral Resolution of Self-Adjoint Elements in a W^*-Algebra

Let h be a self-adjoint element of a W^*-algebra $\mathcal{M}$ and for real λ, let $e(\lambda)=s((\lambda 1-h)^+)$. Let C be the W^*-subalgebra generated by 1 and h. Then by 1.10.4, $e(\lambda)\in C$ and $e(\lambda)$ is the support of $(\lambda 1-h)^+$ in C. Clearly $e(\lambda)\leq e(\mu)$ for $\lambda\leq\mu$.

1.11.1. Lemma. *$\lambda_n\leq\lambda$ $(n=1,2,\ldots)$ and $\lambda_n\to\lambda$ imply $e(\lambda_n)\to e(\lambda)$ in the $s(\mathcal{M},\mathcal{M}_*)$-topology.*

Proof. Let $p=\underset{n}{\mathrm{l.u.b.}}\, e(\lambda_n)$. Then

$$0\leq(\lambda 1-h)^+-(\lambda_n 1-h)^+\leq(\lambda 1-h)-(\lambda_n 1-h)\to 0$$

(uniformly). Since $(\lambda_n 1-h)^+(1-p)=0, (\lambda 1-h)^+(1-p)=0$; hence $e(\lambda)\leq p$, so that $p=e(\lambda)$. q.e.d.

1.11.2. Lemma.
$\lambda\{e(\mu)-e(\lambda)\}\leq\{\mu e(\mu)-(\mu 1-h)^+\}-\{\lambda e(\lambda)-(\lambda 1-h)^+\}\leq\mu\{e(\mu)-e(\lambda)\}$ for $\mu\geq\lambda$.

Proof. $(\lambda 1-h)^+=(\lambda 1-h)e(\lambda)$ and $(\mu 1-h)^+=(\mu 1-h)e(\mu)$. Hence, $\mu e(\mu)-(\mu 1-h)^+=he(\mu)$ and $\lambda e(\lambda)-(\lambda 1-h)^+=he(\lambda)$ and

$$\lambda(e(\mu)-e(\lambda))\leq h(e(\mu)-e(\lambda))\leq\mu(e(\mu)-e(\lambda)).$$

q.e.d.

1.11.3. Theorem. *For any self-adjoint $x\in\mathcal{M}$, there exists a system of projections $\{e(\lambda)\}$ $(-\infty<\lambda<\infty)$ in $\mathcal{M}$, called a resolution of the identity such that*

1. *$\lambda\leq\mu$ implies $e(\lambda)\leq e(\mu)$;*
2. *For a monotone increasing sequence (λ_n) such that $\lambda_n\to\lambda$, $e(\lambda_n)\to e(\lambda)$ in the $s(\mathcal{M},\mathcal{M}_*)$ topology;*
3. *$\lim_{\lambda\to\infty} e(\lambda)=1$ and $\lim_{\lambda\to-\infty} e(\lambda)=0$;*
4. *$x=\int_{-\infty}^{\infty}\lambda\, de(\lambda)=\int_{-\|x\|}^{\|x\|+0}\lambda\, de(\lambda)$,*

where the integral being an abstract Radon-Stieltjes integral with respect to the $s(\mathcal{M},\mathcal{M}_*)$-topology. Such a resolution is unique.

Proof. Clearly $\lambda>\|x\|$ implies $e(\lambda)=1$ and $\lambda<-\|x\|$ implies $e(\lambda)=0$. For any $\delta>0$ and division Δ:

$$-\|x\|-\delta=\lambda_0<\lambda_1<\cdots<\lambda_n=\|x\|+\delta$$

of the interval $[-\|x\|-\delta,\ \|x\|+\delta]$ (with $0<\lambda_i-\lambda_{i-1}<\varepsilon$, $1\leq i\leq n$), we have $m(\Delta) \equiv \sum_{i=1}^{n} \lambda_{i-1}(e(\lambda_i) - e(\lambda_{i-1}))$

$$\leq \sum_{i=1}^{n} [\{\lambda_i e(\lambda_i) - (\lambda_i 1 - x)^+\} - \{\lambda_{i-1} e(\lambda_{i-1}) - (\lambda_{i-1} 1 - x)^+\}]$$

$$= -\{\lambda_0 e(\lambda_0) - (\lambda_0 1 - x)^+\} + \{\lambda_n e(\lambda_n) - (\lambda_n 1 - x)^+\}$$

$$= x \leq \sum_{i=1}^{n} \lambda_i (e(\lambda_i) - e(\lambda_{i-1})) \equiv M(\Delta)$$

and

$$M(\Delta) - m(\Delta) = \sum_{i=1}^{n} (\lambda_i - \lambda_{i-1})(e(\lambda_i) - e(\lambda_{i-1})) \leq \varepsilon 1 .$$

Hence letting $\varepsilon \to 0$, we have

$$x = \int_{-\|x\|-\delta}^{\|x\|+\delta} \lambda \, de(\lambda) = \int_{-\infty}^{\infty} \lambda \, de(\lambda) .$$

We shall show next the uniqueness of $\{e(\lambda)\}$. Suppose that $\{e'(\lambda)\}$ is another resolution of the identity satisfying 1., 2., 3., and $x = \int_{-\infty}^{\infty} \lambda \, de'(\lambda)$. Then

$$(\lambda_0 1 - x) = \int_{-\infty}^{\infty} (\lambda_0 - \lambda) de'(\lambda) = \int_{-\infty}^{\lambda_0} (\lambda_0 - \lambda) de'(\lambda) + \int_{\lambda_0}^{\infty} (\lambda_0 - \lambda) de'(\lambda).$$

Hence, $(\lambda_0 1 - x)^+ = \int_{-\infty}^{\lambda_0} (\lambda_0 - \lambda) de'(\lambda)$, and so

$$e((\lambda_0 1 - x)^+) = e(\lambda_0) = \int_{-\infty}^{\lambda_0 - 0} de'(\lambda) = e'(\lambda_0 - 0) = e'(\lambda_0) . \qquad \text{q.e.d.}$$

1.12. The Polar Decomposition of Elements of a W^*-Algebra

1.12.1. Theorem. *Let $\mathcal{M}$ be a W^*-algebra, and let a be an element of $\mathcal{M}$. a can be decomposed as follows: $a = u|a|$, where $|a| = (a^* a)^{\frac{1}{2}}$ and u is a partial isometry of $\mathcal{M}$ such that $u^* u = s(|a|)$ and $u u^* = s(|a^*|)$. Such a decomposition is unique. This decomposition is called the polar decomposition of a.*

Proof. Put $h(n) = \left(a^* a + \frac{1}{n} 1\right)^{\frac{1}{2}}$ (n a positive integer) and

$$a(n) = a\left(a^* a + \frac{1}{n} 1\right)^{-\frac{1}{2}} ;$$

then $a(n)^* a(n) = \left(a^* a + \frac{1}{n} 1\right)^{-\frac{1}{2}} a^* a \left(a^* a + \frac{1}{n} 1\right)^{-\frac{1}{2}} = \dfrac{a^* a}{a^* a + \frac{1}{n} 1}$;

hence $\|a(n)\| \leq 1$ and $a(n)\left(a^* a + \frac{1}{n} 1\right)^{\frac{1}{2}} = a$. Since $h(n) \to (a^* a)^{\frac{1}{2}}$ (uniformly), for arbitrary $\varepsilon > 0$ there exists an n_0 such that

$$\|h(n) - (a^* a)^{\frac{1}{2}}\| < \varepsilon \quad (n \geq n_0).$$

By the weak compactness of the unit sphere S of $\mathscr{M}$, there exists an accumulation point b of $\{a(n)\}$. Since

$$\{a(n)(a^* a)^{\frac{1}{2}}\} \subset a + \varepsilon S \qquad (n \geq n_0), \qquad b(a^* a) \in a + \varepsilon S.$$

Since ε is arbitrary, $a = b(a^* a)^{\frac{1}{2}}$.

Let p (resp. q) be the support of $(a^* a)^{\frac{1}{2}}$ (resp. $(a a^*)^{\frac{1}{2}}$). Then $a = q a$, since $\{(1-q)a\}\{(1-q)a\}^* = (1-q) a a^* (1-q) = (1-q) a a^* = 0$. Hence

$$a = q a = q b (a^* a)^{\frac{1}{2}}.$$

$$a^* a = (a^* a)^{\frac{1}{2}} p b^* q b p (a^* a)^{\frac{1}{2}} \text{ and so } (a^* a)^{\frac{1}{2}} (p - p b^* q b p)(a^* a)^{\frac{1}{2}} = 0.$$

Since $\|b\| \leq 1$, we conclude that $p = p b^* q b p$. Define $u = q b p$. Then u is a partial isometry having the initial projection p. Moreover

$$a a^* = u (a^* a) u^*$$

and so the final projection is $u p u^* = q$.

Now suppose $a = u|a| = u'|a|$ is another polar decomposition of a. Then $u'^* a = |a| = u'^* u |a|$. Hence $(p - u'^* u)|a| = 0$. Let

$$\mathscr{R} = \{x \mid (p - u'^* u) x = 0, x \in \mathscr{M}\}.$$

Then $\mathscr{R}$ is a σ-closed right ideal. Hence $\mathscr{R} = e \mathscr{M}$ for some projection e. Hence $s(|a|) = p \leq e$. Therefore $(p - u'^* u) p = 0$. On the other hand, $p u'^* u p = u'^* u$. Hence $p = u'^* u$ and so $u' = u$. q.e.d.

1.13. Linear Functionals on a W^*-Algebra

1.13.1. Definition. *A positive linear functional φ on a W^*-algebra $\mathscr{M}$ is said to be normal if it satisfies $\varphi(\underset{\alpha}{\text{l.u.b.}}\, x_\alpha) = \underset{\alpha}{\text{l.u.b.}}\, \varphi(x_\alpha)$ for every uniformly bounded increasing directed set $\{x_\alpha\}$ of positive elements in $\mathscr{M}$.*

1.13.2. Theorem. *Let φ be a positive linear functional on a W^*-algebra $\mathscr{M}$. Then the following conditions are equivalent.*

1. *φ is normal*
2. *φ ic $\sigma(\mathscr{M}, \mathscr{M}_*)$-continuous.*

Proof. 2.⇒1. was proved (cf. 1.7.4). We prove the implication 1.⇒2. Let $\{p_\alpha\}$ be an increasing directed set of projections such that

$$x \to \varphi(x p_\alpha)$$

is $\sigma(\mathscr{M}, \mathscr{M}_*)$-continuous. Let p be the l.u.b. of the $\{p_\alpha\}$. Then p is also a projection (cf. 1.13.4). Therefore,

$$|\varphi(x(p-p_\alpha))| \leq \varphi(x(p-p_\alpha)x^*)^{\frac{1}{2}} \varphi(p-p_\alpha)^{\frac{1}{2}} \leq \varphi(1)^{\frac{1}{2}} \varphi(p-p_\alpha)^{\frac{1}{2}} \quad \text{for } x \in S.$$

(S is the unit sphere of $\mathscr{M}$).
Hence $\varphi(xp)$ is a uniform limit of the directed set $\{\varphi(xp_\alpha)\}$ on S, so that $\varphi(xp)$ is also $\sigma(\mathscr{M}, \mathscr{M}_*)$-continuous on S, and so it is continuous on $\mathscr{M}$. Therefore, there exists a maximal projection p_0 such that $x \to \varphi(x p_0)$ is $\sigma(\mathscr{M}, \mathscr{M}_*)$-continuous.

Suppose $p_0 < 1$ and take a $\psi \in T$, the set of all $\sigma(\mathscr{M}, \mathscr{M}_*)$-continuous positive linear functionals on $\mathscr{M}$, such that $\varphi(1-p_0) < \psi(1-p_0)$. We shall show that there is a non-zero projection $p_1 \leq 1-p_0$ such that $\varphi(p) < \psi(p)$ for all non-zero projections $p\ (\leq p_1) \in \mathscr{M}$.

Assume that this is not true. Then for every non-zero projection $p\ (\leq 1-p_0)$, there exists a non-zero projection q such that $q \leq p$ and $\varphi(q) \geq \psi(q)$. Take a maximal projection q_0 such that $q_0 \leq 1-p_0$ and $\varphi(q_0) \geq \psi(q_0)$; then $q_0 = 1-p_0$. In fact, if $q_0 < 1-p_0$, we can choose another non-zero projection $q_1 < 1-p_0-q_0$ such that $\varphi(q_1) \geq \psi(q_1)$. Hence $\varphi(q_0+q_1) = \varphi(q_0) + \varphi(q_1) \geq \psi(q_0) + \psi(q_1) = \psi(q_0+q_1)$. This contradicts the maximality of q_0. Then $\varphi(1-p_0) \geq \psi(1-p_0)$, a contradiction.

Hence, $\varphi(p) < \psi(p)$ for all non-zero projections $p\ (\leq p_1)$. Therefore $\varphi(q) \leq \psi(q)$ for all projections $q\ (\leq p_1)$. Since the spectrum space of any maximal commutative C^*-subalgebra of $p_1 \mathscr{M} p_1$ is Stonean, every positive element of $p_1 \mathscr{M} p_1$ is a uniform limit of finite positive linear combinations of projections in $p_1 \mathscr{M} p_1$. Hence, $\varphi(a) \leq \psi(a)$ for $a\ (\geq 0) \in p_1 \mathscr{M} p_1$. Hence

$$\begin{aligned} |\varphi(x(p_0+p_1))| &\leq |\varphi(x p_0)| + |\varphi(x p_1)| \\ &\leq |\varphi(x p_0)| + \varphi(1)^{\frac{1}{2}} \varphi(p_1 x^* x p_1)^{\frac{1}{2}} \\ &\leq |\varphi(x p_0)| + \varphi(1)^{\frac{1}{2}} \psi(p_1 x^* x p_1)^{\frac{1}{2}}. \end{aligned}$$

Therefore, $x \to \varphi(x(p_1+p_0))$ is $s(\mathscr{M}, \mathscr{M}_*)$-continuous, and so $\sigma(\mathscr{M}, \mathscr{M}_*)$-continuous (cf. 1.8.10). This contradicts the maximality of p_0. Hence $p_0 = 1$. q.e.d.

This theorem has the following important implication. Let $\mathscr{M}^*$ be the dual of $\mathscr{M}$. We canonically embed $\mathscr{M}_*$ into $\mathscr{M}^*$. Then $\mathscr{M}_*$ is a norm-closed linear subspace of $\mathscr{M}^*$ generated by the set T. On the other hand, by the above theorem, T is the set of all normal positive linear functionals. Since normality is determined by the order property on $\mathcal{M}$ only, the space $\mathcal{M}_*$ is unique — that is, if $\mathcal{M}_{*1}^* = \mathcal{M}_{*2}^* = \mathcal{M}$ for two Banach spaces $\mathcal{M}_{*1}, \mathcal{M}_{*2}$, then $\mathcal{M}_{*1}$ coincides with $\mathcal{M}_{*2}$ when they are canonically embedded into $\mathcal{M}^*$. Hence, $\mathcal{M}$ has only one predual $\mathcal{M}_*$ such that $\mathcal{M}_*^* = \mathcal{M}$.

1.13.3. Corollary. *Let $\mathcal{M}$ be a W^*-algebra. Then $\mathcal{M}$ has a unique predual space $\mathcal{M}_*$ such that the dual of $\mathcal{M}_*$ is $\mathcal{M}$—that is, if $\mathcal{M}_{*1}^* = \mathcal{M}_{*2}^* = \mathcal{M}$ for two Banach spaces $\mathcal{M}_{*1}, \mathcal{M}_{*2}$, then $\mathcal{M}_{*1} = \mathcal{M}_{*2}$, where $(\mathcal{M}_{*i})^*$ is the dual Banach space of $\mathcal{M}_{*i}$ $(i=1,2)$.*

1.13.4. Definition. *Let $\{e_\alpha | \alpha \in \mathbb{I}\}$ be a family of mutually orthogonal projections in $\mathcal{M}$. Then the sum $\sum_{\alpha\in\mathbb{I}} e_\alpha$ is defined as follows: Let $\mathbb{J}$ be any finite subset of $\mathbb{I}$ and put $P_{\mathbb{J}} = \sum_{\alpha\in\mathbb{J}} e_\alpha$. Then $\{P_{\mathbb{J}}\}$ is a uniformly bounded increasing directed set under the order defined by the set-inclusion of subsets $\mathbb{J}$. Hence, this directed set converges to $\operatorname{l.u.b.}_{\mathbb{J}} P_{\mathbb{J}}$ in $\sigma(\mathcal{M}, \mathcal{M}_*)$. Moreover,*

$$\alpha\,\varphi(\operatorname{l.u.b.}_{\mathbb{J}} P_{\mathbb{J}} - P_{\mathbb{J}})^2 = \varphi((\operatorname{l.u.b.}_{\mathbb{J}} P_{\mathbb{J}} - P_{\mathbb{J}})^{\frac{1}{2}} (\operatorname{l.u.b.}_{\mathbb{J}} P_{\mathbb{J}} - P_{\mathbb{J}})^{\frac{1}{2}})$$
$$\leq \varphi(\operatorname{l.u.b.}_{\mathbb{J}} P_{\mathbb{J}} - P_{\mathbb{J}})^{\frac{1}{2}} \|\varphi\|^{\frac{1}{2}} \quad \text{for } \varphi \in T.$$

Hence, $\{P_{\mathbb{J}}\}$ converges to the $\operatorname{l.u.b.}_{\mathbb{J}} P_{\mathbb{J}}$ in $s(\mathcal{M}, \mathcal{M}_*)$, so that $\operatorname{l.u.b.}_{\mathbb{J}} P_{\mathbb{J}}$ is a projection (cf. 1.8.12).

We define $\sum_{\alpha\in\mathbb{I}} e_\alpha = \operatorname{l.u.b.}_{\mathbb{J}} P_{\mathbb{J}}$. From the proof of 1.13.2, we can easily see that normality is equivalent to complete additivity (i.e., for arbitrary $\{e_\alpha | \alpha \in \mathbb{I}\}$ of mutually orthogonal projections in $\mathcal{M}$, $\varphi\left(\sum_{\alpha\in\mathbb{I}} e_\alpha\right) = \sum_{\alpha\in\mathbb{I}} \varphi(e_\alpha)$).

1.13.5. Definition. *We shall call a $\sigma(\mathcal{M}, \mathcal{M}_*)$-continuous linear functional on $\mathcal{M}$ a normal linear functional.*

Concluding remarks on 1.13.

1. Theorem 1.13.2 is due to Dixmier [32].

2. In this section, the reader learns a quite new class of Banach spaces—that is, the existence of a dual Banach space E which has a unique Banach space F such that $F^* = E$. Of course, the reflexivity implies such the property, but one can easily see that a reflexive W^*-algebra is finite-dimensional (cf. [155]).

It seems to be interesting to seek a characterization of a dual Banach space having such a property.

Reference. [32].

1.14. Polar Decomposition of Linear Functionals on a W^*-Algebra

1.14.1. Definition. *Let φ_1, φ_2 be two positive linear functionals on a C^*-algebra $\mathscr{A}$. We say that φ_1 is orthogonal to φ_2 if* $\|\varphi_1-\varphi_2\|=\|\varphi_1\|+\|\varphi_2\|$.

1.14.2. Definition. *Let φ be a normal positive linear functional on a W^*-algebra $\mathscr{M}$. Put $\mathscr{L}=\{x|\varphi(x^*x)=0, x\in\mathscr{M}\}$. $\mathscr{L}$ is a $s(\mathscr{M},\mathscr{M}_*)$-closed left ideal. Hence, it is $\sigma(\mathscr{M},\mathscr{M}_*)$-closed* (cf. 1.8.11), *and so $\mathscr{L}=\mathscr{M}p$ for some projection p.*

p is the greatest of all projections q such that $\varphi(q)=0$.

The projection $1-p$ (denoted by $s(\varphi)$) is called the support of φ. Clearly $\varphi(x)=\varphi(s(\varphi)x)=\varphi(xs(\varphi))=\varphi(s(\varphi)xs(\varphi))$ for $x\in\mathscr{M}$. φ is said to be faithful if $s(\varphi)=1$. Let $\{\varphi_\alpha|\alpha\in\mathbb{I}\}$ be a family of normal positive linear functionals on $\mathscr{M}$. The family is said to be faithful if $\varphi_\alpha(x^*x)=0$ for all $\alpha\in\mathbb{I}$ imply $x=0$ $(x\in\mathscr{M})$.

1.14.3. Theorem. *Let $\mathscr{M}$ be a W^*-algebra and let $\mathscr{M}_*$ be the predual of $\mathscr{M}$. If $f\in\mathscr{M}_*$ and $f^*=f$, then it can be expressed as the difference of two normal positive linear functionals f_1, f_2, $f=f_1-f_2$, which satisfy $\|f\|=\|f_1\|+\|f_2\|$. Such a decomposition is unique. This decomposition is called the orthogonal decomposition of f, and we write $f_1=f^+$ and $f_2=f^-$.*

Proof. Let $\|f\|=1$. By the $\sigma(\mathscr{M},\mathscr{M}_*)$-compactness of $\mathscr{M}^s\cap S$, there exists an element x of $\mathscr{M}^s$ such that $f(x)=1$ and $\|x\|=1$. Put $\mathscr{E}=\{x\,|\,f(x)=1, x\in\mathscr{M}^s\cap S\}$. Then $\mathscr{E}$ is a $\sigma(\mathscr{M},\mathscr{M}_*)$-compact convex set. Let u be an extreme point of $\mathscr{E}$. u is then extreme in $\mathscr{M}^s\cap S$. Hence u is a self-adjoint unitary element. Hence $u=p-p'$, where p is a projection, $p'=1-p$, and $f(p-p')=1$. Put $f_1(x)=f(px)$ and $f_2(x)=-f(p'x)$. Then $(f_1+f_2)(1)=f_1(p)+f_2(p')=1$, and

$$|(f_1+f_2)(x)|=|f(px)-f(p'x)|=|f((p-p')x)|\le\|p-p'\|\;\|f\|\;\|x\|.$$

Hence, $\|f_1+f_2\|=1$, so that $f_1+f_2\ge0$ (cf. 1.5.2).

The norm of f_1 on $p\mathscr{M}p$ is $f_1(p)$. For suppose $f_1(p)<f_1(x)$ for some $x(\|x\|\le1)\in p\mathscr{M}p$. Then

$$\|x-p'\|\le1 \text{ and } f(x-p')=f(x)-f(p')>f(p)-f(p')=1,$$

a contradiction. Hence $f_1\ge0$ on $p\mathscr{M}p$, $f_1=\dfrac{(f+f_1+f_2)}{2}$, and so $f_1^*=f_1$. Hence $f_1(y^*)=f(py^*)=\overline{f(yp)}=\overline{f_1(y)}=\overline{f(py)}$, so that

$$f(yp)=f(yp) \quad \text{for } y\in\mathscr{M},$$

and thus $f(p(yp))=f((yp)p)=f(yp)=f(py)$. Hence,

$$f_1(pyp)=f(pyp)=f(py)=f_1(y),$$

and so f_1 is positive on $\mathscr{M}$. Similarly, f_2 is positive on $\mathscr{M}$ and $f_2(p')=\|f_2\|$. Hence, $\|f_1-f_2\|=\|f_1\|+\|f_2\|$.

Next suppose $f=f_1-f_2=f_1'-f_2'$, $\|f_1\|+\|f_2\|=\|f_1'\|+\|f_2'\|$, and $f_i, f_i'\geq 0$ $(i=1,2)$. Then $f_1(1)+f_2(1)=f_1'(1)+f_2'(1)$ and

$$f_1(1)-f_2(1)=f_1'(1)-f_2'(1).$$

Hence $f_i(1)=f_i'(1)$ *for* $i=1,2$. Since $s(f_1)\leq p$,

$$f(s(f_1))=f_1(s(f_1))=\|f_1\|=\|f_1'\|=f_1'(s(f_1))-f_2'(s(f_1)).$$

Hence $f_2'(s(f_1))=0$ and $f_1'(s(f_1))=\|f_1'\|$. Therefore $s(f_1')\leq s(f_1)$. Now

$$f_1(x)=f(s(f_1)x)=f_1'(s(f_1)x)-f_2'(s(f_1)x)=f_1'(s(f_1)x)$$

and

$$f_1'(s(f_1)x)=f_1'(s(f_1')x)+f_1'\{(s(f_1)-s(f_1'))x\}=f_1'(s(f_1')x)=f_1'(x).$$

Hence $f_1(x)=f_1'(x)$ and $f_2(x)=f_2'(x)$. q.e.d.

1.14.4. Theorem. *Let g be a $\sigma(\mathcal{M},\mathcal{M}_*)$-continuous linear functional on a W^*-algebra $\mathcal{M}$. Then it can be expressed as follows: $g=R_v\varphi$, where φ is a normal positive linear functional on $\mathcal{M}$ with $\|g\|=\|\varphi\|$, and v is a partial isometry of $\mathcal{M}$ having $s(\varphi)$ as the initial projection. Such a decomposition is unique. φ (denoted by $|g|$) is called the absolute value of g. The final projection of v is $s(|g^*|)$. This decomposition is called the polar decomposition of g.*

Proof. We may assume that $\|g\|=1$. Let $\mathscr{E}=\{a\,|\,g(a)=1, a\in S\}$ and let u be an extreme point of $\mathscr{E}$. Then u is extreme in S. Hence, u is a partial isometry (cf. 1.6.4). $\|R_u g\|\leq\|R_u\|\,\|g\|\leq 1$ and $R_u g(1)=g(u)=1$. Hence, $R_u g\geq 0$. Since $uu^*u=u$, $g(u)=g(uu^*u)=R_u g(uu^*)=1$. Hence,

$$uu^*\geq s(R_u g).$$

Let $w=u^*s(R_u g)$. Then $w^*w=s(R_u g)$; w is a partial isometry having $s(R_u g)$ as its initial projection. Moreover,

$$(R_u g)(x)=R_u g(x\,s(R_u g))=g(x\,s(R_u g)u)=g(xw^*)=R_{w^*}g(x)$$

$(x\in\mathcal{M})$.

Hence $R_u g=R_{w^*}g$.

Next, we shall show that $g(x)=g(xp)=g(qx)$ $(x\in\mathcal{M})$, where $p=ww^*$ and $q=w^*w$.

Suppose that for some $x_0(\|x_0\|\leq 1)$, $g(x_0(1-p))=\beta>0$. Then

$$\|nw^*+x_0(1-p)\|=\|\{nw^*+x_0(1-p)\}\{nw+(1-p)x_0^*\}\|^{\frac{1}{2}}$$
$$=\|n^2q+x_0(1-p)x_0^*\|^{\frac{1}{2}}\leq(1+n^2)^{\frac{1}{2}}.$$

On the other hand,

$$\begin{aligned}g(nw^*+x_0(1-p))&=ng(w^*)+g(x_0(1-p))\\&=n+\beta>(n^2+1)^{\frac{1}{2}}\\&\geq\|nw^*+x_0(1-p)\|\end{aligned}$$

for a sufficiently large number n. Hence, $g(x(1-p))=0$ for $x\in\mathcal{M}$. Similarly, $g(x)=g(qx)$ for $x\in\mathcal{M}$. Hence, $g(x)=g(xp)=g(xww^*)=R_{w^*}g(xw)$. Hence $g=R_w\varphi$, where $\varphi=R_{w^*}g\geq 0$ and $\|g\|=\|R_{w^*}g\|=\|\varphi\|$.

Finally, we show the uniqueness of the decomposition. Let $g=R_w\varphi=R_{w'}\varphi'$. Let $w'w'^*=\overline{p}$, $w'^*w'=\overline{q}$. Then $g(x)=\varphi(xw)=\varphi'(xw')$, and $\varphi(x)=\varphi(xq)=\varphi(xw^*w)=\varphi'(xw^*w')=\varphi'(\overline{q}xw^*w')$. Hence $\varphi(1-\overline{q})=0$, and so $q=s(\varphi)\leq\overline{q}$. Similarly $\overline{q}=s(\varphi')\leq q$, and so $q=\overline{q}$.

Put $w'^*w=h+ik$ (h,k self-adjoint). Then $h,k\in q\mathcal{M}q$ since $w'^*w\in q\mathcal{M}q$. $\varphi(w'^*w)=\varphi'(w'^*w')=\varphi'(q)=1=\varphi(h)+i\varphi(k)$. Hence, $\varphi(h)=1$. Since $\|h\|\leq 1$, $h=q$, and since $\|w'^*w\|\leq 1$, $k=0$. Therefore $w'^*w=q$ and $w=w'$. q.e.d.

Concluding remark on 1.14.

Theorem 1.14.3 is due to Grothendieck [66].

References. [152], [209].

1.15. Concrete C^*-Algebras and W^*-Algebras

Let $\mathcal{H}$ be a complex Hilbert space and let $B(\mathcal{H})$ be the algebra of all bounded linear operators on $\mathcal{H}$. We can define various topologies in $B(\mathcal{H})$.

1. The uniform topology in $B(\mathcal{H})$. The uniform topology in $B(\mathcal{H})$ is given by the operator norm $\|a\|$ $(a\in B(\mathcal{H}))$, where $\|a\|=\sup_{\substack{\|\xi\|\leq 1\\ \xi\in\mathcal{H}}}\|a\xi\|$. $B(\mathcal{H})$ is a Banach algebra with this norm. We shall take the adjoint operation $a\to a^*$ as the involution $*$ on $B(\mathcal{H})$ (i.e., $(a\xi,\eta)=(\xi,a^*\eta)$ for $\xi,\eta\in\mathcal{H}$, where $(\ ,\ )$ is the scalar product of $\mathcal{H}$). Then

$$\|a\|^2=\sup_{\|\xi\|=1}\|a\xi\|^2=\sup_{\|\xi\|=1}(a\xi,a\xi)=\sup_{\|\xi\|=1}(a^*a\xi,\xi)\leq\|a^*a\|.$$

Since $\|a^*\|=\|a\|$, $\|a\|^2=\|a^*a\|$. Therefore, $B(\mathcal{H})$ is a C^*-algebra, and so any uniformly closed self-adjoint subalgebra of $B(\mathcal{H})$ is also a C^*-algebra.

2. The strong operator topology in $B(\mathcal{H})$. Let $\xi\in\mathcal{H}$. The function $a\to\|a\xi\|$ is a semi-norm on $B(\mathcal{H})$. The set of all such semi-norms $\{\|a\xi\|\,|\,\xi\in\mathcal{H}\}$ defines a Hausdorff locally convex topology in $B(\mathcal{H})$. This is the strong operator topology.

3. The strongest operator topology. Let $(\xi_i)\subset\mathcal{H}$ be any sequence of elements in $\mathcal{H}$ such that $\sum_{i=1}^{\infty}\|\xi_i\|^2<+\infty$. The function

$$\left(\sum_{i=1}^{\infty}\|a\xi_i\|^2\right)^{\frac{1}{2}}\quad(a\in B(\mathcal{H}))$$

defines a semi-norm on $B(\mathscr{H})$. The set of all such semi-norms

$$\left\{\left(\sum_{i=1}^{\infty}\|a\xi_i\|^2\right)^{\frac{1}{2}}\middle|(\xi_i)\subset\mathscr{H},\ \sum_{i=1}^{\infty}\|\xi_i\|^2<+\infty\right\}$$

defines a Hausdorff locally convex topology in $B(\mathscr{H})$. This is the strongest operator topology.

4. The weak operator topology. For ξ, $\eta\in\mathscr{H}$, the function $|(a\xi,\eta)|$ is a semi-norm on $B(\mathscr{H})$. The set of all semi-norms $\{|(a\xi,\eta)|\,|\,\xi,\eta\in\mathscr{H}\}$ defines a Hausdorff locally convex topology. This is the weak operator topology.

5. The σ-weak operator topology. For (ξ_n), $(\eta_n)\subset\mathscr{H}$ such that $\sum_{n=1}^{\infty}\|\xi_n\|^2<+\infty$, $\sum_{n=1}^{\infty}\|\eta_n\|^2<+\infty$, consider the semi-norm $\left|\sum_{n=1}^{\infty}(a\xi_n,\eta_n)\right|$ on $B(\mathscr{H})$. The set of all such semi-norms will define a Hausdorff locally convex topology, the σ-weak operator topology on $B(\mathscr{H})$.

Let V be the linear space of all continuous linear functionals on $B(\mathscr{H})$ with respect to the weak operator topology. Then the weak operator topology is equivalent to $\sigma(B(\mathscr{H}),V)$. One can easily see that the unit sphere S of $B(\mathscr{H})$ is $\sigma(B(\mathscr{H}),V)$-compact.

Let $B(\mathscr{H})^*$ be the dual of $B(\mathscr{H})$ with the uniform topology. Since the uniform topology is stronger than the weak operator topology, V may be identified with a linear subspace of $B(\mathscr{H})^*$.

Since S is $\sigma(B(\mathscr{H}),V)$-compact and the unit sphere of V is the polar S^0 of S in V, the norm topology of V as a linear subspace of $B(\mathscr{H})^*$ is weaker than the $\tau(V,B(\mathscr{H}))$-topology. Hence, $\sigma(V,B(\mathscr{H}))\leq$ the norm topology $\leq\tau(V,B(\mathscr{H}))$. Therefore, by Mackey's theorem, the dual Banach space of the normed space V coincides with $B(\mathscr{H})$ as a set. Moreover the unit sphere of $B(\mathscr{H})$ is the polar of S^0 since

$$\sup_{\substack{\|\xi\|=\|\eta\|=1\\ \xi,\eta\in\mathscr{H}}}|(a\xi,\eta)|=\|a\|\,.$$

Hence, the dual Banach space of the normed space V coincides with the Banach space $B(\mathscr{H})$, and so the dual of $\bar{V}$ is $B(\mathscr{H})$, where $\bar{V}$ is the norm-completion of V. Hence $B(\mathscr{H})$ is a W^*-algebra. Let $B(\mathscr{H})_*$ be the predual of $B(\mathscr{H})$, and let $\mathscr{N}$ be any $\sigma(B(\mathscr{H}),B(\mathscr{H})_*)$-closed self-adjoint subalgebra of $B(\mathscr{H})$. Then $\mathscr{N}$ is a dual Banach space. Hence it too is a W^*-algebra. In particular, any self-adjoint subalgebra of $B(\mathscr{H})$ closed with respect to the weak operator topology is a W^*-algebra, since $\sigma(B(\mathscr{H}),V)$ is weaker than $\sigma(B(\mathscr{H}),B(\mathscr{H})_*)$.

1.15.1. Proposition. *Let $\mathscr{N}$ be a self-adjoint subalgebra of $B(\mathscr{H})$. The following conditions are equivalent. $\mathscr{N}$ is closed in* 1. *the weak operator topology;* 2. *the σ-weak operator topology;* 3. *the strong operator topology;* 4. *the strongest operator topology;* 5. $\sigma(B(\mathscr{H}),B(\mathscr{H})_*)$.

Proof. Clearly, the weak operator topology $\leq$ the σ-weak operator topology $\leq \sigma(B(\mathscr{H}), B(\mathscr{H})_*) \leq \tau(B(\mathscr{H}), B(\mathscr{H})_*)$ and the weak operator topology $\leq$ the strong operator topology $\leq$ the strongest operator topology $\leq s(B(\mathscr{H}), B(\mathscr{H})_*) \leq \tau(B(\mathscr{H}), B(\mathscr{H})_*)$.

For if $\sum_{n=1}^{\infty} \|\xi_n\|^2 < +\infty$, $\sum_{n=1}^{\infty} \|\eta_n\|^2 < +\infty$ and $f_m(x) = \sum_{n=1}^{m} (x\xi_n, \eta_n)$, then

$$|f(x) - f_m(x)| \leq \sum_{n=m+1}^{\infty} |(x\xi_n, \eta_n)| = \sum_{n=m+1}^{\infty} \|x\| \, \|\xi_n\| \, \|\eta_n\|$$
$$\leq \|x\| \left(\sum_{n=m+1}^{\infty} \|\xi_n\|^2 \right)^{\frac{1}{2}} \left(\sum_{n=m+1}^{\infty} \|\eta_n\|^2 \right)^{\frac{1}{2}},$$

and so $\|f - f_m\| \to 0$ $(m \to \infty)$. Therefore $\tau(B(\mathscr{H}), B(\mathscr{H})_*)$ is the strongest topology.

Now suppose that $\mathscr{N}$ is $\tau(B(\mathscr{H}), B(\mathscr{H})_*)$-closed. Let $\mathscr{N}_1$ be the $\sigma(B(\mathscr{H}), V)$-closure of $\mathscr{N}$ in $B(\mathscr{H})$. Since V is a norm-dense, self-adjoint invariant linear subspace of $B(\mathscr{H})_*$, by Kaplansky density theorem, the unit sphere of $\mathscr{N}$ is $\tau(B(\mathscr{H}), B(\mathscr{H})_*)$-dense in the unit sphere of $\mathscr{N}_1$, but $\mathscr{N}$ is $\tau(B(\mathscr{H}), B(\mathscr{H})_*)$-closed. Hence $\mathscr{N}_1 = \mathscr{N}$. q.e.d.

Henceforward, we shall call a self-adjoint subalgebra which is closed with respect to the weak operator topology, a weakly closed self-adjoint subalgebra of $B(\mathscr{H})$, since there can be no confusion.

1.15.2. Proposition. *Let $\mathscr{N}$ be a weakly closed self-adjoint subalgebra of $B(\mathscr{H})$, and let $\mathscr{N}_*$ be the predual of $\mathscr{N}$. Then* 1. *the weak operator topology, the σ-weak operator topology and $\sigma(\mathscr{N}, \mathscr{N}_*)$ are equivalent on bounded spheres;* 2. *the strong operator topology, the strongest operator topology and $s(\mathscr{N}, \mathscr{N}_*)$ are equivalent on bounded spheres.*

Proof. 1. is clear from the $\sigma(\mathscr{N}, \mathscr{N}_*)$-compactness of the unit sphere of $\mathscr{N}$. On the other hand, $x_\alpha \to x$ in the strong operator topology is equivalent to $(x_\alpha - x)^*(x_\alpha - x) \to 0$ in the weak operator topology. q.e.d.

Next, we shall find a more exact form of $B(\mathscr{H})_*$. Let (ξ_α) be a complete orthonormal system of $\mathscr{H}$. For $a \in B(\mathscr{H})$, put $\|a\|_2 = \left(\sum_\alpha \|a\xi_\alpha\|^2\right)^{\frac{1}{2}}$. Then $\|a\|_2$ does not depend upon the special choice of (ξ_α)—in fact, let (η_β) be another complete orthonormal system; then

$$\sum_\beta \|a\eta_\beta\|^2 = \sum_\beta \| \, |a|\eta_\beta\|^2 = \sum_\beta \sum_\alpha |(|a|\eta_\beta, \xi_\alpha)|^2 = \sum_\alpha \| \, |a|\xi_\alpha\|^2 = \sum_\alpha \|a\xi_\alpha\|^2.$$

Let $H = \{a \mid \|a\|_2 < +\infty, a \in B(\mathscr{H})\}$. Then H is a linear subspace of $B(\mathscr{H})$ and it is a pre-Hilbert space with the norm $\|a\|_2$.

Elements in H are called the operators of Hilbert-Schmidt class. The scalar product of H is $(a,b) = \sum_\alpha (b^* a\xi_\alpha, \xi_\alpha)$. We can easily see the following properties.

1. $a \in H$ implies $a^* \in H$ and $\|a^*\|_2 = \|a\|_2$,
2. $b \in B(\mathscr{H})$, $a \in H$ imply ba, $ab \in H$, $\|ba\|_2 \leq \|b\| \, \|a\|_2$ and $\|ab\|_2 \leq \|a\|_2 \, \|b\|$.

Therefore H is a two-sided ideal of $B(\mathscr{H})$.

$$\|a\xi\|_2 = \sum_\alpha |(a\xi, \xi_\alpha)|^2 = \sum_\alpha |(\xi, a^* \xi_\alpha)|^2 \leq \|\xi\|^2 \sum_\alpha \|a^* \xi_\alpha\|^2 = \|a\|_2^2 \, \|\xi\|^2.$$

Hence $\|a\| \leq \|a\|_2$.

Now we shall show that H is a Hilbert space with the norm $\| \;\|_2$. Let (a_n) be a $\| \;\|_2$-Cauchy sequence in H. Then it is also a $\| \;\|$-Cauchy sequence. Let $a_n \to a_0$ (uniformly). For arbitrary $\varepsilon > 0$, there exists an integer n_0 such that

$$\sum_\alpha \|(a_m - a_n)\xi_\alpha\|^2 = \|a_m - a_n\|_2^2 < \varepsilon \;\; (m, n \geq n_0).$$

For an arbitrary finite subset $(\xi_{\alpha_1}, \xi_{\alpha_2}, \ldots, \xi_{\alpha_l})$, we have

$$\sum_{i=1}^{l} \|(a_m - a_n)\xi_{\alpha_i}\|^2 < \varepsilon \;\; (m, n \geq n_0).$$

Hence

$$\lim_{m \to \infty} \sum_{i=1}^{l} \|(a_m - a_n)\xi_{\alpha_i}\|^2 = \sum_{i=1}^{l} \|(a_0 - a_n)\xi_{\alpha_i}\|^2 \leq \varepsilon \;\; (n > n_0).$$

Therefore, we have $\sum_\alpha \|(a_0 - a_n)\xi_\alpha\|^2 \leq \varepsilon$ $(n \geq 0)$, and so $a_0 \in H$ and $a_n \to a_0$ in H. Therefore H is a Hilbert space.

For $a \in H$, $\sum_\alpha \|a\xi_\alpha\|^2 < +\infty$; hence $a\xi_\alpha = 0$ for all ξ_α except for a countable subset (ξ_{α_i}). Let p_n be the orthogonal projection of $\mathscr{H}$ onto the finite-dimensional subspace of $\mathscr{H}$ spanned by $(\xi_{\alpha_1}, \xi_{\alpha_2}, \ldots, \xi_{\alpha_n})$. Then $\|a - ap_n\|_2^2 = \sum_{j=n+1}^{\infty} \|a\xi_{\alpha_j}\|^2 \to 0$. Hence $\|a - ap_n\| \to 0$. Therefore a is a compact linear operator.

Next, let $h \; (\geq 0) \in B(\mathscr{H})$ and put $Tr(h) = \sum_\alpha (h\xi_\alpha, \xi_\alpha)$; then $Tr(h)$ does not depend on the choice of (ξ_α)—in fact,

$$\sum_\alpha (h\xi_\alpha, \xi_\alpha) = \sum_\alpha \|h^{\frac{1}{2}} \xi_\alpha\|^2 = \|h^{\frac{1}{2}}\|_2^2.$$

An element $a \in B(\mathscr{H})$ is called an operator of trace class if $Tr(|a|) < +\infty$.

Let $T(\mathscr{H})$ be the set of all operators of trace class. Suppose a, $b \in H$, and let $ab = u|ab|$ be the polar decomposition of ab; then

$$\sum_\alpha (|ab|\xi_\alpha, \xi_\alpha) = \sum_\alpha (u^* ab\xi_\alpha, \xi_\alpha) = \sum_\alpha (b\xi_\alpha, a^* u\xi_\alpha) \leq \|b\|_2 \|a^* u\|_2 < +\infty.$$

Hence $ab \in T(\mathscr{H})$. Conversely let $c \in T(\mathscr{H})$. Then $c = v|c| = v|c|^{\frac{1}{2}}|c|^{\frac{1}{2}}$ and $v|c|^{\frac{1}{2}}$, $|c|^{\frac{1}{2}} \in H$, where $v|c|$ is the polar decomposition of c. Therefore, we have, $a \in T(\mathscr{H})$ implies $a^* \in T(\mathscr{H})$; $b \in B(\mathscr{H})$, $a \in T(\mathscr{H})$ imply ba, $ab \in T(\mathscr{H})$.

Now for $a \in T(\mathscr{H})$, define $Tr(a) = \sum_{\alpha}(a\xi_\alpha, \xi_\alpha)$. Then $Tr(a)$ does not depend on the choice of (ξ_α)—in fact, $a = bc$ for $b, c \in H$ and so

$$\sum_{\alpha}(bc\,\xi_\alpha, \xi_\alpha) = \sum_{\alpha}(c\,\xi_\alpha, b^*\xi_\alpha) = (c, b^*).$$

Let u (unitary) $\in B(\mathscr{H})$, $h(\geq 0) \in T(\mathscr{H})$; then

$$Tr(u^* h u) = \|u^* h^{\frac{1}{2}} u\|_2^2 = \|h^{\frac{1}{2}}\|_2^2 = Tr(h).$$

Hence $Tr(u^* a u) = Tr(a)$ for $a \in T(\mathscr{H})$, since Tr is linear. Therefore $Tr(ua) = Tr(u^*(ua)u) = Tr(au)$ and so $Tr(ab) = Tr(ba)$ for $b \in B(\mathscr{H})$ and $a \in T(\mathscr{H})$.

Now, put $\varphi_a(x) = Tr(xa)$ ($x \in B(\mathscr{H})$ and $a \in T(\mathscr{H})$). Then φ_a is a linear functional on $B(\mathscr{H})$.

$$\begin{aligned}|\varphi_a(x)| &= |Tr(xa)| = |Tr(xv|a|)| = |Tr(xv|a|^{\frac{1}{2}}|a|^{\frac{1}{2}})| = (xv|a|^{\frac{1}{2}}, |a|^{\frac{1}{2}}) \\ &\leq \||a|^{\frac{1}{2}}\|_2 \|xv\| \, \||a|^{\frac{1}{2}}\|_2 = \|x\| \, \||a|^{\frac{1}{2}}\|_2^2 = \|x\| \, Tr(|a|).\end{aligned}$$

Moreover $|\varphi_a(v^*)| = Tr(|a|)$. Hence, $\|\varphi_a\| = Tr(|a|)$. The mapping $a \to \varphi_a$ of $T(\mathscr{H})$ into $B(\mathscr{H})^*$ is a linear mapping.

Now put $\|a\|_1 = Tr(|a|)$ $(a \in T(\mathscr{H}))$. Then $T(\mathscr{H})$ is a normed linear space under $\| \;\|_1$. Now let (ζ_n) and (ζ'_n) $(n = 1, 2, \ldots)$ be two arbitrary orthonormal systems of $\mathscr{H}$. For $a \in T(\mathscr{H})$,

$$\begin{aligned}\sum_{n=1}^{\infty} |(a\zeta_n, \zeta'_n)| &= \sum_{n=1}^{\infty} |(v|a|\zeta_n, \zeta'_n)| = \sum_{n=1}^{\infty} |(|a|^{\frac{1}{2}}\zeta_n, |a|^{\frac{1}{2}} v^* \zeta'_n)| \\ &\leq \sum_{n=1}^{\infty} \|a^{\frac{1}{2}}\zeta_n\| \, \||a|^{\frac{1}{2}} v^* \zeta'_n\| \leq \left(\sum_{n=1}^{\infty} \|a^{\frac{1}{2}}\zeta_n\|^2\right)^{\frac{1}{2}} \left(\sum_{n=1}^{\infty} \||a|^{\frac{1}{2}} v^* \zeta'_n\|^2\right)^{\frac{1}{2}} \\ &\leq \|a^{\frac{1}{2}}\|_2 \, \||a|^{\frac{1}{2}} v^*\|_2 \leq \||a|^{\frac{1}{2}}\|_2^2 = Tr(|a|),\end{aligned}$$

where $a = v|a|$ is the polar decomposition of a.

On the other hand, let (η_n) be a complete orthonormal system of the subspace $s(|a|)\mathscr{H}$ of $\mathscr{H}$, and put $\eta'_n = v\eta_n$. Then (η'_n) is an orthonormal system. Moreover,

$$\sum_{n=1}^{\infty} |(a\eta_n, \eta'_n)| = \sum_{n=1}^{\infty} |(v|a|\eta_n, v\eta_n)| = \sum_{n=1}^{\infty} (|a|\eta_n, \eta_n) = Tr(|a|).$$

Hence, $Tr(|a|) = \sup_{(\zeta_n),(\zeta'_n)} \sum_{n=1}^{\infty} |(a\zeta_n, \zeta'_n)|$. Moreover,

$$\|a\|_1 = Tr(|a|) = \||a|^{\frac{1}{2}}\|_2^2 \geq \||a|^{\frac{1}{2}}\|^2 = \||a|\| = \|a\|.$$

Now let (a_j) be a $\| \;\|_1$-Cauchy sequence in $T(\mathscr{H})$; then (a_j) is also $\| \;\|$-Cauchy. Let $a_j \to a_0$ (uniformly). For an arbitrary $\varepsilon > 0$, there exists an integer j_0 such that $\sup_{\substack{(\zeta_n)\\(\zeta'_n)}} \sum_{n=1}^{\infty} |((a_{j_1} - a_{j_2})\zeta_n, \zeta'_n)| < \varepsilon$ $(j_1, j_2, \geq j_0)$.

For an arbitrary positive integer m,

$$\sum_{n=1}^{m} |((a_{j_1}-a_{j_2})\zeta_n, \zeta'_n)| < \varepsilon$$

and so $\lim_{j_1\to\infty} \sum_{n=1}^{m} |((a_{j_1}-a_{j_2})\zeta_n, \zeta'_n)| \leq \varepsilon$. Hence $\sum_{n=1}^{m} |((a_0-a_{j_2})\zeta_n, \zeta'_n)| \leq \varepsilon$ and so $\sum_{n=1}^{\infty} |((a_0-a_{j_2})\zeta_n, \zeta'_n)| \leq \varepsilon$ for arbitrary $(\zeta_n), (\zeta'_n)$. Therefore $a_0-a_{j_2} \in T(\mathscr{H})$ and $\|a_0-a_{j_2}\|_1 \leq \varepsilon$ $(j_2 \geq j_0)$ and so $a_0 \in T(\mathscr{H})$ and $a_j \to a_0$ in $T(\mathscr{H})$. Namely, $T(\mathscr{H})$ is a Banach space. Hence, $\{\varphi_a | a \in T(\mathscr{H})\}$ is a closed linear subspace of $B(\mathscr{H})^*$.

Let a be a bounded operator of finite rank (i.e. $a(\mathscr{H})$ is finite-dimensional) and let $\mathscr{H}_0$ be the null space of a. Then the orthocomplement $\mathscr{H}_0^\perp$ of $\mathscr{H}_0$ is finite-dimensional. Let $(\zeta_1, \zeta_2, \ldots, \zeta_n)$ be a complete orthonormal system of $\mathscr{H}_0^\perp$. Then $Tr(xa) = \sum_{i=1}^{n} (xa\zeta_i, \zeta_i) = \sum_{i=1}^{n} (x\eta_i, \zeta_i)$, where $\eta_i = a\zeta_i$. Hence φ_a is $\sigma(B(\mathscr{H}), V)$-continuous and so $\varphi_\alpha \in V$.

Let $a \in T(\mathscr{H})$ and $a = v|a|$ be the polar decomposition of a. Since $|a|$ is a compact operator, the spectral resolution of $|a|$ implies that $|a| = \sum_{i=1}^{\infty} \lambda_i e_i$, $\lambda_i \geq 0$, $\lambda_i \to 0$, and $\dim(e_i \mathscr{H}) = 1$, where (e_i) is a family of mutually orthogonal projections in $B(\mathscr{H})$ and $\dim(\mathscr{H})$ is the dimension of $\mathscr{H}$. Put $h_n = \sum_{i=1}^{n} \lambda_i e_i$; then

$$\big\| |a| - h_n \big\|_1 = \Big\| \sum_{i=n+1}^{\infty} \lambda_i e_i \Big\|_1 = \sum_{i=n+1}^{\infty} \lambda_i \to 0 \quad (n \to \infty).$$

Hence $\varphi_{|a|} \in \bar{V} = B(\mathscr{H})_*$, and

$$\varphi_a(x) = Tr(xa) = Tr(xv|a|) = \varphi_{|a|}(xv) = R_v \varphi_{|a|}(x).$$

Hence $\varphi_a \in B(\mathscr{H})_*$, and so $\{\varphi_a | a \in T(\mathscr{H})\} \subset B(\mathscr{H})_*$.

On the other hand, for non-zero $\xi, \eta \in \mathscr{H}$, let a be the linear operator of rank one on $\mathscr{H}$ such that $a\eta = \xi$ and $a\zeta = 0$ for all ζ with $(\zeta, \eta) = 0$. Then,

$$Tr(xa) = \left(xa\frac{\eta}{\|\eta\|}, \frac{\eta}{\|\eta\|} \right) = \frac{1}{\|\eta\|^2}(x\xi, \eta).$$

Hence, $V \subset \{\varphi_a | a \in T(\mathscr{H})\}$ and so $\{\varphi_a | a \in T(\mathscr{H})\} = B(\mathscr{H})_*$.

Now we have,

1.15.3. Theorem. *The predual $B(\mathscr{H})_*$ of $B(\mathscr{H})$ may be identified with the Banach space of all trace class operators on $\mathscr{H}$ under the isometric linear mapping $a \to \varphi_a$, where $\varphi_a(x) = Tr(xa)$ $(x \in B(\mathscr{H})$ and $a \in T(\mathscr{H}))$. Moreover, under this identification, positive elements in $B(\mathscr{H})_*$ are identified with positive elements in $T(\mathscr{H})$.*

1.15.4. Corollary. *Let φ be a positive linear functional on $B(\mathscr{H})$. Then the following conditions are equivalent:*

1. *φ is normal;*
2. *There exists a sequence (ξ_n) of elements in $\mathscr{H}$ with $\sum_{n=1}^{\infty} \|\xi_n\|^2 < +\infty$ such that $\varphi(x) = \sum_{n=1}^{\infty} (x\xi_n, \xi_n)$ $(x \in B(\mathscr{H}))$.*

Proof. Clearly, 2. implies 1.

Suppose $\varphi \in B(\mathscr{H})_*$. Then there exists a positive operator h of trace class such that $\varphi(x) = Tr(xh)$ for $x \in B(\mathscr{H})$. Put $h = \sum_{i=1}^{\infty} \lambda_i e_i$, where $\lambda_i \geq 0$, $\sum_{i=1}^{\infty} \lambda_i < +\infty$, $\dim(e_i \mathscr{H}) = 1$ and (e_i) is a family of mutually orthogonal projections. Let (ξ_i') be an orthonormal system of $\mathscr{H}$ such that $e_i \xi_i' = \xi_i'$. Then,

$$Tr(xh) = \sum_{i=1}^{\infty} (xh\xi_i', \xi_i') = \sum_{i=1}^{\infty} \lambda_i (x\xi_i', \xi_i') = \sum_{i=1}^{\infty} (x\sqrt{\lambda_i}\xi_i', \sqrt{\lambda_i}\xi_i')$$

and $\sum_{i=1}^{\infty} \|\sqrt{\lambda_i}\xi_i'\|^2 = \sum_{i=1}^{\infty} \lambda_i < +\infty$. q.e.d.

1.15.5. Corollary. *Let f be a linear functional on $B(\mathscr{H})$. Then the following conditions are equivalent:*

1. *$f \in B(\mathscr{H})_*$;*
2. *There exist two sequences (ξ_n), (η_n) of elements in $\mathscr{H}$ such that $\sum_{n=1}^{\infty} \|\xi_n\|^2 < +\infty$, $\sum_{n=1}^{\infty} \|\eta_n\|^2 < +\infty$, and $f(x) = \sum_{n=1}^{\infty} (x\xi_n, \eta_n)$ $(x \in B(\mathscr{H}))$.*

For the proof, use the polar decomposition of f. q.e.d.

1.15.6. Corollary. *The strongest operator topology on $B(\mathscr{H})$ is equivalent to $s(B(\mathscr{H}), B(\mathscr{H})_*)$. The σ-weak operator topology on $B(\mathscr{H})$ is equivalent to $\sigma(B(\mathscr{H}), B(\mathscr{H})_*)$.*

Concluding remark on 1.15.

Theorem 1.15.3 is due to Dixmier [30] and von Neumann and Schatten [174].

References. [30], [174].

1.16. The Representation Theorems for C^*-Algebras and W^*-Algebras

1.16.1. Definition. *Let $\mathcal{M}_1, \mathcal{M}_2$ be two W^*-algebras and let Φ be a $*$-homomorphism of $\mathcal{M}_1$ into $\mathcal{M}_2$. Φ is said to be a W^*-homomorphism if it is a continuous mapping of $\mathcal{M}_1$ (with the $\sigma(\mathcal{M}_1, \mathcal{M}_{1*})$-topology) into $\mathcal{M}_2$ (with the $\sigma(\mathcal{M}_2, \mathcal{M}_{2*})$-topology).*

1.16.2. Proposition. *Let Φ be a W^*-homomorphism of a W^*-algebra $\mathcal{M}_1$ into another W^*-algebra $\mathcal{M}_2$. Then the image $\Phi(\mathcal{M}_1)$ is $\sigma(\mathcal{M}_2, \mathcal{M}_{2*})$-closed.*

Proof. Let $\mathcal{I} = \{x | \Phi(x) = 0, x \in \mathcal{M}_1\}$. Then $\mathcal{I}$ is a $\sigma(\mathcal{M}_1, \mathcal{M}_{1*})$-closed two-sided ideal in $\mathcal{M}_1$. Hence there exists a central projection z in $\mathcal{M}_1$ such that $\mathcal{I} = \mathcal{M}_1 z$. The restriction of Φ to the W^*-algebra $\mathcal{M}_1(1-z)$ is a $*$-isomorphism. Hence, it is an isometry (cf. 1.2.6). Therefore $\Phi(S)$ is the unit sphere of $\Phi(\mathcal{M}_1)$, where S is the unit sphere of $\mathcal{M}_1$. Hence, $\Phi(S)$ is $\sigma(\mathcal{M}_2, \mathcal{M}_{2*})$-compact and so $\Phi(\mathcal{M}_1)$ is a W^*-subalgebra of $\mathcal{M}_2$. q.e.d.

1.16.3. Definition. *Let $\mathcal{A}$ be a C^*-algebra. A $*$-representation of $\mathcal{A}$ is a $*$-homomorphism π of $\mathcal{A}$ into $B(\mathcal{H})$ ($\mathcal{H}$ some Hilbert space). We shall denote this $*$-representation of $\mathcal{A}$ by $\{\pi, \mathcal{H}\}$.*

1.16.4. Definition. *Let $\mathcal{M}$ be a W^*-algebra. A W^*-representation of $\mathcal{M}$ is a W^*-homomorphism π of $\mathcal{M}$ into $B(\mathcal{H})$ ($\mathcal{H}$ some Hilbert space). We shall denote this W^*-representation of $\mathcal{M}$ by $\{\pi^w, \mathcal{H}\}$ or $\{\pi, \mathcal{H}\}$.*

Let φ be a bounded positive linear functional on a C^*-algebra $\mathcal{A}$. Introduce a conjugate bilinear functional $(x,y) = \varphi(y^* x)$ in $\mathcal{A}$. Let $\mathcal{I} = \{x | \varphi(x^* x) = 0, x \in \mathcal{A}\}$. Then $\mathcal{I}$ is a closed left ideal of $\mathcal{A}$. Define a conjugate bilinear functional on the quotient space $\mathcal{A}/\mathcal{I}$ such that if $x \in x_\varphi$, $y \in y_\varphi$, then $(x_\varphi, y_\varphi) = \varphi(y^* x)$ (here x_φ (resp. y_φ) is the class containing x (resp. y)). The expression (x_φ, y_φ) does not depend on the special choice of the representatives x, y. (x_φ, y_φ) will define a scalar product on $\mathcal{A}/\mathcal{I}$ under which $\mathcal{A}/\mathcal{I}$ is a pre-Hilbert space.

Let $\mathcal{H}_\varphi$ be the completion of $\mathcal{A}/\mathcal{I}$ with respect to the scalar product. Then $\mathcal{H}_\varphi$ is a Hilbert space. Now we construct a $*$-representation of $\mathcal{A}$ on $\mathcal{H}_\varphi$ via φ, denoted by $\{\pi_\varphi, \mathcal{H}_\varphi\}$. Put $\pi_\varphi(a) x_\varphi = (ax)_\varphi$. Then $\pi_\varphi(a)$ is a linear operator on $\mathcal{A}/\mathcal{I}$.

$$\|\pi_\varphi(a) x_\varphi\|^2 = \varphi(x^* a^* a x) \leq \|a^* a\| \, \varphi(x^* x) = \|a\|^2 \, \|x_\varphi\|^2.$$

Hence, $\pi_\varphi(a)$ is a bounded linear operator on the pre-Hilbert space. Hence it can be uniquely extended to a bounded linear operator on $\mathcal{H}_\varphi$, also denoted by $\pi_\varphi(a)$.

Clearly $a \to \pi_\varphi(a)$ is a homomorphism of $\mathcal{A}$ into $B(\mathcal{H}_\varphi)$. Moreover,

$$(\pi_\varphi(a) b_\varphi, c_\varphi) = \varphi(c^* a b) = \varphi((a^* c)^* b) = (b_\varphi, \pi_\varphi(a^*) c_\varphi)$$

$(a, b, c \in \mathcal{A})$. Hence, $\pi_\varphi(a^*) = \pi_\varphi(a)^*$. Therefore $\{\pi_\varphi, \mathcal{H}_\varphi\}$ is a $*$-representation of $\mathcal{A}$.

1.16.5. Definition. *Let $\mathscr{A}$ be a C^*-algebra and let $\mathscr{S}$ be the set of all states on $\mathscr{A}$. $\mathscr{S}$, with the topology $\sigma(\mathscr{S},\mathscr{A})$, is called the state space of $\mathscr{A}$. For arbitrary $\varphi\in\mathscr{S}$, consider the *-representation $\{\pi_\varphi,\mathscr{H}_\varphi\}$. Let $K=\sum_{\varphi\in\mathscr{S}}\oplus\mathscr{H}_\varphi$ be the direct sum of $\{\mathscr{H}_\varphi\}$. Namely $K=\{(\xi_\varphi)_{\varphi\in\mathscr{S}}\,|\,\xi_\varphi\in\mathscr{H}_\varphi$ and $\sum_{\varphi\in\mathscr{S}}\|\xi_\varphi\|^2<+\infty\}$ with the scalar product $((\xi_\varphi),(\eta_\varphi))=\sum_{\varphi\in\mathscr{S}}(\xi_\varphi,\eta_\varphi)$. Put $U(a)=\sum_{\varphi\in\mathscr{S}}\oplus\pi_\varphi(a)$. Then $U(a)\in B(K)$, and the mapping $a\to U(a)$ is a *-representation of $\mathscr{A}$. This *-representation $\{U,K\}$ is called the universal *-representation of $\mathscr{A}$.*

1.16.6. Theorem. *The universal *-representation $\{U,K\}$ of $\mathscr{A}$ is an isometric isomorphism. Therefore, every C^*-algebra is *-isomorphic to a uniformly closed self-adjoint subalgebra of $B(\mathscr{H})$ on some Hilbert space $\mathscr{H}$.*

Proof. Without loss of generality, we can assume that $\mathscr{A}$ has an identity.

$$\begin{aligned}\|U(a)\|^2&=\sup_{\varphi\in\mathscr{S}}\|\pi_\varphi(a)\|^2\geq\sup_{\varphi\in\mathscr{S}}\|\pi_\varphi(a)1_\varphi\|^2\\&=\sup_{\varphi\in\mathscr{S}}\varphi(a^*a)=\|a^*a\|\qquad\text{(cf. 1.5.2)}\\&=\|a\|^2.\end{aligned}$$

On the other hand, $\|\pi_\varphi(a)\|\leq\|a\|$. Hence, $\|U(a)\|\leq\|a\|$ and so $\|a\|=\|U(a)\|$ for $a\in\mathscr{A}$. q.e.d.

Let $\mathscr{M}$ be a W^*-algebra and let φ be a normal state on $\mathscr{M}$. Then for $\xi,\eta\in\mathscr{H}_\varphi$, set $f(x)=(\pi_\varphi(x)\xi,\eta)$ $(x\in\mathscr{M})$. Since $\mathscr{M}_\varphi$ is dense in $\mathscr{H}_\varphi$, there exists two sequences (a_n) and (b_n) in $\mathscr{M}$ such that $\|a_{n\varphi}-\xi\|\to 0$ and $\|b_{n\varphi}-\eta\|\to 0$. Then,

$$\begin{aligned}&|(\pi_\varphi(x)\xi,\eta)-(\pi_\varphi(x)a_{n\varphi},b_{n\varphi})|\\&\leq|(\pi_\varphi(x)(\xi-a_{n\varphi}),\eta)|+|(\pi_\varphi(x)a_{n\varphi},(\eta-b_{n\varphi}))|\\&\leq\|x\|\,\|\xi-a_{n\varphi}\|\,\|\eta\|+\|x\|\,\|a_{n\varphi}\|\,\|\eta-b_{n\varphi}\|\to 0\qquad(n\to\infty).\end{aligned}$$

Hence, $f(x)$ is a uniform limit of sequences $\{f_n(x)\}$ on the unit sphere of $\mathscr{M}$, where $f_n(x)=(\pi_\varphi(x)a_{n\varphi},b_{n\varphi})=\varphi(b_n^*xa_n)$. $f\in\mathscr{M}_*$, since $f_n\in\mathscr{M}_*$. Hence, the mapping $x\to\pi_\varphi(x)$ of $\mathscr{M}$, with the $\sigma(\mathscr{M},\mathscr{M}_*)$-topology, into $B(\mathscr{H}_\varphi)$ with the $\sigma(B(\mathscr{H}),B(\mathscr{H})_*)$-topology, is continuous on bounded spheres, and so it is continuous on $\mathscr{M}$ (i.e., $\{\pi_\varphi,\mathscr{H}_\varphi\}$ is a W^*-representation of $\mathscr{M}$ (say $\{\pi_\varphi^w,\mathscr{H}_\varphi\}$)).

1.16.7. Theorem. *Let $\mathscr{M}$ be a W^*-algebra. Then, it has a faithful W^*-representation $\{\pi,\mathscr{H}\}$ (i.e., $\pi(a)=0$ if and only if $a=0$). Therefore, $\mathscr{M}$ is *-isomorphic to a weakly closed self-adjoint subalgebra of $B(\mathscr{H})$ on some Hilbert space $\mathscr{H}$.*

Proof. Let $\mathscr{S}_m$ be the set of all normal states on $\mathscr{M}$, and consider the W^*-representations $[\{\pi_\varphi^w, \mathscr{H}_\varphi\} | \varphi \in \mathscr{S}_m]$ of $\mathscr{M}$. Set $\mathscr{H} = \sum_{\varphi \in \mathscr{S}_m} \oplus \mathscr{H}_\varphi$ and $\pi(x) = \sum_{\varphi \in \mathscr{S}_m} \oplus \pi_\varphi(x)$ $(x \in \mathscr{M})$. Let F be the set of all finite linear combinations of elements of $\bigcup_{\varphi \in \mathscr{S}_m} \mathscr{H}_\varphi$. F is dense in $\mathscr{H}$. Let $\xi = \sum_{i=1}^n \xi_i$ and $\eta = \sum_{i=1}^n \eta_i$ $(\xi_i, \eta_i \in \mathscr{H}_{\varphi_i}, i = 1, 2, \ldots, n)$. Then $f(x) = (\pi(x)\xi, \eta) = \sum_{i=1}^n (\pi_{\varphi_i}(x)\xi_i, \eta_i)$. Hence, $f \in \mathscr{M}_*$, and so $(\pi(x)\xi', \eta') \in \mathscr{M}_*$ $(\xi', \eta' \in \mathscr{H})$. Therefore, $\{\pi, \mathscr{H}\}$ is a W^*-representation of $\mathscr{M}$ (say $\{\pi^w, \mathscr{H}\}$). Suppose $\pi(a) = 0$. Then $\pi_\varphi(a) = 0$ for all $\varphi \in \mathscr{S}_m$. So $\varphi(a) = 0$ for all $\varphi \in \mathscr{S}_m$. Hence, $a = 0$ (1.7). q.e.d.

Concluding remarks on 1.16.

Theorem 1.16.6 is due to Gelfand-Naimark [55]. They defined a C^*-algebra $\mathscr{A}$ as a Banach *-algebra satisfying the following conditions: 1. $\|x^*x\| = \|x\|^2$ and 2. $1 + x^*x$ has an inverse $(x \in \mathscr{A})$. Then they proved Theorem 1.16.6. They further asked whether axiom 2. was superfluous. This question was solved by Fukamiya, Kelley-Vaught and Kaplansky (cf. Concluding remarks on 1.4). Also they asked whether the axiom 1. could be replaced by $\|x^*x\| = \|x^*\|\,\|x\|$. This question was solved by Glimm and Kadison [61] and Ono [136].

References. [55], [149], [154].

1.17. The Second Dual of a C^*-Algebra

Let $\mathscr{A}$ be a C^*-algebra, $\mathscr{A}^*$ the dual Banach space of $\mathscr{A}$ and let $\mathscr{A}^{**}$ be the second dual of $\mathscr{A}$. Let Γ be the set of all positive linear functionals φ on $\mathscr{A}$ such that $\|\varphi\| \leq 1$. Then Γ is $\sigma(\mathscr{A}^*, \mathscr{A})$-compact. Let $C(\Gamma)$ be the Banach space of all continuous complex valued functions on the compact space Γ with the usual supremum norm. For $a \in \mathscr{A}$ with $a^* = a$, let A be the commutative C^*-subalgebra generated by a. Then there exists a character χ on A such that $\chi(a) = \|a\|$ or $-\|a\|$.

Let $\tilde{\chi}$ be an extension of χ on $\mathscr{A}$ such that $\|\chi\| = \|\tilde{\chi}\|$. $\tilde{\chi}$ is a state. Hence, $\sup_{\varphi \in \Gamma} |\varphi(a)| = \|a\|$. Therefore, the self-adjoint portion $\mathscr{A}^s$ of $\mathscr{A}$ may be isometrically embedded into the real Banach space $C_r(\Gamma)$ of all continuous real-valued functions on Γ. Hence, $\mathscr{A}$ may be topologically embedded into $C(\Gamma)$.

1.17.1. Proposition. *Let g be a bounded linear functional on $\mathscr{A}$. Then g is a finite linear combination of states.*

Proof. Consider $\mathscr{A} \subset C(\Gamma)$. g can be extended to a bounded linear functional $\tilde{g}$ on $C(\Gamma)$. Then by Riesz's theorem, there exists a complex Radon measure μ on Γ such that

$$g(a) = \int_\Gamma \varphi(a)\, d\mu(\varphi) \qquad (a \in \mathscr{A}).$$

Let $\mu = \mu_1 - \mu_2 + i\mu_3 - i\mu_4$, where μ_i $(i = 1,2,3,4)$ are positive Radon measures on Γ. Then $g_i(a) = \int_\Gamma \varphi(a)\, d\mu_i(\varphi)$ for $a \in \mathscr{A}$ $(i = 1,2,3,4)$ are bounded positive linear functionals on $\mathscr{A}$. q.e.d.

1.17.2. Theorem. *Let $\mathscr{A}^{**}$ be the second dual of the C^*-algebra $\mathscr{A}$. Then $\mathscr{A}^{**}$ is a W^*-algebra in a natural manner. Moreover, $\mathscr{A}$ is a C^*-subalgebra of $\mathscr{A}^{**}$ when it is canonically embedded into $\mathscr{A}^{**}$.*

Proof. Let $\{U, K\}$ be the universal $*$-representation of $\mathscr{A}$. Then the mapping $x \to U(x)$ of $\mathscr{A}$ onto $U(\mathscr{A})$ is an isometric $*$-isomorphism. We shall identify $\mathscr{A}$ with $U(\mathscr{A})$. Let $\overline{U(\mathscr{A})}$ be the weak closure of $\mathscr{A}$. $\overline{U(\mathscr{A})}$ is a W^*-algebra.

Let $\overline{U(\mathscr{A})}_*$ be the predual of $\overline{U(\mathscr{A})}$. For $f \in \overline{U(\mathscr{A})}_*$,

$$\|f\| = \sup_{\substack{\|x\| \leq 1 \\ x \in \overline{U(\mathscr{A})}}} |f(x)| = \sup_{\substack{\|x\| \leq 1 \\ x \in \mathscr{A}}} |f(x)|$$

by Kaplansky's density theorem. Therefore, the mapping $f \to f|\mathscr{A}$ is isometric, where $f|\mathscr{A}$ is the restriction of f to $\mathscr{A}$.

Since $\{f|\mathscr{A} \mid f \in \overline{U(\mathscr{A})}_*\}$ contains all states on $\mathscr{A}$, by 1.17.1, it is $\mathscr{A}^*$. Hence $\mathscr{A}^* = \overline{U(\mathscr{A})}_*$ and $(\overline{U(\mathscr{A})}_*)^* = \overline{U(\mathscr{A})}$. So $\overline{U(\mathscr{A})}$ is the second dual of $\mathscr{A}$. q.e.d.

1.17.3. Corollary. *Let $\mathscr{A}$ be a C^*-algebra and let $\mathscr{I}$ be a uniformly closed two-sided ideal in $\mathscr{A}$. Then $\mathscr{I}$ is self-adjoint and the quotient algebra $\mathscr{A}/\mathscr{I}$ is also a C^*-algebra.*

Proof. Let $\mathscr{I}^{00}$ be the bipolar of $\mathscr{I}$ in $\mathscr{A}^{**}$. Then $\mathscr{I}^{00}$ is the $\sigma(\mathscr{A}^{**}, \mathscr{A}^*)$-closure of $\mathscr{I}$ in $\mathscr{A}^{**}$. Hence, it is a $\sigma(\mathscr{A}^{**}, \mathscr{A}^*)$-closed ideal of $\mathscr{A}^{**}$. Therefore, there exists a central projection z in $\mathscr{A}^{**}$ such that $\mathscr{I}^{00} = \mathscr{A}^{**} z$. Since $\mathscr{I} = \mathscr{I}^{00} \cap \mathscr{A} = \mathscr{A}^{**} z \cap \mathscr{A}$, $\mathscr{I}$ is self-adjoint.

$$\mathscr{A}/\mathscr{I} \approx \mathscr{A} + \mathscr{A}^{**} z / \mathscr{A}^{**} z \approx \mathscr{A}(1 - z). \qquad \text{q.e.d.}$$

1.17.4. Corollary. *Let $\mathscr{A}, \mathscr{B}$ be C^*-algebras and let Φ be a $*$-homomorphism of $\mathscr{A}$ into $\mathscr{B}$. Then the image $\Phi(\mathscr{A})$ of $\mathscr{A}$ is a C^*-subalgebra of $\mathscr{B}$. Therefore, the image of a $*$-representation of a C^*-algebra is also a C^*-algebra.*

Proof. Let $\mathscr{I} = \{x \mid \Phi(x) = 0, x \in \mathscr{A}\}$. Then Φ induces a $*$-isomorphism Φ' of the C^*-algebra $\mathscr{A}/\mathscr{I}$ into $\mathscr{B}$. Hence the image is closed, since Φ' is isometric (cf. 1.2.6). q.e.d.

1.17.5. Corollary. *Let $\mathscr{A}$ be a C*-algebra, $\mathscr{B}$ a C*-subalgebra of $\mathscr{A}$, and let $\mathscr{I}$ be a uniformly closed two-sided ideal of $\mathscr{A}$. Then $\mathscr{B}+\mathscr{I}$ is also a C*-subalgebra of $\mathscr{A}$.*

Proof. Consider the quotient C^*-algebra $\mathscr{A}/\mathscr{I}$ and the canonical homomorphism Φ of $\mathscr{A}$ onto $\mathscr{A}/\mathscr{I}$. Then $\Phi(\mathscr{B})=\mathscr{B}+\mathscr{I}$ is uniformly closed. q.e.d.

1.17.6. Proposition. *Let $\mathscr{A}$ be a C*-algebra, $\mathscr{B}$ a C*-subalgebra of $\mathscr{A}$, and let $\mathscr{L}$ be a uniformly closed left ideal satisfying $\mathscr{L}\mathscr{B}\subset\mathscr{L}$. Put*

$$|||x||| = \inf_{y\in\mathscr{L}} \|x+y\| \quad (x\in\mathscr{B}).$$

Then $|||x||| = \inf_{w\in\mathscr{L}\cap\mathscr{B}} \|x+w\|$*—i.e.,* $\mathscr{B}+\mathscr{L}/\mathscr{L}=\mathscr{B}/\mathscr{L}\cap\mathscr{B}$ *as Banach spaces. Moreover* $\mathscr{B}/\mathscr{L}\cap\mathscr{B}$ *is a C*-algebra since* $\mathscr{L}\cap\mathscr{B}$ *is a two-sided ideal in* $\mathscr{B}$.

Proof. Consider $\mathscr{A}^{**}$ and the bipolars, $\mathscr{B}^{00}$ and $\mathscr{L}^{00}$, in $\mathscr{A}^{**}$. $\mathscr{L}^{00}$ is a $\sigma(\mathscr{A}^{**},\mathscr{A}^*)$-closed left ideal. Hence, there exists a projection e in $\mathscr{A}^{**}$ such that $\mathscr{L}^{00}=\mathscr{A}^{**}e$. Let $\mathscr{R}=\mathscr{B}^{00}+\mathbb{C}1$, where $\mathbb{C}$ is the field of complex numbers and 1 is the identity of $\mathscr{A}^{**}$. Since $\mathscr{L}\mathscr{B}\subset\mathscr{L}$, $\mathscr{L}^{00}\mathscr{B}^{00}\subset\mathscr{L}^{00}$ and so $\mathscr{L}^{00}\mathscr{R}\subset\mathscr{L}^{00}$. $\mathscr{R}$ is a $\sigma(\mathscr{A}^{**},\mathscr{A}^*)$-closed self-adjoint algebra of $\mathscr{A}^{**}$. Let u be any unitary in $\mathscr{R}$. Then

$$u^*\mathscr{L}^{00}u\subset\mathscr{L}^{00}=u^*u\mathscr{L}^{00}u^*u\subset u^*\mathscr{L}^{00}u.$$

Hence, $u^*\mathscr{L}^{00}u=\mathscr{L}^{00}$, and so $u^*eu=e$ for every unitary $u\in\mathscr{R}$. Hence, e commutes with $\mathscr{R}$. Clearly the polar $\mathscr{L}^0$ of $\mathscr{L}$ is $R_{(1-e)}\mathscr{A}^*$. So

$$|||x||| = \inf_{y\in\mathscr{L}}\|x+y\| = \sup_{\substack{f\in\mathscr{L}^0\\ \|f\|=1}} |f(x)| = \sup_{\substack{f\in\mathscr{A}^*\\ \|f\|=1}} |f(x(1-e))| = \|x(1-e)\|$$

$(x\in\mathscr{A})$. In particular, the mapping $w\to w(1-e)$ $(w\in\mathscr{B})$ is a *-homomorphism, since $(1-e)$ commutes with $\mathscr{B}$. Clearly $\inf_{y\in\mathscr{L}}\|x+y\|\leq\inf_{y\in\mathscr{B}\cap\mathscr{L}}\|x+y\|$.

On the other hand, if $w\in\mathscr{B}$, $w(1-e)=0$, then $w\in\mathscr{A}^{**}e$. Hence, $w\in\mathscr{L}\cap\mathscr{B}$. Therefore, there exists a *-isomorphism of $\mathscr{B}/\mathscr{B}\cap\mathscr{L}$ into $\mathscr{B}(1-e)$. So $\|w(1-e)\| = \inf_{y\in\mathscr{L}\cap\mathscr{B}}\|x+y\|$. q.e.d.

Now let $\mathscr{M}$ be a W^*-algebra, $\mathscr{M}^*$ the dual of $\mathscr{M}$, and let $\mathscr{M}^{**}$ be the second dual of $\mathscr{M}$. Let $\mathscr{M}_*$ be the predual of $\mathscr{M}$ and let $\mathscr{M}_*^0$ be the polar of $\mathscr{M}_*$ in $\mathscr{M}^{**}$. Then $\mathscr{M}_*^0$ is a $\sigma(\mathscr{M}^{**},\mathscr{M}^*)$-closed two-sided ideal, for $\mathscr{M}_*$ is invariant under L_a, R_a $(a\in\mathscr{M})$.

Let z be the central projection of $\mathscr{M}^{**}$ such that $\mathscr{M}_*^0=\mathscr{M}^{**}z$. Then $\mathscr{M}^{**}=\mathscr{M}^{**}z\oplus\mathscr{M}^{**}(1-z)$. Let $a\in\mathscr{M}^{**}$. Then $a|\mathscr{M}_*\in(\mathscr{M}_*)^*=\mathscr{M}$. Hence there exists an element $b\in\mathscr{M}$ such that $a|\mathscr{M}_*=b$, and so

$$a-b\in\mathscr{M}_*^0=\mathscr{M}^{**}z.$$

Therefore $\mathscr{M}^{**}=\mathscr{M}+\mathscr{M}^{**}z=\mathscr{M}^{**}(1-z)+\mathscr{M}^{**}z$ and

$$\mathscr{M}^*=R_z\mathscr{M}^*\oplus R_{1-z}\mathscr{M}^*=R_z\mathscr{M}^*\oplus\mathscr{M}_*.$$

1.17.7. Proposition. *Let $\mathcal{M}$ be a W^*-algebra, $\mathcal{M}_*$ the predual of $\mathcal{M}$, and let $\mathcal{M}^*$ be the dual of $\mathcal{M}$. Then there exists a linear mapping $R_{(1-z)}$ (z a central projection of $\mathcal{M}^{**}$) of $\mathcal{M}^*$* onto *$\mathcal{M}_*$ satisfying the following conditions:*

1. $R_{1-z}^2 = R_{1-z}$;
2. $\|R_{1-z}f\| \leq \|f\| \quad (f \in \mathcal{M}^*)$;
3. $R_{1-z}f \geq 0$ *if* $f \geq 0 \quad (f \in \mathcal{M}^*)$;
4. *$\mathcal{M}_*$ is a closed subspace of $\mathcal{M}^*$ invariant under L_a, R_a ($a, b \in \mathcal{M}^{**}$).*

1.17.8. Proposition. *Let $\mathcal{M}$ be a W^*-algebra and let $\mathcal{M}^{**}$ be the second dual of $\mathcal{M}$. Then there exists a W^*-homomorphism of $\mathcal{M}^{**}$* onto *$\mathcal{M}$.*

Proof. $\mathcal{M}^{**} = \mathcal{M}^{**}(1-z) + \mathcal{M}^{**}z = \mathcal{M} + \mathcal{M}^{**}z$ and

$$\mathcal{M}(1-z) = \mathcal{M}^{**}(1-z).$$

Consider the mapping $\Phi: x \to x(1-z)$ of $\mathcal{M}$ onto $\mathcal{M}^{**}(1-z)$. Then Φ is a W^*-isomorphism of $\mathcal{M}$ onto $\mathcal{M}^{**}(1-z)$. Let

$$\Phi_1(y) = \Phi^{-1}(y(1-z)) \quad (y \in \mathcal{M}^{**}).$$

Then clearly Φ_1 is a W^*-homomorphism of $\mathcal{M}^{**}$ onto $\mathcal{M}$. q.e.d.

Remark. A positive linear functional belonging to $R_z\mathcal{M}^*$ is said to be singular. Then, it is known that a positive linear functional φ on $\mathcal{M}$ is singular if and only if for any non-zero projection $p \in \mathcal{M}$ there exists a non-zero projection q in $\mathcal{M}$ such that $\varphi(q) = 0$ (cf. [203]).

Concluding remarks on 1.17.

Theorem 1.17.2 is due to Sherman [188]. A complete proof was given by Takeda [200] and Grothendieck [66].

References. [66], [150], [181], [188], [200].

1.18. Commutative W^*-Algebras

Let (Ω, μ) be a measure space with $\mu(\Omega) < +\infty$, let $L^\infty(\Omega, \mu)$ be the C^*-algebra of all essentially bounded μ-measurable functions on Ω, and let $L^1(\Omega, \mu)$ be the Banach space of all μ-integrable functions on Ω. Then, by the Radon-Nikodym theorem, $L^1(\Omega, \mu)^* = L^\infty(\Omega, \mu)$. Hence $L^\infty(\Omega, \mu)$ is a commutative W^*-algebra.

More generally, let (Γ, ν) be a localizable measure space (i.e., a direct sum of finite measure spaces (cf. [182])), and let $L^\infty(\Gamma, \nu)$ be the C^*-algebra of all essentially bounded locally ν-measurable functions on Ω, and let $L^1(\Gamma, \nu)$ be the Banach space of all ν-integrable functions on Γ. Then $L^1(\Gamma, \nu)^* = L^\infty(\Gamma, \nu)$. Hence, $L^\infty(\Gamma, \nu)$ is a commutative W^*-algebra.

1.18.1. Proposition. *Let $\mathcal{M}$ be a commutative W^*-algebra. Then $\mathcal{M}$ is $*$-isomorphic to a W^*-algebra $L^\infty(\Gamma, \nu)$ on some localizable measure space (Γ, ν).*

Proof. Let Ω be the spectrum space of $\mathscr{M}$. Then $\mathscr{M}=C(\Omega)$. Let φ be a normal state on $\mathscr{M}$. By Riesz' representation theorem, φ defines a unique positive Radon measure μ_φ on Ω such that

$$\varphi(a)=\int_\Omega a(t)\,d\mu_\varphi(t) \qquad (a\in\mathscr{M}).$$

Let $\Phi_\varphi: a\to a(t)$ be the mapping of $\mathscr{M}$ into the W^*-algebra $L^\infty(\Omega,\mu_\varphi)$. Φ_φ is a *-homomorphism. Moreover, for $b\in C(\Omega)$,

$$\int_\Omega a(t)b(t)\,d\mu_\varphi(t)=\varphi(ab)$$

and $C(\Omega)$ is dense in $L^1(\Omega,\mu_\varphi)$. Therefore, Φ_φ is a W^*-homomorphism. Hence $\Phi_\varphi(\mathscr{M})$ is a W^*-subalgebra of $L^\infty(\Omega,\mu_\varphi)$, and so $\Phi_\varphi(\mathscr{M})=L^\infty(\Omega,\mu_\varphi)$, since $\Phi_\varphi(\mathscr{M})\supset C(\Omega)$ in $L^\infty(\Omega,\mu_\varphi)$.

Let $\mathscr{I}_\varphi$ be the kernel of Φ_φ. Then $\mathscr{I}_\varphi=\mathscr{M}z$ for some projection z in $\mathscr{M}$. On the other hand, $\varphi(z)=\int_\Omega z(t)\,d\mu_\varphi(t)=0$. Hence, $z\le 1-s(\varphi)$. The converse is also true. Therefore, $z=1-s(\varphi)$.

Φ_φ defines a *-isomorphism of $\mathscr{M}s(\varphi)$ onto $L^\infty(\Omega,\mu_\varphi)$. Let $\{\varphi_\alpha\}_{\alpha\in\mathbb{I}}$ be a maximal family of normal states on $\mathscr{M}$ such that

$$s(\varphi_{\alpha_1})\cdot s(\varphi_{\alpha_2})=0 \qquad (\alpha_1,\alpha_2\in\mathbb{I},\ \alpha_1\neq\alpha_2).$$

Then, $\mathscr{M}=\sum_{\alpha\in\mathbb{I}}\oplus\,\mathscr{M}s(\varphi_\alpha)=\sum_{\alpha\in\mathbb{I}}\oplus\,L^\infty(s(\varphi_\alpha),\mu_{\varphi_\alpha})=L^\infty\Big(\bigcup_\alpha s(\mu_{\varphi_\alpha}),\ \sum_{\alpha\in\mathbb{I}}\oplus\,\mu_\alpha\Big)$

($s(\mu_{\varphi_\alpha})$ is the support of μ_{φ_α}). Therefore, $\mathscr{M}$ is *-isomorphic to

$$L^\infty\Big(\bigcup_{\alpha\in\mathbb{I}} s(\mu_{\varphi_\alpha}),\ \sum_{\alpha\in\mathbb{I}}\oplus\,\mu_{\varphi_\alpha}\Big)$$

on the localizable measure space $\Big(\bigcup_{\alpha\in\mathbb{I}} s(\mu_{\varphi_\alpha}),\ \sum_{\alpha\in\mathbb{I}}\oplus\,\mu_{\varphi_\alpha}\Big)$.

Remark. Let K be a Stonean space. A positive Radon measure μ on K is called normal if for every uniformly bounded increasing directed set $\{f_\alpha\}$ of continuous real valued functions on Ω, $\operatorname{l.u.b.}_\alpha \mu(f_\alpha)=\mu\big(\operatorname{l.u.b.}_\alpha f_\alpha\big)$.

K is called hyper-Stonean if K has a faithful family $\{\mu_\beta\}_{\beta\in\mathbb{J}}$ of normal measures (i.e. for $f\ (\ge 0)\in C(K)$, $\mu_\beta(f)=0$ for all $\beta\in\mathbb{J}$ imply $f=0$). Then, $C(K)$ is a W^*-algebra if and only if K is hyper-Stonean (cf. [31]).

Reference. [182].

1.19. The C^*-Algebra $C(\mathscr{H})$ of all Compact Linear Operators on a Hilbert Space $\mathscr{H}$

Let $\mathscr{H}$ be a Hilbert space, $C(\mathscr{H})$ the C^*-algebra of all compact linear operators on $\mathscr{H}$, $T(\mathscr{H})$ the Banach space of all trace-class operators, $B(\mathscr{H})$ the W^*-algebra of all bounded linear operators on $\mathscr{H}$. $T(\mathscr{H})$ can be identified with the predual $B(\mathscr{H})_*$ of $B(\mathscr{H})$ (cf. 1.15.3). For $x\in C(\mathscr{H})$ and $a\in T(\mathscr{H})$, let $\psi_a(x)=Tr(xa)$. Then ψ_a is a bounded linear functional on $C(\mathscr{H})$.

1.19.1. Proposition. *The mapping $a \to \psi_a$ of $T(\mathscr{H})$ into $C(\mathscr{H})^*$ is an isometric linear mapping of $T(\mathscr{H})$* onto *$C(\mathscr{H})^*$. Therefore, under the mapping $a \to \psi_a$, $T(\mathscr{H})$ can be identified with $C(\mathscr{H})^*$. Hence, we have $C(\mathscr{H})^* = T(\mathscr{H})$ and $T(\mathscr{H})^* = B(\mathscr{H})$.*

Proof. Consider the second dual $C(\mathscr{H})^{**}$ of $C(\mathscr{H})$. Let 1 be the identity of $C(\mathscr{H})^{**}$. Let (p_α) be a uniformly bounded increasing directed set of all projections in $C(\mathscr{H})$. Let $p = \text{l.u.b.}_\alpha\, p_\alpha$ in $C(\mathscr{H})^{**}$. Then the $s(C(\mathscr{H})^{**}, C(\mathscr{H})^*)$-limit of the p_α is p. For $a \in C(\mathscr{H})$, $\|p_\alpha a p_\alpha - a\| \to 0$, since a is a compact operator. Hence $C(\mathscr{H}) \subset pC(\mathscr{H})p$ and so $p=1$. Let φ be a state on $C(\mathscr{H})$. Then $\lim_\alpha \varphi(p_\alpha) = \varphi(1) = 1$. Hence, there exists a sequence of indices (α_n) such that $\varphi(1-p_{\alpha_n}) < 1/n$. Then

$$\begin{aligned}
&\sup_{\substack{\|x\|\le 1\\ x\in C(\mathscr{H})}} |\varphi(x) - \varphi(p_{\alpha_n} x\, p_{\alpha_n})| \\
&\le \sup_{\substack{\|x\|\le 1\\ x\in C(\mathscr{H})}} |\varphi((1-p_{\alpha_n}) x p_{\alpha_n}) + \varphi(p_{\alpha_n} x(1-p_{\alpha_n})) + \varphi((1-p_{\alpha_n}) x (1-p_{\alpha_n})| \\
&\le 3\varphi(1-p_{\alpha_n})^{\frac{1}{2}} \to 0 \qquad (n\to\infty).
\end{aligned}$$

Hence, $\lim \|\varphi - L_{p_{\alpha_n}} R_{p_{\alpha_n}} \varphi\| = 0$.

Since $L_{p_{\alpha_n}} R_{p_{\alpha_n}} \varphi$ is zero on $(1-p_{\alpha_n}) C(\mathscr{H}) p_{\alpha_n} + p_{\alpha_n} C(\mathscr{H})(1-p_{\alpha_n}) + (1-p_{\alpha_n}) C(\mathscr{H})(1-p_{\alpha_n})$, $L_{p_{\alpha_n}} R_{p_{\alpha_n}} \varphi$ can be considered as a positive linear functional on $p_{\alpha_n} C(\mathscr{H}) p_{\alpha_n}$. Since $p_{\alpha_n} C(\mathscr{H}) p_{\alpha_n}$ is finite-dimensional, there exists a positive element $p_{\alpha_n} a_n p_{\alpha_n}$ of $p_{\alpha_n} C(\mathscr{H}) p_{\alpha_n}$ which satisfies

$$\varphi(p_{\alpha_n} x p_{\alpha_n}) = Tr(p_{\alpha_n} a_n p_{\alpha_n} p_{\alpha_n} x p_{\alpha_n}) = Tr(p_{\alpha_n} a_n p_{\alpha_n} x).$$

Moreover,

$$\|L_{p_{\alpha_n}} R_{p_{\alpha_n}} \varphi - L_{p_{\alpha_m}} R_{p_{\alpha_m}} \varphi\| = \|p_{\alpha_n} a_n p_{\alpha_n} - p_{\alpha_m} a_m p_{\alpha_m}\|_1 .$$

Therefore $\{p_{\alpha_n} a_n p_{\alpha_n}\}$ is a Cauchy sequence in $T(\mathscr{H})$. So there exists a trace class operator a_0 such that $\lim_n \varphi(p_{\alpha_n} x p_{\alpha_n}) = \varphi(x) = Tr(a_0 x)$ for $x \in C(\mathscr{H})$, and $\|\varphi\| = \|a_0\|_1 = Tr(|a_0|)$. q.e.d.

Concluding remarks on 1.19.

The problem of extending Theorem 1.19.1 to more general Banach spaces is interesting. Grothendieck [65] has solved this problem in some special cases. In general, the second dual of a Banach algebra is again a Banach algebra (cf. [7]). Hence, $C(E)^{**}$ (E, a Banach space) is a Banach algebra.

References. [30], [174].

1.20. The Commutation Theorem of von Neumann

Let $\mathscr{H}$ be a Hilbert space, $B(\mathscr{H})$ the set of all bounded linear operators, and let $\mathscr{L}$ be a subset of $B(\mathscr{H})$. We denote by $\mathscr{L}'$ the set of elements of $B(\mathscr{H})$ commuting with all elements of $\mathscr{L}$, and we call $\mathscr{L}'$ the commutant of $\mathscr{L}$.

Put $(\mathscr{L}')'=\mathscr{L}''$ (the bicommutant of $\mathscr{L}$), $(\mathscr{L}'')'=\mathscr{L}'''$, ... It is clear that $\mathscr{L}'$ is a subalgebra of $B(\mathscr{H})$ containing the identity, $\mathscr{L}''\supset\mathscr{L}$, and $\mathscr{L}_1\subset\mathscr{L}_2$ implies $\mathscr{L}_1'\supset\mathscr{L}_2'$. Therefore $\mathscr{L}_1''\subset\mathscr{L}_2''$. Hence, $\mathscr{L}'\supset(\mathscr{L}'')'=\mathscr{L}'''$. On the other hand, $\mathscr{L}'\subset(\mathscr{L}')''=\mathscr{L}'''$. Hence, $\mathscr{L}'=\mathscr{L}'''=\mathscr{L}^{(5)}=\cdots$ and $\mathscr{L}\subset\mathscr{L}''=\mathscr{L}^{(4)}=\cdots$. If $\mathscr{L}$ is a self-adjoint set, $\mathscr{L}'$ is a self-adjoint subalgebra. Let $\mathscr{H}_1$ and $\mathscr{H}_2$ be two Hilbert spaces and let $\mathscr{H}_1\odot\mathscr{H}_2$ be the algebraic tensor product of $\mathscr{H}_1$ and $\mathscr{H}_2$. Then there exists a unique pre-Hilbert space structure on $\mathscr{H}_1\odot\mathscr{H}_2$ satisfying

$$(\xi_1\otimes\xi_2,\eta_1\otimes\eta_2)=(\xi_1,\eta_1)(\xi_2,\eta_2)\qquad(\xi_1,\eta_1\in\mathscr{H}_1\text{ and }\xi_2,\eta_2\in\mathscr{H}_2).$$

The Hilbert space (denoted by $\mathscr{H}_1\otimes\mathscr{H}_2$) obtained by the completion of $\mathscr{H}_1\odot\mathscr{H}_2$ is called the tensor product of $\mathscr{H}_1$ and $\mathscr{H}_2$.

Let $a_1,b_1\in B(\mathscr{H}_1)$ and $a_2,b_2\in B(\mathscr{H}_2)$. The algebraic tensor product $a_1\otimes a_2$ defines a unique continuous linear operator on $\mathscr{H}_1\otimes\mathscr{H}_2$ such that $(a_1\otimes a_2)\xi_1\otimes\xi_2=a_1\xi_1\otimes a_2\xi_2$. $a_1\otimes a_2$ is bilinear in a_1 and a_2, $(a_1b_1)\otimes(a_2b_2)=(a_1\otimes a_2)(b_1\otimes b_2)$, and $(a_1\otimes a_2)^*=a_1^*\otimes a_2^*$.

Let $(\eta_\alpha)_{\alpha\in\mathbb{I}}$ be a complete orthonormal system of $\mathscr{H}_2$. The mapping $\xi_1\to\xi_1\otimes\eta_\alpha$ is an isometry u_α of $\mathscr{H}_1$ onto a closed subspace $\mathscr{H}^\alpha$ of $\mathscr{H}_1\otimes\mathscr{H}_2$. The $\{\mathscr{H}^\alpha\}_{\alpha\in\mathbb{I}}$ are mutually orthogonal, and $\mathscr{H}_1\otimes\mathscr{H}_2=\sum_{\alpha\in\mathbb{I}}\oplus\mathscr{H}^\alpha$. u_α^* is a linear mapping of $\mathscr{H}_1\otimes\mathscr{H}_2$ onto $\mathscr{H}_1$ such that

$$u_\alpha^*(\mathscr{H}_1\otimes\mathscr{H}_2\ominus\mathscr{H}^\alpha)=0,$$

where $\mathscr{H}_1\otimes\mathscr{H}_2\ominus\mathscr{H}^\alpha$ is the ortho-complement of $\mathscr{H}^\alpha$ in $\mathscr{H}_1\otimes\mathscr{H}_2$. It is also an isometry on $\mathscr{H}^\alpha$, and $u_\alpha u_\alpha^*$ is the projection e_α of $\mathscr{H}_1\otimes\mathscr{H}_2$ onto $\mathscr{H}^\alpha$. Let $a\in B(\mathscr{H}_1\otimes\mathscr{H}_2)$. Then $u_\alpha^*au_\beta\in B(\mathscr{H}_1)$. Let $a_{\alpha\beta}=u_\alpha^*au_\beta$. a is perfectly determined by the matrix $(a_{\alpha\beta})$. In fact, if $a_{\alpha\beta}=b_{\alpha\beta}$ for $\alpha,\beta\in\mathbb{I}$, $u_\alpha^*au_\beta=u_\alpha^*bu_\beta$, and so $e_\alpha ae_\beta=e_\alpha be_\beta$. Hence $a=b$.

Moreover $(\lambda a)_{\alpha\beta}=\lambda a_{\alpha\beta}$ (λ a complex number), $(a+b)_{\alpha\beta}=a_{\alpha\beta}+b_{\alpha\beta}$, $(a^*)_{\alpha\beta}=(a_{\beta\alpha})^*$ and $(ab)_{\alpha\beta}\xi_1=u_\alpha^*(ab)u_\beta\xi_1=u_\alpha^*a\left(\sum_{\gamma\in\mathbb{I}}u_\gamma u_\gamma^*\right)bu_\beta\xi_1=\sum_{\gamma\in\mathbb{I}}a_{\alpha\gamma}b_{\gamma\beta}\xi_1$ for $\xi_1\in\mathscr{H}$.

$$\begin{aligned}(a_1\otimes 1_{\mathscr{H}_2})_{\alpha\beta}\xi_1&=u_\alpha^*(a_1\otimes 1_{\mathscr{H}_2})u_\beta\xi_1=u_\alpha^*(a_1\otimes 1_{\mathscr{H}_2})\xi_1\otimes\eta_\beta\\&=u_\alpha^*(a_1\xi_1\otimes\eta_\beta)=\delta_{\alpha\beta}a_1\xi_1\end{aligned}$$

($a_1\in B(\mathscr{H}_1)$, $\xi_1\in\mathscr{H}_1$, $1_{\mathscr{H}_1}$ (resp. $1_{\mathscr{H}_2}$) is the identity of $B(\mathscr{H}_1)$ (resp. $B(\mathscr{H}_2)$). Hence $(a_1\otimes 1_{\mathscr{H}_2})_{\alpha\beta}=\delta_{\alpha\beta}a_1$.

1.20.1. Lemma. *If* $a\in B(\mathscr{H}_1\otimes\mathscr{H}_2)$ *commutes with* $u_\alpha u_\beta^*$ ($\alpha,\beta\in\mathbb{I}$), *a is of the form* $a_1\otimes 1_{\mathscr{H}_2}$ ($a_1\in B(\mathscr{H}_1)$).

Proof. $a_{\alpha\beta}=u_\alpha^* a u_\beta = u_\gamma^* u_\gamma u_\alpha^* a u_\beta = u_\gamma^* a u_\gamma u_\alpha^* u_\beta$. Since $u_\alpha^* u_\beta = 0$ for $\alpha \neq \beta$ and $u_\alpha^* u_\alpha = 1_{\mathscr{H}_1}$, $a_{\alpha\beta}=0$ for $\alpha\neq\beta$ and $a_{\alpha\alpha}=u_\gamma^* a u_\gamma$ for $\gamma\in\mathbb{I}$. Hence $a_{\alpha\beta}=\delta_{\alpha\beta}a_1$ with $a_1\in B(\mathscr{H}_1)$. q.e.d.

1.20.2. Lemma. *For a subset D of $B(\mathscr{H}_1)$ containing 0, let $\mathscr{M}_D$ be the set of all elements a of $B(\mathscr{H}_1\otimes\mathscr{H}_2)$ such that $a_{\alpha\beta}\in D$. Then $(D\otimes 1_{\mathscr{H}_2})'=\mathscr{M}_{D'}$ and $(D\otimes 1_{\mathscr{H}_2})''=D''\otimes 1_{\mathscr{H}_2}$. Moreover, if D contains $1_{\mathscr{H}_1}$, $(\mathscr{M}_D)'=D'\otimes 1_{\mathscr{H}_2}$ and $(\mathscr{M}_D)''=\mathscr{M}_{D''}$.*

Proof. Let $a_1\otimes 1_{\mathscr{H}_2}$ $(a_1\in D)$ and $b\in(D\otimes 1_{\mathscr{H}_2})'$. Then

$$\{(a_1\otimes 1_{\mathscr{H}_2})b\}_{\alpha\beta}=a_1 b_{\alpha\beta}=\{b(a_1\otimes 1_{\mathscr{H}_2})\}_{\alpha\beta}=b_{\alpha\beta}a_1 .$$

Hence $(D\otimes 1_{\mathscr{H}_2})'\subset\mathscr{M}_{D'}$. The converse is clear. Hence, $(D\otimes 1_{\mathscr{H}_2})'=\mathscr{M}_{D'}$. Further, $(u_\gamma u_\delta^*)_{\alpha\beta}=u_\alpha^* u_\gamma u_\delta^* u_\beta = 0$ $(\alpha\neq\gamma$ or $\beta\neq\delta)$ or $=1$ $(\alpha=\gamma$ and $\beta=\delta)$. Therefore, $u_\gamma u_\delta^*\in\mathscr{M}_{D'}$. Hence $(D\otimes 1_{\mathscr{H}_2})''=(\mathscr{M}_{D'})'\subset B(\mathscr{H}_1)\otimes 1_{\mathscr{H}_2}$, and $(D\otimes 1_{\mathscr{H}_2})''\subset(D'\otimes 1_{\mathscr{H}_2})'=\mathscr{M}_{D''}$, so that $(D\otimes 1_{\mathscr{H}_2})''=D''\otimes 1_{\mathscr{H}_2}$.

Finally, suppose $1_{\mathscr{H}_1}\in D$. Then $u_\gamma u_\delta^*\in\mathscr{M}_D$, $(\mathscr{M}_D)'\subset B(\mathscr{H}_1)\otimes 1_{\mathscr{H}_2}$, and $\mathscr{M}_D\supset D\otimes 1_{\mathscr{H}_2}$. Hence, $(\mathscr{M}_D)'=D'\otimes 1_{\mathscr{H}_2}$; $(\mathscr{M}_D)''=(D'\otimes 1_{\mathscr{H}_2})'=\mathscr{M}_{D''}$. q.e.d.

1.20.3. Theorem. *Let $\mathscr{M}$ be a self-adjoint subalgebra of $B(\mathscr{H})$ on some Hilbert space $\mathscr{H}$, and suppose that $\mathscr{M}$ contains the identity of $B(\mathscr{H})$. Then the following conditions are equivalent.*

1. *$\mathscr{M}$ is weakly closed;*
2. *$\mathscr{M}''=\mathscr{M}$.*

Proof. Clearly, 2. implies 1. Now suppose that $\mathscr{M}$ is weakly closed and $\mathscr{M}\subsetneqq\mathscr{M}''$. Then there exists a linear functional f on $\mathscr{M}''$ continuous with respect to the weak operator topology, such that $f(\mathscr{M})=0$ and $f(a)\neq 0$ for some $a\in\mathscr{M}''$.

Let $f(x)=\sum_{i=1}^n (x\xi_i,\xi_i')$ $(x\in\mathscr{M}'', \xi_i, \xi_i'\in\mathscr{H}$ $(i=1,2,\dots,n))$. Let $(\eta_i\,|\,i=1,2,\dots,n)$ be a complete orthonormal system in a n-dimensional Hilbert space $\mathscr{H}_1$. Put $\mathscr{H}^i=\mathscr{H}\otimes\eta_i$, and let u_i be an isometry $\mathscr{H}$ onto $\mathscr{H}^i$ such that $\xi\to\xi\otimes\eta_i$ $(\xi\in\mathscr{H})$. Then

$$f(x)=\sum_{i=1}^n (x\xi_i,\xi_i')=\sum_{i=1}^n (x u_i^* u_i\xi_i, u_i^* u_i\xi_i')$$

$$=\sum_{i=1}^n (u_i x u_i^* u_i\xi_i, u_i\xi_i')=((x\otimes 1_{\mathscr{H}_1})\xi,\xi') \quad (x\in\mathscr{M}''),$$

where $\xi=\sum_{i=1}^n u_i\xi_i$, $\xi'=\sum_{i=1}^n u_i\xi_i'\in\mathscr{H}\otimes\mathscr{H}_1$. Therefore, $((\mathscr{M}\otimes 1_{\mathscr{H}_1})\xi,\xi')=0$ and $((a\otimes 1_{\mathscr{H}_1})\xi,\xi')\neq 0$. Let $\mathscr{K}$ be the closed subspace of $\mathscr{H}\otimes\mathscr{H}_1$

generated by the set $\{(\mathscr{M}\otimes 1_{\mathscr{H}_1})\xi\}$, and let e be the projection of $\mathscr{H}\otimes\mathscr{H}_1$ onto $\mathscr{X}$.

Since $\mathscr{X}$ is invariant under the *-algebra $\mathscr{M}\otimes 1_{\mathscr{H}_1}$, e belongs to $(\mathscr{M}\otimes 1_{\mathscr{H}_1})'$. On the other hand, $\mathscr{M}''\otimes 1_{\mathscr{H}_1}=(\mathscr{M}\otimes 1_{\mathscr{H}_1})''$, and $\mathscr{M}\otimes 1_{\mathscr{H}_1}$ contains the identity. Therefore, $e(a\otimes 1_{\mathscr{H}_1})\xi=(a\otimes 1_{\mathscr{H}_1})e\xi=(a\otimes 1_{\mathscr{H}_1})\xi$. Hence, $(a\otimes 1_{\mathscr{H}_1})\xi\in\mathscr{X}$. Therefore, there exists a sequence (a_n) of $\mathscr{M}$ such that $\|(a_n\otimes 1_{\mathscr{H}_1})\xi-(a\otimes 1_{\mathscr{H}_1})\xi\|\to 0$ $(n\to\infty)$. Hence, $((\mathscr{M}\otimes 1_{\mathscr{H}_1})\xi,\xi')=0$ implies $((a\otimes 1_{\mathscr{H}_1})\xi,\xi')=0$, a contradiction. q.e.d.

1.20.4. Corollary. *Let $\mathscr{M}$ and $\mathscr{N}$ be two weakly closed self-adjoint subalgebras of $B(\mathscr{H})$ containing the identity $1_{\mathscr{H}}$. Then $R(\mathscr{M},\mathscr{N})=(\mathscr{M}'\cap\mathscr{N}')'$, where $R(\mathscr{M},\mathscr{N})$ is the weakly closed self-adjoint subalgebra of $B(\mathscr{H})$ generated by $\mathscr{M}$ and $\mathscr{N}$.*

1.20.5. Proposition. *Let $\mathscr{N}$ be a weakly closed self-adjoint algebra containing $1_{\mathscr{H}}$ on a Hilbert space $\mathscr{H}$, $\mathscr{Z}=\mathscr{N}\cap\mathscr{N}'$, t_{ij} $(i,j=1,2,\ldots,n)$ elements in $\mathscr{N}$, t'_{ij} $(i,j=1,2,\ldots,n)$ elements in $\mathscr{N}'$. Then the following conditions are equivalent:*

1. $\sum_{k=1}^{n} t_{ik}t'_{kj}=0$ $(i,j=1,2,\ldots,n)$;
2. *there exist elements z_{ij} $(i,j=1,2,\ldots,n)$ in $\mathscr{Z}$ such that*

$$\sum_{k=1}^{n} t_{ik}z_{kj}=0,\qquad \sum_{k=1}^{n} z_{ik}t'_{kj}=t'_{ij}\quad (i,j=1,2,\ldots,n).$$

Proof. It is easy to prove that $2.\Rightarrow 1.$ We shall prove $1.\Rightarrow 2.$ Suppose $\sum_{k=1}^{n} t_{ik}t'_{kj}=0$ $(i,j=1,2,\ldots,n)$. Let $\mathscr{K}$ be an n-dimensional Hilbert space. Let $t=(t_{ij})_{i,j=1,2,\ldots,n}\in\mathscr{N}\otimes B(\mathscr{K})$ and $t'=(t'_{ij})_{i\,j=1,2,\ldots,n}\in\mathscr{N}'\otimes B(\mathscr{K})$. Then $tt'=0$. Put $\mathscr{I}'=\{x'\,|\,tx'=0, x'\in\mathscr{N}'\otimes B(\mathscr{K})\}$. Then $\mathscr{I}'$ is a σ-closed right ideal of $\mathscr{N}'\otimes B(\mathscr{K})$. Hence there exists a projection $z'=(z_{ij})_{i,j=1,\ldots,n}$ such that $\mathscr{I}'=z'(\mathscr{N}'\otimes B(\mathscr{K}))$. Since $(\mathscr{N}'\otimes 1_{\mathscr{K}})\mathscr{I}'\subset\mathscr{I}'$, $z_{ij}\in\mathscr{N}$, and so $z_{ij}\in\mathscr{N}\cap\mathscr{N}'$ $(i,j=1,2,\ldots,n)$. Since $tz'=0$, $\sum_{k=1}^{n} t_{ik}z_{kj}=0$ $(i,j=1,2,\ldots,n)$. Since $z't'=t'$, $\sum_{k=1}^{n} z_{ik}t'_{kj}=t'_{ij}$ $(i,j=1,2,\ldots,n)$. q.e.d.

1.21. *-Representations of C^*-Algebras, 1

1.21.1. Definition. *Let $\{\pi,\mathscr{H}\}$ be a *-representation of a C^*-algebra $\mathscr{A}$. $\{\pi,\mathscr{H}\}$ is said to be cyclic if there exists an element ξ in $\mathscr{H}$ such that $[\pi(\mathscr{A})\xi]=\mathscr{H}$ ($[\pi(\mathscr{A})\xi]$ is the closed subspace of $\mathscr{H}$ generated by $\{\pi(a)\xi\,|\,a\in\mathscr{A}\}$). Such an element ξ is called a cyclic vector of $\{\pi,\mathscr{H}\}$. An element η in $\mathscr{H}$ is called a separating vector of $\{\pi,\mathscr{H}\}$, if $[\pi(\mathscr{A})'\eta]=\mathscr{H}$.*

1.21.2. Definition. *Let $\{\pi_1, \mathscr{H}_1\}$, $\{\pi_2, \mathscr{H}_2\}$ be two *-representations of a C^*-algebra $\mathscr{A}$. $\{\pi_1, \mathscr{H}_1\}$ is said to be equivalent to $\{\pi_2, \mathscr{H}_2\}$ if there exists an unitary operator U of $\mathscr{H}_1$* onto *$\mathscr{H}_2$ such that $\pi_2(a)U = U\pi_1(a)$ for $a \in \mathscr{A}$.*

We shall identify equivalent *-representations.

1.21.3. Definition. *Let $\{\pi_\alpha, \mathscr{H}_\alpha\}_{\alpha\in\mathrm{II}}$ be a family of *-representations of a C^*-algebra $\mathscr{A}$. The sum of the $\{\pi_\alpha, \mathscr{H}_\alpha\}_{\alpha\in\mathrm{II}}$, $\{\pi, \mathscr{H}\} = \sum_{\alpha\in\mathrm{II}} \{\pi_\alpha, \mathscr{H}_\alpha\}$, is a *-representation $\{\pi, \mathscr{H}\}$ of $\mathscr{A}$ defined as follows: $\mathscr{H} = \sum_{\alpha\in\mathrm{II}} \oplus \mathscr{H}_\alpha$ and $\pi(a) = \sum_{\alpha\in\mathrm{II}} \oplus \pi_\alpha(a)$ $(a \in \mathscr{A})$.*

1.21.4. Definition. *Let $\{\pi, \mathscr{H}\}$ be a *-representation of $\mathscr{A}$ and let E' be a projection in $\pi(\mathscr{A})'$. Consider the *-representation $\{\pi_1, \mathscr{H}_1\}$ defined as follows. $\mathscr{H}_1 = E'\mathscr{H}$ and $\pi_1(a) = \pi(a)E'$ $(a \in \mathscr{A})$. This representation, denoted by $\{\pi E', E'\mathscr{H}\}$, is said to be a sub*-representation of $\{\pi, \mathscr{H}\}$.*

We define analogously a cyclic W^*-representation, the equivalence, a sum and a subrepresentation for W^*-representations of a W^*-algebra.

1.21.5. Proposition. *Let $\{\pi, \mathscr{H}\}$ be a *-representation of a C^*-algebra $\mathscr{A}$. Suppose that $\{\pi, \mathscr{H}\}$ has a cyclic vector ξ and put $\varphi(x) = (\pi(x)\xi, \xi)$ $(x \in \mathscr{A})$. φ is a bounded positive linear functional on $\mathscr{A}$, and $\{\pi, \mathscr{H}\}$ is equivalent to $\{\pi_\varphi, \mathscr{H}_\varphi\}$.*

Proof. It is clear that φ is bounded positive. Consider the mapping $\pi(a)\xi \to a_\varphi$ $(a \in \mathscr{A})$. $\|\pi(a)\xi\|^2 = \varphi(a^* a) = \|a_\varphi\|^2$. Hence, this mapping can be uniquely extended to a unitary operator U of $\mathscr{H}$ onto $\mathscr{H}_\varphi$. Further, $\pi_\varphi(b)U\pi(a)\xi = (ba)_\varphi = U\pi(b)\pi(a)\xi$. Hence $U\pi(b) = \pi_\varphi(b)U$ $(b \in \mathscr{A})$. q.e.d.

1.21.6. Proposition. *Let φ be a non-zero bounded positive linear functional on a C^*-algebra $\mathscr{A}$. The *-representation $\{\pi_\varphi, \mathscr{H}_\varphi\}$ of $\mathscr{A}$ is cyclic.*

Proof. Let $\{\pi_\varphi^w, \mathscr{H}_\varphi\}$ be the W^*-representation of $\mathscr{A}^{**}$ obtained from $\{\pi_\varphi, \mathscr{H}_\varphi\}$. $\{\pi_\varphi^w, \mathscr{H}_\varphi\}$ is equivalent to the *-representation $\{\pi_\varphi, \mathscr{H}_\varphi\}$ of $\mathscr{A}^{**}$ constructed via the $\sigma(\mathscr{A}^{**}, \mathscr{A}^*)$-continuous positive linear functional φ. Hence $1_\varphi \in \mathscr{H}_\varphi$ and $[\pi_\varphi(\mathscr{A})1_\varphi] = [\pi_\varphi(\mathscr{A}^{**})1_\varphi] = \mathscr{H}$. So 1_φ is a cyclic vector. q.e.d.

1.21.7. Proposition. *Let $\{\pi, \mathscr{H}\}$ be a *-representation of a C^*-algebra $\mathscr{A}$ such that $\pi(a)\xi = 0$ (all $a \in \mathscr{A}$) imply $\xi = 0$ (called a nowhere trivial *-representation). Then $\{\pi, \mathscr{H}\}$ is equivalent to a sum of cyclic *-representations, constructed via bounded positive linear functionals.*

Proof. Let $\{\xi_\alpha\}_{\alpha\in\mathrm{II}}$ be a maximal family of non-zero vectors in $\mathscr{H}$ such that $[\pi(\mathscr{A})\xi_\alpha]$ is orthogonal to $[\pi(\mathscr{A})\xi_\beta]$ $(\alpha, \beta \in \mathrm{II},\ \alpha \neq \beta)$. Then $\sum_{\alpha\in\mathrm{II}} \oplus [\pi(\mathscr{A})\xi_\alpha] = \mathscr{H}$. In fact, suppose there exists a non-zero vector ξ

in $\mathscr{H}$ such that ξ is orthogonal to $[\pi(\mathscr{A})\xi_\alpha]$ for all $\alpha\in\mathbb{I}$. Then $(\pi(\mathscr{A})\xi, \pi(\mathscr{A})\xi_\alpha)=(\xi, \pi(\mathscr{A})\xi_\alpha)=0$, and $[\pi(\mathscr{A})\xi]\neq(0)$. This contradicts the maximality $\{\xi_\alpha\}$. Let E'_α be the orthogonal projection of $\mathscr{H}$ onto $[\pi(\mathscr{A})\xi_\alpha]$. $\pi(a)E'_\alpha\mathscr{H}\subset[\pi(\mathscr{A})\xi_\alpha]$. Hence, $E'_\alpha\pi(a)E'_\alpha=\pi(a)E'_\alpha$ and $\pi(a)E'_\alpha=E'_\alpha\pi(a)$ (if $a^*=a$), and so $\pi(a)E'_\alpha=E'_\alpha\pi(a)$ $(a\in\mathscr{A})$. Therefore, $\{\pi E'_\alpha, E'_\alpha\mathscr{H}\}$ is a sub*-representation of $\{\pi,\mathscr{H}\}$, and

$$\{\pi,\mathscr{H}\} = \sum_{\alpha\in\mathbb{I}}\{\pi E'_\alpha, E'_\alpha\mathscr{H}\}. \qquad \text{q.e.d.}$$

1.21.8. Definition. *Let $\{\pi,\mathscr{H}\}$ be a *-representation of a C*-algebra $\mathscr{A}$. $\{\pi,\mathscr{H}\}$ is said to be irreducible if every non-zero element in $\mathscr{H}$ is a cyclic vector of $\{\pi,\mathscr{H}\}$.*

1.21.9. Proposition. *Let $\{\pi,\mathscr{H}\}$ be a *-representation of the C*-algebra $\mathscr{A}$. The following conditions are equivalent:*

1. *$\{\pi,\mathscr{H}\}$ is irreducible;*
2. *$\overline{\pi(\mathscr{A})}=B(\mathscr{H})$, where $\overline{\pi(\mathscr{A})}$ is the weak closure of $\pi(\mathscr{A})$ in $B(\mathscr{H})$.*

Proof. Suppose $\{\pi,\mathscr{H}\}$ is irreducible. Then $\overline{\pi(\mathscr{A})}$ contains $1_{\mathscr{H}}$. Consider $\pi(\mathscr{A})'$, a W^*-algebra. If $\pi(\mathscr{A})'$ contains a non-trivial projection E', then $\pi(\mathscr{A})E'\mathscr{H}\subseteqq E'\mathscr{H}\neq\mathscr{H}$, a contradiction. Hence $\pi(\mathscr{A})'=(\mathbb{C}1_{\mathscr{H}})$ ($\mathbb{C}$, the field of complex numbers). Hence $\overline{\pi(\mathscr{A})}=\pi(\mathscr{A})''=B(\mathscr{H})$.

Conversely, suppose there exists a non-zero element ξ_0 in $\mathscr{H}$ such that $[\pi(\mathscr{A})\xi_0]\subsetneqq\mathscr{H}$, and let E' be the orthogonal projection of $\mathscr{H}$ onto $[\pi(\mathscr{A})\xi_0]$. Then E' is in $\pi(\mathscr{A})'$. Hence, $\pi(\mathscr{A})'\neq(\mathbb{C}1_{\mathscr{H}})$ and so $\pi(\mathscr{A})''\subsetneqq B(\mathscr{H})$. q.e.d.

1.21.10. Theorem. *Let φ be a state on the C*-algebra $\mathscr{A}$. Then the following conditions are equivalent:*

1. *$\{\pi_\varphi,\mathscr{H}_\varphi\}$ is irreducible;*
2. *φ is extreme in the state space $\mathscr{S}$ of $\mathscr{A}$.*

Proof. One may assume that $\mathscr{A}$ has an identity.

Suppose that $\{\pi_\varphi,\mathscr{H}_\varphi\}$ is irreducible and $\varphi=(\varphi_1+\varphi_2)/2$ $(\varphi_1,\varphi_2\in\mathscr{S})$. Since $|\varphi_1(b^*a)|\leq\varphi_1(b^*b)^{\frac{1}{2}}\varphi_1(a^*a)^{\frac{1}{2}}\leq 2\varphi(b^*b)^{\frac{1}{2}}\varphi(a^*a)^{\frac{1}{2}}=2\|a_\varphi\|\,\|b_\varphi\|$, $B(a_\varphi,b_\varphi)=\varphi_1(b^*a)$ defines a continuous positive definite conjugate bilinear functional on $\mathscr{H}_\varphi$. Hence, there exists a bounded positive linear operator H on $\mathscr{H}_\varphi$ such that $B(a_\varphi,b_\varphi)=(Ha_\varphi,b_\varphi)$ $(a,b\in\mathscr{A})$.

One easily sees that $H\in\pi(\mathscr{A})'$. Hence, $H=\lambda 1_{\mathscr{H}_\varphi}$, and so $\varphi_1=\varphi_2=\varphi$.

Conversely, suppose that $\pi_\varphi(\mathscr{A})'\neq(\mathbb{C}1_{\mathscr{H}})$. Take a non-trivial projection P' in $\pi_\varphi(\mathscr{A})'$, and set $\varphi_1(a)=\dfrac{(\pi_\varphi(a)P'1_\varphi,1_\varphi)}{\lambda_1}$ and

$$\varphi_2(a)=\frac{(\pi_\varphi(a)(1_{\mathscr{H}_\varphi}-P')1_\varphi,1_\varphi)}{\lambda_2} \qquad (a\in\mathscr{A},\ \lambda_1=(P'1_\varphi,1_\varphi),$$

and $\lambda_2=((1_{\mathscr{H}_\varphi}-P')1_\varphi, 1_\varphi))$. Then $\varphi_1, \varphi_2 \in \mathscr{S}$, and $\lambda_1\varphi_1+\lambda_2\varphi_2=\varphi$, and $\lambda_1+\lambda_2=1$.

Assume that $\varphi=\varphi_1$. Then

$$\left\|\pi_\varphi(a)P'\frac{1}{\sqrt{\lambda_1}}1_\varphi\right\| = \left\|\frac{P'}{\sqrt{\lambda_1}}\pi_\varphi(a)1_\varphi\right\| = \left\|\frac{P'}{\sqrt{\lambda_1}}a_\varphi\right\| = \|a_\varphi\|.$$

Hence $\dfrac{P'}{\sqrt{\lambda_1}}$ is an isometry on $\mathscr{H}_\varphi$. This is a contradiction, so $\varphi\neq\varphi_1$. Similarly, $\varphi_2\neq\varphi$. q.e.d.

1.21.11. Definition. *An extreme state on a C^*-algebra is called a pure state.*

1.21.12. Theorem. *Let $\mathscr{A}$ be a C^*-algebra. Then $\mathscr{A}$ has a faithful family of irreducible *-representations $\{\pi_\alpha, \mathscr{H}_\alpha\}_{\alpha\in\mathbb{I}}$ (i.e., $\pi_\alpha(x)=0$ for all $\alpha\in\mathbb{I}$ imply $x=0$).*

Proof. Let $\mathfrak{F}$ be the set of all positive linear functionals φ on $\mathscr{A}$ such that $\|\varphi\|\leq 1$. Then $\mathfrak{F}$ is a $\sigma(\mathscr{A}^*, \mathscr{A})$-compact convex set. Let $\mathfrak{F}_0$ be the set of all extreme points in $\mathfrak{F}$; then $\varphi=0$ or $\|\varphi\|=1$ $(\varphi\in\mathfrak{F}_0)$. If $\varphi\neq 0$, φ is a pure state on $\mathscr{A}$. Consider all irreducible *-representations $\{\pi_\varphi, \mathscr{H}_\varphi\}_{\varphi\in\mathfrak{F}_0, \varphi\neq 0}$ of $\mathscr{A}$. Then $\pi_\varphi(a)=0$ for all φ $(\neq 0)\in\mathfrak{F}_0$ implies $a=0$. q.e.d.

1.21.13. Proposition. *Let $\mathscr{A}$ be a C^*-algebra, and let Φ be a bounded linear mapping of $\mathscr{A}$ into the W^*-algebra $B(\mathscr{H})$. Then Φ can be uniquely extended to a continuous linear mapping $\tilde{\Phi}$ of $\mathscr{A}^{**}$, with the topology $\sigma(\mathscr{A}^{**}, \mathscr{A}^*)$, into $B(\mathscr{H})$, with the topology $\sigma(B(\mathscr{H}), B(\mathscr{H})_*)$.*

In particular, let $\{\pi, \mathscr{H}\}$ be a *-representation of $\mathscr{A}$. Then, the extended mapping, denoted by $\{\tilde{\pi}, \mathscr{H}\}$, is a W^*-representation $\{\pi^w, \mathscr{H}\}$ of $\mathscr{A}^{**}$. Conversely, let $\{\pi^w, \mathscr{H}\}$ be a W^*-representation of $\mathscr{A}^{**}$ on the Hilbert space $\mathscr{H}$. The restriction of π on $\mathscr{A}$ is a *-representation $\{\pi, \mathscr{H}\}$ of $\mathscr{A}$, and its unique extension $\{\tilde{\pi}, \mathscr{H}\}$ to the W^*-representation of $\mathscr{A}^{**}$ coincides with $\{\pi^w, \mathscr{H}\}$. Hence, there exists a one-to-one correspondence between *-representations of $\mathscr{A}$ and W^*-representations of $\mathscr{A}^{**}$.

Proof. Let Φ^* (resp. Φ^{**}) be the dual (resp. the second dual) of Φ. Φ^{**} is a continuous linear mapping of $\mathscr{A}^{**}$, with the $\sigma(\mathscr{A}^{**}, \mathscr{A}^*)$-topology, into $B(\mathscr{H})^{**}$, with the $\sigma(B(\mathscr{H})^{**}, B(\mathscr{H})^*)$-topology.

By 1.17.8, $B(\mathscr{H})^{**}=B(\mathscr{H})+B(\mathscr{H})^{**}z$, where $B(\mathscr{H})^{**}z=B(\mathscr{H})_*^0$ in $B(\mathscr{H})^{**}$, and $B(\mathscr{H})^{**}z\cap B(\mathscr{H})=(0)$. The canonical projection Ψ_1 of $B(\mathscr{H})^{**}$ onto $B(\mathscr{H})$ is a W^*-homomorphism. Now set $\tilde{\Phi}(x)=\Psi_1\Phi^{**}(x)$ $(x\in\mathscr{A}^{**})$. Then $\tilde{\Phi}$ is a continuous mapping of $\mathscr{A}^{**}$, with the $\sigma(\mathscr{A}^{**}, \mathscr{A}^*)$-topology, into $B(\mathscr{H})$, with the $\sigma(B(\mathscr{H}), B(\mathscr{H})_*)$-topology.

Moreover, $\tilde{\Phi}(a)=\Psi_1\Phi^{**}(a)=\Psi_1\Phi(a)=\Phi(a)$ $(a\in\mathscr{A})$. Hence, $\tilde{\Phi}$ is an extension of Φ. By the continuity of $\tilde{\Phi}$ and the density of $\mathscr{A}$, this extension is unique.

Next, suppose Φ is a *-representation. For $x, y\in\mathscr{A}^{**}$, there exist directed sets $\{x_\alpha\}$, $\{y_\beta\}$ in $\mathscr{A}$ such that $x_\alpha\to x$, $y_\beta\to y$ in the $\sigma(\mathscr{A}^{**},\mathscr{A}^*)$-topology. Hence,

$$\tilde{\Phi}(x_\alpha y)=\lim_\beta \tilde{\Phi}(x_\alpha y_\beta)=\lim_\beta \Phi(x_\alpha y_\beta)=\lim_\beta \Phi(x_\alpha)\Phi(y_\beta)=\Phi(x_\alpha)\tilde{\Phi}(y)$$

and $\tilde{\Phi}(xy)=\lim_\alpha \tilde{\Phi}(x_\alpha y)=\lim_\alpha \Phi(x_\alpha)\tilde{\Phi}(y)=\tilde{\Phi}(x)\tilde{\Phi}(y)$. Analogously we have $\tilde{\Phi}(x^*)=\tilde{\Phi}(x)^*$. Hence $\tilde{\Phi}$ is a W^*-representation of $\mathscr{A}^{**}$. The remainder of the proof is clear. q.e.d.

1.21.14. Definition. *Let $\{\pi^w,\mathscr{H}\}$ be a W^*-representation of a W^*-algebra $\mathscr{M}$. The kernel $\mathscr{I}=\{a\,|\,\pi(a)=0,\ a\in\mathscr{M}\}$ is a $\sigma(\mathscr{M},\mathscr{M}_*)$-closed two-sided ideal. Hence there exists a unique central projection z in $\mathscr{M}$ such that $\mathscr{I}=\mathscr{M}z$. Put $1-z=s(\pi)$. $s(\pi)$ is called the support of $\{\pi^w,\mathscr{H}\}$. The restriction of π to $\mathscr{M}s(\pi)$ is one-to-one.*

1.21.15. Definition. *Let $\{\pi_1^w,\ \mathscr{H}_1\}$, $\{\pi_2^w,\ \mathscr{H}_2\}$ be two W^*-representations of $\mathscr{M}$. If $s(\pi_1)=s(\pi_2)$, we say that $\{\pi_1^w,\ \mathscr{H}_1\}$ is quasi-equivalent to $\{\pi_2^w,\ \mathscr{H}_2\}$.*

Clearly the quasi-equivalence is an equivalence relation, so that by this relation we can classify W^*-representations of $\mathscr{M}$ into quasi-equivalence classes.

Let $\mathscr{D}(\mathscr{M})$ be the family of all quasi-equivalence classes of the W^*-representations of $\mathscr{M}$. Then, to each element $\rho\in\mathscr{D}(\mathscr{M})$, there corresponds a unique non-zero central projection $c(\rho)$ of $\mathscr{M}$ such that $c(\rho)=s(\pi)$ $(\{\pi,\mathscr{H}\}\in\rho)$.

Conversely, let z be a non-zero central projection of $\mathscr{M}$. Then $\mathscr{M}z$ is a W^*-algebra, so that it has a faithful W^*-representation $\{\pi,\mathscr{H}\}$. Then the mapping $a\to\pi(az)$ of $\mathscr{M}$ into $B(\mathscr{H})$ is a W^*-representation of $\mathscr{M}$. Hence, there exists a one-to-one correspondence between all elements of $\mathscr{D}(\mathscr{M})$ and all non-zero central projections of $\mathscr{M}$. For convenience, we shall add an imaginary 0-element to $\mathscr{D}(\mathscr{M})$, and we shall denote by $\mathscr{D}'(\mathscr{M})$ the set $\mathscr{D}(\mathscr{M})\cup(0)$.

There is a one-to-one correspondence between $\mathscr{D}'(\mathscr{M})$ and the set $\mathscr{Z}^p$ of all central projections of $\mathscr{M}$. Since $\mathscr{Z}^p$ is a complete Boolean algebra, we can canonically introduce its Boolean algebra structure into $\mathscr{D}'(\mathscr{M})$. Also, we can regard every element ρ of $\mathscr{D}(\mathscr{M})$ as a W^*-homomorphism of $\mathscr{M}$ onto $\mathscr{M}c(\rho)$. Therefore, within the quasi-equivalence, the W^*-representation theory of $\mathscr{M}$ can be completely reduced to the structure theory (ideal theory) of the W^*-algebra $\mathscr{M}$.

Now let $\mathscr{A}$ be a C^*-algebra, $\{\pi,\mathscr{H}\}$ a *-representation of $\mathscr{A}$. Then there corresponds a unique W^*-representation $\{\tilde{\pi},\mathscr{H}\}$ of $\mathscr{A}^{**}$. Using

$\{\tilde{\pi}, \mathscr{H}\}$, we define the quasi-equivalence and the support of $\{\pi, \mathscr{H}\}$ (i.e., $s(\pi)=s(\tilde{\pi})$).

1.21.16. Theorem. *Let $\mathscr{A}$ be a C^*-algebra, $\tilde{\mathscr{A}}$ the C^*-subalgebra of $\mathscr{A}^{**}$ generated by $\mathscr{A}$ and 1, and let $\{\pi_i, \mathscr{H}_i\}$ $(i=1,2,\ldots,n)$ be a finite family of mutually inequivalent irreducible *-representations of $\mathscr{A}$.*

1. Let $T_1 \in B(\mathscr{H}_1), \ldots, T_n \in B(\mathscr{H}_n)$, and let $E_1, E_2, \ldots, E_n$ be finite-dimensional projections of $B(\mathscr{H}_1), B(\mathscr{H}_2), \ldots, B(\mathscr{H}_n)$, and $\varepsilon>0$. Then there exists an element $a \in \mathscr{A}$ such that $E_j \pi_j(a)=E_j T_j$, $\pi_j(a) E_j=T_j E_j$ $(1 \leq j \leq n)$ and $\|a\|-\varepsilon<\max\limits_{1 \leq j \leq n}(\|T_j\|)$.

2. Let $T_1 \in B(\mathscr{H}_1), \ldots, T_n \in B(\mathscr{H}_n)$ be self-adjoint elements, and $E_1, \ldots, E_n$ finite dimensional projections of $B(\mathscr{H}_1), \ldots, B(\mathscr{H}_n)$, and let $\varepsilon>0$. Then there exists a self-adjoint element h of $\mathscr{A}$ such that $E_j \pi_j(h)=E_j T_j$, $\pi_j(h) E_j=T_j E_j$ $(1 \leq j \leq n)$, and $\|h\|-\varepsilon<\max\limits_{1 \leq j \leq n}(\|T_j\|)$.

3. Let $T_1 \in B(\mathscr{H}_1), \ldots, T_n \in B(\mathscr{H}_n)$ of unitaries, and $E_1, \ldots, E_n$ finite-dimensional projections of $B(\mathscr{H}_1), \ldots, B(\mathscr{H}_n)$. Then there exists a unitary element u of $\tilde{\mathscr{A}}$ such that $\tilde{\pi}(u) E_j=T_j E_j$ $(1 \leq j \leq n)$.

Proof. $\mathscr{A}^{**} s(\pi_j)$ is *-isomorphic to $\tilde{\pi}_j(\mathscr{A}^{**})=\overline{\pi_j(\mathscr{A})}=B(\mathscr{H}_j)$ under $\tilde{\pi}_j$. Hence $s(\pi_j)$ is a minimal central projection in the center of $\mathscr{A}^{**}$, and so $s(\pi_j) s(\pi_k)=0$ or $s(\pi_j) s(\pi_k)=s(\pi_j)=s(\pi_k)$. Suppose that $s(\pi_j) s(\pi_k)=s(\pi_j)=s(\pi_k)$ for $j \neq k$. $\tilde{\pi}_j(x s(\pi_j))=\pi_j(x)$ and $\tilde{\pi}_k(x s(\pi_k))=\pi_k(x)$ $(x \in \mathscr{A})$.

The mapping $\pi_j(x) \rightarrow \pi_k(x)$ $(x \in \mathscr{A})$ can be uniquely extended to a W^*-homomorphism Φ of $B(\mathscr{H}_j)$ onto $B(\mathscr{H}_k)$. Put $\varphi(Y)=(\Phi(Y) \xi, \xi)$ $(Y \in B(\mathscr{H}_j)$, $\|\xi\|=1$, and $\xi \in \mathscr{H}_k)$. Then φ is a normal pure state on $B(\mathscr{H}_j)$. Hence $\varphi(Y)=Tr(K Y)$ for $Y \in B(\mathscr{H}_j)$, where K is a positive element of trace-class. Since φ is a pure state, K must be a one-dimensional projection. Take an element η of $\mathscr{H}_j$ such that $K \eta=\eta$ and $\|\eta\|=1$. Then $\varphi(Y)=Tr(K Y)=(Y \eta, \eta)$ $(Y \in B(\mathscr{H}_j))$. Hence, we have

$$\psi(x)=(\pi_k(x) \xi, \xi)=(\Phi(\pi_j(x)) \xi, \xi)=(\pi_j(x) \eta, \eta)=\varphi(x).$$

Since $\{\pi_j, \mathscr{H}_j\}$ and $\{\pi_k, \mathscr{H}_k\}$ are irreducible, they are equivalent to $\{\pi_\varphi, \mathscr{H}_\varphi\}$ and $\{\pi_\psi, \mathscr{H}_\psi\}$ respectively. Hence $\{\pi_j, \mathscr{H}_j\}$ is equivalent to $\{\pi_k, \mathscr{H}_k\}$, a contradiction. Hence, $s(\pi_j) s(\pi_k)=0$.

Consider the sum $\{\pi, \mathscr{H}\}=\sum_{j=1}^n\{\pi_j, \mathscr{H}_j\}$. Then $s(\pi)=\sum_{j=1}^n s(\pi_j)$. Hence, $\tilde{\pi}(\mathscr{A}^{**})=\sum_{j=1}^n \oplus B(\mathscr{H}_j)$ and so $\mathscr{A}^{**} s(\pi)$ is *-isomorphic to $\sum_{j=1}^n \oplus B(\mathscr{H}_j)$ under $\tilde{\pi}$.

Put $E=\sum_{j=1}^{n} E_j$. Then E is a projection belonging to $\sum_{j=1}^{n} \oplus B(\mathscr{H}_j)$. Hence, there exists a unique projection p in $\mathscr{A}^{**}$ such that $\tilde{\pi}(p)=E$ and $p\leq s(\pi)$. Set $\mathfrak{F}=\{y\mid\tilde{\pi}(y)E=E\tilde{\pi}(y)=0,\ y\in\mathscr{A}^{**}\}$. Then $\mathfrak{F}=(1-p)\mathscr{A}^{**}(1-p)$, and so $\mathfrak{F}$ is $\sigma(\mathscr{A}^{**},\mathscr{A}^{*})$-closed.

$$\begin{aligned}\left(\sum_{j=1}^{n}\oplus B(\mathscr{H}_j)\right)E+E\left(\sum_{j=1}^{n}\oplus B(\mathscr{H}_j)\right) &= \sum_{j=1}^{n}\oplus B(\mathscr{H}_j)E_j+\sum_{j=1}^{n}\oplus E_jB(\mathscr{H}_j)\\ &= \sum_{j=1}^{n}\oplus(E_jB(\mathscr{H}_j)E_j\\ &\quad +(1_{\mathscr{H}_j}-E_j)B(\mathscr{H}_j)E_j)\\ &\quad +\sum_{j=1}^{n}\oplus E_jB(\mathscr{H}_j)(1_{\mathscr{H}_j}-E_j).\end{aligned}$$

Let $Q_1^j, Q_2^j, \ldots, Q_{m_j}^j$ $(j=1,2,\ldots,n)$ be a family of mutually orthogonal one-dimensional projections of $B(\mathscr{H}_j)$ such that $\sum_{l=1}^{m_j} Q_l^j=E_j$. Then $B(\mathscr{H}_j)E_j=\sum_{l=1}^{m_j} B(\mathscr{H}_j)Q_l^j$ and $B(\mathscr{H}_j)Q_l^j\cap B(\mathscr{H}_j)Q_r^j=(0)$ $(r\neq l)$. Moreover,

$$\|YQ_l^j\|=\|Q_l^jY^*YQ_l^j\|^{\frac{1}{2}}=(Q_l^jY^*YQ_l^j\xi,\xi)^{\frac{1}{2}}=\|Y\xi\|$$

$(\xi\in\mathscr{H}_j,\ \|\xi\|=1,$ and $Q_l^j\xi=\xi)$.

Hence the Banach space $B(\mathscr{H}_j)Q_l^j$ is a Hilbert space. Hence $B(\mathscr{H}_j)E_j$ and so $E_jB(\mathscr{H}_j)$ are reflexive Banach spaces. Therefore,

$$\left(\sum_{j=1}^{n}\oplus B(\mathscr{H}_j)\right)E+E\left(\sum_{j=1}^{n}\oplus B(\mathscr{H}_j)\right)$$

is again reflexive. Since $\mathscr{A}^{**}p+p\mathscr{A}^{**}$ is isometric to

$$\left(\sum_{j=1}^{n}\oplus B(\mathscr{H}_j)\right)E+E\left(\sum_{j=1}^{n}\oplus B(\mathscr{H}_j)\right)$$

under $\tilde{\pi}$, $\mathscr{A}^{**}p+p\mathscr{A}^{**}$ is reflexive.

Let $\mathfrak{F}^0$ be the polar of $\mathfrak{F}$ in $\mathscr{A}^*$. $\mathfrak{F}^{00}$ (in $\mathscr{A}^{**}$) $=\mathfrak{F}$, since

$$\mathfrak{F}=(1-p)\mathscr{A}^{**}(1-p)$$

is $\sigma(\mathscr{A}^{**},\mathscr{A}^*)$-closed. Hence by the general theory of Banach spaces, the dual of $\mathfrak{F}^0$ is $\mathscr{A}^{**}/\mathfrak{F}=\mathscr{A}^{**}p+p\mathscr{A}^{**}$, and so $\mathfrak{F}^0$ is reflexive. Hence, the unit sphere of $\mathfrak{F}^0$ is $\sigma(\mathscr{A}^*,\mathscr{A}^{**})$-compact, and so it is $\sigma(\mathscr{A}^*,\mathscr{A})$-compact. Hence $\mathfrak{F}^0$ is $\sigma(\mathscr{A}^*,\mathscr{A})$-closed. Therefore, $(\mathfrak{F}^{00}\cap\mathscr{A})^0=\mathfrak{F}^0$, and so $\mathfrak{F}^{0*}=\mathscr{A}/\mathfrak{F}^{00}\cap\mathscr{A}=\mathscr{A}^{**}/\mathfrak{F}^{00}$, since $\mathfrak{F}^0$ is reflexive. Hence, $\|ap+pa\|=\inf_{x\in\mathfrak{F}^{00}\cap\mathscr{A}}\|a+x\|$, and $\mathscr{A}^{**}p+p\mathscr{A}^{**}=\{ap+pa\mid a\in\mathscr{A}\}$. Since $\sum_{j=1}^{n}(T_jE_j+E_jT_j)\in\left(\sum_{j=1}^{n}\oplus B(\mathscr{H}_j)\right)E+E\left(\sum_{j=1}^{n}\oplus B(\mathscr{H}_j)\right)$, there exists an element a of $\mathscr{A}$ such that

$$\begin{aligned}\pi(a)E+\pi(a)E &= \sum_{j=1}^{n} (\pi_j(a)E_j+E_j\pi(a))\\ &= \sum_{j=1}^{n} 2E_j\pi(a)E_j+(1_{\mathscr{H}_j}-E_j)\pi_j(a)E_j+E_j\pi_j(a)(1_{\mathscr{H}_j}-E_j))\\ &= \sum_{j=1}^{n} (2E_jT_jE_j+(1_{\mathscr{H}_j}-E_j)T_jE_j+E_jT_j(1_{\mathscr{H}_j}-E_j))\end{aligned}$$

and $\|a\|-\varepsilon<\max_{1\leq j\leq n}\{\|T_j\|\}$. Hence, $E_j\pi_j(a)E_j=E_jT_jE_j$, $(1_{\mathscr{H}_j}-E_j)\pi_j(a)E_j=(1_{\mathscr{H}_j}-E_j)T_jE_j$; $E_j\pi_j(a)(1_{\mathscr{H}_j}-E_j)=E_jT_j(1_{\mathscr{H}_j}-E_j)$. Hence, $E_j\pi_j(a)=E_jT_j$ and $\pi_j(a)E_j=T_jE_j$ $(1\leq j\leq n)$.

Next, suppose that the T_j are self-adjoint. Then by the above results, there exists an element a of $\mathscr{A}$ such that $E_j\pi(a)=E_jT_j$, $\pi_j(a)E_j=T_jE_j$ $(1\leq j\leq n)$. Therefore, $\pi_j(a^*)E_j=T_jE_j$ and $E_j\pi_j(a^*)=E_jT_j$. Hence, $\pi_j\left(\frac{a+a^*}{2}\right)E_j=T_jE_j$ and $E_j\pi_j\left(\frac{a+a^*}{2}\right)=E_jT_j$ $(1\leq j\leq n)$.

Finally, suppose the T_j are unitary. There exist finite-dimensional projections P_j and unitary U_j in $B(\mathscr{H}_j)$ such that $E_j\leq P_j$, $U_jP_j=P_jU_j$, and $U_jE_j=T_jE_j$. There exist self-adjoint elements H_j in $B(\mathscr{H}_j)$ such that $H_jP_j=P_jH_j$ and $\exp(iH_j)=U_j$.

Take a self-adjoint element h of $\mathscr{A}$ such that $\pi_j(h)P_j=H_jP_j$; then $P_j\pi_j(h)=P_jH_j$. Hence

$$\pi_j(\exp ih)P_j=\exp i\pi_j(h)P_j=\exp i(H_j)P_j.$$

Hence,

$$\pi_j(\exp ih)E_j=(\exp iH_j)E_j=U_jE_j$$

and $\pi_j(\exp ih)E_j=T_jE_j$. q.e.d.

1.21.17. Corollary. *Every irreducible *-representaion $\{\pi,\mathscr{H}\}$ of a C^*-algebra $\mathscr{A}$ is algebraically irreducible—i.e., $\pi(\mathscr{A})\xi=\mathscr{H}$ for every non-zero $\xi\in\mathscr{H}$.*

1.21.18. Corollary. *Let φ be a pure state of a C^*-algebra $\mathscr{A}$ and let $\mathscr{L}=\{x\,|\,\varphi(x^*x)=0,\ x\in\mathscr{A}\}$. Then the quotient Banach space $\mathscr{A}/\mathscr{L}$ is a Hilbert space with the scalar product defined by $\varphi(y^*x)$, and $\mathscr{L}$ is a maximal left ideal of $\mathscr{A}$. Furthermore the null space of φ is $\mathscr{L}+\mathscr{L}^*$ $(\mathscr{L}^*=\{x^*\,|\,x\in\mathscr{L}\})$.*

Proof. Let $\mathscr{L}^{00}$ be the bipolar of $\mathscr{L}$ in $\mathscr{A}^{**}$. Then it is a σ-closed maximal left ideal of $\mathscr{A}^{**}$. Hence there exists a minimal projection e in $\mathscr{A}^{**}$ with $\mathscr{L}^{00}=\mathscr{A}^{**}(1-e)$. By the previous results, $\mathscr{A}/\mathscr{L}$ is isometrically isomorphic to $\mathscr{A}^{**}/\mathscr{L}^{00}=\mathscr{A}^{**}e=$ the Hilbert space. Since $(\mathscr{L}+\mathscr{L}^*)^{00}=\mathscr{A}^{**}(1-e)+(1-e)\mathscr{A}^{**}$, $\mathscr{A}^{**}/(\mathscr{L}+\mathscr{L}^*)^{00}=e\mathscr{A}^{**}e=$ a one-dimensional space. Hence the null space of φ is $\mathscr{L}+\mathscr{L}^*$. q.e.d.

1.21.19. Proposition. *Let $\mathscr{A}$ be a C^*-algebra and let $\mathscr{L}$ be a regular maximal left ideal (i.e., there exists an element x_0 in $\mathscr{A}$ with $xx_0 - x \in \mathscr{L}$ for $x \in \mathscr{A}$) of $\mathscr{A}$. Then there exists a unique pure state φ on $\mathscr{A}$ with $\mathscr{L} = \{x \mid \varphi(x^* x) = 0, x \in \mathscr{A}\}$.*

Proof. Let $\mathscr{L}^{00} = \mathscr{A}^{**} p$. Since $\mathscr{L}$ is closed in $\mathscr{A}$, $p < 1$ and so $(1-p)\mathscr{A}^{**}(1-p) \neq (0)$. Take a normal state ψ on $\mathscr{A}^{**}$ with $\psi(1-p) = 1$, and set $\varphi(x) = \psi((1-p)x(1-p))$ $(x \in \mathscr{A})$. Then clearly

$$\mathscr{L} \subset \{x \mid \varphi(x^* x) = 0,\ x \in \mathscr{A}\}.$$

Since $\mathscr{L}$ is maximal, $\mathscr{L} = \{x \mid \varphi(x^* x) = 0, x \in \mathscr{A}\}$. Now let Γ be the set of all positive linear functionals f on $\mathscr{A}$ with $f(\mathscr{L}) = 0$ and $\|f\| \leq 1$. Then Γ is a compact convex subset of $\mathscr{A}^*$ in the $\sigma(\mathscr{A}^*, \mathscr{A})$-topology. One can easily see that any extreme point in Γ is again extreme in the set of all positive linear functionals g on $\mathscr{A}$ with $\|g\| \leq 1$. Hence there exists a pure state φ_0 on $\mathscr{A}$ with $\varphi_0(\mathscr{L}) = 0$. By 1.21.18, the null space of φ_0 is $\mathscr{L} + \mathscr{L}^*$. Hence $\varphi = \varphi_0$. q.e.d.

Concluding remarks on 1.21.

Theorem 1.21.12 is due to Segal [180]; Theorem 1.21.16 is due to Kadison [85].

References. [56], [85], [115], [138], [180], [202].

1.22. Tensor Products of C^*-Algebras and W^*-Algebras

We shall first of all state some facts concerning the tensor product of Banach spaces. Let E and F be two Banach spaces, $E \odot F$ the algebraic tensor product of E and F. A norm β on $E \odot F$ is said to be a cross norm if $\beta(x \otimes y) = \|x\| \, \|y\|$ for every $x \in E$, $y \in F$. $E \otimes_\beta F$ denotes the completion of $E \odot F$ with respect to β. The "least cross norm" λ is obtained by the natural algebraic imbedding of $E \odot F$ into $B(E^*, F)$, where $B(E^*, F)$ is the Banach space of all bounded linear operators of E^* into F. If under this imbedding, $T^u \in B(E^*, F)$ corresponds to a tensor product $u = \sum_{j=1}^n x_j \otimes y_j$, then for $x^* \in E^*$ $T^u x^* = \sum_{j=1}^n \langle x_j, x^* \rangle y_j$. We define $\lambda(u) = \|T^u\|$. Hence, $\lambda(u) = \sup\limits_{\substack{x^* \in E^* \\ y^* \in F^* \\ \|x^*\| = \|y^*\| = 1}} \left| \sum_{j=1}^n \langle x_j, x^* \rangle \langle y_j, y^* \rangle \right|$.

The greatest cross norm γ is defined by $\gamma(u) = \inf \sum_{j=1}^n \|x_j\| \, \|y_j\|$, where the inf is taken over all representations of u. γ is a cross norm and $\gamma \geq \lambda$. Analogously, consider the algebraic tensor product $E^* \odot F^*$ of the duals E^*, F^*. Elements of $E^* \odot F^*$ can be considered as linear functionals on $E \odot F$. Let α be a norm on $E \odot F$, and set $\alpha^*(f) = \sup_{\alpha(x) \leq 1} |\langle x, f \rangle|$ $(f \in E^* \odot F^*)$. If α^* is finite on $E^* \odot F^*$, it defines a norm on $E^* \odot F^*$. We call α^* the dual norm of α. If $\lambda \leq \alpha \leq \gamma$, α^* is also a cross norm on $E^* \odot F^*$. Moreover $\gamma^* = \lambda$. Therefore $\lambda^* \geq \alpha^* \geq \lambda$ if $\lambda \leq \alpha \leq \gamma$. λ is the least cross norm among all cross norms having again cross norms as dual norms.

Now let $C_0(\Omega)$ be the commutative C^*-algebra of all complex valued continuous functions vanishing at infinity on a locally compact Hausdorff space Ω. Let F be a Banach space, and let $C_0(\Omega, F)$ be the Banach space of all F-valued continuous functions $a(t)$ vanishing at infinity on Ω, with the norm $\|a\| = \sup_{t \in \Omega} \|a(t)\|$. For $f \in C_0(\Omega)$, $x \in F$, the mapping $\Phi: f \otimes x \to f(t)x$ of $C_0(\Omega) \odot F$ into $C_0(\Omega, F)$ can be uniquely extended to a linear mapping $C_0(\Omega) \odot F$ into $C_0(\Omega, F)$. Grothendieck [65] showed that $\|\Phi(w)\| = \lambda(w)$ $(w \in C_0(\Omega) \odot F)$, and so Φ can be uniquely extended to an isometric linear mapping of $C_0(\Omega) \otimes_\lambda F$ onto $C_0(\Omega, F)$. We shall identify $C_0(\Omega) \otimes_\lambda F$ with $C_0(\Omega, F)$ under the mapping Φ.

If F is a C^*-algebra, $C_0(\Omega, F)$ becomes a C^*-algebra with the product as pointwise multiplication. Next, let $L^\infty(\Omega, \mu)$ be the commutative W^*-algebra of all essentially bounded μ-locally measurable functions on a localizable measure space Ω. The predual of $L^\infty(\Omega, \mu)$ is $L^1(\Omega, \mu)$, all μ-integrable functions on Ω.

Let F be a Banach space, and let $L^1(\Omega, \mu, F)$ be the Banach space of all F-valued μ-Bochner integrable functions $g(t)$ on Ω with the norm $\|g\| = \int_\Omega \|g(t)\| \, d\mu(t)$. For $f \in L^1(\Omega, \mu)$, $x \in F$, the mapping $\Psi: f \otimes x \to f(t)x$ of $L^1(\Omega, \mu) \odot F$ into $L^1(\Omega, \mu, F)$ can be uniquely extended to a linear mapping of $L^1(\Omega, \mu) \odot F$ into $L^1(\Omega, \mu, F)$.

Grothendieck [65] showed that $\|\Psi(w)\| = \gamma(w)$ for $w \in L^1(\Omega, \mu) \odot F$, and that Ψ can be uniquely extended to an isometric linear mapping of $L^1(\Omega, \mu) \otimes_\gamma F$ onto $L^1(\Omega, \mu, F)$.

Consider $L^\infty(\Omega, \mu) \otimes_\lambda F^*$. Since $\gamma^* = \lambda$, we have

$$L^\infty(\Omega, \mu) \otimes_\lambda F^* = L^\infty(\Omega, \mu) \otimes_{\gamma^*} F^*,$$

where γ is the greatest cross norm on $L^1(\Omega, \mu) \odot F$. Since $L^1(\Omega, \mu)$ has the metric approximation property (cf. [65]), the canonical mapping of $L^1(\Omega, \mu) \otimes_\gamma F$ into the Banach space $J(L^\infty(\Omega, \mu), F^*)$, of all integrable bilinear functionals on $L^\infty(\Omega, \mu) \times F^*$, is an isometry. The dual of $L^\infty(\Omega, \mu) \otimes_\lambda F^*$ is $J(L^\infty(\Omega, \mu), F^*)$. Hence, $\lambda^* = \gamma$ on $L^1(\Omega, \mu) \odot F$, where λ is the least cross norm on $L^\infty(\Omega, \mu) \odot F^*$, and the dual norm λ^* is considered in $L^\infty(\Omega, \mu)^* \odot F^{**}$ $(\supset L^1(\Omega, \mu) \odot F)$. For more information on these items, we shall refer to [65], [174].

Let $\mathscr{A}$ and $\mathfrak{B}$ be two C^*-algebras. The algebraic tensor product $\mathscr{A} \odot \mathfrak{B}$ can be considered as a *-algebra. The algebraic tensor product $\mathscr{A}^* \odot \mathfrak{B}^*$ is a set of linear functionals on $\mathscr{A} \odot \mathfrak{B}$.

A positive linear functional ρ on $\mathscr{A} \odot \mathfrak{B}$ is defined as follows: $\rho(x^* x) \geq 0$ for $x \in \mathscr{A} \odot \mathfrak{B}$. A C^*-norm α on $\mathscr{A} \odot \mathfrak{B}$ is defined as a norm satisfying $\alpha(x^* x) = \alpha(x)^2$ and $\alpha(xy) \leq \alpha(x)\alpha(y)$ $(x, y \in \mathscr{A} \odot \mathfrak{B})$.

1.22.1. Proposition. *Let φ and ψ be positive elements of $\mathscr{A}^*$ and $\mathfrak{B}^*$ respectively. Then $\varphi \otimes \psi$ is a positive linear functional on $\mathscr{A} \odot \mathfrak{B}$.*

Proof. Let $x = \sum_{j=1}^{n} a_j \otimes b_j \in \mathscr{A} \odot \mathfrak{B}$. Then

$$\varphi \otimes \psi(x^* x) = \sum_{i,j=1}^{n} \varphi(a_i^* a_j)\psi(b_i^* b_j).$$

For any family of complex numbers $(\lambda_1, \lambda_2, \ldots, \lambda_n)$,

$$\sum_{i,j=1}^{n} \psi(b_i^* b_j)\bar{\lambda}_i \lambda_j = \psi\left(\left(\sum_{j=1}^{n} \lambda_j b_j\right)^* \left(\sum_{j=1}^{n} \lambda_j b_j\right)\right) \geq 0.$$

Hence, the matrix $(\psi(b_i^* b_j))_{i,j=1,2,\ldots,n}$ is positive, and so it is a positive linear combination of one-dimensional projections. Since any one-dimensional projection is of the form $(\bar{\lambda}_i \lambda_j)_{i,j=1,2,\ldots,n}$, we have $\sum_{i,j=1}^{n} \varphi(a_i^* a_j)\psi(b_i^* b_j) \geq 0$. q.e.d.

Now let $Q = \{L_{y^*} R_y \varphi \otimes \psi \mid y \in \mathscr{A} \odot \mathfrak{B}, \varphi \in \mathscr{S}_{\mathscr{A}}, \psi \in \mathscr{S}_{\mathfrak{B}}\}$, where $\mathscr{S}_{\mathscr{A}}$ (resp. $\mathscr{S}_{\mathfrak{B}}$) is the state space of $\mathscr{A}$ (resp. $\mathfrak{B}$). Q is a set of positive linear functionals on $\mathscr{A} \odot \mathfrak{B}$.

Since every element of $\mathscr{A}^*$ (resp. $\mathfrak{B}^*$) is a linear combination of positive elements, every element of $\mathscr{A}^* \odot \mathfrak{B}^*$ is a linear combination of elements of the form $\varphi \otimes \psi$ $(\varphi \in \mathscr{S}_{\mathscr{A}}, \psi \in \mathscr{S}_{\mathfrak{B}})$. Therefore, if α is a norm on $\mathscr{A} \odot \mathfrak{B}$ such that $\alpha^*(\varphi \otimes \psi)$ is finite $(\varphi \in \mathscr{S}_{\mathscr{A}}, \psi \in \mathscr{S}_{\mathfrak{B}})$, then α^* is finite.

1.22.2. Proposition. *There exists a least C^*-norm α_0 among all C^*-norm α on $\mathscr{A} \odot \mathfrak{B}$ such that α^* is finite. α_0 is a cross norm and $\lambda \leq \alpha_0 \leq \gamma$.*

Proof. Let α be a C^*-norm on $\mathscr{A} \odot \mathfrak{B}$ such that α^* is finite. The completion $\mathscr{A} \otimes_\alpha \mathfrak{B}$ is a C^*-algebra.

Every element of $\mathscr{A}^* \odot \mathfrak{B}^*$ can be extended uniquely to a bounded linear functional (denoted by the same notation) on $\mathscr{A} \otimes_\alpha \mathfrak{B}$. Let $\mathscr{D}$ be the C^*-algebra obtained from $\mathscr{A} \otimes_\alpha \mathfrak{B}$ by adjoining an identity.

Every bounded positive linear functional ρ on $\mathscr{A} \otimes_\alpha \mathfrak{B}$ can be extended to a positive linear functional (denoted by the same notation) on $\mathscr{D}$ such that $\rho(1) = \|\rho\|$.

Now, $L_{y^*} R_y \varphi \otimes \psi(\alpha(x^* x)1 - x^* x) \geq 0$ $(\varphi \in \mathscr{S}_{\mathscr{A}}, \psi \in \mathscr{S}_{\mathfrak{B}})$. Hence,

$$\alpha(x^* x) = \alpha(x)^2 \geq \frac{\varphi \otimes \psi(y^* x^* x y)}{\varphi \otimes \psi(y^* y)}$$

if $\varphi \otimes \psi(y^* y) \neq 0$. On the other hand, consider the *-representation $\{\pi_{\varphi\otimes\psi}, \mathscr{H}_{\varphi\otimes\psi}\}$ of $\mathscr{A} \otimes_\alpha \mathfrak{B}$.

$$\sup_{\substack{\varphi\otimes\psi(y^*y)\neq 0 \\ y\in\mathscr{A}\odot\mathfrak{B}}} \frac{\varphi \otimes \psi(y^* x^* x y)}{\varphi \otimes \psi(y^* y)} = \|\pi_{\varphi\otimes\psi}(x)\|^2 .$$

Let $\alpha_0(x) = \sup_{\substack{\varphi\in\mathscr{S}_{\mathscr{A}} \\ \varphi\in\mathscr{S}_{\mathfrak{B}}}} \|\pi_{\varphi\otimes\psi}(x)\|$ $(x\in\mathscr{A}\odot\mathfrak{B})$. α_0 is a C^*-norm on $\mathscr{A}\odot\mathfrak{B}$, and $\alpha(x) \geq \alpha_0(x)$ $(x\in\mathscr{A}\odot\mathfrak{B})$.

We shall show next that α_0 is a cross norm and $\alpha_0 \geq \lambda$. Consider the *-representations $\{\pi_\varphi, \mathscr{H}_\varphi\}$ and $\{\pi_\psi, \mathscr{H}_\psi\}$ of $\mathscr{A}$ and $\mathfrak{B}$, respectively. One can easily see that the mapping $a_\varphi \otimes b_\psi \to (a\otimes b)_{\varphi\otimes\psi}$ of $\mathscr{H}_\varphi \otimes \mathscr{H}_\psi$ into $\mathscr{H}_{\varphi\otimes\psi}$ extends to a unitary operator U of $\mathscr{H}_\varphi \otimes \mathscr{H}_\psi$ onto $\mathscr{H}_{\varphi\otimes\psi}$.

Moreover, $U(\pi_\varphi(a)\otimes 1_{\mathscr{H}_\psi} \cdot 1_{\mathscr{H}_\varphi} \otimes \pi_\psi(b))U^* = \pi_{\varphi\otimes\psi}(a\otimes b)$. Hence, $\alpha_0(a\otimes b) \leq \|a\|\,\|b\|$.

On the other hand,

$$\alpha_0(a\otimes b)^2 = \alpha_0(a^* a \otimes b^* b) \geq \sup_{(\varphi,\psi)\in\mathscr{S}_{\mathscr{A}}\times\mathscr{S}_{\mathfrak{B}}} |\langle a^* a \otimes b^* b, \varphi\otimes\psi\rangle| = \|a\|^2 \|b\|^2 .$$

Hence, $\alpha_0(a\otimes b) = \|a\|\,\|b\|$.

Next, for $x\in\mathscr{A}\odot\mathfrak{B}$,

$$\sup_{\substack{\|a\|\leq 1 \\ \|b\|\leq 1 \\ (\varphi,\psi)\in\mathscr{S}_{\mathscr{A}}\times\mathscr{S}_{\mathfrak{B}}}} |\varphi\otimes\psi(x\, a\otimes b)| \leq \alpha_0(x\, a\otimes b) \leq \alpha_0(x)\alpha_0(a\otimes b) \leq \alpha_0(x).$$

Consider the second duals $\mathscr{A}^{**}$ and $\mathfrak{B}^{**}$. Then for $f\in\mathscr{A}^{**}$, $g\in\mathfrak{B}^{**}$, we have the polar decompositions $f = R_U|f|$ and

$$g = R_V|g| \quad (U\in\mathscr{A}^{**},\ V\in\mathfrak{B}^{**}).$$

By the Kaplansky density theorem, there exist directed sets (a_α) and (b_β) such that $a_\alpha \to U$ in the $\tau(\mathscr{A}^{**},\mathscr{A}^*)$-topology, $b_\beta \to V$ in the $\tau(\mathfrak{B}^{**},\mathfrak{B}^*)$-topology $(a_\alpha\in\mathscr{A}, b_\beta\in\mathfrak{B}$ and $\|a_\alpha\|, \|b_\beta\| \leq 1)$. Hence,

$$\lambda(x) = \sup_{\substack{(f,g)\in\mathscr{A}^*\times\mathfrak{B}^* \\ \|f\|=\|g\|=1}} |\langle x, f\otimes g\rangle| = \sup_{\substack{(\varphi,\psi)\in\mathscr{S}_{\mathscr{A}}\times\mathscr{S}_{\mathfrak{B}} \\ \|a\|\leq 1, \|b\|\leq 1}} |\langle x, R_a\varphi \otimes R_b\psi\rangle|$$

$(x\in\mathscr{A}\odot\mathfrak{B})$. Therefore, $\alpha_0(x) \geq \lambda(x)$ for $x\in\mathscr{A}\odot\mathfrak{B}$. Moreover, α_0 is a cross norm; hence, $\alpha_0 \leq \gamma$. q.e.d.

1.22.3. Proposition. *Let $\mathscr{A}$ be a commutative C^*-algebra, and let $\mathfrak{B}$ be a C^*-algebra. Then, the C^*-norm α_0 on $\mathscr{A}\odot\mathfrak{B}$ coincides with λ, and $\mathscr{A}\otimes_\lambda\mathfrak{B} = C_0(\Omega,\mathfrak{B})$ (Ω is the spectrum space of $\mathscr{A}$).*

Proof. Let $\mathscr{A} = C_0(\Omega)$. Under the mapping Φ, $\mathscr{A}\otimes_\lambda\mathfrak{B} = C_0(\Omega,\mathfrak{B})$. Moreover,

$$\begin{aligned}\Phi(a_1\otimes b_1\, a_2\otimes b_2) &= \Phi(a_1 a_2 \otimes b_1 b_2) = (a_1 a_2)(t) b_1 b_2 = (a_1(t) b_1)(a_2(t) b_2) \\ &= \Phi(a_1\otimes b_1)\Phi(a_2\otimes b_2),\end{aligned}$$

and $\Phi(a^* \otimes b^*)=\overline{a(t)}b^*=\Phi(a\otimes b)^*$ $(a_1,a_2,a\in\mathscr{A};\ b_1,b_2,b\in\mathfrak{B})$. Hence, Φ is a *-isomorphism of $\mathscr{A}\odot\mathfrak{B}$ into $C_0(\Omega,\mathfrak{B})$. Hence, λ is a C^*-norm on $\mathscr{A}\odot\mathfrak{B}$, and so $\lambda=\alpha_0$, for λ^* is finite. Moreover the C^*-algebra $\mathscr{A}\otimes_\lambda\mathfrak{B}$ is identified with the C^*-algebra $C_0(\Omega,\mathfrak{B})$ under Φ. q.e.d.

1.22.4. Lemma. *Let $\mathscr{A}$, $\mathfrak{B}$ be two commutative C^*-algebras and suppose that $\mathfrak{B}$ has an identity. Then there exists only one C^*-norm α on $\mathscr{A}\odot\mathfrak{B}$, and $\alpha=\lambda$. Further, $\mathscr{A}\otimes_\lambda\mathfrak{B}=C_0(\Omega_1\times\Omega_2)$, where Ω_1 (resp. Ω_2) is the spectrum space of $\mathscr{A}$ (resp. $\mathfrak{B}$).*

Proof. Let α be a C^*-norm on $\mathscr{A}\odot\mathfrak{B}$. By the uniqueness of the C^*-norm on a C^*-algebra (cf. 1.2.6) we have $\alpha(a\otimes 1)=\|a\|$ $(a\in\mathscr{A})$. Let χ be a character of $\mathscr{A}\otimes_\alpha\mathfrak{B}$. For $h(\geq 0)\in\mathfrak{B}$, set $f_h(a)=\chi(a\otimes h)$. $0\leq f_h\leq f_{\|h\|1}$. Since f_1 is a character on $\mathscr{A}$, we have $\chi(a\otimes h)=f_1(a)g(h)$. g can be extended uniquely to a state $\tilde{g}$ on $\mathfrak{B}$.

Let $a_0\in\mathscr{A}$ with $f_1(a_0)=1$; then

$$\begin{aligned}f_1(a_0)\tilde{g}(h_1)\,f_1(a_0)\tilde{g}(h_2)&=\chi(a_0\otimes h_1)\chi(a_0\otimes h_2)=\chi(a_0^2\otimes h_1h_2)\\&=f_1(a_0)^2\,\tilde{g}(h_1h_2)\qquad(h_1,h_2\in\mathfrak{B}).\end{aligned}$$

Hence every character on $\mathscr{A}\otimes_\alpha\mathfrak{B}$ is a product of two characters on $\mathscr{A}$ and $\mathfrak{B}$. Let Γ be the spectrum space of $\mathscr{A}\otimes_\alpha\mathfrak{B}$. By the above results, Γ is a subset of $\Omega_1\times\Omega_2$. Moreover Γ is closed in $\Omega_1\times\Omega_2$. Suppose that $\Gamma\subsetneqq\Omega_1\times\Omega_2$. There exists an open set $G_1\times G_2$ in $\Omega_1\times\Omega_2$ such that $G_1\times G_2\cap\Gamma=(\emptyset)$, where G_1 (resp. G_2) is an open set in Ω_1 (resp. Ω_2). Hence, there exists a non-zero continuous function a (resp. b) with a compact support on Ω_1 (resp. Ω_2) such that $s(a)\subset G_1$ and $s(b)\subset G_2$. Then $\chi(a\otimes b)=0$ for all $\chi\in\Gamma$ and so $a\otimes b=0$ in $\mathscr{A}\otimes_\alpha\mathfrak{B}$, a contradiction. Hence $\Gamma=\Omega_1\times\Omega_2$. Therefore,

$$\alpha(y)=\sup_{(\chi_1,\chi_2)\in\Omega_1\times\Omega_2}|\chi_1\otimes\chi_2(y)|$$

for $y\in\mathscr{A}\odot\mathfrak{B}$. Hence, $\mathscr{A}\otimes_\alpha\mathfrak{B}=C_0(\Omega_1\times\Omega_2)$ under the canonical identification.

On the other hand,

$$\begin{aligned}\lambda(y)&=\sup_{\substack{\|a\|\leq 1\\ \|b\|\leq 1\\ (\varphi,\psi)\in\mathscr{S}_{\mathscr{A}}\times\mathscr{S}_{\mathfrak{B}}}}|\langle y,R_a\varphi\otimes R_b\psi\rangle|=\sup_{\substack{\|a\|\leq 1\\ \|b\|\leq 1\\ (\chi_1,\chi_2)\in\Omega_1\times\Omega_2}}|\langle y,R_a\chi_1\otimes R_b\chi_2\rangle|\\&=\sup_{\substack{\|a\|\leq 1\\ \|b\|\leq 1\\ (\chi_1,\chi_2)\in\Omega_1\times\Omega_2}}|\chi_1\otimes\chi_2(ya\otimes b)|=\sup_{(\chi_1,\chi_2)\in\Omega_1\times\Omega_2}|\chi_1\otimes\chi_2(y)|=\alpha(y)\\&\qquad\qquad(y\in\mathscr{A}\odot\mathfrak{B}).\end{aligned}$$

Hence, $\alpha=\lambda$. q.e.d.

1.22.5. Proposition. *Let $\mathscr{A}$ be a commutative C^*-algebra, and let $\mathfrak{B}$ be a C^*-algebra with an identity. Then, there exists only one C^*-norm α on $\mathscr{A}\odot\mathfrak{B}$, and $\alpha=\lambda$.*

Proof. Let α be a C^*-norm on $\mathscr{A} \odot \mathfrak{B}$ and let χ_1 be an arbitrary character of $\mathscr{A}$. Set $\overline{\chi}_1(a \otimes 1) = \chi_1(a)$ $(a \in \mathscr{A})$; then $\overline{\chi}_1$ is a character on $\mathscr{A} \otimes 1$. Let φ be a state on $\mathscr{A} \otimes_\alpha \mathfrak{B}$ such that $\varphi = \overline{\chi}_1$ on $\mathscr{A} \otimes 1$. By the same reasoning as in the proof of 1.22.4, $\varphi = \chi_1 \otimes \psi$ (ψ on a state on $\mathfrak{B}$). Let $\mathfrak{F} = \{\varphi \mid \varphi = \overline{\chi}_1$ on $\mathscr{A} \otimes 1, \varphi \in \mathscr{S}_{\mathscr{A} \otimes_\alpha \mathfrak{B}}\}$. Then $\mathfrak{F}$ is a $\sigma((\mathscr{A} \otimes_\alpha \mathfrak{B})^*, \mathscr{A} \otimes_\alpha \mathfrak{B})$-compact convex set. Let $\mathscr{E} = \{\psi \mid \varphi = \chi_1 \otimes \psi, \varphi \in \mathfrak{F}\}$. $\mathscr{E}$ is a $\sigma(\mathfrak{B}^*, \mathfrak{B})$-compact convex set. Let h be an arbitrary self-adjoint element of $\mathfrak{B}$. Let C be a maximal commutative C^*-subalgebra of $\mathfrak{B}$ containing h. By 1.22.4, $\mathscr{A} \otimes_\alpha C = \mathscr{A} \otimes_\lambda C$. Let χ_2 be a character of C such that $|\chi_2(h)| = \|h\|$. $\chi_1 \otimes \chi_2$ is a pure state on $\mathscr{A} \otimes_\alpha C$. Let ρ be an extended state on $\mathscr{A} \otimes_\alpha \mathfrak{B}$. Then $\rho = \chi_1 \otimes \psi$ and $\psi = \chi_2$ on C. Hence, $\sup_{\psi \in \mathscr{E}} |\psi(h)| = \|h\|$ $(h \in \mathfrak{B}^s)$. Therefore, $\mathscr{E} = \mathscr{S}_{\mathfrak{B}}$ by the separation theorem in locally convex spaces. Hence $\Omega_1 \times \mathscr{S}_{\mathfrak{B}} \subset \mathscr{S}_{\mathscr{A} \otimes_\alpha \mathfrak{B}}$, and so

$$\mathscr{S}_{\mathscr{A}} \times \mathscr{S}_{\mathfrak{B}} \subset \mathscr{S}_{\mathscr{A} \otimes_\alpha \mathfrak{B}}.$$

Hence, α^* is finite, and $\alpha \geq \alpha_0$ by 1.22.2.

Conversely, let φ be a pure state on $\mathscr{A} \otimes_\alpha \mathfrak{B}$. $\{\pi_\varphi, \mathscr{H}_\varphi\}$ is irreducible, and $\pi_\varphi(\mathscr{A} \otimes 1)$ belongs to the center of $\pi_\varphi(\mathscr{A} \otimes_\alpha \mathfrak{B})''$. Hence, the restriction of φ to $\mathscr{A} \otimes 1$ is a character, and so $\varphi \in \mathfrak{F}$. Therefore

$$\alpha(y)^2 = \alpha(y^* y) = \sup_{(\chi_1, \psi) \in \Omega_1 \times \mathscr{S}_{\mathfrak{B}}} |\langle y^* y, \chi_1 \otimes \psi \rangle| \leq \alpha_0(y^* y) = \alpha_0(y)^2.$$

Hence $\alpha = \alpha_0 = \lambda$. q.e.d.

1.22.6. Theorem. *Let $\mathscr{A}$, $\mathfrak{B}$ be two C^*-algebras with identities. Then the C^*-norm α_0 on $\mathscr{A} \odot \mathfrak{B}$ is the least C^*-norm among all C^*-norms α on $\mathscr{A} \odot \mathfrak{B}$.*

Proof. Let α be a C^*-norm on $\mathscr{A} \odot \mathfrak{B}$. By the uniqueness of the C^*-norm on a C^*-algebra, $\alpha(a \otimes 1) = \|a\|$ and

$$\alpha(1 \otimes b) = \|b\| \qquad (a \in \mathscr{A}, b \in \mathfrak{B}).$$

Let ξ be a pure state on $\mathscr{A}$ and set $\overline{\xi}(a \otimes 1) = \xi(a)$. Let φ be a state on $\mathscr{A} \otimes_\alpha \mathfrak{B}$ such that $\varphi = \overline{\xi}$ on $\mathscr{A} \otimes 1$. Then

$$\varphi(a \otimes b) = \xi(a) \psi(b) \qquad (a \in \mathscr{A}, b \in \mathfrak{B}),$$

where ψ is a state on $\mathfrak{B}$. This is easily seen by a line of reasoning similar to that used in the proof of 1.22.4. Set $\mathfrak{F} = \{\varphi \mid \varphi = \overline{\xi}$ on $\mathscr{A} \otimes 1, \varphi \in \mathscr{S}_{\mathscr{A} \otimes_\alpha \mathfrak{B}}\}$ and $\mathscr{E} = \{\psi \mid \varphi = \xi \otimes \psi, \varphi \in \mathfrak{F}\}$. $\mathscr{E}$ is a $\sigma(\mathfrak{B}^*, \mathfrak{B})$-compact convex set. For an arbitrary self-adjoint element $b \in \mathfrak{B}$, let C be a maximal commutative C^*-subalgebra of $\mathfrak{B}$ containing b. By 1.22.5, $\mathscr{A} \otimes_\alpha C = \mathscr{A} \otimes_\lambda C$. Let χ_2 be a character of C such that $|\chi_2(b)| = \|b\|$. $\xi \otimes \chi_2$ is a pure state of $\mathscr{A} \otimes_\alpha C$. Hence there exists a state φ which extends $\xi \otimes \chi_2$ on $\mathscr{A} \otimes_\alpha \mathfrak{B}$. Let $\varphi = \xi \otimes \psi$. Then $\psi = \chi_2$ on C. Hence, $\sup_{\psi \in \mathscr{E}} |\psi(b)| = \|b\|$ $(b \in \mathfrak{B}^s)$, and so $\mathscr{E} = \mathscr{S}_{\mathfrak{B}}$. Therefore $\xi \otimes \mathscr{S}_{\mathfrak{B}} \subset \mathscr{S}_{\mathscr{A} \otimes_\alpha \mathfrak{B}}$ for an arbitrary pure state ξ, and so α^* is finite. Hence $\alpha \geq \alpha_0$. q.e.d.

Remark. Takesaki [204] showed that there exists a pair of C^*-algebras $\mathscr{A}$, $\mathfrak{B}$ as follows: $\mathscr{A}\odot\mathfrak{B}$ has a C^*-cross norm α_1 which is different from α_0. Also, he proved that there exists a class of non-commutative, infinite-dimensional C^*-algebras $\mathscr{A}$ such that $\mathscr{A}\odot\mathfrak{B}$ has a unique C^*-norm (i.e., α_0) for every C^*-algebra $\mathfrak{B}$. It would be an interesting problem to find a characterization of the C^*-algebras $\mathscr{A}$ such that $\mathscr{A}\odot\mathfrak{B}$ has a unique C^*-norm for every C^*-algebra $\mathfrak{B}$.

1.22.7. Proposition. *Let $\mathscr{A}$ and $\mathfrak{B}$ be C^*-algebras and let α be a C^*-norm on $\mathscr{A}\odot\mathfrak{B}$ which satisfies $\alpha(a\otimes b)\leq k\|a\|\,\|b\|$ ($a\in\mathscr{A}$, $b\in\mathfrak{B}$, k is a fixed positive number). Then $\alpha_0\leq\alpha\leq\gamma$.*

Proof. Consider a faithful *-representation $\{\pi,\mathscr{H}\}$ of $\mathscr{A}\otimes_\alpha\mathfrak{B}$ without a trivial part. The linear mapping $a\to\pi(a\otimes b)$ of $\mathscr{A}$ into $B(\mathscr{H})$ is bounded. Hence, by 1.21.13, it can be extended uniquely to a continuous linear mapping (denoted by π_b^1) of $\mathscr{A}^{**}$, with the $\sigma(\mathscr{A}^{**},\mathscr{A}^*)$-topology into $B(\mathscr{H})$, with the $\sigma(B(\mathscr{H}),B(\mathscr{H})_*)$-topology. Analogously, the linear mapping, $b\to\pi(a\otimes b)$ of $\mathfrak{B}$ into $B(\mathscr{H})$, can be extended to a continuous mapping π_a^2 of $\mathfrak{B}^{**}$ into $B(\mathscr{H})$. Let 1 (resp. 1′) be the identity of $\mathscr{A}^{**}$ (resp. $\mathfrak{B}^{**}$). Let $\{a_\alpha\}$ (resp. $\{b_\beta\}$) be a directed set in $\mathscr{A}$, (resp. $\mathfrak{B}$) such that $0\leq\alpha_\alpha\leq 1$ $(0\leq b_\beta\leq 1')$ which converges to 1 in the $\tau(\mathscr{A}^{**},\mathscr{A}^*)$-topology (converges 1′ in the $\tau(\mathfrak{B}^{**},\mathfrak{B}^*)$-topology). Then a_α converges to 1 in the $s(\mathscr{A}^{**},\mathscr{A}^*)$-topology and so $a_\alpha^2\to 1$ in the $s(\mathscr{A}^{**},\mathscr{A}^*)$-topology.

$$\begin{aligned}\pi^1_{b_1 b_2}(1)&=\lim_\alpha \pi^1_{b_1 b_2}(a_\alpha^2)=\lim_\alpha \pi(a_\alpha^2\otimes b_1 b_2)\\ &=\lim_\alpha \pi(a_\alpha\otimes b_1)\pi(a_\alpha\otimes b_2)=\lim_\alpha \pi^1_{b_1}(a_\alpha)\pi^1_{b2}(a_\alpha).\end{aligned}$$

Since π_b^1 is a continuous mapping of $\mathscr{A}^{**}$, with the $\sigma(\mathscr{A}^{**},\mathscr{A}^*)$-topology into $B(\mathscr{H})$ with the $\sigma(B(\mathscr{H}),B(\mathscr{H})_*)$-topology, π_b^1 is continuous with respect to the $\tau(\mathscr{A}^{**},\mathscr{A}^*)$ and $\tau(B(\mathscr{H}),B(\mathscr{H})_*)$-topologies. Hence, $\pi_b^1(a_\alpha)\to\pi_b^1(1)$ in the $s(B(\mathscr{H}),B(\mathscr{H})_*)$-topology and so

$$\pi^1_{b_1}(a_\alpha)\pi^1_{b_2}(a_\alpha)\to\pi^1_{b_1}(1)\pi^1_{b_2}(1).$$

Hence, we have $\pi^1_{b_1 b_2}(1)=\pi^1_{b_1}(1)\pi^1_{b_2}(1)$. Analogously, we have

$$\pi^1_{b^*}(1)=\pi^1_b(1)^*.$$

Therefore, $b\to\pi_b^1(1)$ is a *-representation of $\mathfrak{B}$ and $a\to\pi_a^2(1')$ is a *-representation of $\mathscr{A}$.

Moreover,

$$\begin{aligned}\pi_a^2(1')\pi_b^1(1)&=\lim_{\alpha,\beta}\pi(a\otimes b_\beta)\pi(a_\alpha\otimes b)=\lim_\alpha\ \lim_\beta \pi(a a_\alpha\otimes b_\beta b)\\ &=\lim_\alpha \pi(a a_\alpha\otimes b)=\pi(a\otimes b),\end{aligned}$$

and $\pi_b^1(1)\pi_a^2(1') = \lim_{\alpha,\beta} \pi(a_\alpha a \otimes b b_\beta) = \pi(a \otimes b)$. Hence,

$$\pi_a^2(1')\pi_b^1(1) = \pi_b^1(1)\pi_a^2(1') \qquad (a \in \mathscr{A},\ b \in \mathfrak{B}).$$

Let $\mathscr{A}_1 = \mathscr{A} + \mathbb{C}1$ and $\mathfrak{B}_1 = \mathfrak{B} + \mathbb{C}1$, ($\mathbb{C}$ the field of complex numbers). Then the *-representation $\pi_a^2(1')$ of $\mathscr{A}$ (resp. $\pi_b^1(1)$ of $\mathfrak{B}$) can be extended to a *-representation of $\mathscr{A}_1$ (resp. $\mathfrak{B}_1$) such that $\pi_1^2(1') = \pi_1^1(1) = 1_{\mathscr{H}}$. The mapping $x \otimes y \to \pi_x^2(1)\pi_y^1(1')$ can be uniquely extended to a *-homomorphism Φ of $\mathscr{A}_1 \odot \mathfrak{B}_1$ into $B(\mathscr{H})$.

Suppose $\Phi\left(\sum_{i=1}^n x_i \otimes y_i\right) = 0$ $(x_i \in \mathscr{A}_1,\ y_i \in \mathfrak{B}_1)$. Then

$$\Phi\left(\left(\sum_{i=1}^n (x_i \otimes y_i)\right) y\right) = 0 \ (y \in \mathscr{A} \odot \mathfrak{B}), \qquad \left(\sum_{i=1}^n x_i \otimes y_i\right) y \in \mathscr{A} \odot \mathfrak{B},$$

and Φ is one-to-one on $\mathscr{A} \odot \mathfrak{B}$. Hence $\left(\sum_{i=1}^n x_i \otimes y_i\right) y = 0$ in $\mathscr{A} \odot \mathfrak{B}$.

Now let $\varphi \in \mathscr{S}_{\mathscr{A}}$, $\psi \in \mathscr{S}_{\mathfrak{B}}$, and let $\tilde{\varphi}$ (resp. $\tilde{\psi}$) be the unique state extension of φ (resp. ψ) on $\mathscr{A}_1$ (resp. $\mathfrak{B}_1$). Since

$$\{\pi_{\tilde{\varphi} \otimes \tilde{\psi}}, \mathscr{H}_{\tilde{\varphi} \otimes \tilde{\psi}}\} = \{\pi_{\tilde{\varphi}} \otimes \pi_{\tilde{\psi}}, \mathscr{H}_{\tilde{\varphi}} \otimes \mathscr{H}_{\tilde{\psi}}\},$$

there exist directed sets $\{a_\alpha\}$ (resp. $\{b_\beta\}$) in $\mathscr{A}$ (resp. $\mathfrak{B}$) such that

$$\pi_{\tilde{\varphi}}(a_\alpha) \to 1_{\mathscr{H}_{\tilde{\varphi}}}$$

strongly (resp. $\pi_{\tilde{\varphi}}(b_\beta) \to 1_{\mathscr{H}_{\tilde{\psi}}}$ strongly). Hence,

$$\begin{aligned} &\tilde{\varphi} \otimes \tilde{\psi}\left(\left(\sum_{i=1}^n x_i \otimes y_i\right)(a_\alpha \otimes b_\alpha)\right) \\ &\qquad = \left(\pi_{\tilde{\varphi} \otimes \tilde{\psi}}\left(\sum_{i=1}^n x_i \otimes y_i\right) \pi_{\tilde{\varphi} \otimes \tilde{\psi}}(a_\alpha \otimes b_\alpha)(1 \otimes 1)_{\tilde{\varphi} \otimes \tilde{\psi}}, (1 \otimes 1)_{\tilde{\varphi} \otimes \tilde{\psi}}\right) \\ &\qquad \to \tilde{\varphi} \otimes \tilde{\psi}\left(\sum_{i=1}^n x_i \otimes y_i\right) = 0. \end{aligned}$$

On the other hand, $\{\tilde{\varphi} \mid \varphi \in \mathscr{S}_{\mathscr{A}}\}$ (resp. $\{\tilde{\psi} \mid \psi \in \mathscr{S}_{\mathfrak{B}}\}$) is $\sigma(\mathscr{A}_1^*, \mathscr{A}_1)$-dense in $\mathscr{S}_{\mathscr{A}_1}$ (resp. $\sigma(\mathfrak{B}_1^*, \mathfrak{B}_1)$-dense in $\mathscr{S}_{\mathfrak{B}_1}$). Hence, $\xi \otimes \eta\left(\sum_{i=1}^n x_i \otimes y_i\right) = 0$ for $\xi \in \mathscr{S}_{\mathscr{A}_1}$, $\eta \in \mathscr{S}_{\mathfrak{B}_1}$, and so $\sum_{i=1}^n x_i \otimes y_i = 0$. Therefore, Φ is a *-isomorphism of $\mathscr{A}_1 \odot \mathfrak{B}_1$ into $B(\mathscr{H})$. Set $\tilde{\alpha}(w) = \|\Phi(w)\|$ $(w \in \mathscr{A}_1 \odot \mathfrak{B}_1)$. Then $\tilde{\alpha}(y) = \alpha(y)$ for $y \in \mathscr{A} \odot \mathfrak{B}$. Since $\tilde{\alpha}$ is a C^*-norm on $\mathscr{A}_1 \odot \mathfrak{B}_1$, $\tilde{\alpha} \geq \alpha_0$ by 1.22.6, and so $\alpha \geq \alpha_0$. Moreover

$$\begin{aligned} \alpha(a \otimes b) = \|\Phi(a \otimes b)\| &= \|\pi_a^2(1')\pi_b^1(1)\| \leq \|\pi_a^2(1')\| \, \|\pi_b^1(1)\| \\ &\leq \|a\| \, \|b\| \qquad (a \in \mathscr{A},\ b \in \mathfrak{B}). \end{aligned}$$

Hence $\alpha_0 \leq \alpha \leq \gamma$. q.e.d.

1.22.8. Definition. *Let $\mathscr{A}$, $\mathfrak{B}$ be two C^*-algebras. We shall call the C^*-algebra $\mathscr{A} \otimes_{\alpha_0} \mathfrak{B}$ the C^*-tensor product of $\mathscr{A}$ and $\mathfrak{B}$, or more simply, the tensor product of the C^*-algebras $\mathscr{A}$ and $\mathfrak{B}$ (denoted by $\mathscr{A} \otimes \mathfrak{B}$).*

1.22.9. Proposition. *Let $\{\pi_1, \mathscr{H}_1\}$ (resp. $\{\pi_2, \mathscr{H}_2\}$) be a *-representation of $\mathscr{A}$ (resp. $\mathfrak{B}$), and set*

$$\pi_1 \otimes \pi_2(a \otimes b) = (\pi_1(a) \otimes 1_{\mathscr{H}_2})(1_{\mathscr{H}_1} \otimes \pi_2(b)) \qquad (a \in \mathscr{A},\, b \in \mathfrak{B}).$$

*Then $\pi_1 \otimes \pi_2$ can be uniquely extended to a *-representation (denoted by $\{\pi_1 \otimes \pi_2, \mathscr{H}_1 \otimes \mathscr{H}_2\}$) of $\mathscr{A} \otimes \mathfrak{B}$ on $\mathscr{H}_1 \otimes \mathscr{H}_2$. Moreover, if $\{\pi_1, \mathscr{H}_1\}$ and $\{\pi_2, \mathscr{H}_2\}$ are faithful, then $\{\pi_1 \otimes \pi_2, \mathscr{H}_1 \otimes \mathscr{H}_2\}$ is faithful.*

Proof. Consider the mapping $\Phi\colon \sum_{i=1}^{n} a_i \otimes b_i \to \sum_{i=1}^{n} (a_i \otimes 1_{\mathscr{H}_2})(1_{\mathscr{H}_1} \otimes b_i)$ of $\mathscr{A} \odot \mathfrak{B}$ into $B(\mathscr{H}_1 \otimes \mathscr{H}_2)$. One can easily see that Φ is a *-homomorphism. For $\xi_1^i \in \mathscr{H}_1$, $\xi_2^i \in \mathscr{H}_2$ $(i = 1, 2, \ldots, n)$, set

$$\varphi(x) = \left(\Phi(x) \sum_{i=1}^{n} \xi_1^i \otimes \xi_2^i,\ \sum_{i=1}^{n} \xi_1^i \otimes \xi_2^i\right) \qquad (x \in \mathscr{A} \odot \mathfrak{B}).$$

Then $\varphi \in \mathscr{A}^* \odot \mathfrak{B}^*$ and is positive. Hence, $\alpha_0(x^* x) \geq \dfrac{\varphi(x^* x)}{\left\| \sum_{i=1}^{n} \xi_1^i \otimes \xi_2^i \right\|^2}$, and so $\alpha_0(x) \geq \|\Phi(x)\|$. Hence Φ can be uniquely extended to a *-homomorphism of $\mathscr{A} \otimes_\alpha \mathfrak{B}$ into $B(\mathscr{H}_1 \otimes \mathscr{H}_2)$.

Next, suppose that $\{\pi_1, \mathscr{H}_1\}$ and $\{\pi_2, \mathscr{H}_2\}$ are faithful. Let $V_i =$ the convex span of $\{\varphi_i \mid \varphi_i(x) = (x\xi_i, \xi_i),\ \xi_i \in \mathscr{H}_i,\ \|\xi_i\| = 1\ (i = 1, 2)\}$. Then $\overline{V}_1 = \mathscr{S}_{\mathscr{A}}$ and $\overline{V}_2 = \mathscr{S}_{\mathfrak{B}}$, when $\{\pi_1, \mathscr{H}_1\}$ and $\{\pi_2, \mathscr{H}_2\}$ are faithful. ($\overline{V}_1$ (resp. $\overline{V}_2$) is the $\sigma(\mathscr{A}^*, \mathscr{A})$-closure (resp. $\sigma(\mathfrak{B}^*, \mathfrak{B})$-closure) of V_1 (resp. V_2)).

Put $\alpha(x) = \|\Phi(x)\|$ $(x \in \mathscr{A} \odot \mathfrak{B})$. Then $\mathscr{S}_{\mathscr{A}} \times \mathscr{S}_{\mathfrak{B}} \subset (\mathscr{A} \otimes_\alpha \mathfrak{B})^*$. Hence $\alpha_0 \leq \alpha$. Hence, $\alpha_0 = \alpha$. q.e.d.

We now define the tensor product of W^*-algebras. Let $\mathscr{M}$ and $\mathscr{N}$ be two W^*-algebras, and let $\mathscr{M}_*$ and $\mathscr{N}_*$ be their respective preduals. Consider the C^*-tensor product $\mathscr{M} \otimes_{\alpha_0} \mathscr{N}$. Let α_0^* be the dual norm of α_0 in $\mathscr{M}^* \odot \mathscr{N}^*$. Since $\mathscr{M}_* \subset \mathscr{M}^*$ and $\mathscr{N}_* \subset \mathscr{N}^*$, we can consider the norm α_0^* on $\mathscr{M}_* \odot \mathscr{N}_*$. $\mathscr{M}_* \otimes_{\alpha_0^*} \mathscr{N}_*$ is a closed subspace of $(\mathscr{M} \otimes_{\alpha_0} \mathscr{N})^*$. Since $\mathscr{M}_* \odot \mathscr{N}_*$ is invariant under R_x, L_x $(x \in \mathscr{M} \odot \mathscr{N})$, $\mathscr{M}_* \otimes_{\alpha_0^*} \mathscr{N}_*$ is also invariant under R_x, L_x $(x \in \mathscr{M} \otimes_{\alpha_0} \mathscr{N})$. Hence, the polar $\mathscr{I}$, of $\mathscr{M}_* \otimes_{\alpha_0^*} \mathscr{N}_*$ in the second dual of $\mathscr{M} \otimes_{\alpha_0} \mathscr{N}$ is a two-sided ideal, and $(\mathscr{M}_* \otimes_{\alpha_0^*} \mathscr{N}_*)^* = (\mathscr{M} \otimes_{\alpha_0} \mathscr{N})^{**}/\mathscr{I}$. The canonical mapping

$$\mathscr{M} \otimes_{\alpha_0} \mathscr{N} \to (\mathscr{M} \otimes_{\alpha_0} \mathscr{N} + \mathscr{I})/\mathscr{I}$$

is a *-isomorphism, and so is an isometry. Hence, the C^*-algebra $\mathscr{M} \otimes_{\alpha_0} \mathscr{N}$ can be considered as a C^*-subalgebra of the W^*-algebra $(\mathscr{M} \otimes_{\alpha_0} \mathscr{N})^{**}/\mathscr{I}$.

Hence $(\mathscr{M}_* \otimes_{\alpha_0^*} \mathscr{N}_*)^* = (\mathscr{M} \otimes_{\alpha_0} \mathscr{N})^{**}/\mathscr{I}$ is a W^*-algebra and the C^*-algebra $\mathscr{M} \otimes_\alpha \mathscr{N}$ is σ-dense in it.

1.22.10. Definition. *The W^*-algebra $(\mathscr{M}_* \otimes_{\alpha_0^*} \mathscr{N}_*)^*$, more simply denoted by $\mathscr{M} \bar{\otimes} \mathscr{N}$, is called the W^*-tensor product or simply the tensor product of the W^*-algebras $\mathscr{M}$ and $\mathscr{N}$.*

1.22.11. Proposition. *Let $\{\pi_1^w, \mathscr{H}_1\}$ (resp. $\{\pi_2^w, \mathscr{H}_2\}$) be a W^*-representation of the W^*-algebra $\mathscr{M}$ (resp. $\mathscr{N}$). Put*

$$\pi_1 \otimes \pi_2(x \otimes y) = (\pi_1(x) \otimes 1_{\mathscr{H}_2})(1_{\mathscr{H}_1} \otimes \pi_2(y)) \quad (x \in \mathscr{M},\, y \in \mathscr{N}).$$

Then $\pi_1 \otimes \pi_2$ can be uniquely extended to a W^*-representation on $\mathscr{H}_1 \otimes \mathscr{H}_2$ (denoted by $\{(\pi_1 \otimes \pi_2)^w, \mathscr{H}_1 \otimes \mathscr{H}_2\}$) of $\mathscr{M} \bar{\otimes} \mathscr{N}$. Moreover, if $\{\pi_1^w, \mathscr{H}_1\}$ and $\{\pi_2^w, \mathscr{H}_2\}$ are faithful, $\{(\pi_1 \otimes \pi_2)^w, \mathscr{H}_1 \otimes \mathscr{H}_2\}$ is faithful.

Proof. First of all, consider the *-representation $\{\pi_1 \otimes \pi_2, \mathscr{H}_1 \otimes \mathscr{H}_2\}$ of the C^*-algebra $\mathscr{M} \otimes \mathscr{N}$. Then $(\pi_1 \otimes \pi_2(x)\xi_1 \otimes \xi_2, \eta_1 \otimes \eta_2) = f_1 \otimes f_2(x)$ $(x \in \mathscr{M} \otimes \mathscr{N}, \xi_1, \eta_1 \in \mathscr{H}_1, \xi_2, \eta_2 \in \mathscr{H}_2, f_1(a) = (\pi_1(a)\xi_1, \eta_1)$ and $f_2(b) = (\pi_2(b)\xi_2, \eta_2)$ for $a \in \mathscr{M}, b \in \mathscr{N})$.

Consider the mappings $\pi_1 \otimes \pi_2 : \mathscr{M} \otimes \mathscr{N} \to B(\mathscr{H}_1 \otimes \mathscr{H}_2)$ and

$$(\pi_1 \otimes \pi_2)^* : (\mathscr{M} \otimes \mathscr{N})^* \leftarrow B(\mathscr{H}_1 \otimes \mathscr{H}_2)^*.$$

By the previous considerations, one can easily see that

$$(\pi_1 \otimes \pi_2)^* (B(\mathscr{H}_1 \otimes \mathscr{H}_2)_*) \subset \mathscr{M}_* \otimes_{\alpha_0^*} \mathscr{N}_*.$$

Let $(\pi_1 \otimes \pi_2)_0^*$ be the restriction of $(\pi_1 \otimes \pi_2)^*$ to $B(\mathscr{H}_1 \otimes \mathscr{H}_2)_*$, and let $((\pi_1 \otimes \pi_2)_0^*)^*$ be its dual. $((\pi_1 \otimes \pi_2)_0^*)^*$ is a continuous mapping of $(\mathscr{M}_* \otimes_{\alpha_0^*} \mathscr{N}_*)^* = \mathscr{M} \bar{\otimes} \mathscr{N}$ with the $\sigma(\mathscr{M} \bar{\otimes} \mathscr{N}, (\mathscr{M} \bar{\otimes} \mathscr{N})_*)$-topology into $B(\mathscr{H}_1 \otimes \mathscr{H}_2)$, with the $\sigma(B(\mathscr{H}_1 \otimes \mathscr{H}_2), B(\mathscr{H}_1 \otimes \mathscr{H}_2)_*)$-topology. Clearly $((\pi_1 \otimes \pi_2)_0^*)^* = \pi_1 \otimes \pi_2$ on $\mathscr{M} \otimes \mathscr{N}$. Since $\pi_1 \otimes \pi_2$ is a *-homomorphism, one can easily show that $((\pi_1 \otimes \pi_2)_0^*)^*$ is a W^*-representation of $\mathscr{M} \bar{\otimes} \mathscr{N}$.

Now suppose π_1, π_2 are faithful, and let $\mathscr{I}$ be the kernel of $\{(\pi_1 \otimes \pi_2)^w, \mathscr{H}_1 \otimes \mathscr{H}_2\}$. If $\mathscr{I} \neq (0)$, there exists a positive element h (>0) in $\mathscr{I}$. On the other hand, by 1.15.5, for $f \in \mathscr{M}_*$, $g \in \mathscr{N}_*$ there exists sequences $(\xi_1^i), (\eta_1^i) \subset \mathscr{H}_1$, $(\xi_2^i), (\eta_2^i) \subset \mathscr{H}_2$ such that $\sum_{i=1}^\infty \|\xi_j^i\|^2 < +\infty$,

$$\sum_{i=1}^\infty \|\eta_j^i\|^2 < +\infty \quad (j = 1, 2), \qquad f(a) = \sum_{i=1}^\infty (\pi_1(a)\xi_1^i, \eta_1^i)$$

and $g(b) = \sum_{i=1}^\infty (\pi_2(b)\xi_2^i, \eta_2^i)$ $(a \in \mathscr{M}, b \in \mathscr{N})$. Then

$$\begin{aligned} |f \otimes g(h)| &= \left| \sum_{i,j=1}^\infty (\pi_1 \otimes \pi_2(h)\xi_1^i \otimes \xi_2^i, \eta_1^i \otimes \eta_2^i) \right| \\ &\leq \sum_{i,j=1}^\infty |(\pi_1 \otimes \pi_2(h)\xi_1^i \otimes \xi_2^i, \eta_1^i \otimes \eta_2^i)|. \end{aligned}$$

Hence, $h = 0$, a contradiction, and so $(\pi_1 \otimes \pi_2)^w$ is faithful. q.e.d.

1.22.12. Proposition. *Let $L^\infty(\Omega,\mu)$ be the commutative W^*-algebra of all essentially bounded locally μ-measurable functions on a localizable measure space Ω. Let $\mathcal{M}$ be a W^*-algebra. Then*

$$L^1(\Omega,\mu)\otimes_{\alpha_0}\mathcal{M}_*=L^1(\Omega,\mu)\otimes_\gamma\mathcal{M}_*=L^1(\Omega,\mu,\mathcal{M}_*),$$

where $L^1(\Omega,\mu,\mathcal{M}_*)$ is the Banach space of all $\mathcal{M}_*$-valued Bochner μ-integrable functions on Ω.

1.22.13. Theorem. *Let $\mathcal{M}$ be a W^*-algebra with the separable predual $\mathcal{M}_*$, and let $L^\infty(\Omega,\mu,\mathcal{M})$ be the Banach space of all $\mathcal{M}$-valued essentially bounded weakly * μ-locally measurable functions on a localizable measure space Ω. Then, $L^\infty(\Omega,\mu,\mathcal{M})$ is a W^*-algebra under the pointwise multiplication, and its predual is $L^1(\Omega,\mu,\mathcal{M}_*)$.*

Moreover the mapping $f\otimes a\to f(t)a$ $(f\in L^\infty(\Omega,\mu), a\in\mathcal{M})$ can be uniquely extended to a *-isomorphism Φ of the W^*-algebra

$$L^\infty(\Omega,\mu)\bar{\otimes}\mathcal{M} \text{ onto } L^\infty(\Omega,\mu,\mathcal{M}).$$

Proof. Since $\mathcal{M}_*$ is separable, by the Dunford-Pettis theorem [43], for any $x\in(L^1(\Omega,\mu)\otimes_\gamma\mathcal{M}_*)^*$, there exists a unique $\mathcal{M}$-valued essentially bounded weakly* locally μ-measurable function $g^x(t)$ on Ω such that

$$x(\eta\otimes\xi)=\int_\Omega\langle g^x(t),\xi\rangle\eta(t)d\mu(t)$$

and ess. sup $\|g^x(t)\|=\|x\|$ $(\eta\in L^1(\Omega,\mu), \xi\in\mathcal{M}_*)$.

Under the mapping $x\to g^x$, $(L^1(\Omega,\mu)\otimes_\gamma\mathcal{M}_*)^*$ is isometrically isomorphic to $L^\infty(\Omega,\mu,\mathcal{M})$. Therefore, by the Dunford-Pettis theorem and Grothendieck's theorem, the dual of $L^1(\Omega,\mu,\mathcal{M}_*)$ is $L^\infty(\Omega,\mu,\mathcal{M})$, for $L^1(\Omega,\mu)\otimes_\gamma\mathcal{M}_*=L^1(\Omega,\mu,\mathcal{M}_*)$.

Since (Ω,μ) is localizable, $L^\infty(\Omega,\mu,\mathcal{M})=\sum_{\alpha\in\mathbb{I}}\oplus L^\infty(\Omega_\alpha,\mu_\alpha,\mathcal{M})$ with $\mu_\alpha(\Omega_\alpha)<+\infty$. Hence we may assume that $\mu(\Omega)<+\infty$. Take $x,y\in L^\infty(\Omega,\mu,\mathcal{M})$. For $\xi\in\mathcal{M}_*$, $R_{y(t)}\xi\in\mathcal{M}_*$, consider a vector valued function $t\to R_{y(t)}\xi$ on Ω. For $a\in\mathcal{M}$, we have

$$(R_{y(t)}\xi)(a)=\xi(ay(t))=L_a\xi(y(t)).$$

Hence $(R_{y(t)}\xi)(a)=\langle y(t),g(t)\rangle$, where $g(t)=L_a\xi$ for all $t\in\Omega$.

Since $g(t)\in L^1(\Omega,\mu,\mathcal{M}_*)$, $(R_{y(t)}\xi)(a)$ is μ-measurable for all $a\in\mathcal{M}$, so that $t\to R_{y(t)}\xi$ is weakly μ-measurable. Since $\mathcal{M}_*$ is separable, $t\to R_{y(t)}\xi$ is strongly μ-measurable. Hence for $\eta\in L^1(\Omega,\mu)$, $t\to\eta(t)R_{y(t)}\xi$ is strongly μ-measurable. Since the set of all finite linear combinations of $\mathcal{M}_*$-valued functions of the forms $\eta(t)\xi$ is norm-dense in $L^1(\Omega,\mu,\mathcal{M}_*)$, and since the function $\eta(t)R_{y(t)}\xi$ belongs to $L^1(\Omega,\mu,\mathcal{M}_*)$, for $h\in L^1(\Omega,\mu,\mathcal{M}_*)$, there exists a sequence (h_n) such that $h_n\to h$ in $L^1(\Omega,\mu,\mathcal{M}_*)$ and $R_{y(t)}h_n(t)\in L^1(\Omega,\mu,\mathcal{M}_*)$ $(n\geq 1)$.

Then $\|R_{y(t)}h_n(t)-R_{y(t)}h(t)\| \leq \|y\| \, \|h_n(t)-h(t)\|$. Since there exists a subsequence (n_j) of (n) such that $\|h_{n_j}(t)-h(t)\| \to 0$ (μ—a.e.), the above inequality implies the strong measurability of the function $R_{y(t)}h(t)$. Since $\|R_{y(t)}h(t)\| \leq \|y\| \, \|h(t)\|$, $R_{y(t)}h(t)$ belongs to $L^1(\Omega,\mu,\mathcal{M}_*)$, so that $\langle x(t), R_{y(t)}h(t)\rangle = \langle x(t)y(t), h(t)\rangle$ is measurable. Therefore $x(t)y(t)$ is weakly* μ-measurable and $\|x(t)y(t)\| \leq \|x\| \, \|y\|$.

Moreover, $\langle x(t)^*, \eta(t)\xi\rangle = \langle x(t), \overline{\eta(t)}\xi^*\rangle$ ($\eta \in L^1(\Omega,\mu)$, $\xi \in \mathcal{M}_*$). Hence, if we define $x^*(t) = x(t)^*$, then $x^* \in L^\infty(\Omega,\mu,\mathcal{M})$, and moreover $\|x^* x\| = \text{ess. sup}\,\|x(t)^* x(t)\| = \|x\|^2$. Therefore $L^\infty(\Omega,\mu,\mathcal{M})$ is a C^*-algebra, and it is the dual of $L^1(\Omega,\mu,\mathcal{M}_*)$. Hence $L^\infty(\Omega,\mu,\mathcal{M})$ is a W^*-algebra.

On the other hand, $L^\infty(\Omega,\mu)\otimes_{\alpha_0}\mathcal{M} = L^\infty(\Omega,\mu)\otimes_\lambda\mathcal{M}$ and

$$L^1(\Omega,\mu)\otimes_{\alpha_0^*}\mathcal{M}_* = L^1(\Omega,\mu)\otimes_{\lambda^*}\mathcal{M}_* = L^1(\Omega,\mu)\otimes_\gamma\mathcal{M}_* = L^1(\Omega,\mu,\mathcal{M}_*).$$

Hence, $L^\infty(\Omega,\mu)\overline{\otimes}\mathcal{M} = L^1(\Omega,\mu,\mathcal{M}_*)^* = L^\infty(\Omega,\mu,\mathcal{M})$. The mapping

$$\Phi: f\otimes a \to f(t)a \qquad (f\in L^\infty(\Omega,\mu),\ a\in\mathcal{M})$$

can be extended uniquely to a $*$-isomorphism of the W^*-algebra $L^\infty(\Omega,\mu)\overline{\otimes}\mathcal{M}$ onto the W^*-algebra $L^\infty(\Omega,\mu,\mathcal{M})$. Hence,

$$L^\infty(\Omega,\mu)\overline{\otimes}\mathcal{M} = L^\infty(\Omega,\mu,\mathcal{M})$$

as W^*-algebras under the isomorphism Φ. q.e.d.

1.22.14. Proposition. *Let $\mathcal{M}$ be a W^*-algebra, and let $(v_\beta \mid \beta\in\mathrm{II})$ be a family of partial isometries in $\mathcal{M}$ such that $v_{\beta_0} = e_{\beta_0}$, $v_\beta^* v_\beta = e_{\beta_0}$ (β_0 fixed, $\beta\in\mathrm{II}$), $\{v_\beta v_\beta^* \mid \beta\in\mathrm{II}\}$ is a family of mutually orthogonal projections, and $\sum_{\beta\in\mathrm{II}} v_\beta v_\beta^* = 1$. Let $\mathcal{M}_1 = \left\{x \,\middle|\, x = \sum_{\beta\in\mathrm{II}} v_\beta x_{\beta_0} v_\beta^*,\ x_{\beta_0}\in e_{\beta_0}\mathcal{M}e_{\beta_0}\right\}$ and let $\mathcal{M}_2$ be the W^*-subalgebra of $\mathcal{M}$ generated by $\{v_\beta v_\gamma^* \mid \beta,\gamma\in\mathrm{II}\}$. Then $\mathcal{M}_1$ is the W^*-subalgebra composed of all elements of $\mathcal{M}$ commuting with $\mathcal{M}_2$, and $\mathcal{M}$ is $*$-isomorphic to $\mathcal{M}_1\overline{\otimes}\mathcal{M}_2$. Moreover, $\mathcal{M}_2$ is $*$-isomorphic to a W^*-algebra $B(\mathcal{H})$, with* $\dim(\mathcal{H}) = \mathrm{Card}(\mathrm{II})$.

Proof. Let $\{\pi, \mathcal{H}_1\}$ be a faithful W^*-representation of $\mathcal{M}$ such that $\pi(1) = 1_{\mathcal{H}_1}$. We shall identify $\mathcal{M}$ with $\pi(\mathcal{M})$. Put $e_{\beta_0}\mathcal{H}_1 = \mathcal{H}_2$ and $\mathcal{H}_3 = l^2(\mathrm{II})$, where $l^2(\mathrm{II})$ is the Hilbert space of all square summable complex valued functions on II.

Let u_β be the unitary mapping of $\mathcal{H}_2$ onto $v_\beta v_\beta^*\mathcal{H}_1$ defined by v_β. Then we can define canonically a unitary mapping of $\mathcal{H}_1$ onto $\mathcal{H}_2\otimes\mathcal{H}_3$. We shall identify the two Hilbert spaces by this unitary mapping. Then by the consideration of 1.20, the weakly closed self-adjoint algebra generated by $\{u_\beta u_\gamma^* \mid \beta,\gamma\in\mathrm{II}\}$ is the algebra $1_{\mathcal{H}_2}\otimes B(\mathcal{H}_3)$, so that $\mathcal{M}_2 = 1_{\mathcal{H}_2}\otimes B(\mathcal{H}_3)$.

Since $\mathcal{M}_1$ commutes with $\mathcal{M}_2$, $\mathcal{M}_1 \subset (1_{\mathcal{H}_2} \otimes B(\mathcal{H}_3))' = B(\mathcal{H}_2) \otimes 1_{\mathcal{H}_3}$. Moreover, for $a \in \mathcal{M}$, $a_{\beta\gamma} = u_\beta^* a u_\gamma = v_\beta^* a v_\gamma$. Hence,

$$\begin{aligned}\mathcal{M} \subset \{(e_{\beta_0} \mathcal{M} e_{\beta_0})' \otimes 1_{\mathcal{H}_3}\}' &= \{(e_{\beta_0} \mathcal{M} e_{\beta_0} \otimes 1_{\mathcal{H}_3})' \cap (1_{\mathcal{H}_2} \otimes B(\mathcal{H}_3))'\}' \\ &= R(e_{\beta_0} \mathcal{M} e_{\beta_0} \otimes 1_{\mathcal{H}_3}, 1_{\mathcal{H}_2} \otimes B(\mathcal{H}_3)),\end{aligned}$$

$\mathcal{M}_1 = (e_{\beta_0} \mathcal{M} e_{\beta_0}) \otimes 1_{\mathcal{H}_3}$, and $1_{\mathcal{H}_2} \otimes B(\mathcal{H}_3) = \mathcal{M}_2$. Hence by 1.22.11, $\mathcal{M} = \mathcal{M}_1 \bar{\otimes} \mathcal{M}_2$. q.e.d.

Concluding remarks on 1.22.

1. Theorem 1.22.6 is due to Takesaki [204].

2. A C^*-algebra is said to be simple if it has no proper closed two-sided ideal. One can easily see that the C^*-tensor product of simple C^*-algebras with identity is again simple from Theorem 1.22.6 (cf. [204]).

3. The following problem would be interesting: Let $\mathcal{M}$ and $\mathcal{N}$ be W^*-algebras and let $\mathcal{M}_*$ and $\mathcal{N}_*$ be their preduals. If a cross norm α on $\mathcal{M}_* \odot \mathcal{N}_*$ satisfies that the dual of $\mathcal{M}_* \otimes_\alpha \mathcal{N}_*$ is a W^*-algebra, then can we conclude that $\alpha = \alpha_0^*$?

4. The tensor product of C^*-algebras was studied first by Turumaru ([212], [213], [214]).

References. [69], [154], [204], [212], [213], [214], [221].

1.23. The Inductive Limit and Infinite Tensor Product of C^*-Algebras

Let $\mathbb{I}$ be a directed set of indices, and let $\{\mathcal{A}_\alpha \mid \alpha \in \mathbb{I}\}$ be a family of C^*-algebras with identity. Suppose for $\alpha, \beta \in \mathbb{I}$, with $\alpha \leq \beta$, there exists a $*$-isomorphism $\Phi_{\beta,\alpha}$ of $\mathcal{A}_\alpha$ into $\mathcal{A}_\beta$ such that $\Phi_{\beta,\alpha}(1_\alpha) = 1_\beta$ (1_α (resp. 1_β) is the identity of $\mathcal{A}_\alpha$ (resp. $\mathcal{A}_\beta$)).

Moreover, suppose that the family of these $*$-isomorphisms $\{\Phi_{\beta,\alpha}\}$ satisfies the following condition: $\Phi_{\gamma,\beta} \Phi_{\beta,\alpha} = \Phi_{\gamma,\alpha}$ ($\alpha \leq \beta \leq \gamma$). Let $\mathfrak{F}$ be the set of all functions $\{x_\alpha\}$ on $\mathbb{I}$ such that $x_\alpha \in \mathcal{A}_\alpha$. Then $\mathfrak{F}$ is a $*$-algebra under the pointwise addition and multiplication and $\{x_\alpha\}^* = \{x_\alpha^*\}$.

Let $\mathcal{L}$ be a subset of all elements $\{x_\alpha\}$ in $\mathfrak{F}$ satisfying the following conditions: There exists an index γ in $\mathbb{I}$ (depending on $\{x_\alpha\}$) such that $x_\alpha = \Phi_{\alpha,\gamma}(x_\gamma)$ for all $\alpha (\geq \gamma) \in \mathbb{I}$. Let $\{x_\alpha\} \in \mathcal{L}$ and $\beta \geq \alpha \geq \gamma$. Then $x_\beta = \Phi_{\beta,\gamma}(x_\gamma) = \Phi_{\beta,\alpha} \Phi_{\alpha,\gamma}(x_\gamma) = \Phi_{\beta,\alpha}(x_\alpha)$.

For $x = \{x_\alpha\}$, $y = \{y_\alpha\} \in \mathcal{L}$, there exist γ_1, γ_2 such that

$$\Phi_{\alpha,\gamma_1}(x_{\gamma_1}) = x_\alpha \qquad (\alpha \geq \gamma_1)$$

and $\Phi_{\alpha,\gamma_2}(y_{\gamma_2}) = y_\alpha$ ($\alpha \geq \gamma_2$). Take $\gamma_3 \geq \gamma_1, \gamma_2$. Then $\Phi_{\alpha,\gamma_3}(x_{\gamma_3}) = x_\alpha$ and $\Phi_{\alpha,\gamma_3}(y_{\gamma_3}) = y_\alpha$ ($\alpha \geq \gamma_3$). Hence $\Phi_{\alpha,\gamma_3}(x_{\gamma_3} + y_{\gamma_3}) = x_\alpha + y_\alpha$ ($\alpha \geq \gamma_3$), and so $x + y \in \mathcal{L}$. Analogously one can easily see that $\mathcal{L}$ is a $*$-algebra. Moreover, $\|x_\alpha\| = \|\Phi_{\alpha,\gamma}(x_\gamma)\| = \|x_\gamma\|$ $(\alpha \geq \gamma)$. Hence $\lim_\alpha \|x_\alpha\|$ exists. Put $\|x\| = \lim_\alpha \|x_\alpha\|$ $(x \in \mathscr{L})$. Then $\| \ \|$ is a C^*-semi-norm on $\mathscr{L}$. Put

$$\mathscr{I} = \{x \mid \|x\| = 0, x \in \mathscr{L}\}.$$

$\mathscr{I}$ is a self-adjoint two-sided ideal of $\mathscr{L}$.

Consider the quotient *-algebra $\mathscr{L}/\mathscr{I}$. The semi-norm $\| \ \|$ on $\mathscr{L}$ is constant on each class $y + \mathscr{I}$ $(y \in \mathscr{L})$, and so defines a C^*-norm on $\mathscr{L}/\mathscr{I}$, also denoted by $\| \ \|$.

Let $\mathscr{A}$ be the C^*-algebra obtained by the completion of $\mathscr{L}/\mathscr{I}$ under $\| \ \|$. Set $\mathscr{L}_\beta = \{x \mid x = \{x_\alpha\} \in \mathscr{L}$ and $x_\alpha = \Phi_{\alpha,\beta}(x_\beta)$ $(\alpha \geq \beta)\}$. $\mathscr{L}_\beta$ is a *-subalgebra of $\mathscr{L}$.

Let x_β be an arbitrary element of $\mathscr{A}_\beta$. We shall define an element $x = \{x_\alpha\}$ in $\mathfrak{F}$ as follows: if $\alpha \geq \beta$, $x_\alpha = \Phi_{\alpha,\beta}(x_\beta)$ and if $\alpha \not\geq \beta$, $x_\alpha = 0$. Then $x \in \mathscr{L}_\beta$. Hence the mapping $x_\beta \to x = \{x_\alpha\}$, of $\mathscr{A}_\beta$ into $\mathscr{L}_\beta$ induces a *-isomorphism Φ_β of $\mathscr{A}_\beta$ onto a C^*-subalgebra $\mathscr{L}_\beta + \mathscr{I}/\mathscr{I}$ (denoted by $\tilde{\mathscr{A}}_\beta$) of $\mathscr{A}$.

Moreover, for $y_\beta \in \mathscr{A}_\beta$ and $\gamma \geq \beta$,

$$\Phi_\gamma \Phi_{\gamma,\beta}(y_\beta) = \{x_\alpha\} + \mathscr{I}(\{x_\alpha\} \in \mathscr{L}_\gamma \text{ and } x_\gamma = \Phi_{\gamma,\beta}(y_\beta)).$$

$x_\alpha = \Phi_{\alpha,\gamma}(x_\gamma)$ $(\alpha \geq \gamma)$. Hence, we have $x_\alpha = \Phi_{\alpha,\gamma}\Phi_{\gamma,\beta}(y_\beta) = \Phi_{\alpha,\beta}(y_\beta)$. On the other hand, $\Phi_\beta(y_\beta) = \{x'_\alpha\} + \mathscr{I}$ $(\{x'_\alpha\} \in \mathscr{L}_\beta$ and $x'_\beta = y_\beta)$. Hence,

$$x'_\alpha = \Phi_{\alpha,\beta}(y_\beta) \qquad (\alpha \geq \beta),$$

and so $x_\alpha = x'_\alpha$ for $\alpha \geq \gamma$. Hence $\{x_\alpha\} - \{x'_\alpha\} \in \mathscr{I}$. Therefore, we have $\Phi_\gamma \Phi_{\gamma,\beta} = \Phi_\beta$ $(\gamma \geq \beta)$.

Clearly $\mathscr{L}_{\beta_1} \subset \mathscr{L}_{\beta_2}$ if $\beta_1 \leq \beta_2$, and $\bigcup_{\alpha \in \Pi} \mathscr{L}_\alpha = \mathscr{L}$. Hence

$$\tilde{\mathscr{A}}_{\beta_1} \subset \tilde{\mathscr{A}}_{\beta_2} \qquad (\beta_1 \leq \beta_2),$$

and the uniform closure of $\bigcup_{\alpha \in \Pi} \tilde{\mathscr{A}}_\alpha$ is $\mathscr{A}$.

1.23.1. Definition. *The C^*-algebra $\mathscr{A}$, denoted by*

$$\lim_\alpha \{\mathscr{A}_\alpha; \Phi_{\beta,\alpha} \mid (\beta,\alpha) \in \Pi \times \Pi \text{ and } \beta \geq \alpha\},$$

*is called the inductive limit of $\{\mathscr{A}_\alpha \mid \alpha \in \Pi\}$ defined by the family of *-isomorphisms $\{\Phi_{\beta,\alpha}\}$.*

Then we have

1.23.2. Proposition. *Let $\mathscr{A} = \lim_\alpha \{\mathscr{A}_\alpha; \Phi_{\beta,\alpha} \mid (\beta,\alpha) \in \Pi \times \Pi, \beta \geq \alpha\}$. Then there exists a directed set, by inclusion (i.e., $\tilde{\mathscr{A}}_\alpha \subseteq \tilde{\mathscr{A}}_\beta$ is equivalent to $\alpha \leq \beta$) of C^*-subalgebras $\{\tilde{\mathscr{A}}_\alpha \mid \alpha \in \Pi\}$ of $\mathscr{A}$ containing the identity which satisfies the following conditions: 1. for $\alpha \in \Pi$ there exists a *-isomorphism Φ_α of $\mathscr{A}_\alpha$ onto $\tilde{\mathscr{A}}_\alpha$ such that $\Phi_\beta \Phi_{\beta,\alpha} = \Phi_\alpha$ $(\beta \geq \alpha)$; 2. the uniform closure of $\bigcup_{\alpha \in \Pi} \tilde{\mathscr{A}}_\alpha$ is $\mathscr{A}$.*

Conversely, let $\mathscr{D}$ be a C-algebra with identity having a directed set, by inclusion, of C*-subalgebras $\{\mathscr{D}_\alpha | \alpha \in \mathbb{I}\}$ of $\mathscr{D}$ containing the identity. Suppose $\mathscr{D}$ satisfies the following conditions:* 1. *for $\alpha \in \mathbb{I}$ there exists a *-isomorphism Ψ_α of $\mathscr{A}_\alpha$ onto $\mathscr{D}_\alpha$ such that $\Psi_\beta \Phi_{\beta,\alpha} = \Psi_\alpha$ $(\beta \geq \alpha)$;* 2. *the uniform closure of $\bigcup_{\alpha \in \mathbb{I}} \mathscr{D}_\alpha$ is $\mathscr{D}$. Then, there exists a *-isomorphism Λ of $\mathscr{A}$ onto $\mathscr{D}$ such that $\Lambda(\tilde{\mathscr{A}}_\alpha) = \mathscr{D}_\alpha$ and $\Lambda = \Psi_\alpha \Phi_\alpha^{-1}$ on $\tilde{\mathscr{A}}_\alpha$* $(a \in \mathbb{I})$.

Proof. Clearly, $\Lambda_\alpha = \Psi_\alpha \Phi_\alpha^{-1}$ defines a *-isomorphism of $\tilde{\mathscr{A}}_\alpha$ onto $\mathscr{D}_\alpha$. Let $\beta \geq \alpha$ and $a_\alpha \in \tilde{\mathscr{A}}_\alpha$. Then

$$\begin{aligned}\Lambda_\beta(a_\alpha) &= \Psi_\beta \Phi_\beta^{-1}(a_\alpha) = \Psi_\beta \Phi_\beta^{-1}(\Phi_\alpha \Phi_\alpha^{-1}(a_\alpha)) = \Psi_\beta \Phi_\beta^{-1} \Phi_\beta \Phi_{\beta,\alpha}(\Phi_\alpha^{-1}(a_\alpha)) \\ &= \Psi_\alpha \Phi_\alpha^{-1}(a_\alpha).\end{aligned}$$

Hence $\Lambda_\alpha = \Lambda_\beta$ on $\tilde{\mathscr{A}}_\alpha$.

For $b \in \bigcup_{\alpha \in \mathbb{I}} \tilde{\mathscr{A}}_\alpha$, there exists some α such that $b \in \tilde{\mathscr{A}}_\alpha$. Define $\Lambda(b) = \Lambda_\alpha(b)$; $\Lambda(b)$ is well-defined. Λ defines a *-isomorphism of $\bigcup_\alpha \tilde{\mathscr{A}}_\alpha$ onto $\bigcup_\alpha \mathscr{D}_\alpha$. Moreover, Λ is an isometry, since $\tilde{\mathscr{A}}_\alpha$ and $\mathscr{D}_\alpha$ are C^*-algebras. Hence, Λ can be uniquely extended to a *-isomorphism Λ of $\mathscr{A}$ onto $\mathscr{D}$. q.e.d.

1.23.3. Corollary. *Let $\mathscr{A}$ be a C*-algebra with identity, and let $\{\mathscr{A}_\alpha | \alpha \in \mathbb{I}\}$ be a directed set, by inclusion, of C*-subalgebras of $\mathscr{A}$ containing the identity. Suppose that $\mathscr{A}$ is the uniform closure of $\bigcup_{\alpha \in \mathbb{I}} \mathscr{A}_\alpha$. For $\alpha \leq \beta$, let $\Phi_{\beta,\alpha}$ be the identity mapping of $\mathscr{A}_\alpha$ into $\mathscr{A}_\beta$. Then $\mathscr{A}$ is *-isomorphic to $\varinjlim_\alpha \{\mathscr{A}_\alpha; \Phi_{\beta,\alpha} | (\beta,\alpha) \in \mathbb{I} \times \mathbb{I}, \beta \geq \alpha\}$.*

1.23.4. Proposition. *Let $\mathscr{A} = \varinjlim_\alpha \{\mathscr{A}_\alpha; \Phi_{\beta,\alpha} | (\beta,\alpha) \in \mathbb{I} \times \mathbb{I}, \beta \geq \alpha\}$,*

$$\mathfrak{B} = \varinjlim_\alpha \{\mathfrak{B}_\alpha; \Psi_{\beta,\alpha} | (\beta,\alpha) \in \mathbb{I} \times \mathbb{I}, \beta \geq \alpha\}.$$

*Suppose that there exists a *-isomorphism Λ_α of $\mathscr{A}_\alpha$* onto $\mathfrak{B}_\alpha$ $(\alpha \in \mathbb{I})$. If $\Lambda_\beta \Phi_{\beta,\alpha} = \Psi_{\beta,\alpha} \Lambda_\alpha$ $(\alpha, \beta(\beta \geq \alpha) \in \mathbb{I})$, *then $\mathscr{A}$ is *-isomorphic to $\mathfrak{B}$.*

Proof. Consider the directed sets of C^*-algebras $\{\tilde{\mathscr{A}}_\alpha\}$ (resp. $\{\tilde{\mathfrak{B}}_\alpha\}$) and the *-isomorphisms $\{\Phi_\alpha\}$ (resp. $\{\Psi_\alpha\}$). Then $\Phi_\beta \Phi_{\beta,\alpha} = \Phi_\alpha$ and $\Psi_\beta \Psi_{\beta,\alpha} = \Psi_\alpha$ $(\beta \geq \alpha)$. Consider a *-isomorphism $\Psi_\alpha \Lambda_\alpha \Phi_\alpha^{-1}$ of

$$\tilde{\mathscr{A}}_\alpha \text{ onto } \tilde{\mathfrak{B}}_\alpha \qquad (\alpha \in \mathbb{I}).$$

Then for $a_\alpha \in \tilde{\mathscr{A}}_\alpha$,

$$\begin{aligned}\Psi_\beta \Lambda_\beta \Phi_\beta^{-1}(a_\alpha) &= \Psi_\beta \Lambda_\beta \Phi_\beta^{-1}(\Phi_\alpha \Phi_\alpha^{-1}(a_\alpha)) = \Psi_\beta \Lambda_\beta \Phi_{\beta,\alpha}(\Phi_\alpha^{-1}(a_\alpha)) \\ &= \Psi_\beta \Psi_{\beta,\alpha} \Lambda_\alpha \Phi_\alpha^{-1}(a_\alpha) = \Psi_\alpha \Lambda_\alpha \Phi_\alpha^{-1}(a_\alpha).\end{aligned}$$

Hence, $\Psi_\beta \Lambda_\beta \Phi_\beta^{-1} = \Psi_\alpha \Lambda_\alpha \Phi_\alpha^{-1}$ on $\tilde{\mathscr{A}}_\alpha$ $(\beta \geq \alpha)$. Therefore, we can easily define a *-isomorphism Λ of $\mathscr{A}$ onto $\mathfrak{B}$ such that $\Lambda = \Psi_\alpha \Lambda_\alpha \Phi_\alpha^{-1}$ on $\tilde{\mathscr{A}}_\alpha$. q.e.d.

1.23.5. Proposition *Let* $\mathscr{A} = \lim_n \{\mathscr{A}_n; \Phi_{m,n} \mid m \geq n, m, n = 1, 2, \ldots\}$ *and* $\mathfrak{B} = \lim_n \{\mathfrak{B}_n; \Psi_{m,n} \mid m \geq n, m, n = 1, 2, \ldots\}$, *and each* $\mathscr{A}_n$, $\mathfrak{B}_n$ *is *-isomorphic to* $B(\mathscr{H}_n)$ $(\dim(\mathscr{H}_n) < +\infty)$. *Then* $\mathscr{A}$ *is *-isomorphic to* $\mathfrak{B}$.

Proof. Consider $\{\tilde{\mathscr{A}}_n\}$ and $\{\tilde{\mathfrak{B}}_n\}$. By induction, we shall define a *-isomorphism Λ_n of $\tilde{\mathscr{A}}_n$ onto $\tilde{\mathfrak{B}}_n$. For $n=1$, we take an arbitrary *-isomorphism Λ_1 of $\tilde{\mathscr{A}}_1$ onto $\tilde{\mathfrak{B}}_1$. Suppose Λ_n $(n \leq n_0)$ is defined. We can write: $\tilde{\mathscr{A}}_{n_0+1} = \tilde{\mathscr{A}}_{n_0} \otimes \tilde{\mathscr{A}}'_{n_0}$ and $\tilde{\mathfrak{B}}_{n_0+1} = \tilde{\mathfrak{B}}_{n_0} \otimes \tilde{\mathfrak{B}}'_{n_0}$. Take a *-isomorphism Δ_{n_0+1} of $1 \otimes \tilde{\mathscr{A}}'_n$ onto $1 \otimes \tilde{\mathfrak{B}}'_{n_0}$, and set $\Lambda_{n_0+1} = \Lambda_{n_0} \otimes \Delta_{n_0+1}$. Then $\Lambda_n = \Lambda_m$ on $\tilde{\mathscr{A}}_n$ $(n \leq m)$. Therefore, we can define a *-isomorphism Λ of $\mathscr{A}$ onto $\mathfrak{B}$ such that $\Lambda = \Lambda_n$ on $\tilde{\mathscr{A}}_n$. q.e.d.

1.23.6. Definition. *Let* $\mathscr{A}$ *be a* C^**-algebra with identity.* $\mathscr{A}$ *is said to be uniformly hyperfinite if there exists a directed set, by inclusion, of* C^**-subalgebras* $\{\mathscr{A}_\alpha \mid \alpha \in \mathbb{I}\}$ *containing the identity such that the uniform closure of* $\bigcup_{\alpha \in \mathbb{I}} \mathscr{A}_\alpha$ *is* $\mathscr{A}$, *and* $\mathscr{A}_\alpha$ *is *-isomorphic to* $B(\mathscr{H}_\alpha)$

$$(\dim(\mathscr{H}_\alpha) < +\infty, \alpha \in \mathbb{I}).$$

1.23.7. Proposition. *Let* $\mathscr{A}$ *be a separable, uniformly hyperfinite* C^**-algebra. Then there exists an increasing sequence (i.e.,* $\mathscr{A}_n \subset \mathscr{A}_m$ $(n \leq m)$*) of* C^**-subalgebras* $\mathscr{A}_n$ *containing the identity such that* $\mathscr{A}_n$ *is *-isomorphic to* $B(\mathscr{H}_n)$ $(\dim(\mathscr{H}_n) < +\infty)$, *and the uniform closure of* $\bigcup_{n=1}^{\infty} \mathscr{A}_n$ *is* $\mathscr{A}$.

Proof. Let $\mathscr{A}$ = the uniform closure of $\bigcup_{\alpha \in \mathbb{I}} \mathscr{A}_\alpha$. We shall choose by induction an increasing sequence $\{\mathscr{A}_{\alpha_n}\}$. Let (a_n) be a sequence which is dense in $\mathscr{A}$. There exists an $\mathscr{A}_{\alpha_1}$ and $b_1 \in \mathscr{A}_{\alpha_1}$ such that $\|a_1 - b_1\| < 1$. Suppose $\mathscr{A}_{\alpha_n}$ $(n \leq n_0)$ is defined. Then there exists a family

$$\{\mathscr{A}_{\alpha_{n_0+1,1}}, \mathscr{A}_{\alpha_{n_0+1,2}}, \ldots, \mathscr{A}_{\alpha_{n_0+1,n_0+1}}\} \qquad (\alpha_{n_0+1,i} \in \mathbb{I})$$

and $b_{n_0+1,i} \in \mathscr{A}_{\alpha_{n_0+1,i}}$ $(1 \leq i \leq n_0+1)$ such that

$$\|a_i - b_{n_0+1,i}\| < \frac{1}{n_0+1} \qquad (1 \leq i \leq n_0+1).$$

Take an α such that $\alpha_1, \alpha_2, \ldots, \alpha_{n_0}, \alpha_{n_0+1,1}, \ldots, \alpha_{n_0+1,n_0+1} \leq \alpha$ and set $\alpha_{n_0+1} = \alpha$. Then $\{\mathscr{A}_{\alpha_n}\}$ is an increasing sequence and the uniform closure of $\bigcup_{n=1}^{\infty} \mathscr{A}_{\alpha_n}$ is $\mathscr{A}$. q.e.d.

Remark. In general, the isomorphism classes of uniformly hyperfinite C^*-algebras $\mathscr{A}$ $\left(=\text{the uniform closure of } \bigcup_{\alpha\in\mathbb{I}} \mathscr{A}_\alpha\right)$ depend on

$$\dim(\mathscr{H}_\alpha) \qquad (\alpha\in\mathbb{I}).$$

Glimm [58] gave the complete classification of isomorphism classes of separable, uniformly hyperfinite C^*-algebras.

1.23.8. Proposition. *Let $\mathscr{A}$ be a C^*-algebra and let $\{\mathscr{A}_\alpha|\alpha\in\mathbb{I}\}$ be a directed set, by inclusion, of C^*-subalgebras of $\mathscr{A}$ such that the uniform closure of $\bigcup_{\alpha\in\mathbb{I}}\mathscr{A}_\alpha$ is $\mathscr{A}$.*

Suppose that $\mathscr{A}_\alpha$ $(\alpha\in\mathbb{I})$ is simple (i.e., no proper closed two-sided ideal). Then $\mathscr{A}$ is again simple.

Proof. Let $\mathscr{I}$ be a proper closed two-sided ideal of $\mathscr{A}$. Then $\mathscr{A}_\alpha\cap\mathscr{I}=(0)$. Consider the canonical *-homomorphism $\Phi:\mathscr{A}\to\mathscr{A}/\mathscr{I}$; then Φ is isometric on $\mathscr{A}_\alpha$. Hence it is isometric on $\mathscr{A}$, and so $\mathscr{I}=(0)$. q.e.d.

1.23.9. Corollary. *Let $\mathscr{A}=\lim_{\alpha\in\mathbb{I}}\{\mathscr{A}_\alpha;\Phi_{\beta,\alpha}|(\beta,\alpha)\in\mathbb{I}\times\mathbb{I},\beta\geq\alpha\}$. Suppose that $\mathscr{A}_\alpha$ $(\alpha\in\mathbb{I})$ is simple. Then $\mathscr{A}$ is again simple.*

Now let $\mathscr{A}=\lim_{\alpha}\{\mathscr{A}_\alpha;\Phi_{\beta,\alpha}|(\beta,\alpha)\in\mathbb{I}\times\mathbb{I},\beta\geq\alpha\}$. Let $\{\varphi_\alpha|\alpha\in\mathbb{I}\}$ be a family of states on $\mathscr{A}_\alpha$ such that $\varphi_\beta(\Phi_{\beta,\alpha}(x_\alpha))=\varphi_\alpha(x_\alpha)$ $(x_\alpha\in\mathscr{A}_\alpha$ and $\beta\geq\alpha)$. Then we define a state $\tilde{\varphi}_\alpha$ on $\tilde{\mathscr{A}}_\alpha$ as follows: $\tilde{\varphi}_\alpha(a_\alpha)=\varphi_\alpha(\Phi_\alpha^{-1}(a_\alpha))$ $(a_\alpha\in\tilde{\mathscr{A}}_\alpha)$. Then

$$\begin{aligned}\tilde{\varphi}_\beta(a_\alpha)&=\varphi_\beta(\Phi_\beta^{-1}(a_\alpha))=\varphi_\beta(\Phi_\beta^{-1}\Phi_\alpha\Phi_\alpha^{-1}(a_\alpha))=\varphi_\beta(\Phi_\beta^{-1}\Phi_\beta\Phi_{\beta,\alpha}\Phi_\alpha^{-1}(a_\alpha))\\&=\varphi_\beta(\Phi_{\beta,\alpha}\Phi_\alpha^{-1}(a_\alpha))=\varphi_\alpha(\Phi_\alpha^{-1}(a_\alpha))=\tilde{\varphi}_\alpha(a_\alpha)\qquad(a_\alpha\in\tilde{\mathscr{A}}_\alpha).\end{aligned}$$

Hence, by using the family $\{\tilde{\varphi}_\alpha|\alpha\in\mathbb{I}\}$, we can define a linear functional φ on $\bigcup_{\alpha\in\mathbb{I}}\tilde{\mathscr{A}}_\alpha$ such that $\varphi=\tilde{\varphi}_\alpha$ on $\tilde{\mathscr{A}}_\alpha$. Since $\|\tilde{\varphi}_\alpha\|=1$, $\|\varphi\|=1$ on $\bigcup_{\alpha\in\mathbb{I}}\tilde{\mathscr{A}}_\alpha$. Hence φ can be uniquely extended to a state φ on $\mathscr{A}$.

1.23.10. Definition. *The above φ is called the inductive limit of $\{\varphi_\alpha;\Phi_{\beta,\alpha}|(\beta,\alpha)\in\mathbb{I}\times\mathbb{I},\beta\geq\alpha\}$, and is denoted by*

$$\varphi=\lim_{\alpha}\{\varphi_\alpha;\Phi_{\beta,\alpha}\,|\,(\beta,\alpha)\in\mathbb{I}\times\mathbb{I},\beta\geq\alpha\}.$$

Now let $\mathbb{I}$ be a set of indices, $\{\mathscr{A}_\alpha|\alpha\in\mathbb{I}\}$ be a family of C^*-algebras with identity 1_α, and let $\mathbb{J}$ be the set of all finite subsets of $\mathbb{I}$. $\mathbb{J}$ is a directed set by inclusion. For $\gamma\in\mathbb{J}$, consider the C^*-algebra $\mathfrak{B}_\gamma=\bigotimes_{\alpha\in\gamma}\mathscr{A}_\alpha$. Let $\gamma_1\leq\gamma_2$; then $\mathfrak{B}_{\gamma_2}=\mathfrak{B}_{\gamma_1}\otimes\bigotimes_{\alpha\in\gamma_2-\gamma_1}\mathscr{A}_\alpha$. Φ_{γ_2,γ_1} will be defined as the mapping $x\to x\otimes 1$ of $\mathfrak{B}_{\gamma_1}$ into $\mathfrak{B}_{\gamma_2}$, where 1 is the identity of $\bigotimes_{\alpha\in\gamma_2-\gamma_1}\mathscr{A}_\alpha$. Then $\Phi_{\gamma_3,\gamma_2}\Phi_{\gamma_2,\gamma_1}=\Phi_{\gamma_3,\gamma_1}$ $(\gamma_1\leq\gamma_2\leq\gamma_3)$.

1.23.11. Definition. $\mathscr{A}=\lim\limits_{\gamma}\{\mathfrak{B}_\gamma;\Phi_{\gamma',\gamma}|(\gamma',\gamma)\in\mathbb{J}\times\mathbb{J},\gamma'\geq\gamma\}$, *denoted by* $\bigotimes\limits_{\alpha\in\mathbb{I}}\mathscr{A}_\alpha$, *is called the infinite tensor product of C^*-algebras* $\{\mathscr{A}_\alpha|\alpha\in\mathbb{I}\}$.

Suppose that γ contains only one element α; then we shall denote $\tilde{\mathfrak{B}}_\gamma$ by $\tilde{\mathscr{A}}_\alpha$. Let $\bigodot\limits_{\alpha\in\mathbb{I}}\mathscr{A}_\alpha=\mathscr{A}_0$ be the algebraical $*$-subalgebra of $\mathscr{A}$ generated by $\{\tilde{\mathscr{A}}_\alpha|\alpha\in\mathbb{I}\}$. Then the uniform closure of $\mathscr{A}_0$ is $\mathscr{A}$.

1.23.12. Proposition. *Let* $\mathbb{I}=\bigcup\limits_{\beta\in\mathbb{K}}\mathbb{I}_\beta$ *and* $\mathbb{I}_{\beta_1}\cap\mathbb{I}_{\beta_2}=(\emptyset)$ *if*

$$\beta_1\neq\beta_2\qquad(\beta_1,\beta_2\in\mathbb{K}).$$

Then $\bigotimes\limits_{\beta\in\mathbb{K}}\bigotimes\limits_{\alpha\in\mathbb{I}_\beta}\mathscr{A}_\alpha=\bigotimes\limits_{\alpha\in\mathbb{K}}\mathscr{A}_\alpha$.

Proof. Clearly $\bigodot\limits_{\beta\in\mathbb{K}}\bigodot\limits_{\alpha\in\mathbb{I}_\beta}\mathscr{A}_\alpha=\bigodot\limits_{\alpha\in\mathbb{I}}\mathscr{A}_\alpha$ and so $\bigotimes\limits_{\beta\in\mathbb{K}}\bigotimes\limits_{\alpha\in\mathbb{I}_\beta}\mathscr{A}_\alpha=\bigotimes\limits_{\alpha\in\mathbb{K}}\mathscr{A}_\alpha$. q.e.d.

Now let $\varphi_\alpha\in\mathscr{S}_{\mathscr{A}_\alpha}$ and for $\gamma\in\mathbb{J}$, set $\psi_\gamma=\bigotimes\limits_{\alpha\in\gamma}\varphi_\alpha$. Then ψ_γ is a state on $\mathfrak{B}_\gamma$. If $\gamma_1=(\alpha_1,\alpha_2,\dots,\alpha_n)$ and $\gamma_2=(\alpha_1,\alpha_2,\dots,\alpha_n,\alpha_{n+1},\dots,\alpha_m)$, then

$$\begin{aligned}\psi_{\gamma_2}(\Phi_{\gamma_2,\gamma_1}(a_{\alpha_1}\otimes a_{\alpha_2}\otimes\cdots\otimes a_{\alpha_n}))&=\psi_{\gamma_2}(a_{\alpha_1}\otimes a_{\alpha_2}\otimes\cdots\otimes a_{\alpha_n}\otimes 1_{\alpha_{n+1}}\otimes\cdots\otimes 1_{\alpha_m})\\&=\varphi_{\alpha_1}(a_{\alpha_1})\varphi_{\alpha_2}(a_{\alpha_2})\dots\varphi_{\alpha_n}(a_{\alpha_n})\\&=\psi_{\gamma_1}(a_{\alpha_1}\otimes a_{\alpha_2}\otimes\cdots\otimes a_{\alpha_n})\end{aligned}$$

$(a_{\alpha_i}\in\mathscr{A}_{\alpha_i},1\leq i\leq n)$. Hence, $\psi_{\gamma_2}(\Phi_{\gamma_2,\gamma_1}(x))=\psi_{\gamma_1}(x)$ for $x\in\mathfrak{B}_{\gamma_1}$. Hence we have $\lim\limits_{\gamma}\{\psi_\gamma;\Phi_{\gamma',\gamma}|(\gamma',\gamma)\in\mathbb{J}\times\mathbb{J},\gamma'\geq\gamma\}$.

1.23.13. Definition. $\lim\limits_{\gamma}\{\psi_\gamma;\Phi_{\gamma',\gamma}|(\gamma',\gamma)\in\mathbb{J}\times\mathbb{J},\gamma'\geq\gamma\}$, *denoted by* $\bigotimes\limits_{\alpha\in\mathbb{I}}\varphi_\alpha$, *is called the infinite product state of* $\{\varphi_\alpha|\varphi\in\mathbb{I}\}$ *on* $\bigotimes\limits_{\alpha\in\mathbb{I}}\mathscr{A}_\alpha$.

Remark. If φ_α is a factorial state for $\alpha\in\mathbb{I}$, then $\bigotimes\limits_{\alpha\in\mathbb{I}}\varphi_\alpha$ is again a factorial state (cf. chapter 2, chapter 4).
Inductive limit of C^*-algebras was studied first by Takeda [201].
References. [69], [130], [201].

1.24. Radon-Nikodym Theorems in W^*-Algebras

In this section, we shall give two generalizations of Radon-Nikodym theorem to general W^*-algebras.

1.24.1. Proposition. *Let $\mathscr{M}$ be a W^*-algebra and $\varphi(\geq 0)\in\mathscr{M}_*$. Suppose that $R_a\varphi$ is self-adjoint for some $a\in\mathscr{M}$. Then*

$$|(R_a\varphi)(h)|\leq\|a\|\,\varphi(h)\qquad(h\in\mathscr{M},h\geq 0).$$

Proof.

$$(R_a\varphi)^*(x)=\overline{R_a\varphi(x^*)}=\overline{\varphi(x^*a)}=\overline{\varphi((a^*x)^*)}=\varphi(a^*x)=(R_a\varphi)(x)$$
$$=\varphi(xa)\qquad (x\in\mathcal{M}).$$

Hence, $\varphi(a^*x)=\varphi(xa)$, so that $\varphi(xa^2)=\varphi(xaa)=\varphi(a^*xa)$. Therefore $R_{a^2}\varphi\geq 0$ and so $\varphi(xa^4)=\varphi((a^2)^*xa^2)$. In a similar manner one can prove that $\varphi(xa^{2^{n+1}})=\varphi((a^{2^n})^*x(a^{2^n}))$ $(x\in\mathcal{M})$. Then for $h\geq 0$,

$$|\varphi(ha)|=|\varphi(h^{\frac{1}{2}}h^{\frac{1}{2}}a)|\leq\varphi(h)^{\frac{1}{2}}\varphi(a^*ha)^{\frac{1}{2}}=\varphi(h)^{\frac{1}{2}}\varphi(ha^2)^{\frac{1}{2}}$$
$$\leq\varphi(h)^{\frac{1}{2}}\{\varphi(h)^{\frac{1}{2}}\varphi((a^2)^*ha^2)^{\frac{1}{2}}\}^{\frac{1}{2}}=\varphi(h)^{\frac{1}{2}}\varphi(h)^{\frac{1}{4}}\varphi(ha^4)^{\frac{1}{4}}$$
$$=\varphi(h)^{\frac{1}{2}+\frac{1}{4}}\varphi(ha^4)^{\frac{1}{4}}\leq\cdots\leq\varphi(h)^{\sum_{i=1}^{n}\frac{1}{2^i}}\varphi(ha^{2^n})^{\frac{1}{2^n}}$$
$$=\varphi(h)^{1-\frac{1}{2^n}}\varphi(ha^{2^n})^{\frac{1}{2^n}}\leq\varphi(h)^{1-\frac{1}{2^n}}(\|\varphi\|\,\|h\|\,\|a\|^{2^n})^{\frac{1}{2^n}}\to\|a\|\,\varphi(h)$$
$$(n\to\infty).$$

Hence, $|\varphi(ha)|\leq\|a\|\,\varphi(h)$. q.e.d.

1.24.2. Lemma. *Let $\varphi(\geq 0)\in\mathcal{M}_*$, and let $R_b\varphi=R_v|R_b\varphi|$ be the polar decomposition of $R_b\varphi$ $(b\in\mathcal{M})$. Then the absolute value $|R_b\varphi|$ of $R_b\varphi$ is bounded by $\|b\|\,\varphi$—i.e., $|R_b\varphi|\leq\|b\|\,\varphi$.*

Proof. Since $|R_b\varphi|=R_{v^*}R_b\varphi$, $|R_b\varphi|(x)=\varphi(xv^*b)$, so that we have

$$|\varphi(hv^*b)|\leq\|v^*b\|\,\varphi(h)\leq\|b\|\,\varphi(h)\qquad (h\in\mathcal{M},\ h\geq 0).\qquad\text{q.e.d.}$$

1.24.3. Theorem. *Let ψ be a normal positive linear functional on $\mathcal{M}$ such that $\psi\leq\varphi$. Then there exists a positive element t_0 of $\mathcal{M}$, with $0\leq t_0\leq 1$, such that $\psi(x)=\varphi(t_0xt_0)$ $(x\in\mathcal{M})$.*

Proof. Consider the W^*-algebra $s(\varphi)\mathcal{M}s(\varphi)$. Let $\tilde{\varphi}$ be the restriction of φ to $s(\varphi)\mathcal{M}s(\varphi)$ and consider the W^*-representation $\{\pi_{\tilde{\varphi}}^{w},\mathcal{H}_{\tilde{\varphi}}\}$ of $s(\varphi)\mathcal{M}s(\varphi)$. This representation is faithful. For simplicity, we shall identify $s(\varphi)\mathcal{M}s(\varphi)$ with its image under $\pi_{\tilde{\varphi}}$. Then $\varphi(x)=(x\xi,\xi)$ for $x\in s(\varphi)\mathcal{M}s(\varphi)$, where $(,)$ is the scalar product of $\mathcal{H}_{\tilde{\varphi}}$, and ξ is the image of $s(\varphi)$ in $\mathcal{H}_{\tilde{\varphi}}$.

Let $\{s(\varphi)\mathcal{M}s(\varphi)\}'$ be the commutant of $s(\varphi)\mathcal{M}s(\varphi)$ in $\mathcal{H}_{\tilde{\varphi}}$. Then $[s(\varphi)\mathcal{M}s(\varphi)\xi]=[\{s(\varphi)\mathcal{M}s(\varphi)\}'\xi]=\mathcal{H}_{\tilde{\varphi}}$, where $[()]$ is the closed linear subspace of $\mathcal{H}_{\tilde{\varphi}}$ generated by $()$—i.e., ξ is a cyclic and separating vector.

Let $\tilde{\psi}$ be the restriction of ψ on $s(\varphi)\mathcal{M}s(\varphi)$. Then $\tilde{\psi}\leq\tilde{\varphi}$, and so there exists a positive element h_0', with $\|h_0'\|\leq 1$, of $\{s(\varphi)\mathcal{M}s(\varphi)\}'$ such that $\tilde{\psi}(x)=(xh_0'\xi,h_0'\xi)$ $(x\in s(\varphi)\mathcal{M}s(\varphi))$. Now consider the σ-continuous linear functional f' on the W^*-algebra $\{s(\varphi)\mathcal{M}s(\varphi)\}'$, where

$$f'(y')=(y'h_0'\xi,\xi)\qquad (y'\in\{s(\varphi)\mathcal{M}s(\varphi)\}').$$

Then $f'=R_{h'_0}g'$, where $g'(y')=(y'\xi,\xi)$ for $y'\in\{s(\varphi)\mathcal{M}s(\varphi)\}'$. Since $g'\geq 0$, $|f'|\leq\|h'_0\|g'$ by 1.24.2, so that there exists a positive element $t_0\in s(\varphi)\mathcal{M}s(\varphi)$, $0\leq t_0\leq 1$, such that $|f'|(y')=(y't_0\xi,\xi)$. Then

$$|f'|(y')=R_{v'^*}f'(y')=f'(y'v'^*)=g'(y'v'^*h'_0),$$

where $R_{v'}|f'|=f'$ is the polar decomposition of f'. Hence,

$$(y't_0\xi,\xi)=(y'v'^*h'_0\xi,\xi)\qquad(y'\in\{s(\varphi)\mathcal{M}s(\varphi)\}').$$

Since $[\{s(\varphi)\mathcal{M}s(\varphi)\}'\xi]=\mathcal{H}_{\tilde{\varphi}}$, we have $t_0\xi=v'^*h'_0\xi$, and so

$$v't_0\xi=v'v'^*h'_0\xi.$$

On the other hand,

$$\begin{aligned}(y'v'v'^*h'_0\xi,\xi)&=|f'|(y'v')=R_{v'}|f'|(y')\\&=f(y')=(y'h'_0\xi,\xi)\qquad(y'\in\{s(\varphi)\mathcal{M}s(\varphi)\}').\end{aligned}$$

Hence, $v'v'^*h'_0\xi=h'_0\xi$, and so $v't_0\xi=h'_0\xi$. Therefore,

$$\begin{aligned}\tilde{\psi}(x)&=(xh'_0\xi,h'_0\xi)=(xv't_0\xi,v't_0\xi)\\&=(xv'^*v't_0\xi,t_0\xi)=(xv'^*h'_0\xi,t_0\xi)\\&=(xt_0\xi,t_0\xi)=(t_0xt_0\xi,\xi)=\tilde{\varphi}(t_0xt_0)\qquad(x\in s(\varphi)\mathcal{M}s(\varphi)).\end{aligned}$$

Now we have

$$\begin{aligned}\psi(x)&=\psi(s(\varphi)xs(\varphi))=\tilde{\psi}(s(\varphi)xs(\varphi))\\&=\tilde{\varphi}(t_0s(\varphi)xs(\varphi)t_0)=\varphi(t_0xt_0)\qquad(x\in\mathcal{M}).\end{aligned}$$

q.e.d.

1.24.4. Proposition. *Let $\mathcal{M}$ be a W^*-algebra, and let φ,ψ be two normal positive linear functionals on $\mathcal{M}$ such that $\psi\leq\varphi$. Then there exists a positive element h_0 in $\mathcal{M}$, with $0\leq h_0\leq 1$, such that*

$$\psi(x)=\tfrac{1}{2}\varphi(h_0x+xh_0)\qquad(x\in\mathcal{M}).$$

Proof. Let $A_1=\{h\in\mathcal{M}\mid\|h\|\leq 1,h^*=h\}$. A_1 is convex and $\sigma(\mathcal{M},\mathcal{M}_*)$-compact. Hence, under the mapping $h\to\frac{1}{2}(L_h\varphi+R_h\varphi)$, the image $\mathfrak{F}$ of A_1 is convex and $\sigma(\mathcal{M}_*,\mathcal{M})$-compact.

Suppose $\psi\notin\mathfrak{F}$; then $\psi^0\not\supseteq\mathfrak{F}^0$, where ψ^0 and $\mathfrak{F}^0$ are the polars of ψ and $\mathfrak{F}$ respectively in the self-adjoint portion of $\mathcal{M}$. Hence, there exists an element x_0 in $\mathcal{M}^s$ such that $|f(x_0)|\leq 1$ $(f\in\mathfrak{F})$ and $|\psi(x_0)|>1$.

On the other hand, let $x_0=x_0^+-x_0^-$. Then

$$\tfrac{1}{2}\varphi\{(e-(1-e))x_0+x_0(e-(1-e))\}=\varphi(x_0^++x_0^-)\geq\psi(x_0^++x_0^-),$$

where $e=s(x^+)$. Hence, $\psi(x_0^++x_0^-)\leq 1$. Similarly

$$\tfrac{1}{2}\varphi(((1-e)-e)x_0+x_0((1-e)-e))=-\varphi(x_0^++x_0^-)\leq-\psi(x_0^++x_0^-).$$

Hence, $-\psi(x_0^+ + x_0^-) \geq -1$. Therefore,

$$-1 \leq -\psi(x_0^+ + x_0^-) \leq \psi(x_0) = \psi(x_0^+) - \psi(x_0^-) \leq \varphi(x_0^+ + x_0^-) \leq 1.$$

Hence $|\psi(x_0)| \leq 1$, a contradiction. Hence

$$\psi(x) = \tfrac{1}{2}\varphi(h_1 x + x h_1) \qquad (x \in \mathscr{M} \text{ and some } h_1 \in A_1).$$

Moreover, since $\psi \geq 0$,

$$\begin{aligned}\psi(1) = \varphi(h_1^+ - h_1^-) &\geq \psi(p-(1-p)) = \tfrac{1}{2}\varphi\big(h_1(p-(1-p)) + (p-(1-p))h_1\big)\\ &= \varphi(h_1^+ + h_1^-),\end{aligned}$$

where $p = s(h_1^+)$. Hence $\psi(h_1^-) = 0$, so that

$$\psi(x) = \tfrac{1}{2}\varphi(h_1^+ x + x h_1^+) \qquad (x \in \mathscr{M}). \qquad \text{q.e.d.}$$

1.24.5. Proposition. *Let $\mathscr{M}$ be a W^*-algebra, $\mathscr{N}$ a W^*-subalgebra of $\mathscr{M}$. Let f be a $\sigma(\mathscr{N}, \mathscr{N}_*)$-continuous linear functional on $\mathscr{N}$. Then f can be extended to a $\sigma(\mathscr{M}, \mathscr{M}_*)$-continuous linear functional $\tilde{f}$ on $\mathscr{M}$ with $\|\tilde{f}\| = \|f\|$.*

Proof. First of all, suppose φ be a normal positive linear functional on $\mathscr{N}$. Let $\mathscr{N}^0$ be the polar of $\mathscr{N}$ in $\mathscr{M}_*$. Then $\mathscr{N}$ is the dual of $\mathscr{M}_*/\mathscr{N}^0$. Hence, there exists a self-adjoint element $g \in \mathscr{M}_*$ such that $g = \varphi$ on $\mathscr{N}$. Put $g = g^+ - g^-$ and $\varphi_1 = g^+ + g^-$. Then $\varphi \leq \varphi_1$ on $\mathscr{N}$. Hence by 1.24.3, there exists a positive element h, with $0 \leq h \leq 1$, in $\mathscr{N}$ such that $\varphi(y) = \varphi_1(hyh)$ $(y \in \mathscr{N})$. Set $\tilde{\varphi}(x) = \varphi_1(hxh)$ $(x \in \mathscr{M})$; then $\tilde{\varphi} \geq 0$ and $\tilde{\varphi} = \varphi$ on $\mathscr{N}$, and $\|\tilde{\varphi}\| = \tilde{\varphi}(1) = \varphi_1(h1h) = \varphi_1(hs(h)1s(h)h) = \varphi_1(heh) = \|\varphi\|$ where e is the identity of $\mathscr{N}$.

Next let $f = R_u|f|$ be the polar decomposition. Then there exists a normal positive linear functional φ' on $\mathscr{M}$ such that $|f| = \varphi'$ on $\mathscr{N}$ and $\||f|\| = \|\varphi'\|$. Put $\tilde{f} = R_u \varphi'$; then $\|\tilde{f}\| = \|R_u \varphi'\| = \|f\|$ and $\tilde{f} = f$ on $\mathscr{N}$. q.e.d.

Concluding remark on 1.24.

If $\mathscr{M}$ is a semi-finite algebra, we may have a more general Radon-Nikodym theorem (cf. [44], [184]).

References. [154], [158].

2. Classification of W^*-Algebras

2.1. Equivalence of Projections and the Comparability Theorem

2.1.1. Definition. *Let p, q be two projections of a W^*-algebra $\mathcal{M}$. If there exists a partial isometry u in $\mathcal{M}$ such that $u^*u=p$ and $uu^*=q$, then p is said to be equivalent to q and denote this by $p\sim q$. If there exists a projection q_1 ($\leq q$) equivalent to p, we write this by $p\prec q$ or $q\succ p$. We write $p\sim q$ or $p\prec q$ by $p\precsim q$.*

One can easily see that the relation "$\sim$" satisfies the conditions of equivalence, and the relation "$\prec$" is reflexive and transitive.

Let $p\prec q$, and let z be a central projection in $\mathcal{M}$. Then $pz\prec qz$. Let $(p_\alpha)_{\alpha\in\mathbb{I}}$ (resp. $(q_\alpha)_{\alpha\in\mathbb{I}}$) be a family of mutually orthogonal projections such that $p_\alpha\sim q_\alpha$ ($\alpha\in\mathbb{I}$). Then $\sum_{\alpha\in\mathbb{I}} p_\alpha$ is equivalent to $\sum_{\alpha\in\mathbb{I}} q_\alpha$. In fact, let u_α be a partial isometry giving the equivalence $p_\alpha\sim q_\alpha$; then, by using the $s(\mathcal{M},\mathcal{M}_*)$-topology, we can define an infinite sum $\sum_{\alpha\in\mathbb{I}} u_\alpha$ $(=u)$ and easily see that $u^*u=p$ and $uu^*=q$.

Clearly $p\sim q$ implies $p\prec q$ and $p\succ q$.

2.1.2. Proposition. *If $p\prec q$ and $p\succ q$, then $p\sim q$.*

Proof. We have $p\sim q_1\leq q$ and $q\sim p_1\leq p$. The relations $q_1\leq q$ and $q\sim p_1$ imply $q_1\sim p_2\leq p_1$. Hence $p\sim q_1\sim p_2$. Let u be a partial isometry giving $p\sim p_2$ and let $p^{(0)}=p=u^*u$, $p^{(1)}=p_1,\ldots,p^{(2n)}=u^n u^{n*}$ and $p^{(2n+1)}=\{(up_1)^n\}\{(up_1)^n\}^*$ $(n=1,2,\ldots)$. Then $p\geq p_1\geq p_2$ implies

$$upu^*=uu^*=p^{(2)}\geq up_1u^*=p^{(3)}\geq up_2u^*=p^{(4)}.$$

Analogously $p^{(2n)}\geq p^{(2n+1)}\geq p^{(2n+2)}$. Therefore the sequence $\{p^{(n)}\}$ is decreasing.

Put $e=\bigwedge_n p^{(n)}$. Then

$$\begin{array}{l} p=e+(p^{(0)}-p^{(1)})+(p^{(1)}-p^{(2)})+(p^{(2)}-p^{(3)})+\cdots; \\ \quad\ \ \| \qquad\quad\ \wr \qquad\qquad\qquad \| \qquad\qquad\qquad \wr \\ p_1=e+(p^{(2)}-p^{(3)})+(p^{(1)}-p^{(2)})+(p^{(4)}-p^{(5)})+\cdots. \end{array}$$

Hence $p\sim p_1\sim q$. q.e.d.

2.1.3. Comparability theorem. *Let p and q be projections in $\mathcal{M}$. Then there exists a central projection z in $\mathcal{M}$ such that $pz \succ qz$ and $pz' \prec qz'$, where $z' = 1 - z$.*

Proof. If $q\mathcal{M}p = (0)$, then $c(q)c(p) = 0$ by 1.10.7. Hence we take $c(p)$ as the z. If $q\mathcal{M}p \neq (0)$, we take an extreme point u in the unit sphere of $q\mathcal{M}p$. Then, by 1.6.5, u is a partial isometry and $(q-f)\mathcal{M}(p-e) = (0)$, where $u^*u = e$ and $uu^* = f$. Let z be the central support of $p-e$. Then $q - f \leq z'$ and $p - e \leq z$. Hence $zp = ze + z(p-e) \succ zf = zq$ and

$$z'p = z'e \prec z'f + z'(q-f) = z'q. \qquad \text{q.e.d.}$$

2.1.4. Proposition. *Let $a \in \mathcal{M}$, and let p (resp. q) be the left (resp. right) support of a. Then $p \sim q$.*

Proof. Let $a = v|a|$ be the polar decomposition of a. Then $v^*v = s(|a|) = p$ and $vv^* = s(|a^*|) = q$. q.e.d.

2.1.5. Proposition. *Let p, q be two projections in $\mathcal{M}$. Then*

$$p \vee q - q \sim p - q \wedge p.$$

Proof. Let $p' = 1 - p$ and $q' = 1 - q$. Then

$$s(|pq'|) = s(q'pq') = 1 - (q + p' \wedge q').$$

Since $(p \vee q) + (p' \wedge q') = 1$, $1 - (q + q' \wedge p') = (p \vee q) - q$. On the other hand, $s(|q'p|) = 1 - (p' + p \wedge q) = p - p \wedge q$. Hence $(p \vee q) - q \sim p - p \wedge q$. q.e.d.

2.1.6. Definition. *A W^*-algebra $\mathcal{M}$ is called a factor if its center consists of the scalar multiplies of identity only.*

2.1.7. Corollary. *Let $\mathcal{M}$ be a factor, and let p, q be projections in $\mathcal{M}$. Then $p \prec q$ or $p \succ q$.*

2.1.8. Definition. *A W^*-algebra $\mathcal{M}$ is said to be countably decomposable if every family of mutually orthogonal non-zero projections in $\mathcal{M}$ is at most countable. A projection p in $\mathcal{M}$ is said to be countably decomposable if $p\mathcal{M}p$ is countably decomposable.*

2.1.9. Proposition. *Let $\mathcal{M}$ be a W^*-algebra with the separable predual $\mathcal{M}_*$. Then $\mathcal{M}$ is countably decomposable.*

Proof. Let $(\varphi_n | n = 1, 2, \ldots)$ be a countable family of normal states which is norm dense in the set of all normal states on $\mathcal{M}$. Let $\varphi(a) = \sum_{n=1}^{\infty} \frac{1}{2^n} \varphi_n(a)$ $(a \in \mathcal{M})$. φ is faithful on $\mathcal{M}$. Let $(e_\alpha)_{\alpha \in \mathbb{I}}$ be a family of mutually orthogonal non-zero projections in $\mathscr{M}$. Then

$$\varphi\left(\sum_{\alpha\in\mathbb{I}} e_\alpha\right) = \sum_{\alpha\in\mathbb{I}} \varphi(e_\alpha) < \infty.$$

Hence $\varphi(e_\alpha)=0$ for all $\alpha\in\mathbb{I}$, except for a countable number of indices α. q.e.d.

2.1.10. Proposition. *Let $\mathscr{N}$ be a weakly closed self-adjoint algebra on a separable Hilbert space $\mathscr{H}$. Then $\mathscr{N}_*$ is separable.*

Proof. Let $(\xi_n | n=1,2,\ldots)$ be a set of elements which is dense in $\mathscr{H}$ and set $f_{m,n}(a)=(a\xi_m,\xi_n)$ $(a\in\mathscr{N})$. Then $\{f_{m,n}|m,n=1,2,\ldots\}$ is total in $\mathscr{N}_*$—i.e., $f_{m,n}(x)=0$ $(x\in\mathscr{N}; m,n=1,2,3,\ldots)$ imply $x=0$. Hence $\mathscr{N}_*$ is separable. q.e.d.

Let $\mathscr{M}$ be a W^*-algebra, and let a be a positive element in $\mathscr{M}$. For an arbitrary non-zero projection e in $\mathscr{M}$, set $M_e(a)=\|eae\|$,

$$m_e(a)=\|a\|-\|e(\|a\|1-a)e\|$$

and $\omega_e(a)=M_e(a)-m_e(a)$ $(a\in\mathscr{M})$. When $e=1$, we simply write $M(a)$, $m(a)$, $\omega(a)$. Further, define $\omega_0(a)=0$. Let $\mathfrak{F}$ be a family of projections in $\mathscr{M}$ commuting with a. We set $\omega_{\mathfrak{F}}(a)=\sup_{e\in\mathfrak{F}}\omega_e(a)$.

2.1.11. Lemma. *Let $\mathscr{M}$ be a W^*-algebra, Z its center, and let a be a positive element of $\mathscr{M}$. Then there exists a central projection z and a unitary element in $\mathscr{M}$ such that $\omega_z(\frac{1}{2}(a+u^*au))\leq\frac{3}{4}\omega(a)$* and

$$\omega_{z'}(\tfrac{1}{2}(a+u^*au))\leq\tfrac{3}{4}\omega(a),$$

where $z'=1-z$.

Proof. Let $n(a)=\frac{1}{2}(M(a)+m(a))$. There exist spectral projections p, q of a such that $p+q=1: M_p(a)\leq n(a)$ and $m_q(a)\geq n(a)$. Let z be a central projection such that $pz\prec qz$ and $pz'\succ qz'$ $(z'=1-z)$. Let v (resp. w) be a partial isometry giving the equivalence $pz\sim q_1\leq qz$ (resp. $qz'\sim p_1\leq pz'$). Set $u=v+v^*+w+w^*+(1-pz-q_1-qz'-p_1)$. Then u is unitary. Moreover,

$$az\geq m(a)pz+n(a)qz=m(a)pz+n(a)q_1+n(a)(qz-q_1).$$

Hence,

$$(u^*au)z\geq m(a)q_1+n(a)pz+n(a)(qz-q_1),$$

and so

$$\begin{aligned}\tfrac{1}{2}(a+u^*au)z &\geq \tfrac{1}{2}(m(a)+n(a))(pz+q_1)+n(a)(qz-q_1)\\ &\geq \tfrac{1}{2}(m(a)+n(a))(pz+q_1+qz-q_1)=(M(a)-\tfrac{3}{4}\omega(a))z.\end{aligned}$$

On the other hand, it is clear that $\frac{1}{2}(a+u^*au)z\leq M(a)z$, and so $\omega_z(\frac{1}{2}(a+u^*au))\leq\frac{3}{4}\omega(a)$. Similarly, we have $\omega_{z'}(\frac{1}{2}(a+u^*au))\leq\frac{3}{4}\omega(a)$. q.e.d.

2.1.12. Lemma. *Let $a\ (\geq 0) \in \mathcal{M}$ and let $\mathfrak{F} = \{z_1, z_2, \ldots, z_n\}$ be a finite family of mutually orthogonal central projections, with the sum* 1, *in $\mathcal{M}$. Then there exist a finite family $\mathfrak{F}' = \{c_1, c_2, \ldots, c_m\}$ of mutually orthogonal central projections, with the sum* 1, *and a unitary element u in $\mathcal{M}$ such that $\omega_{\mathfrak{F}'}(\frac{1}{2}(a + u^* a u)) \leq \frac{3}{4} \omega_{\mathfrak{F}}(a)$.*

Proof. Since $\mathcal{M} = \sum_{i=1}^{n} \oplus \mathcal{M} z_i$, the proof can be reduced to the case of $\mathfrak{F} = \{1\}$. Hence by 2.1.11, there exists a family $\mathfrak{F}$ satisfying the above conditions. q.e.d.

Let x be an element of $\mathcal{M}$, and let $K(x)$ be the uniformly closed convex subset of $\mathcal{M}$ generated by $\{u^* x u \mid u \in \mathcal{M}^u\}$ ($\mathcal{M}^u$ the group of all unitary elements in $\mathcal{M}$). Let C be the set of all real valued functions on $\mathcal{M}^u$ such that $f(u) \geq 0\ (u \in \mathcal{M}^u)$, $f(u) = 0$ for all $u \in \mathcal{M}^u$, except for a finite number of points, and $\sum_{u \in \mathcal{M}^u} f(u) = 1$.

For $f \in C$ and $x \in \mathcal{M}$, set $f \cdot x = \sum_{u \in \mathcal{M}^u} f(u) u x u^*$. Then for $g, f \in C$, we have

$$\begin{aligned} g \cdot (f \cdot x) &= \sum_{u \in \mathcal{M}^u} g(u) u (f \cdot x) u^* = \sum_u g(u) u \Big(\sum_v f(v) v x v^* \Big) u^* \\ &= \sum_{u,v} g(u) f(v) (uv) x (uv)^* = \sum \Big(\sum_{w = u \cdot v} g(u) f(v) \Big) w x w^* \\ &= \sum_w \sum_u g(u) f(u^{-1} w) w x w^* = g * f \cdot x, \end{aligned}$$

where $g * f$ is the convolution multiplication of g and f. Hence $g * f \in C$.

2.1.13. Lemma. *Let a be a positive element of $\mathcal{M}$ and $\varepsilon > 0$. Then there exist an $f \in C$ and a central element c in $\mathcal{M}$ such that $\|f \cdot a - c\| < \varepsilon$.*

Proof. For an arbitrary positive integer r, there exists a finite family $\mathfrak{F} = \{z_{1,r}, z_{2,r}, \ldots, z_{n_r,r}\}$ of mutually orthogonal central projections with the sum 1 and an $f_r \in C$ such that $\omega_{\mathfrak{F}}(f_r \cdot a) \leq (\frac{3}{4})^r \omega(a)$ by 2.1.12. Hence, we can choose a family $\{\mu_{1,r}, \ldots, \mu_{n_r,r}\}$ of real numbers such that

$$\Big\| f_r \cdot a - \sum_{i=1}^{n_r} \mu_{i,r} z_{i,r} \Big\| \leq (\tfrac{3}{4})^r \omega(a). \qquad \text{q.e.d.}$$

2.1.14. Lemma. *Let $a_1, a_2, \ldots, a_n$ be a finite family of positive elements in $\mathcal{M}$ and $\varepsilon > 0$. Then there exist an $f \in C$ and central elements $c_1, c_2, \ldots, c_n$ in $\mathcal{M}$ such that $\|f \cdot a_i - c_i\| < \varepsilon\ (1 \leq i \leq n)$.*

Proof. Suppose that this is true for n. Take a $g \in C$ and a central element c_{n+1} such that $\|g * f \cdot a_{n+1} - c_{n+1}\| < \varepsilon$. Then

$$\|g * f \cdot a_i - c_i\| < \varepsilon \qquad (1 \leq i \leq n). \qquad \text{q.e.d.}$$

2.1.15. Lemma. *Let* $(a_i|i=1,2,\ldots)$ *be a sequence of positive elements in* $\mathcal{M}$. *Then there exist a sequence* $(c_i|i=1,2,\ldots)$ *of central elements in* $\mathcal{M}$ *and a sequence* $(f_j|j=1,2,\ldots)$ *in* C *such that*

$$\|f_j \cdot a_i - c_i\| \to 0 \qquad (j\to\infty) \qquad (i=1,2,\ldots).$$

Proof. For each positive integer j, take a $g_j \in C$ and central elements $c_{1,j}, c_{2,j}, \ldots, c_{n,j}$ such that $\|g_j \cdot a_i - c_{i,j}\| < 2^{-j}$ $(1\leq i\leq j)$.

Set $f_j = g_j * g_{j-1} * \cdots * g_1$. Then,

$$\|f_{j+1}\cdot a_i - c_{i,j}\| = \|g_{j+1}(f_j\cdot a_i) - g_{j+1}\cdot c_{i,j}\| \leq \|f_j\cdot a_i - c_{i,j}\| \leq 2^{-j} \qquad (1\leq i\leq j).$$

Hence $\|f_{j+1}\cdot a_i - f_j\cdot a_i\| \leq 2^{-j+1}$. Therefore $(f_j\cdot a_i)$ is a Cauchy sequence for each i, and so $(c_{i,j})$ is also Cauchy for each i.

Let $c_i = \lim_{j\to\infty} c_{i,j}$. Then $\|f_j\cdot a_i - c_i\| \to 0$ $(j\to\infty)$ $(i=1,2,\ldots)$. q.e.d.

2.1.16. Theorem. *Let* x *be an element of a* W^**-algebra* $\mathcal{M}$, *and let* $K(x)$ *be the uniformly closed convex subset of* $\mathcal{M}$ *generated by* $\{u^* x u | u \in \mathcal{M}^u\}$, *where* $\mathcal{M}^u$ *is the group of all unitary elements in* $\mathcal{M}$. *Then* $K(x) \cap Z \neq (\emptyset)$, *where* Z *is the center of* $\mathcal{M}$.

Proof. Let $x = x_1 - x_2 + i x_3 - i x_4$ $(x_1, x_2, x_3, x_4 \geq 0)$. By 2.1.15, there exist a sequence (f_j) in C and central elements c_1, c_2, c_3, c_4 in $\mathcal{M}$ such that $\|f_j\cdot x_i - c_i\| \to 0$ $(j\to\infty)$ $(i=1,2,3,4)$. Hence $c_1 - c_2 + i c_3 - i c_4 \in K(x) \cap Z$. q.e.d.

References. [29], [94].

2.2. Classification of W^*-Algebras

2.2.1. Definition. *Let* p *be a projection of a* W^**-algebra* $\mathcal{M}$. p *is said to be finite if for a projection* p_1 *in* $\mathcal{M}$, $p_1 \leq p$ *and* $p_1 \sim p$ *imply* $p_1 = p$; p *is said to be purely infinite if it does not contain any non-zero finite projection;* p *is said to be infinite if it is not finite.*

One can easily see that along with a finite projection, all smaller ones are also finite. A W^*-algebra is said to be finite if its identity is finite; a W^*-algebra is said to be purely infinite if its identity is purely infinite.

Let $\{z_\alpha\}_{\alpha\in\mathbb{I}}$ be a set of mutually orthogonal finite central projections of a W^*-algebra $\mathcal{M}$ and let $z = \sum_{\alpha\in\mathbb{I}} z_\alpha$. Suppose $z \sim p \leq z$. Then $p z_\alpha \sim z_\alpha$. Hence $p z_\alpha = z_\alpha$ $(\alpha\in\mathbb{I})$, so that $p = z$. Hence z is also finite. Therefore there exists a unique maximal finite central projection z_1 in $\mathcal{M}$. Similarly, one can see that there exists a unique maximal purely infinite central projection z_3 in $\mathcal{M}$.

Set $z_2=1-z_1-z_3$. If $z_2\neq 0$, any non-zero central projection z $(\leq z_2)$ contains a non-zero finite projection and z is infinite.

2.2.2. Definition. *$\mathcal{M}$ is said to be semi-finite if $z_3=0$; properly infinite if $z_1=0$; properly infinite and semi-finite if $z_1=z_3=0$.*

2.2.3. Theorem. *Any W^*-algebra $\mathcal{M}$ is uniquely decomposed into a direct sum of three algebras which are finite, properly infinite and semi-finite; and purely infinite, respectively.*

2.2.4. Proposition. *Let $\mathcal{M}$ be a properly infinite W^*-algebra. Then there exists a sequence (p_n) of mutually orthogonal, equivalent projections in $\mathcal{M}$ with $p_n\sim 1$ and $\sum_{n=1}^{\infty} p_n=1$.*

Proof. It is enough to show that there exists a non-zero central projection z in $\mathcal{M}$ such that there exists a sequence (p_n) of mutually orthogonal, equivalent projections in $\mathcal{M}z$ with $p_n\sim z$ and $\sum_{n=1}^{\infty} p_n=z$.

Take a projection $q_1<1$ with $q_1\sim 1$, and let v be a partial isometry with $v^*v=1$ and $vv^*=q_1$. Then $v^{*n}v^n=1$, and so $q_n=v^n v^{*n}$ is a projection for each n and $q_n\geq q_{n+1}$. Put $r_n=q_n-q_{n+1}$. Then

$$(v^n r_1)^*(v^n r_1)=r_1(v^{n*}v^n)r_1=r_1$$

and

$$(v^n r_1)(v^n r_1)^*=v^n(q_1-q_2)v^{n*}=q_{n+1}-q_{n+2}=r_{n+1} \qquad (n=1,2,\ldots).$$

Hence $0\neq r_1\sim r_2\sim\cdots$ and (r_n) is a family of mutually orthogonal projections.

Now let $(q_\alpha)_{\alpha\in\mathbb{I}}$ be a maximal family of mutually orthogonal, equivalent projections such that $(r_n)\subset(q_\alpha)_{\alpha\in\mathbb{I}}$. Then by the comparability theorem, there exists a central projection z in $\mathcal{M}$ such that

$$\Big(1-\sum_{\alpha\in\mathbb{I}} q_\alpha\Big)z\precsim r_1 z\neq 0.$$

Put $q_0=\Big(1-\sum_{\alpha\in\mathbb{I}} q_\alpha\Big)$; then $z=q_0z+\sum_{\alpha\in\mathbb{I}} q_\alpha z$. Since $\mathrm{Card}(\mathbb{I})\geq\aleph_0$, we can divide the set $\mathbb{I}$ as follows: $\mathbb{I}=\bigcup_{j=1}^{\infty}\mathbb{I}_j$, $\mathbb{I}_j\cap\mathbb{I}_k=(\emptyset)$ $(j\neq k)$ and

$$\mathrm{Card}(\mathbb{I}_j)=\mathrm{Card}(\mathbb{I}) \qquad (j=1,2,\ldots).$$

Put $e_1=q_0z+\sum_{\alpha\in\mathbb{I}_1} q_\alpha z$ and $e_j=\sum_{\alpha\in\mathbb{I}_j} q_\alpha z$ $(j=2,\ldots)$. Then clearly

$e_j\sim z$ $(j=1,2,\ldots)$ and $\sum_{j=1}^{\infty} e_j=z$. q.e.d.

2.2.5. Proposition. *Let $\mathcal{M}$ be a W^*-algebra. Then $\mathcal{M}$ can be written as follows: $\mathcal{M}=\sum_{\alpha\in\mathbb{I}}\oplus\mathcal{M}_\alpha$, where $\mathcal{M}_\alpha$ is countably decomposable or it contains an uncountable family $(e_\beta|\beta\in\mathbb{J})$ of mutually orthogonal, equivalent, countably decomposable projections in $\mathcal{M}_\alpha$ with $\sum_{\beta\in\mathbb{J}} e_\beta=1_\alpha$ (1_α is the identity of $\mathcal{M}_\alpha$).*

Proof. For a normal state φ on $\mathcal{M}$, the support $s(\varphi)$ is countably decomposable. Let $(p_\gamma)_{\gamma\in\mathbb{K}}$ be a maximal family of mutually orthogonal, equivalent projections, with $p_\gamma\sim s(\varphi)$ ($\gamma\in\mathbb{K}$), in $\mathcal{M}$. Then there exists a central projection z in $\mathcal{M}$ such that $q=\left(1-\sum_{\gamma\in\mathbb{K}} p_\gamma\right)z\prec s(\varphi)z\neq 0$. Let $\{v_\beta|\beta\in(0)\cup\mathbb{K}\}$ be a family of partial isometries in $\mathcal{M}$ with $v_0^*v_0\leq s(\varphi)z$, $v_0v_0^*=q$, $v_\gamma^*v_\gamma=s(\varphi)z$ and $v_\gamma v_\gamma^*=p_\gamma z$ ($\gamma\in\mathbb{K}$). Put

$$\varphi_\beta(x)=\varphi(v_\beta^* x v_\beta)\qquad (x\in\mathcal{M}z,\ \beta\in(0)\cup\mathbb{K}).$$

If $\mathrm{Card}(\mathbb{K})\leq\aleph_0$, set $(0)\cup\mathbb{K}=\{\beta_1,\beta_2,\dots\}$ and define

$$\psi(x)=\sum_{n=1}^{\infty}\frac{1}{2^n}\varphi_{\beta_n}(u_{\beta_n}^* x v_{\beta_n})\qquad (x\in\mathcal{M}z).$$

Then one can easily see that ψ is faithful on $\mathcal{M}z$, so that $\mathcal{M}z$ is countably decomposable. If $\mathrm{Card}(\mathbb{K})>\aleph_0$, we divide $\mathbb{K}$ as follows: $\mathbb{K}=\bigcup_{j\in\mathbb{H}}\mathbb{K}_j$, $\mathbb{K}_{j_1}\cap\mathbb{K}_{j_2}=(\emptyset)$ ($j_i\neq j_2$), $\mathrm{Card}(\mathbb{K}_j)=\aleph_0$. Then $\mathrm{Card}(\mathbb{K})=\mathrm{Card}(\mathbb{H})$. Put $q_j=\sum_{\beta\in\mathbb{K}_j} p_\beta z$ ($j\neq$ some fixed j_0) and $q_{j_0}=q+\sum_{\beta\in\mathbb{K}_{j_0}} p_\beta z$. Then $(q_j|j\in\mathbb{H})$ is a uncountable family of mutually orthogonal, equivalent, countably decomposable projections with $\sum_{j\in\mathbb{H}} q_j=z$. q.e.d.

2.2.6. Definition. *A non-zero projection p in a W^*-algebra $\mathcal{M}$ is said to be abelian if $p\mathcal{M}p$ is commutative.*

2.2.7. Proposition. *Let p be an abelian projection of a W^*-algebra $\mathcal{M}$ and let q be a projection of $\mathcal{M}$ such that $p\sim q$. Then q is again abelian.*

Proof. Let $u^*u=p$ and $uu^*=q$. Then,

$$\begin{aligned} qxqqyq &= uu^*xuu^*yuu^* = upu^*xupu^*yupu^* \\ &= u(pu^*yup)(pu^*xup)u^* = qyqqxq \end{aligned}$$

$(x,y\in\mathcal{M})$. q.e.d.

2.2.8. Proposition. *An abelian projection is finite.*

Proof. Let $p_1\sim p$, $p_1\leq p$, and let $u^*u=p_1$, $uu^*=p$. Then $pup_1=u$, and so $u,u^*\in p\mathcal{M}p$. Hence $u^*u=uu^*$. q.e.d.

2.2.9. Definition. *A W^*-algebra is said to be of type* I *(or discrete) if every non-zero central projection contains an abelian projection; type* II *if it is semi-finite and does not contain any abelian projection; type* III *if it is purely infinite. A W^*-algebra having no discrete W^*-algebra as a direct summand is said to be continuous. A finite type* II *W^*-algebra is said to be of type* II_1. *A properly infinite type* II *W^*-algebra is said to be of type* II_∞.

Since types I, II, and III are mutually exclusive and moreover they are preserved under the taking of sums of mutually orthogonal central projections, we have

2.2.10. Proposition. *Any W^*-algebra is uniquely decomposed into a direct sum of three algebras, respectively, of types* I, II, *and* III.

2.2.11. Proposition. *Let p be a projection of a W^*-algebra $\mathcal{M}$. Then the center of the W^*-algebra $p\mathcal{M}p$ is Zp, where Z is the center of $\mathcal{M}$.*

Proof. Let e be a central projection in $p\mathcal{M}p$. Then

$$(p-e)p\mathcal{M}pe=(p-e)\mathcal{M}e=0.$$

Hence $(p-e)c(e)=0$, and so $pc(e)=ec(e)=e\in Zp$. q.e.d.

2.2.12. Corollary. *Let p be an abelian projection in $\mathcal{M}$. Then $p\mathcal{M}p=Zp$ (Z the center of $\mathcal{M}$).*

2.2.13. Proposition. *Let $\mathcal{M}$ be a W^*-algebra with no abelian projections and let p be a projection in $\mathcal{M}$. Then p can be written as follows: $p=f+g$ with f, g mutually orthogonal, equivalent projections.*

Proof. Consider a maximal set of projections $\{f_\alpha\}$, $\{g_\alpha\}$ which are mutually orthogonal, $\leq p$ and with $f_\alpha \sim g_\alpha$. Let $f=\sum_\alpha f_\alpha$ and $g=\sum_\alpha g_\alpha$.

Then f and g are mutually orthogonal and equivalent.

If $q=p-f-g$ is not zero, we observe that, by hypothesis, $q\mathcal{M}q$ is not commutative, and so there exists a projection $q_1 \leq q$ such that q_1 is not in the center of $q\mathcal{M}q$. Hence $q_1\mathcal{M}(q-q_1)\neq(0)$. Take an extreme point u of the unit sphere of $q_1\mathcal{M}(q-q_1)$. Then $0\neq uu^*\leq q_1$ and $u^*u\leq q-q_1$. This contradicts the maximality. Hence $f+g=p$ and $f\sim g$. q.e.d.

2.2.14. Proposition. *Let $\mathcal{M}$ be a countably decomposable type* III *W^*-algebra, and let p, q be projections such that $c(p)=c(q)$. Then $p\sim q$.*

Proof. By using the comparability theorem, we may assume $p\neq 0$ and $p\prec q$. Let z_0 be a maximal central projection such that $pz_0\succ qz_0$.

Suppose $z_0 \neq 1$ and let $(q_\alpha z'_0)_{\alpha \in \mathrm{I\!I}}$ $(z'_0 = 1 - z_0)$ be a maximal family of mutually orthogonal, equivalent projections such that $q_\alpha z'_0 \sim p z'_0$, $q_\alpha z'_0 \leq q z'_0$. Then there exists a non-zero central projection z, such that $\left(q z'_0 - \sum_{\alpha \in \mathrm{I\!I}} q_\alpha z'_0\right) z \prec p z'_0 z \neq 0$. Since $\mathrm{Card}(\mathrm{I\!I}) \leq \aleph_0$,

$$\sum_{\alpha \in \mathrm{I\!I}} q_\alpha z'_0 z + \left(q - \sum_{\alpha \in \mathrm{I\!I}} q_\alpha\right) z'_0 z \prec p z'_0 z$$

by 2.2.4. Hence $q z'_0 z \prec p z'_0 z \neq 0$ a contradiction, and so $z_0 = 1$. q.e.d.

2.2.15. An example. *Let $\mathscr{N}$ be a commutative W^*-algebra and let $\mathscr{H}$ be an n-dimensional Hilbert space ($n < +\infty$). Then the W^*-algebra $\mathscr{N} \otimes B(\mathscr{H})$ is a finite type* I *W^*-algebra.*

Proof. $\mathscr{N} \otimes B(\mathscr{H})$ is $*$-isomorphic to the algebra $\mathscr{N}_n$ of all $n \times n$ matrices over $\mathscr{N}$. Let $a = (a_{ij}) \in \mathscr{N}_n$, and set $\Phi(a) = \frac{1}{n} \sum_{i=1}^{n} a_{ii}$. Then Φ is a $\mathscr{N}$-valued linear functional on $\mathscr{N}_n$.

One can easily see that Φ satisfies the following conditions:

1. $\Phi(a^* a) \geq 0$, and $\Phi(a^* a) = 0$ if and only if $a = 0$;
2. $\Phi(a^* a) = \Phi(a a^*)$;
3. $\Phi(z a) = z \Phi(a)$ $(a \in \mathscr{N}_n, z \in Z)$ where Z is the center on $\mathscr{N}_n$.

Now suppose $q \sim p$ and $q \leq p$, where p and q are projections in $\mathscr{N}_n$. Then there exists an element u in $\mathscr{N}_n$ such that $u^* u = q$ and $u u^* = p$. Hence $\Phi(q) = \Phi(p)$, and so $\Phi(p - q) = 0$. Hence $p = q$. Therefore, $\mathscr{N}_n$ is finite.

$\mathscr{N} \otimes 1_{\mathscr{H}}$ is the center of $\mathscr{N} \otimes B(\mathscr{H})$. Let e be a one-dimensional projection on $\mathscr{H}$. Then $(1 \otimes e) \mathscr{N} \otimes B(\mathscr{H}) (1 \otimes e) = \mathscr{N} \otimes e$. Hence $1 \otimes e$ is an abelian projection. Clearly $c(1 \otimes e) = 1 \otimes 1_{\mathscr{H}}$, and so $\mathscr{N} \otimes B(\mathscr{H})$ is of type I. q.e.d.

References. [29], [94].

2.3. Type I W^*-Algebras

Let $\mathscr{M}$ be a type I W^*-algebra, and let z be a non-zero central projection of $\mathscr{M}$. Then there exists an abelian projection p $(\leq z)$. Let $(p_\alpha)_{\alpha \in \mathrm{I\!I}}$ be a maximal family of mutually orthogonal, equivalent projections such that $p_\alpha \sim p$ $(\alpha \in \mathrm{I\!I})$. Put $q = \sum_{\alpha \in \mathrm{I\!I}} p_\alpha$. It is impossible that $p \prec 1 - q$. Hence there exists a non-zero central projection z_1 such that $(1 - q) z_1 \sim q_1 \leq p z_1$. If $q_1 = 0$, $z_1 = \sum_{\alpha \in \mathrm{I\!I}} p_\alpha z_1$; otherwise $q_1 = c(q_1) p$ and

$c(q_1) = \sum_{\alpha \in \mathbb{I}} c(q_1) p_\alpha + (1-q) z_1$. Therefore, in any case, z contains a non-zero central projection z_0 such that z_0 is a sum of mutually orthogonal projections which are equivalent to $p z_0$.

Now let $(e_\alpha | \alpha \in \mathbb{J}_1)$ (resp. $(f_\beta | \beta \in \mathbb{J}_2)$) be a family of mutually orthogonal, equivalent abelian projections with the sum z_0. Suppose $\mathrm{Card}(\mathbb{J}_1) = n < +\infty$. Let $\{u_\alpha | \alpha \in \mathbb{J}_1\}$ be a family of partial isometries in $\mathcal{M} z_0$ such that $u_{\alpha_0} = e_{\alpha_0}$, $u_\alpha^* u_\alpha = e_{\alpha_0}$ and $u_\alpha u_\alpha^* = e_\alpha$ ($\alpha \in \mathbb{J}_1$; α_0 a fixed index in $\mathbb{J}_1$). Then $\mathcal{M} z_0 = \mathcal{N} \overline{\otimes} B(\mathcal{H}) = \mathcal{N} \otimes B(\mathcal{H})$ by 1.22.14 ($\dim(\mathcal{H}) = n$, $\mathcal{N}$ is *-isomorphic to $e_{\alpha_0} \mathcal{M} e_{\alpha_0}$). Hence, by 2.2.15, $\mathcal{M} z_0$ is finite and so $\mathrm{Card}(\mathbb{J}_2)$ also must be finite. Therefore, $\mathbb{J}_1$ and $\mathbb{J}_2$ are finite or infinite at the same time. By the comparability theorem, $e_\alpha z \prec f_\beta z$ and $e_\alpha z' \succ f_\beta z'$ for some central projection $z (z' = 1 - z)$.

Let $e_\alpha z \sim k \leq f_\beta z$. Since $k, f_\beta z - k \in f_\beta \mathcal{M} f_\beta = f_\beta Z f_\beta$ (Z the center of $\mathcal{M}$), $k \mathcal{M} (f_\beta z - k) = k f_\beta \mathcal{M} f_\beta (f_\beta z - k) = (0)$. Hence $c(k) c(f_\beta z - k) = 0$. Since $c(k) = c(e_\alpha z) = z z_0$, $f_\beta z = k$. Analogously $e_\alpha z' \sim f_\beta z'$, so that $e_\alpha \sim f_\beta$. Hence, if $\mathbb{J}_1$ and $\mathbb{J}_2$ are finite, $\mathrm{Card}(\mathbb{J}_1) = \mathrm{Card}(\mathbb{J}_2)$.

Next suppose $\mathbb{J}_1$ and $\mathbb{J}_2$ are infinite. Take a normal state φ_{α_0} on $e_{\alpha_0} \mathcal{M} e_{\alpha_0}$. Then $s(\varphi_{\alpha_0}) \leq e_{\alpha_0}$. Hence by the above consideration, $s(\varphi_{\alpha_0}) = e_{\alpha_0} z_{\alpha_0}$ for some central projection z_{α_0} ($\leq z_0$) of $\mathcal{M}$.

$$\sum_{\alpha \in \mathbb{J}_1} e_\alpha z_{\alpha_0} = \sum_{\beta \in \mathbb{J}_2} f_\beta z_{\alpha_0} = z_{\alpha_0}.$$

Let $(v_\alpha | \alpha \in \mathbb{J}_1)$ be a family of partial isometries such that $v_{\alpha_0} = e_{\alpha_0}$, $v_\alpha^* v_\alpha = e_{\alpha_0}$ and $v_\alpha v_\alpha^* = e_\alpha$ ($\alpha \in \mathbb{J}_1$). Put $\varphi_\alpha(x) = \varphi_{\alpha_0}(v_\alpha^* x v_\alpha)$ ($x \in e_\alpha \mathcal{M} e_\alpha$). Let $\mathbb{J}_2^\alpha$ be a set composed of all indices β ($\in \mathbb{J}_2$) such that $f_\beta e_\alpha z_{\alpha_0} \neq 0$. Since $\varphi_\alpha(e_\alpha z_{\alpha_0} e_\alpha) = \sum_{\beta \in \mathbb{J}_2} \varphi_\alpha(e_\alpha f_\beta z_{\alpha_0} e_\alpha)$ and $s(\varphi_\alpha) = e_\alpha z_{\alpha_0}$, $\mathbb{J}_2^\alpha$ is countable. If $\beta \in \mathbb{J}_2 - \bigcup_{\alpha \in \mathbb{J}_1} \mathbb{J}_2^\alpha$, $f_\beta e_\alpha z_{\alpha_0} = 0$ for all α. Hence $f_\beta z_{\alpha_0} = 0$, and so $c(f_\beta) z_{\alpha_0} = z_{\alpha_0} = 0$, a contradiction. Hence $\mathbb{J}_2 = \bigcup_{\alpha \in \mathbb{J}_1} \mathbb{J}_2^\alpha$, and so $\mathrm{Card}(\mathbb{J}_2) \leq \mathrm{Card}(\mathbb{J}_1) \aleph_0 = \mathrm{Card}(\mathbb{J}_1)$. Similarly $\mathrm{Card}(\mathbb{J}_1) \leq \mathrm{Card}(\mathbb{J}_2)$, and so $\mathrm{Card}(\mathbb{J}_1) = \mathrm{Card}(\mathbb{J}_2) = n$.

2.3.1. Definition. *Let z be a central projection of a type* I *W^*-algebra $\mathcal{M}$. Suppose there exists a family $\{p_\alpha | \alpha \in \mathbb{I}\}$ of mutually orthogonal, equivalent abelian projections such that $z = \sum_{\alpha \in \mathbb{I}} p_\alpha$. Then the cardinal number n of $\mathbb{I}$ does not depend on a choice of $(p_\alpha | \alpha \in \mathbb{I})$. Such a central projection z is called a n-homogeneous central projection. If $z = 1$, $\mathcal{M}$ is said to be of type* I_n *(or n-homogeneous). A sum of mutually orthogonal, n-homogeneous central projections is also n-homogeneous.*

Therefore, we have

2.3.2. Theorem. *A type* I *W^*-algebra $\mathscr{M}$ can be decomposed uniquely into a direct sum of type* I_n *W^*-algebras $\mathscr{M}_n$ ($n\in\mathbb{K}$), where $\mathbb{K}$ is a family of mutually distinct cardinal numbers.*

2.3.3. Theorem. *Let $\mathscr{M}$ be a type* I_n *W^*-algebra (n a cardinal number), and let Z be the center of $\mathscr{M}$. Then $\mathscr{M}$ is *-isomorphic to $Z\bar{\otimes}B(\mathscr{H})$* ($\dim(\mathscr{H})=n$).

Proof. Let $\{e_\alpha|\alpha\in\mathbb{I}\}$ be a family of mutually orthogonal equivalent abelian projections in $\mathscr{M}$ such that $\sum_{\alpha\in\mathbb{I}} e_\alpha=1$. Let $\{v_\alpha|\alpha\in\mathbb{I}\}$ be a family of partial isometries in $\mathscr{M}$ such that $v_{\alpha_0}=e_{\alpha_0}$, $v_\alpha^* v_\alpha=e_{\alpha_0}$ and $v_\alpha v_\alpha^*=e_\alpha$ ($\alpha\in\mathbb{I}$, α_0 a fixed index in $\mathbb{I}$). Let $\mathscr{M}_1=\left\{x\,|\,x=\sum_{\alpha\in\mathbb{I}} v_\alpha x_{\alpha_0} v_\alpha^*, x_{\alpha_0}\in e_{\alpha_0}\mathscr{M}e_{\alpha_0}\right\}$. Since $e_{\alpha_0}\mathscr{M}e_{\alpha_0}=Ze_{\alpha_0}$, $x=\sum_{\alpha\in\mathbb{I}} v_\alpha z e_{\alpha_0} v_\alpha^*=\sum_{\alpha\in\mathbb{I}} z v_\alpha e_{\alpha_0} v_\alpha^*=\sum_{\alpha\in\mathbb{I}} z e_\alpha=z$ for some $z\in Z$. Hence $\mathscr{M}_1=Z$. By 1.22.14, $\mathscr{M}=Z\bar{\otimes}B(\mathscr{H})$ with $\dim(\mathscr{H})=n$. q.e.d.

2.3.4. Corollary. *Let $\mathscr{M}_1$ (resp. $\mathscr{M}_2$) be a type I_{n_1} (resp. I_{n_2}) W^*-algebra, and let Z_1 (resp. Z_2) be the center of $\mathscr{M}_1$ (resp. $\mathscr{M}_2$). Then $\mathscr{M}_1$ is *-isomorphic to $\mathscr{M}_2$ if and only if $n_1=n_2$ and Z_1 is *-isomorphic to Z_2.*

References. [94], [97], [98].

2.4. Finite W^*-Algebras

2.4.1. Proposition. *Let $\mathscr{M}$ be a finite W^*-algebra and let Z be the center of $\mathscr{M}$. Then $\mathscr{M}$ is countably decomposable if and only if Z is countably decomposable.*

Proof. Suppose that Z is countably decomposable. Then we shall show that $\mathscr{M}$ has a faithful normal state, and so it is countably decomposable.

Let φ be a normal state on $\mathscr{M}$, and let $\{e_1, e_2, \ldots, e_n\}$ be a maximal family of mutually orthogonal, equivalent projections in $\mathscr{M}$ with $e_i\sim s(\varphi)$ $(1\le i\le n)$. Then there exists a non-zero central projection z_0 such that $\left(1-\sum_{i=1}^n e_i\right)z_0\prec s(\varphi)z_0\neq 0$. Let $\{v_i|i=1,2,\ldots,n+1\}$ be a family of partial isometries such that $v_i^* v_i=s(\varphi)z_0$, $v_i v_i^*=e_i z_0$ $(1\le i\le n)$, and $v_{n+1}^* v_{n+1}\le s(\varphi)z_0$, $v_{n+1}v_{n+1}^*=\left(1-\sum_{i=1}^n e_i\right)z_0$. Put $e_{n+1}=1-\sum_{i=1}^n e_i$ and set $\psi(x)=\sum_{i=1}^{n+1}\varphi(v_i^* e_i x e_i v_i)$ $(x\in\mathscr{M})$. If $\psi(x^*x)=0$, $xe_iv_is(\varphi)=0$ $(1\leq i\leq n+1)$, and so $xv_iv_i^*=0$ $(1\leq i\leq n+1)$. Hence $xz_0=0$. Therefore ψ is faithful on $\mathscr{M}z_0$. Let $\{z_n|n=1,2,\ldots\}$ be a maximal family of mutually orthogonal, non-zero central projections such that for each n there exists a normal state φ_n which is faithful on $\mathscr{M}z_n$. Then by the above result, $\sum_{n=1}^{\infty} z_n=1$. Set $\xi(x)=\sum_{n=1}^{\infty}\frac{1}{2^n}\varphi_n(x)$ $(x\in\mathscr{M})$. Then clearly ξ is faithful on $\mathscr{M}$. The reverse part of the theorem is clear. q.e.d.

2.4.2. Proposition. *Let $\mathscr{M}$ be a finite W^*-algebra, and let p,p_1,q,q_1 be projections in $\mathscr{M}$ satisfying the following conditions: $p_1\leq p$, $q_1\leq q$, $p_1\sim q_1$ and $p\sim q$. Then $p-p_1$ is equivalent to $q-q_1$.*

Proof. By the comparability theorem, there exists a central projection z in $\mathscr{M}$ such that $(p-p_1)z\prec(q-q_1)z$ and

$$(p-p_1)z'\succ(q-q_1)z' \qquad (z'=1-z).$$

If $(p-p_1)z\precsim(q-q_1)z$, there exists a projection q_2 with

$$(p-p_1)z\sim q_2<(q-q_1)z.$$

Then $pz=(p-p_1)z+p_1z\sim q_2+q_1z<(q-q_1)z+q_1z=qz$. Hence $q_2+q_1z\sim qz$, a contradiction. Hence $(p-p_1)z\sim(q-q_1)z$. Analogously $(p-p_1)z'\sim(q-q_1)z'$, and so $p-p_1\sim q-q_1$. q.e.d.

2.4.3. Definition. *Let $\mathscr{A}$ be a C^*-algebra, and let φ be a positive linear functional on $\mathscr{A}$. φ is said to be tracial if $\varphi(a^*a)=\varphi(aa^*)$ $(a\in\mathscr{A})$.*

2.4.4. Theorem. *Let $\mathscr{M}$ be a finite W^*-algebra. Then $\mathscr{M}$ has at least one tracial state.*

Proof. We shall show first that for an arbitrary non-zero projection p of $\mathscr{M}$ and positive integer n, there exist a non-zero projection p_0 in $\mathscr{M}$ and a faithful normal state φ_0 on $p_0\mathscr{M}p_0$ such that $p_0\leq p$ and $\varphi_0(a^*a)\leq\left(1+\frac{1}{n}\right)\varphi_0(aa^*)$ $(a\in p_0\mathscr{M}p_0)$.

Let φ be a normal state on $\mathscr{M}$ with $s(\varphi)\leq p$. We may assume that $s(\varphi)=p$. If for arbitrary two projections e,f, with $e\sim f$, in $p\mathscr{M}p$, $\varphi(e)=\varphi(f)$, one can easily see that $\varphi(a^*a)=\varphi(aa^*)$ $(a\in p\mathscr{M}p)$. Hence we can take this φ as a φ_0. Otherwise, let $\{e_\alpha,f_\alpha|\alpha\in\mathbb{I}\}$ be a maximal family of projections in $p\mathscr{M}p$ such that $e_\alpha\sim f_\alpha$, $\varphi(e_\alpha)>\varphi(f_\alpha)$ $(\alpha\in\mathbb{I})$ and $(e_\alpha)_{\alpha\in\mathbb{I}}$ (resp. $(f_\alpha)_{\alpha\in\mathbb{I}}$) are mutually orthogonal.

Put $e_1=\sum_{\alpha\in\mathbb{I}}e_\alpha$ and $f_1=\sum_{\alpha\in\mathbb{I}}f_\alpha$. Since $\varphi(e_1)>\varphi(f_1)$, $f_1<p$. By 2.4.2, $p-e_1$ is equivalent to $p-f_1$, $\varphi(e)\leq\varphi(f)$ $(e\leq p-e_1, f\leq p-f_1$, and $e\sim f)$.

Set $\mu_0=\inf\mu$, where μ moves over all real numbers μ satisfying $\varphi(e)\leq\mu\varphi(f)$ for all e,f with $e\leq p-e_1$, $f\leq p-f_1$ and $e\sim f$. Clearly $\mu_0\geq\varphi(p-e_1)>0$. For $\varepsilon>0$ $(\mu_0-\varepsilon>0)$, there exist non-zero projections e_2 $(\leq p-e_1)$ and f_2 $(\leq p-f_1)$ in $p\mathscr{M}p$ such that $(\mu_0-\varepsilon)\varphi(f_2)\leq\varphi(e_2)$ and $e_2\sim f_2$.

We shall show that there exist non-zero projections e_3 $(\leq e_2)$ and f_3 $(\leq f_2)$, with $e_3\sim f_3$, in $p\mathscr{M}p$ such that for arbitrary two projections e,f with $e\leq e_3$, $f\leq f_3$ and $e\sim f$, in $\mathscr{M}$, $(\mu_0-\varepsilon)\varphi(f)\leq\varphi(e)$. In fact, if this is not true, there exists a family $\{e_\beta,f_\beta|\beta\in\mathbb{J}\}$ of projections in $p\mathscr{M}p$ such that $e_\beta\leq e_2$, $f_\beta\leq f_2$, $e_\beta\sim f_\beta$ $(\beta\in\mathbb{J})$, $(e_\beta)_{\beta\in\mathbb{J}}$ (resp. $(f_\beta)_{\beta\in\mathbb{J}}$) are mutually orthogonal, $\sum_{\beta\in\mathbb{I}} e_\beta=e_2$ $\left(\text{resp. } \sum_{\beta\in\mathbb{I}} f_\beta=f_2\right)$, and

$$(\mu_0-\varepsilon)\varphi(f_\beta)>\varphi(e_\beta).$$

Hence $(\mu_0-\varepsilon)\varphi(f_2)>\varphi(e_2)$, a contradiction.

Let v be a partial isometry with $v^*v=f_3$ and $vv^*=e_3$. Set $\psi(x)=\varphi(vxv^*)$ $(x\in f_3\mathscr{M}f_3)$. Let r, s be arbitrary two equivalent projections in $f_3\mathscr{M}f_3$. Then

$$(\mu_0-\varepsilon)\varphi(r)\leq\varphi(vsv^*)=\psi(s)\leq\mu_0\varphi(r).$$

By putting $r=s$, we have

$$(\mu_0-\varepsilon)\varphi(r)\leq\psi(r)\leq\mu_0\varphi(r).$$

Hence $\psi(s)\leq\mu_0\varphi(r)\leq\dfrac{\mu_0}{\mu_0-\varepsilon}\psi(r)$, and so

$$\psi(x^*x)\leq\frac{\mu_0}{\mu_0-\varepsilon}\psi(xx^*)\qquad(x\in f_3\mathscr{M}f_3).$$

Take ε such as $\dfrac{\mu_0}{\mu_0-\varepsilon}\leq 1+\dfrac{1}{n}$ and set $\xi(x)=\dfrac{\psi(x)}{\psi(f_3)}$ $(x\in f_3\mathscr{M}f_3)$. Since $\psi(f_3)\geq(\mu_0-\varepsilon)\varphi(f_3)>0$, ξ is well-defined and is a normal state on $f_3\mathscr{M}f_3$. Since φ is faithful on $f_3\mathscr{M}f_3$, ψ is again faithful on $f_3\mathscr{M}f_3$. By putting $f_3=p_0$, $\xi=\varphi_0$, we have the required result.

Now let $\{p_i|i=1,2,\dots,m\}$ be a maximal family of mutually orthogonal projections in $\mathscr{M}$ such that $p_i\sim p_0$ $(1\leq i\leq m)$. Then there exists a non-zero central projection z_0 such that $\left(1-\sum_{i=1}^m p_i\right)z_0\prec p_0z_0\neq 0$. Let $\{v_i|i=1,2,\dots,m+1\}$ be a family of partial isometries with $v_i^*v_i=p_0z_0$ $v_iv_i^*=p_iz_0$ $(1\leq i\leq m)$, and $v_{m+1}^*v_{m+1}\leq p_0z_0$ and $v_{m+1}v_{m+1}^*=\left(1-\sum_{i=1}^m p_i\right)z_0$.

Put $\varphi_n(x) = \sum_{i=1}^{m+1} \varphi_0(v_i^* x v_i)$ $(x \in \mathcal{M})$. Then

$$\begin{aligned}\varphi_n(x^* x) &= \sum_{i=1}^{m+1} \varphi_0(v_i^* x^* x v_i)\\ &= \sum_{i=1}^{m+1} \sum_{j=1}^{m+1} \varphi_0(v_i^* x^* v_j v_j^* x v_i)\\ &\leq \sum_{i=1}^{m+1} \sum_{j=1}^{m+1} \left(1 + \frac{1}{n}\right) \varphi_0(v_j^* x v_i v_i^* x^* v_j),\end{aligned}$$

since $v_j^* x v_i \in p_0 \mathcal{M} p_0$ $(1 \leq i, j \leq m+1)$. Hence

$$\begin{aligned}\varphi_n(x^* x) &\leq \left(1 + \frac{1}{n}\right) \sum_{j=1}^{m+1} \varphi_0(v_j^* x z_0 x^* v_j)\\ &= \left(1 + \frac{1}{n}\right) \sum_{j=1}^{m+1} \varphi_0(v_j^* x x^* v_j)\\ &= \left(1 + \frac{1}{n}\right) \varphi_n(x x^*) \qquad (x \in \mathcal{M}).\end{aligned}$$

Since φ_0 is faithful on $p_0 \mathcal{M} p_0$, $\varphi_n(1) \neq 0$. Put $\psi_n(x) = \dfrac{\varphi_n(x)}{\varphi_n(1)}$ $(x \in \mathcal{M})$; then ψ_n is a normal state on $\mathcal{M}$ satisfying the inequality

$$\psi_n(x^* x) \leq \left(1 + \frac{1}{n}\right) \psi_n(x x^*) \qquad (x \in \mathcal{M}).$$

Since the state space $\mathcal{S}$ of $\mathcal{M}$ is $\sigma(\mathcal{M}^*, \mathcal{M})$-compact, a sequence $\{\psi_n | n = 1, 2, \ldots\}$ has an accumulation point ψ_0 in $\mathcal{S}$. Then

$$\psi_0(x^* x) \leq \left(1 + \frac{1}{n}\right) \psi_0(x x^*) \qquad (x \in \mathcal{M})$$

for every positive integer n. Hence $\psi_0(x^* x) \leq \psi_0(x x^*)$, and so $\psi_0(x x^*) = \psi_0((x^*)^* (x^*)) \leq \psi_0(x^* x)$. Hence $\psi_0(x^* x) = \psi_0(x x^*)$ $(x \in \mathcal{M})$. q.e.d.

2.4.5. Proposition. *Let $\mathcal{M}$ be a finite W^*-algebra, a an element of $\mathcal{M}$, and let C_a be the $\sigma(\mathcal{M}, \mathcal{M}_*)$-closed convex subset of $\mathcal{M}$ generated by $\{u^* a u | u \in \mathcal{M}^u\}$ ($\mathcal{M}^u$ the unitary group of $\mathcal{M}$). Then $C_a \cap Z$ consists of a single element, where Z is the center of $\mathcal{M}$.*

Proof. $C_a \cap Z \neq (\emptyset)$ by 2.1.16. Let $a = a_1 + i a_2$ $(a_1, a_2 \in \mathcal{M}^s)$. Clearly $C_a \cap Z \subset C_{a_1} \cap Z + i C_{a_2} \cap Z$. Hence we may assume that a is self-adjoint. Since $C_{\|a\| 1 + a} \cap Z = \|a\| 1 + C_a \cap Z$, we may assume that $a \geq 0$.

Suppose that $C_a \cap Z$ contains two distinct central elements c_1, c_2. Then there exists a non-zero central projection z and a positive number λ such that $c_1 z > c_2 z + \lambda z$ or $c_2 z > c_1 z + \lambda z$. It suffices to assume that $c_1 z > c_2 z + \lambda z$. By passing to $\mathcal{M} z$, we may assume that $z = 1$. By the reasoning in the proof of 2.4.4, there exists a normal state ψ_n $(n = 1, 2, \ldots)$ on $\mathcal{M}$ such that $\psi_n(x^* x) \le \left(1 + \frac{1}{n}\right) \psi_n(x x^*)$ $(x \in \mathcal{M})$. Since

$$u^* a u = (a^{\frac{1}{2}} u)^* (a^{\frac{1}{2}} u) \qquad (u \in \mathcal{M}^u),$$

$\psi_n(u^* a u) \le \left(1 + \frac{1}{n}\right) \psi_n(a)$ and $\psi_n(a) \le \left(1 + \frac{1}{n}\right) \psi_n(u^* a u)$. Hence $\psi_n(y_1) \le \left(1 + \frac{1}{n}\right)^2 \psi_n(y_2)$ $(y_1, y_2 \in C_a)$, and so $\psi_n(c_1) \le \left(1 + \frac{1}{n}\right)^2 \psi_n(c_2)$.

On the other hand $\psi_n(c_1) \ge \psi_n(c_2) + \lambda$, and so

$$\left(1 + \frac{1}{n}\right)^2 \psi_n(c_2) \ge \psi_n(c_2) + \lambda \qquad (n = 1, 2, \ldots),$$

a contradiction. q.e.d.

2.4.6. Theorem. *Let $\mathcal{M}$ be a finite W^*-algebra and let Z be its center. Then there exists a unique linear mapping $\natural$ (called the $\natural$-operation): $x \to x^\natural$ of $\mathcal{M}$ onto Z satisfying the following conditions:*

1. $\|x^\natural\| \le \|x\|$;
2. $1^\natural = 1$;
3. $(z x)^\natural = z x^\natural$;
4. $(x^* x)^\natural = (x x^*)^\natural$;
5. $(x^* x)^\natural \ge 0$, *and* $(x^* x)^\natural = 0$ *if and only if* $x = 0$;
6. $x \to x^\natural$ *is* $\sigma(\mathcal{M}, \mathcal{M}_*)$ *and* $s(\mathcal{M}, \mathcal{M}_*)$*-continuous.* $(x \in \mathcal{M}, z \in Z)$.

Proof. Put $x^\natural = C_x \cap Z$ $(x \in \mathcal{M})$. It is clear that the mapping $x \to x^\natural$ is linear; $\|x^\natural\| \leqq \|x\|$; $1^\natural = 1$; $(z x)^\natural = z x^\natural$.

Since $C_{u^* x u} \cap Z = u^* C_x u \cap Z = C_x \cap Z$ for $u \in \mathcal{M}^u$, $(x^* x)^\natural = (x x^*)^\natural$. Clearly $(x^* x)^\natural \ge 0$. Now let $(h_\alpha)_{\alpha \in \mathbb{I}}$ be a uniformly bounded, increasing directed set of positive elements in $\mathcal{M}$. Clearly $(\underset{\alpha}{\text{l.u.b.}}\, h_\alpha)^\natural \ge \underset{\alpha}{\text{l.u.b.}}\, h_\alpha^\natural$. Suppose that $(\underset{\alpha}{\text{l.u.b.}}\, h_\alpha)^\natural > \underset{\alpha}{\text{l.u.b.}}\, h_\alpha^\natural$. Then there exist a non-zero central projection z_1 in $\mathcal{M}$ and positive number λ such that

$$(\underset{\alpha}{\text{l.u.b.}}\, h_\alpha)^\natural z_1 > (\underset{\alpha}{\text{l.u.b.}}\, h_\alpha^\natural) z_1 + \lambda z_1 .$$

Consider a sequence $\{\psi_n\}$ of normal states on $\mathscr{M} z_1$ such that

$$\psi_n(y^* y) \leq \left(1 + \frac{1}{n}\right) \psi_n(y y^*) \qquad (y \in \mathscr{M} z_1,\ n = 1, 2, \ldots).$$

Then

$$\psi_n\Big((\operatorname*{l.u.b.}_{\alpha} h_\alpha)^\natural z_1\Big) \leq \left(1 + \frac{1}{n}\right) \psi_n\Big((\operatorname*{l.u.b.}_{\alpha} h_\alpha) z_1\Big) = \left(1 + \frac{1}{n}\right) \operatorname*{l.u.b.}_{\alpha} \psi_n(h_\alpha z_1)$$

$$\leq \left(1 + \frac{1}{n}\right)^2 \operatorname*{l.u.b.}_{\alpha} \psi_n(h_\alpha^\natural z_1) = \left(1 + \frac{1}{n}\right)^2 \psi_n\Big(\operatorname*{l.u.b.}_{\alpha} h_\alpha^\natural z_1\Big).$$

Hence $\psi_n(\operatorname*{l.u.b.}_{\alpha} h_\alpha^\natural z_1) + \lambda \leq \left(1 + \frac{1}{n}\right)^2 \psi_n(\operatorname*{l.u.b.}_{\alpha} h_\alpha^\natural z)$ $(n = 1, 2, \ldots)$, a contradiction. Hence $(\operatorname*{l.u.b.}_{\alpha} h_\alpha)^\natural = \operatorname*{l.u.b.}_{\alpha} h_\alpha^\natural$, and so for an arbitrary normal state φ on $\mathscr{M}$, a state $\varphi^\natural$ on $\mathscr{M}: x \to \varphi(x^\natural)\ (x \in \mathscr{M})$ is again normal. Hence the mapping $x \to x^\natural\ (x \in \mathscr{M})$ is $\sigma(\mathscr{M}, \mathscr{M}_*)$-continuous.

Since $\varphi(x^{\natural *} x^\natural) = \varphi((x^{\natural *} x)^\natural) = \varphi^\natural(x^{\natural *} x) \leq \varphi^\natural(x^{\natural *} x^\natural)^{\frac{1}{2}} \varphi^\natural(x^* x)^{\frac{1}{2}}$, $\varphi(x^{\natural *} x^\natural) \leq \varphi^\natural(x^* x) = \varphi((x^* x)^\natural)$ $(\varphi \in \mathscr{S})$. Hence $x^{*\natural} x^\natural \leq (x^* x)^\natural$ $(x \in \mathscr{M})$, and so the $\sigma(\mathscr{M}, \mathscr{M}_*)$-continuity of the $\natural$-operation implies its $s(\mathscr{M}, \mathscr{M}_*)$-continuity.

Now let $\mathscr{L} = \{x \mid (x^* x)^\natural = 0,\ x \in \mathscr{M}\}$. Then $\mathscr{L}$ is a $\sigma(\mathscr{M}, \mathscr{M}_*)$-closed two-sided ideal. Hence there exists a central projection z such that $\mathscr{L} = \mathscr{M} z$. Since $z^\natural = z$, $z = 0$. Hence $(x^* x)^\natural = 0$ implies $x = 0$.

Finally we prove the uniqueness. Suppose a linear mapping $\natural'$ of $\mathscr{M}$ onto Z satisfies the conditions 2., 3., 4., and $(x^* x)^{\natural'} \geq 0$ $(x \in \mathscr{M})$. Then $\|x^{\natural'}\|^2 = \|(x^{\natural'})^* (x^{\natural'})\| \leq \|(x^* x)^{\natural'}\|$. Since $\|x^* x\| 1 - (x^* x) \geq 0$, $\|x^* x\| 1 \geq (x^* x)^{\natural'}$. Hence $\|(x^* x)^{\natural'}\| \leq \|x\|^2$, and so $\natural'$ is uniformly continuous. For $h \in \mathscr{M}$, let K_h be the uniformly closed convex subset of $\mathscr{M}$ generated by $\{u^* h u \mid u \in \mathscr{M}^u\}$. Then $h^{\natural'} = (K_h)^{\natural'} = (K_h \cap Z)^{\natural'} = h^\natural$ by 4., and so $h^{\natural'} = h^\natural$. q.e.d.

2.4.7. Corollary. *Let $\mathscr{M}$ be a finite W^*-algebra. Then $\mathscr{M}$ has a faithful family of normal, tracial states.*

Proof. For an arbitrary normal state φ on $\mathscr{M}$, set $\varphi^\natural(x) = \varphi(x^\natural)$ $(x \in \mathscr{M})$. Then $\varphi^\natural$ is a normal, tracial state on $\mathscr{M}$. Let $\mathfrak{F} = \{\varphi^\natural \mid \varphi$, all normal states on $\mathscr{M}\}$. Then clearly $\mathfrak{F}$ is a faithful family of normal, tracial states on $\mathscr{M}$. q.e.d.

References. [29], [82], [154].

2.5. Traces and Criterions of Types

2.5.1. Definition. *Let $\mathscr{A}$ be a C*-algebra and $\mathscr{A}^+$ the positive portion of $\mathscr{A}$. A functional τ on $\mathscr{A}^+$, with values ≥ 0, finite or infinite, is called a trace (on $\mathscr{A}$) if it satisfies the following conditions:*

1. *if $a, b \in \mathscr{A}^+$, $\tau(a+b)=\tau(a)+\tau(b)$;*
2. *if $a \in \mathscr{A}^+$ and λ is a non-negative real number, $\tau(\lambda a)=\lambda\tau(a)$ (we define $0\cdot(+\infty)=0$);*
3. *if $a \in \mathscr{A}^+$ and u is unitary in $\mathscr{A}$, $\tau(u^* a u)=\tau(a)$.*

A trace τ is said to be faithful if $\tau(a)=0$ implies $a=0$; finite if $\tau(a)<+\infty$ ($a \in \mathscr{A}^+$); semi-finite if for every non-zero $a \in \mathscr{A}^+$, there exists a non-zero element b in $\mathscr{A}^+$ with $\tau(b)<+\infty$ and $b \leq a$.

Let $\mathscr{M}$ be a W-algebra and let τ be a trace on $\mathscr{M}$. τ is said to be normal if for every uniformly bounded increasing directed set $\{a_\alpha\} \subset \mathscr{A}^+$, $\tau(\operatorname{l.u.b.}_\alpha a_\alpha)=\operatorname{l.u.b.}\tau(a_\alpha)$.*

2.5.2. Proposition. *Let $\mathscr{M}$ be a W*-algebra and let τ be a trace on $\mathscr{M}$. The set of $a \in \mathscr{M}^+$ with $\tau(a)<+\infty$ is the positive portion of a two-sided ideal $\mathscr{I}$ of $\mathscr{M}$. There exists a unique linear functional $\tilde{\tau}$ on $\mathscr{I}$ which coincides with τ on $\mathscr{I} \cap \mathscr{M}^+$, and one has $\tilde{\tau}(ax)=\tilde{\tau}(xa)$ ($a \in \mathscr{I}, x \in \mathscr{M}$). Finally if τ is normal, for every $a \in \mathscr{I}$, the linear functional $x \to \tilde{\tau}(ax)$ on $\mathscr{M}$ is $\sigma(\mathscr{M}, \mathscr{M}_*)$-continuous.*

Proof. Set $\mathfrak{F}=\{a \mid \tau(a)<+\infty, a \in \mathscr{M}^+\}$, and $\mathscr{I}_1=\{b \mid b^* b \in \mathfrak{F}, b \in \mathscr{M}\}$. For $b \in \mathscr{I}_1$, a unitary $u \in \mathscr{M}$, $(bu)^*(bu)=u^* b^* b u \in \mathfrak{F}$; $(ub)^*(ub)=b^* b \in \mathfrak{F}$. Hence $bu, ub \in \mathscr{I}_1$. If $b, c \in \mathscr{I}_1$, $(b+c)^*(b+c) \leq 2(b^* b+c^* c) \in \mathfrak{F}$. Hence, $b+c \in \mathscr{I}_1$. Since any element in $\mathscr{M}$ is a linear combination of unitary elements, bx, xb belong again to $\mathscr{I}_1 (x \in \mathscr{M}, b \in \mathscr{I}_1)$ so that $\mathscr{I}_1$ is a two-sided ideal. Let $\mathscr{I}_1 \cdot \mathscr{I}_1$ be the set of all finite linear combinations of elements of the form bc $(b, c \in \mathscr{I}_1)$. Then $\mathscr{I}_1 \cdot \mathscr{I}_1$ is also a two-sided ideal. For $a \in \mathfrak{F}$, $a^{\frac{1}{2}} \in \mathscr{I}_1$, and so $a \in \mathscr{I}_1 \cdot \mathscr{I}_1$. Hence $\mathfrak{F} \subset (\mathscr{I}_1 \cdot \mathscr{I}_1) \cap \mathscr{M}^+$. Conversely, let $c \in \mathscr{I}_1 \cdot \mathscr{I}_1$. Then c is a sum of elements $a^* b$ with $a, b \in \mathscr{I}_1$. By the equality

$$4a^* b=(a+b)^*(a+b)-(a-b)^*(a-b)-i(a+ib)^*(a+ib)+i(a-ib)^*(a-ib),$$

if $c \geq 0$, it is bounded by an element of the form $\sum_{i=1}^{n} a_i^* a_i$ $(a_i \in \mathscr{I}_1)$. Hence $c^{\frac{1}{2}} \in \mathscr{I}_1$, and so $(\mathscr{I}_1 \cdot \mathscr{I}_1) \cap \mathscr{M}^+=\mathfrak{F}$.

Let $a \in \mathscr{I}_1 \cdot \mathscr{I}_1$ and put $a=a_1+i a_2$ $(a_1, a_2 \in \mathscr{M}^s)$. Then $a_1, a_2 \in \mathscr{I}_1 \cdot \mathscr{I}_1$. There exist spectral projections p, q of a_1 such that $p+q=1$, $p a_1 \geq 0$, $q a_1 \leq 0$. $p a_1$, $-q a_1 \in (\mathscr{I}_1 \cdot \mathscr{I}_1) \cap \mathscr{M}^+=\mathfrak{F}$, and so a is a finite linear combination of elements in $\mathfrak{F}$. Hence $\mathscr{I}_1 \cdot \mathscr{I}_1$ is equal to the set of all finite linear combinations of elements in $\mathfrak{F}$. Hence we can take $\mathscr{I}_1 \cdot \mathscr{I}_1$ as the $\mathscr{I}$ in the theorem.

For $a\in\mathscr{I}$, suppose there exist two expressions as follows:

$$a=a_1-a_2=b_1-b_2 \quad (a_1,a_2,b_1,b_2\in\mathscr{I}\cap\mathscr{M}^+).$$

Then $a_1+b_2=a_2+b_1$, and so

$$\tau(a_1+b_2)=\tau(a_1)+\tau(b_2)=\tau(a_2+b_1)=\tau(a_2)+\tau(b_1).$$

Hence $\tau(a_1)-\tau(a_2)=\tau(b_1)-\tau(b_2)$. Therefore τ can be extended uniquely to a linear functional $\tilde{\tau}$ on $\mathscr{I}$. If $a\in\mathscr{I}$ and $u\in\mathscr{M}^u$, then $\tilde{\tau}(u^*au)=\tilde{\tau}(a)$. Hence $\tilde{\tau}(au)=\tilde{\tau}(uauu^*)=\tilde{\tau}(ua)$, and so $\tilde{\tau}(ax)=\tilde{\tau}(xa)$ $(x\in\mathscr{M},\ a\in\mathscr{I})$.

Now suppose that τ is normal. For $a\in\mathscr{I}$, set $\varphi(x)=\tilde{\tau}(ax)(x\in\mathscr{M})$. We shall show that φ is $\sigma(\mathscr{M},\mathscr{M}_*)$-continuous. We may assume that $a\geq 0$. For $x\geq 0$, $\varphi(x)=\tilde{\tau}(ax^{\frac{1}{2}}x^{\frac{1}{2}})=\tilde{\tau}(x^{\frac{1}{2}}ax^{\frac{1}{2}})\geq 0$. Hence $\varphi\geq 0$. Let (x_α) be a uniformly bounded increasing directed set of positive elements in $\mathscr{M}$. $\operatorname{l.u.b.}_\alpha a^{\frac{1}{2}}x_\alpha a^{\frac{1}{2}} = a^{\frac{1}{2}}(\operatorname{l.u.b.}_\alpha x_\alpha)\, a^{\frac{1}{2}}$. Since $a^{\frac{1}{2}}\in\mathscr{I}_1$, $x_\alpha^{\frac{1}{2}}a^{\frac{1}{2}}\in\mathscr{I}_1$. Hence $a^{\frac{1}{2}}x_\alpha a^{\frac{1}{2}}\in\mathscr{I}$. Let $x_\alpha^{\frac{1}{2}}a^{\frac{1}{2}}=v|x_\alpha^{\frac{1}{2}}a^{\frac{1}{2}}|$ be the polar decomposition of $x_\alpha^{\frac{1}{2}}a^{\frac{1}{2}}$. Then $x_\alpha^{\frac{1}{2}}ax_\alpha^{\frac{1}{2}}=va^{\frac{1}{2}}x_\alpha a^{\frac{1}{2}}v^*$. Hence

$$\tilde{\tau}(a^{\frac{1}{2}}x_\alpha a^{\frac{1}{2}})=\tilde{\tau}(v^*va^{\frac{1}{2}}x_\alpha a^{\frac{1}{2}})=\tilde{\tau}(va^{\frac{1}{2}}x_\alpha a^{\frac{1}{2}}v^*)=\tilde{\tau}(x_\alpha^{\frac{1}{2}}ax_\alpha^{\frac{1}{2}})=\tilde{\tau}(ax_\alpha).$$

Hence $x\to\tilde{\tau}(ax)$ $(x\in\mathscr{M})$ is normal. q.e.d.

Now let $\mathscr{M}$ be a semi-finite W^*-algebra, p a non-zero finite projection in $\mathscr{M}$, and let $\{p_\alpha|\alpha\in\mathbb{I}\}$ be a maximal family of mutually orthogonal equivalent projections with $p_\alpha\sim p\,(\alpha\in\mathbb{I})$. Then there exists a non-zero central projection z such that $\left(1-\sum_{\alpha\in\mathbb{I}}p_\alpha\right)z\prec pz\neq 0$. Let $p_0=\left(1-\sum_\alpha p_\alpha\right)z$. Then $z=p_0+\sum_{\alpha\in\mathbb{I}}p_\alpha z$. Let $\{v_\alpha|\alpha\in\mathbb{I}\}$ be a family of partial isometries with $v_\alpha^*v_\alpha=pz$ and $v_\alpha v_\alpha^*=p_\alpha z$, and let v_0 be a partial isometry with $v_0^*v_0\leq pz$ and $v_0v_0^*=p_0$. Take a normal tracial state φ on $zp\mathscr{M}zp$, and define a functional ψ on $\mathscr{M}^+z$ as follows: $\psi(h)=\sum_{\alpha\in(0)\cup\mathbb{I}}\varphi(v_\alpha^*hv_\alpha)$ $(h\in\mathscr{M}^+z)$. Then $\sum_{\alpha\in(0)\cup\mathbb{I}}v_\alpha v_\alpha^*=z$. Put $a_{\gamma,\beta}=v_\gamma^*av_\beta$. Then

$$\begin{aligned}
\psi(a^*a)&=\sum_{\alpha\in(0)\cup\mathbb{I}}\varphi(v_\alpha^*a^*av_\alpha)\\
&=\sum_{\alpha\in(0)\cup\mathbb{I}}\ \sum_{\gamma\in(0)\cup\mathbb{I}}\varphi(v_\alpha^*a^*v_\gamma v_\gamma^*av_\alpha)\\
&=\sum_{\alpha\in(0)\cup\mathbb{I}}\ \sum_{\gamma\in(0)\cup\mathbb{I}}\varphi(a_{\gamma\alpha}^*a_{\gamma\alpha})\\
&=\sum_{\gamma\in(0)\cup\mathbb{I}}\ \sum_{\alpha\in(0)\cup\mathbb{I}}\varphi(a_{\gamma\alpha}a_{\gamma\alpha}^*)\\
&=\sum_{\gamma\in(0)\cup\mathbb{I}}\ \sum_{\alpha\in(0)\cup\mathbb{I}}\varphi(v_\gamma^*av_\alpha v_\alpha^*a^*v_\gamma)\\
&=\sum_{\gamma\in(0)\cup\mathbb{I}}\varphi(v_\gamma^*aa^*v_\gamma)\\
&=\psi(aa^*)\quad(a\in\mathscr{M}z).
\end{aligned}$$

Hence ψ is a trace on $\mathcal{M}z$. By defining $\psi=0$ on $\mathcal{M}^+(1-z)$, ψ can be extended uniquely to a trace on $\mathcal{M}^+$. One can easily see that ψ is a normal semi-finite trace on $\mathcal{M}$. If $\psi(a^* a)=0$ for some

$$a\in\mathcal{M}z,\ \varphi(v_\alpha^* a^* a v_\alpha)=0 \qquad (\alpha\in(0)\cup\mathbb{I}).$$

By taking all normal tracial states φ on $zp\mathcal{M}zp$, we have $av_\alpha=0$, and so $ap_\alpha z=0\,(\alpha\in\mathbb{I}\cup(0))$. Hence $az=0$. Therefore $\mathcal{M}z$ has a faithful family $(\tau_\beta|\beta\in\mathbb{J})$ of semi-finite normal traces—i.e., for $a\in\mathcal{M}^+z$, $\tau_\beta(a)=0\,(\beta\in\mathbb{J})$ imply $a=0$. From this one can easily see that a semi-finite W^*-algebra has a faithful family of semi-finite normal traces. Now we have

2.5.3. Lemma. *Let e be a projection of a semi-finite W^*-algebra. Then e is finite if and only if there exists a faithful family $(\tau_\beta)_{\beta\in\mathbb{J}}$ of semi-finite traces such that $\tau_\beta(e)<+\infty$ $(\alpha\in\mathbb{J})$.*

Proof. By the above result, it is clear that if e is finite, there exists a faithful family of semi-finite normal traces satisfying the condition. Conversely let e be a projection satisfying the condition. Suppose $e\sim e_1\leq e$. Then $\tau_\beta(e-e_1)=0$ $(\beta\in\mathbb{J})$. Hence $e-e_1=0$. q.e.d.

2.5.4. Theorem. *Let $\mathcal{M}$ be a W^*-algebra. Then $\mathcal{M}$ is finite (resp. semi-finite) if and only if for any non-zero, positive element h in $\mathcal{M}$, there exists a normal finite (resp. semi-finite) trace τ such that $\tau(h)\neq 0$. $\mathcal{M}$ is properly infinite (resp. purely infinite) if and only if there is no normal finite trace (resp. semi-finite) on $\mathcal{M}^+$ except for the identical zero trace.*

2.5.5. Corollary. *Let p and q be finite projections in a W^*-algebra $\mathcal{M}$. Then $p\vee q$ is again finite.*

2.5.6. Theorem. *Let $\mathcal{M}$ be a W^*-algebra. Then,* 1. *$\mathcal{M}$ is finite if and only if the *-operation is $s(\mathcal{M},\mathcal{M}_*)$-continuous on bounded spheres;* 2. *$\mathcal{M}$ is semi-finite if and only if there exists an increasing directed set $(e_\alpha)_{\alpha\in\mathbb{I}}$ of projections such that $\operatorname{l.u.b.}_\alpha e_\alpha=1$ and the *-operation is $s(\mathcal{M},\mathcal{M}_*)$-continuous on bounded spheres of $\mathcal{M}e_\alpha$ $(\alpha\in\mathbb{I})$;* 3. *$\mathcal{M}$ is purely infinite if and only if for an arbitrary non-zero projection e in $\mathcal{M}$, the *-operation is not $s(\mathcal{M},\mathcal{M}_*)$-continuous on bounded spheres of $\mathcal{M}e$.*

Proof. Let e be a finite projection in $\mathcal{M}$ and let $c(e)$ be the central support of e. Then $\mathcal{M}c(e)$ is semi-finite. Let $\{\psi_\alpha|\alpha\in\mathbb{I}\}$ be a faithful family of normal semifinite traces of $\mathcal{M}c(e)$ such that $\psi_\alpha(e)<+\infty$ $(\alpha\in\mathbb{I})$. Let $s(\psi_\alpha)$ be the support of ψ_α—i.e., $s(\psi_\alpha)=1-p_\alpha$, where p_α is the greatest projection in $\mathcal{M}$ such that $\psi_\alpha(p_\alpha)=0$. Then $s(\psi_\alpha)$ is a central projection. For any finite subset $\mathbb{J}$ of $\mathbb{I}$, put $z_{\mathbb{J}}=\bigvee_{\alpha\in\mathbb{J}} s(\psi_\alpha)$. Then $\operatorname{l.u.b.}_{\mathbb{J}} z_{\mathbb{J}}=c(e)$.

Now suppose that $\{x_\beta e|\beta\in\mathbb{K}\}$ $(\|x_\beta e\|\leq 1)$ converges to 0 in the $s(\mathcal{M},\mathcal{M}_*)$-topology. The linear span $\mathcal{I}_\alpha$ of $\{h|\psi_\alpha(h)<+\infty, h\in\mathcal{M}^+\}$ is a $\sigma(\mathcal{M},\mathcal{M}_*)$-dense ideal of $\mathcal{M}c(e)$. Hence $\{L_a\tilde{\psi}_\alpha|a\in\mathcal{I}_\alpha,\alpha\in\mathbb{I}\}$ is a total set of $\sigma(\mathscr{M}c(e), (\mathscr{M}c(e))_*)$-continuous linear functionals on $\mathscr{M}c(e)$—viz. $L_a\psi_\alpha(x)=0$ $(a\in\mathscr{I}_\alpha, \alpha\in\mathbb{I})$ imply $x=0$.

$$\begin{aligned}|(L_a\tilde{\psi}_\alpha)((x_\beta e)(x_\beta e)^*)| &= |\tilde{\psi}_\alpha(a x_\beta e x_\beta^*)| \\ &= |\tilde{\psi}_\alpha(e x_\beta^* a x_\beta e)| = |Le\tilde{\psi}_\alpha(e x_\beta^* a x_\beta e)| \\ &\leq (Le\tilde{\psi}_\alpha(e x_\beta^* x_\beta e))^{\frac{1}{2}}(Le\tilde{\psi}_\alpha(e x_\beta^* a^* a x_\beta e))^{\frac{1}{2}} \\ &\to 0 \quad (\alpha\in\mathbb{I}),\end{aligned}$$

since $Le\tilde{\psi}_\alpha$ is normal. Hence $\{(x_\beta e s(\psi_\alpha))^*\}$ converges to 0 in the $s(\mathscr{M},\mathscr{M}_*)$-topology $(\alpha\in\mathbb{I})$, and so $\{(x_\beta e)^*\}$ converges to 0 in the $s(\mathscr{M},\mathscr{M}_*)$-topology.

Next, suppose that e is an infinite projection in a W^*-algebra $\mathscr{M}$, and let v be a partial isometry with $vv^*=e$ and $v^*v<e$. Write $e_i=(v^*)^i(v)^i$ and $u_i=e_i v$ $(i=1,2,\ldots)$. Then $\{e_i-e_{i+1}\}$ is a family of mutually orthogonal non-zero projections, and

$$(u_i-u_{i+1})^*(u_i-u_{i+1})=e_{i+1}-e_{i+2}$$

and

$$(u_i-u_{i+1})(u_i-u_{i+1})^*=e_i-e_{i+1}.$$

Hence $e_i-e_{i+1}\sim e_{i+1}-e_{i+2}$ $(i=1,2,\ldots)$. Let w_i be a partial isometry such that $w_i^*w_i=e_i-e_{i+1}$ and $w_iw_i^*=e_1-e_2$. Since $\{e_i-e_{i+1}\}$ are mutually orthogonal, $\{w_i\}$ converges to 0 in the $s(\mathscr{M},\mathscr{M}_*)$-topology, but $w_iw_i^*=e_1-e_2$. Hence $\{w_i^*\}$ does not converge to 0 in the $s(\mathscr{M},\mathscr{M}_*)$-topology. q.e.d.

2.5.7. Proposition. *Let $\mathscr{M}$ be a semi-finite W^*-algebra. Then there exists a faithful normal semi-finite trace on $\mathscr{M}$.*

Proof. Let $\{\tau_\alpha\}_{\alpha\in\mathbb{I}}$ be a maximal family of normal semi-finite traces whose supports are mutually orthogonal. Then $\sum_{\alpha\in\mathbb{I}} s(\tau_\alpha)=1$. Define $\varphi(h)=\sum_{\alpha\in\mathbb{I}}\tau_\alpha(hs(\tau_\alpha))$ for $h(\geq 0)\in\mathscr{M}$. Then one can easily see that φ is a faithful normal semi-finite trace on $\mathscr{M}$. q.e.d.

References. [37], [151], [154].

2.6. Types of Tensor Products of W^*-Algebras

2.6.1. Proposition. *Let $\mathscr{M}$ and $\mathscr{N}$ be two W^*-algebras, and let $\mathscr{M}\mathbin{\bar{\otimes}}\mathscr{N}$ be the W^*-tensor product of $\mathscr{M}$ and $\mathscr{N}$. $\mathscr{M}\mathbin{\bar{\otimes}}\mathscr{N}$ is finite if and only if both of $\mathscr{M}$ and $\mathscr{N}$ are finite; $\mathscr{M}\mathbin{\bar{\otimes}}\mathscr{N}$ is semi-finite if both of $\mathscr{M}$ and $\mathscr{N}$ are semi-finite.*

Proof. If $\mathcal{M} = \sum_{\alpha \in \mathbb{I}} \oplus \mathcal{M}_\alpha$ and $\mathcal{N} = \sum_{\beta \in \mathbb{J}} \oplus \mathcal{N}_\beta$, $\mathcal{M} \overline{\otimes} \mathcal{N} = \sum_{\substack{\alpha \in \mathbb{I} \\ \beta \in \mathbb{J}}} \oplus \mathcal{M}_\alpha \overline{\otimes} \mathcal{N}_\beta$ by 1.22.11. Let $\mathcal{M}$ and $\mathcal{N}$ be two finite W^*-algebras with faithful normal tracial states φ and ψ, respectively. Then $\varphi \otimes \psi$ is a normal tracial state on $\mathcal{M} \overline{\otimes} \mathcal{N}$.

Since $\{L_a \varphi \mid a \in \mathcal{M}\}$ (resp. $\{L_b \psi \mid b \in \mathcal{N}\}$) is total in $\mathcal{M}_*$ (resp. $\mathcal{N}_*$), $\{L_a \varphi \otimes L_b \psi \mid a \in \mathcal{M}, b \in \mathcal{N}\}$ is total in $\mathcal{M}_* \otimes_{\alpha_0} \mathcal{N}_*$. Hence $\varphi \otimes \psi$ is faithful, and so $\mathcal{M} \overline{\otimes} \mathcal{N}$ is again finite. Therefore one can easily see that the first part of 2.6.1 is true.

Next, let $\mathcal{M}$ and $\mathcal{N}$ be arbitrary two W^*-algebras, and let e (resp. f) be a finite projection in $\mathcal{M}$ (resp. $\mathcal{N}$). Then

$$(e \otimes f)(\mathcal{M} \overline{\otimes} \mathcal{N})(e \otimes f) = (e \mathcal{M} e) \overline{\otimes} (f \mathcal{N} f).$$

Hence from the above results, one can easily see that $e \otimes f$ is again finite in $\mathcal{M} \overline{\otimes} \mathcal{N}$. If $\mathcal{M}$ and $\mathcal{N}$ are semi-finite, there exist increasing directed sets $(e_\alpha)_{\alpha \in \mathbb{I}}$ and $(f_\beta)_{\beta \in \mathbb{J}}$ of finite projections in $\mathcal{M}$ and $\mathcal{N}$ respectively such that $e_\alpha \to 1$ in the $s(\mathcal{M}, \mathcal{M}_*)$-topology and $f_\beta \to 1$ in the $s(\mathcal{N}, \mathcal{N}_*)$-topology. Hence $e_\alpha \otimes f_\beta \to 1 \otimes 1$ in the $s(\mathcal{M} \overline{\otimes} \mathcal{N}, \mathcal{M}_* \otimes_{\alpha_0} \mathcal{N}_*)$-topology, and so $\mathcal{M} \overline{\otimes} \mathcal{N}$ is again semi-finite. q.e.d.

2.6.2. Proposition. *Let $\mathcal{M}$ and $\mathcal{N}$ be two W^*-algebras and let $\mathcal{M} \overline{\otimes} \mathcal{N}$ be the W^*-tensor product of $\mathcal{M}$ and $\mathcal{N}$. $\mathcal{M} \overline{\otimes} \mathcal{N}$ is of type* I *if both of $\mathcal{M}$ and $\mathcal{N}$ are of type* I.

Proof. Let $\mathcal{M}$ (resp. $\mathcal{N}$) be a type I_m (resp. I_n) W^*-algebra (m, n cardinals). Then there exists a family $(e_\alpha)_{\alpha \in \mathbb{I}}$ (resp. $(f_\beta)_{\beta \in \mathbb{J}}$) of mutually orthogonal equivalent, maximal abelian projections in $\mathcal{M}$ (resp. $\mathcal{N}$) such that $\sum_{\alpha \in \mathbb{I}} e_\alpha = 1$, $\sum_{\beta \in \mathbb{J}} f_\beta = 1$. Then $\{e_\alpha \otimes f_\beta\}_{\substack{\alpha \in \mathbb{I} \\ \beta \in \mathbb{J}}}$ is a family of mutually orthogonal, equivalent projections with the sum 1. Since

$$e_\alpha \otimes f_\beta (\mathcal{M} \overline{\otimes} \mathcal{N}) e_\alpha \otimes f_\beta = (e_\alpha \mathcal{M} e_\alpha) \overline{\otimes} (f_\beta \mathcal{N} f_\beta), \qquad e_\alpha \otimes f_\beta$$

is abelian. Hence $\mathcal{M} \overline{\otimes} \mathcal{N}$ is a type I_{mn} W^*-algebra. From this result, one can easily see that 2.6.2 is true. q.e.d.

2.6.3. Proposition. *Let $\mathcal{M}$ and $\mathcal{N}$ be two W^*-algebras and let $\mathcal{M} \overline{\otimes} \mathcal{N}$ be the W^*-tensor product of $\mathcal{M}$ and $\mathcal{N}$. $\mathcal{M} \overline{\otimes} \mathcal{N}$ is of type* II *if $\mathcal{M}$ and $\mathcal{N}$ are semi-finite and one of them is of type* II.

Proof. Let $\mathcal{M}$ (resp. $\mathcal{N}$) be a finite W^*-algebra (resp. continuous finite W^*-algebra) having a faithful normal tracial state φ (resp. ψ). Then $\mathcal{M} \overline{\otimes} \mathcal{N}$ is continuous. In fact, since $\mathcal{N}$ is continuous, there exists a decreasing sequence (e_n) of projections such that $e_n - e_{n+1} \sim e_{n+1}$, $c(e_n) = 1$ $(n = 1, 2, \ldots)$. Then $\varphi \otimes \psi(1 \otimes e_n) = \psi(e_n) = 2^{-n+1} \psi(e_1) \to 0$ $(n \to \infty)$. If $\mathcal{M} \overline{\otimes} \mathcal{N}$ has a type I W^*-algebra as a direct summand, there exists an abelian projection p in $\mathscr{M} \mathbin{\bar{\otimes}} \mathscr{N}$. Clearly $p \prec 1 \otimes e_n$, and so $\varphi \otimes \psi(p) \leq \varphi \otimes \psi(1 \otimes e_n) \to 0$ $(n \to \infty)$, a contradiction. From this result, one can easily see that 2.6.3 is true. q.e.d.

2.6.4. Theorem. *Let $\mathscr{M}$ and $\mathscr{N}$ be two W^*-algebras, and let $\mathscr{M} \mathbin{\bar{\otimes}} \mathscr{N}$ be the W^*-tensor product of $\mathscr{M}$ and $\mathscr{N}$. $\mathscr{M} \mathbin{\bar{\otimes}} \mathscr{N}$ is of type* III *if one of them is of type* III.

Proof. It suffices to assume that $\mathscr{N}$ is of type III. Let φ be an arbitrary normal state on $\mathscr{M}$. For $x \in \mathscr{M} \mathbin{\bar{\otimes}} \mathscr{N}$, put $T(x,f) = \varphi \otimes f(x)$ $(f \in \mathscr{N}_*)$. Then $|\varphi \otimes f(x)| \leq \|\varphi \otimes f\| \, \|x\| = \|\varphi\| \, \|f\| \, \|x\|$. Since $(\mathscr{N}_*)^* = \mathscr{N}$, there exists a unique element $\Phi_\varphi(x)$ in $\mathscr{N}$ such that $(\varphi \otimes f)(x) = f(\Phi_\varphi(x))$. Put $P_\varphi(x) = 1 \otimes \Phi_\varphi(x)$ $(x \in \mathscr{M} \mathbin{\bar{\otimes}} \mathscr{N})$. Then the mapping $x \to P_\varphi(x)$ of $\mathscr{M} \mathbin{\bar{\otimes}} \mathscr{N}$ into $1 \otimes \mathscr{N}$ is linear, and satisfies the following conditions:

1. $P_\varphi(1 \otimes 1) = 1 \otimes 1$;
2. $\|P_\varphi(x)\| \leq \|x\|$;
3. $P_\varphi(h) \geq 0$ $(h \geq 0)$;
4. $P_\varphi(axb) = a P_\varphi(x) b$;
5. $P_\varphi(x)^* P_\varphi(x) \leq P_\varphi(x^* x)$;
6. if $P_\varphi(x^* x) = 0$ for all normal states φ on $\mathscr{M}$, then $x = 0$;
7. P_φ is σ and s-continuous $(x, h \in \mathscr{M} \mathbin{\bar{\otimes}} \mathscr{N}, a, b \in 1 \otimes \mathscr{N})$.

In fact, 1. and 2. are clear. 3. If $h \geq 0$, $0 \leq \varphi \otimes \psi(h) = \psi(\Phi_\varphi(h))$ $(\psi \geq 0)$. Hence $P_\varphi(h) \geq 0$. 4. Put $a = 1 \otimes a_1$, $b = 1 \otimes b_1$ $(a_1, b_1 \in \mathscr{N})$. Then

$$\varphi \otimes f(axb) = \varphi \otimes L_{a_1} R_{b_1} f(x) = L_{a_1} R_{b_1} f(\Phi_\varphi(x)) = f(a_1 \Phi_\varphi(x) b_1).$$

Hence $P_\varphi(axb) = a P_\varphi(x) b$. 5. For $\psi(\geq 0) \in \mathscr{N}_*$,

$$\begin{aligned}\varphi \otimes \psi(P_\varphi(x)^* P_\varphi(x)) &= \varphi \otimes \psi(P_\varphi(x^* P_\varphi(x))) = \varphi \otimes \psi(x^* P_\varphi(x)) \\ &\leq \varphi \otimes \psi(x^* x)^{\frac{1}{2}} \varphi \otimes \psi(P_\varphi(x)^* P_\varphi(x))^{\frac{1}{2}}.\end{aligned}$$

Hence $\varphi \otimes \psi(P_\varphi(x)^* P_\varphi(x)) \leq \varphi \otimes \psi(x^* x) = \varphi \otimes \psi(P_\varphi(x^* x))$. Hence $\psi(\Phi_\varphi(x)^* \Phi_\varphi(x)) \leq \psi(\Phi_\varphi(x^* x))$, and so $P_\varphi(x)^* P_\varphi(x) \leq P_\varphi(x^* x)$. 6. If $P_\varphi(x^* x) = 0$ for all normal states on $\mathscr{M}, g \otimes f(x^* x) = 0$ $(g \in \mathscr{M}_*, f \in \mathscr{N}_*)$. Hence $x^* x = 0$, and so $x = 0$. 7. The σ-continuity of P_φ is clear, and the σ-continuity of P_φ and 5. imply the s-continuity of P_φ.

Now we shall show the following general lemma.

2.6.5. Lemma. *Let $\mathscr{U}$ be a W^*-algebra and let $\mathscr{V}$ be a W^*-subalgebra of $\mathscr{U}$, containing the identity of $\mathscr{U}$. Suppose that there exists a family of mappings $\{Q_\alpha \mid \alpha \in \mathrm{II}\}$ of $\mathscr{U}$ into $\mathscr{V}$ satisfying the following conditions:* 1. $Q_\alpha(1) = 1$; 2. $\|Q_\alpha(x)\| \leq \|x\|$; 3. $Q_\alpha(h) \geq 0$ $(h \geq 0)$; 4. $Q_\alpha(axb) = a Q_\alpha(x) b$; 5. $Q_\alpha(x)^* Q_\alpha(x) \leq Q_\alpha(x^* x)$; 6. *if* $Q_\alpha(x^* x) = 0$ *for all* $\alpha \in \mathrm{II}$, *then* $x = 0$; 7. *Q_α is σ and s-continuous $(x, h \in \mathscr{U}, a, b \in \mathscr{V})$. Then if $\mathscr{V}$ is of type* III, *$\mathscr{U}$ is again of type* III.

Proof. Suppose that there exists a non-zero central projection z in $\mathscr{U}$ such that $\mathscr{U}z$ is semi-finite. Let e be a non-zero finite projection in $\mathscr{U}z$. Take a Q_{α_0} with $Q_{\alpha_0}(e)\neq 0$. Since $Q_{\alpha_0}(e)>0$, there exists a non-zero projection p in $\mathscr{V}$ such that $\lambda p<Q_{\alpha_0}(e)$ for some positive number λ.

Let $(x_\beta)_{\beta\in\mathbb{J}}$ $(\|x_\beta\|\leq 1, x_\beta\in p\mathscr{V}p)$ be a directed set such that $x_\beta\to 0$ in the $s(\mathscr{U},\mathscr{U}_*)$-topology. Then $x_\beta e\to 0$ in the $s(\mathscr{U},\mathscr{U}_*)$-topology. Hence by 2.5.6, $(x_\beta e)^*=e x_\beta^*\to 0$ in the $s(\mathscr{U},\mathscr{U}_*)$-topology, and so

$$Q_{\alpha_0}(e x_\beta^*)=Q_{\alpha_0}(e)x_\beta^*\to 0$$

in the $s(\mathscr{U},\mathscr{U}_*)$-topology. Hence $\{pQ_{\alpha_0}(e)p+(1-p)\}^{-1}pQ_{\alpha_0}(e)x_\beta^*=x_\beta^*$ converges to 0 in the $s(\mathscr{U},\mathscr{U}_*)$-topology. Therefore the *-operation is $s(\mathscr{V},\mathscr{V}_*)$-continuous on bounded spheres of $p\mathscr{V}p$. But, $p\mathscr{V}p$ is of type III, and so this is a contradiction by 2.5.6. q.e.d.

Remark. In general, a linear mapping of a C^*-algebra $\mathscr{A}$ onto its C^*-subalgebra $\mathscr{B}$, satisfying the above conditions 1., 2., 3., 4., and 5., is called a conditional expectation [215]. A conditional expectation is clearly a projection of $\mathscr{A}$ onto $\mathscr{B}$, with norm 1. Conversely, Tomiyama [211] proved that a projection of $\mathscr{A}$ onto $\mathscr{B}$, with norm 1 is automatically a conditional expectation.

Now we obtain the following diagram concerning the types of tensor products of W^*-algebras.

2.6.6. Theorem. Im $\overline{\otimes}$ In $=$ Imn; I $\overline{\otimes}$ I $=$ I; (I *or* II) $\overline{\otimes}$ II $=$ II; *(arbitrary)* $\overline{\otimes}$ III $=$ III; *(finite)* $\overline{\otimes}$ *(finite)* $=$ *(finite)*; *(semi-finite)* $\overline{\otimes}$ *(semi-finite)* $=$ *(semi-finite)*; *(arbitrary)* $\overline{\otimes}$ *(properly infinite)* $=$ *(properly infinite)*; *(arbitrary)* $\overline{\otimes}$ *(purely infinite)* $=$ *(purely infinite)*; *(discrete)* $\overline{\otimes}$ *(discrete)* $=$ *(discrete)*; *(arbitrary)* $\overline{\otimes}$ *(continuous)* $=$ *(continuous)*.

2.6.7. Proposition. *Let $\mathscr{M}_1$ and $\mathscr{M}_2$ be two W^*-algebras, and let Z_1 and Z_2 be the centers of $\mathscr{M}_1$ and $\mathscr{M}_2$, respectively. The center of $\mathscr{M}_1\overline{\otimes}\mathscr{M}_2$ is the σ-closure of $Z_1\odot Z_2$ (i.e., $Z_1\overline{\otimes}Z_2$) in $\mathscr{M}_1\overline{\otimes}\mathscr{M}_2$. In particular, if $\mathscr{M}_1$ and $\mathscr{M}_2$ are factors, $\mathscr{M}_1\overline{\otimes}\mathscr{M}_2$ is again a factor.*

Proof. Let $\{\pi_1,\mathscr{H}_1\}$, $\{\pi_2,\mathscr{H}_2\}$ be faithful W^*-representations of $\mathscr{M}_1$ and $\mathscr{M}_2$, respectively. Then

$$R(\pi_1(Z_1)\otimes 1_{\mathscr{H}_2}, 1_{\mathscr{H}_1}\otimes\pi_2(Z_2))\subset$$

the center $\mathscr{C}$ of $R(\pi_1(\mathscr{M}_1)\otimes 1_{\mathscr{H}_2}, 1_{\mathscr{H}_1}\otimes\pi_2(\mathscr{M}_2))$.

On the other hand,

$$\pi_1(\mathscr{M}_1)\otimes 1_{\mathscr{H}_2}\subset R(\pi_1(\mathscr{M}_1)\otimes 1_{\mathscr{H}_2}, 1_{\mathscr{H}_1}\otimes\pi_2(\mathscr{M}_2))\subset\mathscr{C}';$$

$$\pi_1(\mathscr{M}_1)'\otimes 1_{\mathscr{H}_2}\subset R(\pi_1(\mathscr{M}_1)'\otimes 1_{\mathscr{H}_2}, 1_{\mathscr{H}_1}\otimes\pi_2(\mathscr{M}_2)')\subset\mathscr{C}'.$$

Hence

$$R(\pi_1(\mathscr{M}_1)\otimes 1_{\mathscr{H}_2}, \pi_1(\mathscr{M}_1)'\otimes 1_{\mathscr{H}_2})=R(\pi_1(Z_1)'\otimes 1_{\mathscr{H}_2})\subset\mathscr{C}'.$$

Analogously $R(1_{\mathscr{H}_1} \otimes \pi_2(Z_2)') \subset \mathscr{C}'$. Hence

$$R(\pi_1(Z_1)' \otimes 1_{\mathscr{H}_2}, 1_{\mathscr{H}_1} \otimes \pi_2(Z_2)') \subset \mathscr{C}',$$

and so

$$\mathscr{C} \subset R(\pi_1(Z_1)' \otimes 1_{\mathscr{H}_2}, 1_{\mathscr{H}_1} \otimes \pi_2(Z_2)')' = R(\pi_1(Z_1) \otimes 1_{\mathscr{H}_2}, 1_{\mathscr{H}_1} \otimes \pi_2(Z_2))$$

(cf. 2.8). q.e.d.

Concluding remark on 2.6.

It would be preferable to find a proof of 2.6.7 without use of the representation theorem (i.e. space-free).

References. [37], [117], [151], [154].

2.7. *-Representations of C^*-Algebras and W^*-Algebras, 2

In this section, we shall consider only nowhere trivial *-representations.

2.7.1. Definition. *Let $\{\pi, \mathscr{H}\}$ be a *-representation of a C^*-algebra $\mathscr{A}$, and let E' be a non-zero projection in the commutant $\pi(\mathscr{A})'$. Then the subrepresentation $\{\pi E', E'\mathscr{H}\}$ of $\{\pi, \mathscr{H}\}$ is called a induction of $\{\pi, \mathscr{H}\}$. Similarly we shall define an induction of a W^*-representation of a W^*-algebra.*

2.7.2. Definition. *Let $\{\pi, \mathscr{H}\}$ be a *-representation of a C^*-algebra $\mathscr{A}$, and let $\mathscr{K}$ be a Hilbert space. A mapping $x \to \pi(x) \otimes 1_{\mathscr{K}}$ of $\mathscr{A}$ into $B(\mathscr{H} \otimes \mathscr{K})$ is again a *-representation of $\mathscr{A}$. This is called an amplification of $\{\pi, \mathscr{H}\}$. Similarly we shall define an amplification of a W^*-representation.*

2.7.3. Proposition. *Let $\{\pi, \mathscr{H}\}$ be a *-representation of a C^*-algebra $\mathscr{A}$, and let E'_1, E'_2 be two non-zero projections in the commutant $\pi(\mathscr{A})'$. $\{\pi E'_1, E'_1\mathscr{H}\}$ is equivalent to $\{\pi E'_2, E'_2\mathscr{H}\}$ if and only if $E'_1 \sim E'_2$ in $\pi(\mathscr{A})'$.*

Proof. Suppose that $\{\pi E'_1, E'_1\mathscr{H}\}$ is equivalent to $\{\pi E'_2, E'_2\mathscr{H}\}$. Then there exists a unitary mapping V' of $E'_1\mathscr{H}$ onto $E'_2\mathscr{H}$ such that $V'\pi(x)E'_1 = \pi(x)E'_2V'$ $(x \in \mathscr{A})$. By defining $V'(1_{\mathscr{H}} - E'_1)\mathscr{H} = 0$, V' can be uniquely extended to a partial isometry on $\mathscr{H}$ such that $V'^*V' = E'_1$ and $V'V'^* = E'_2$. Then $V'\pi(x)E'_1 = V'E'_1\pi(x) = V'\pi(x) = \pi(x)V'$. Hence $V' \in \pi(\mathscr{A})'$. The remainder of 2.7.3 is clear. q.e.d.

2.7.4. Proposition. *Let $\{\pi_1^w, \mathscr{H}_1\}$, $\{\pi_2^w, \mathscr{H}_2\}$ be two W^*-representations of a W^*-algebra $\mathscr{M}$ with $k(\pi_1) \subset k(\pi_2)$ ($k(\pi_i)$, the kernels of π_i $(i=1,2)$). Then $\{\pi_2^w, \mathscr{H}_2\}$ is equivalent to an induction of an amplification of $\{\pi_1^w, \mathscr{H}_1\}$.*

Proof. Consider the representation $\{\pi_1+\pi_2, \mathscr{H}_1\oplus\mathscr{H}_2\}$ of $\mathscr{M}$. Let E'_1 (resp. E'_2) be the orthogonal projection of $\mathscr{H}_1\oplus\mathscr{H}_2$ onto $\mathscr{H}_1$ (resp. $\mathscr{H}_2$). Then $E'_1, E'_2\in(\pi_1+\pi_2)(\mathscr{M})'$, $(\pi_1+\pi_2)(\mathscr{M})''=(\pi_1+\pi_2)(\mathscr{M})$, and $c(E'_1)=1_{\mathscr{H}_1\oplus\mathscr{H}_2}$. By the comparability theorem, there exists a family of mutually orthogonal central projections $(z_\alpha)_{\alpha\in\mathrm{II}}$ in $(\pi_1+\pi_2)(\mathscr{M})'$ with the sum $1_{\mathscr{H}_1\oplus\mathscr{H}_2}$ as follows: for each $\alpha\in\mathrm{II}$ there exists a family of mutually orthogonal projections $(E'_{\alpha,\beta})_{\beta\in\mathrm{II}_\alpha}$ in $(\pi_1+\pi_2)(\mathscr{M})'$ with $E'_{\alpha,\beta}\precsim E'_1 z_\alpha(\beta\in\mathrm{II}_\alpha)$ and $\sum_{\beta\in\mathrm{II}_\alpha} E'_{\alpha,\beta}=E'_2 z_\alpha$. Take a cardinal number $\aleph$ such that $\aleph\geq\mathrm{Card}(\mathrm{II}_\alpha)$ $(\alpha\in\mathrm{II})$, and let $\mathscr{K}$ be a $\aleph$-dimensional Hilbert space. Then

$$\begin{aligned}\{\pi_2,\mathscr{H}_2\}&=\{(\pi_1+\pi_2)E'_2, E'_2(\mathscr{H}_1\oplus\mathscr{H}_2)\}\\&=\sum_{\alpha\in\mathrm{II}}\sum_{\beta\in\mathrm{II}_\alpha}\{(\pi_1+\pi_2)E'_{\alpha,\beta}, E'_{\alpha,\beta}(\mathscr{H}_1\oplus\mathscr{H}_2)\}\\&=\sum_{\alpha\in\mathrm{II}}\{(\pi_1\otimes 1_{\mathscr{K}})F'_\alpha, F'_\alpha\,\mathscr{H}\otimes\mathscr{K}\},\end{aligned}$$

where F'_α is a projection in the commutant $\pi_1\otimes 1_{\mathscr{K}}(\mathscr{M})'$. Since $z_{\alpha_1}\cdot z_{\alpha_2}=0\,(\alpha_1,\alpha_2\in\mathrm{II},\alpha_1\neq\alpha_2)$, $F'_{\alpha_1}\cdot F'_{\alpha_2}=0$. Hence

$$\sum_{\alpha\in\mathrm{II}}\{(\pi_1\otimes 1_{\mathscr{K}})F'_\alpha, F'_\alpha(\mathscr{H}\otimes\mathscr{K})\}=\{(\pi_1\otimes 1_{\mathscr{K}})\Big(\sum_{\alpha\in\mathrm{II}}F'_\alpha\Big), \Big(\sum_{\alpha\in\mathrm{II}}F'_\alpha\Big)(\mathscr{H}\otimes\mathscr{K})\}.$$

q.e.d.

2.7.5. Definition. *A weakly closed self-adjoint subalgebra on a Hilbert space $\mathscr{H}$ is called a W^*-algebra on a Hilbert space $\mathscr{H}$.*

Let $\mathscr{N}$ be a W^*-algebra, containing $1_{\mathscr{H}}$, on a Hilbert space $\mathscr{H}$, and let e (resp. e') be a projection in $\mathscr{N}$ (resp. $\mathscr{N}'$). We shall consider the W^*-algebras $e\mathscr{N}e$ and $\mathscr{N}'e$ (resp. $\mathscr{N}e'$ and $e'\mathscr{N}'e'$) as W^*-algebras, containing $1_{e\mathscr{H}}$ (resp. $1_{e'\mathscr{H}}$), on a Hilbert space $e\mathscr{H}$ (resp. $e'\mathscr{H}$). By $(e\mathscr{N}e)'$ (resp. $(e'\mathscr{N}'e')'$) we shall denote the commutant of $e\mathscr{N}e$ (resp. $(e'\mathscr{N}'e')$) on the space $e\mathscr{H}$ (resp. $e'\mathscr{H}$).

2.7.6. Proposition. *Let $\mathscr{N}$ be a W^*-algebra, containing $1_{\mathscr{H}}$, on a Hilbert space $\mathscr{H}$, and let e (resp. e') be a projection in $\mathscr{N}$ (resp. $\mathscr{N}'$). Then $(e\mathscr{N}e)'=\mathscr{N}'e$ and $(e'\mathscr{N}'e')'=\mathscr{N}e'$.*

Proof. Clearly $(\mathscr{N}'e)'\supset(e\mathscr{N}e)$. Let $a\in(\mathscr{N}'e)'$ and let b be a bounded linear operator on $\mathscr{H}$ such that $b=a$ on $e\mathscr{H}$ and $b=0$ on $(1_{\mathscr{H}}-e)\mathscr{H}$. Then for $x'\in\mathscr{N}'$, $bx'=ebex'=ax'e=x'eae=x'b$. Hence $b\in e\mathscr{N}e$, and so $(\mathscr{N}'e)'=e\mathscr{N}e$. Similarly $(\mathscr{N}e')'=e'\mathscr{N}'e'$. q.e.d.

2.7.7. Notation. *Let $\mathscr{N}$ be a W^*-algebra, containing $1_{\mathscr{H}}$, on a Hilbert space $\mathscr{H}$. For $\xi\in\mathscr{H}$, put $\varphi_\xi(x)=(x\xi,\xi)\,(x\in\mathscr{N})$. φ_ξ is a normal positive linear functional on $\mathscr{N}$.* If $\|\xi\|=1$, *φ_ξ is a normal state on $\mathscr{N}$ (called a vector state).*

2.7.8. Lemma. *Let $\mathcal{N}$ be a W^*-algebra, containing $1_{\mathcal{H}}$, on a Hilbert space $\mathcal{H}$. If ψ is a normal state on $\mathcal{N}$ such that $s(\psi)=P_{[\mathcal{N}'\xi]}$ (some $\xi\in\mathcal{H}$) and $\psi\geq\varphi_\xi$, then there is a vector η in $\mathcal{H}$ with $\psi=\varphi_\eta$ and $[\mathcal{N}\eta]=[\mathcal{N}\xi]$ ($P_{[\mathcal{N}'\xi]}$ is the orthogonal projection of $\mathcal{H}$* onto $[\mathcal{N}'\xi]$).

Proof. Since $\varphi_\xi\leq\psi$, there exists a positive element h in $\mathcal{N}$ with $0\leq h\leq 1$ such that for $x\in\mathcal{N}$, $\varphi_\xi(x)=\psi(hxh)$ by 1.24.3. If e is the support projection of h, then $\varphi_\xi(1-e)=\psi(h(1-e)h)=0$. Therefore

$$1-e\leq 1-s(\varphi_\xi)=1-P_{[\mathcal{N}'\xi]}=1-s(\psi),$$

that is $s(\psi)\leq e$.

Let $h=\int_0^1\lambda\, de_\lambda$ be the spectral decomposition of h, and let $p_n=\int_{1/n}^1 de_\lambda$ and $t_n=\int_{1/n}^1\frac{1}{\lambda}\,de_\lambda$. Then $p_n\to 1-e_{+0}=e$ in the strong operator topology. Hence $\psi(p_n)\to\psi(e)$, and so $\psi(p_n-p_m)\to 0$ $(n,m\to\infty)$. Let $\eta_n=t_n\xi$. Then

$$\begin{aligned}\|\eta_n-\eta_m\|^2&=((t_n-t_m)\xi,(t_n-t_m)\xi)\\&=\varphi_\xi((t_n-t_m)^2)=\psi(h(t_n-t_m)^2h)\\&=\psi((p_n-p_m)^2)=\psi(p_n-p_m)\end{aligned}$$

$(n\geq m)$. The sequence (η_n) is therefore a Cauchy sequence of elements in $\mathcal{H}$, convergent to the vector η (say). Since $p_nxp_n\to exe$ $(x\in\mathcal{N})$, we have $\psi(exe)=\lim_n\psi(p_nxp_n)$. But

$$\psi(exe)=\psi(s(\psi)exes(\psi))=\psi(s(\psi)xs(\psi))=\psi(x),$$

so that

$$\begin{aligned}\psi(x)&=\lim_n\psi(p_nxp_n)=\lim_n\psi(ht_nxt_nh)=\lim_n\varphi_\xi(t_nxt_n)=\lim_n\varphi_{t_n\xi}(x)\\&=\lim_n\varphi_{\eta_n}(x)=\varphi_\eta(x)\qquad(x\in\mathcal{N}).\end{aligned}$$

Because $\eta_n=t_n\xi\in[\mathcal{N}\xi]$, $\eta\in[\mathcal{N}\xi]$, and so $[\mathcal{N}\eta]\subset[\mathcal{N}\xi]$. Also

$$h\eta=\lim_n ht_n\xi=\lim_n p_n\xi=e\xi=es(\psi)\xi=s(\psi)\xi=\xi.$$

Consequently $\xi\in\mathcal{N}\eta$, and so $[\mathcal{N}\xi]\subset[\mathcal{N}\eta]$. Hence $[\mathcal{N}\xi]=[\mathcal{N}\eta]$. q.e.d.

2.7.9. Theorem. *Let $\mathcal{N}$ be a W^*-algebra, containing $1_{\mathcal{H}}$, on a Hilbert space $\mathcal{H}$, ξ an element in $\mathcal{H}$ and φ a normal state on $\mathcal{N}$ such that $s(\varphi)\leq P_{[\mathcal{N}'\xi]}$. Then there exists an element η in $[\mathcal{N}\xi]$ with $\varphi=\varphi_\eta$ and if $s(\varphi)=P_{[\mathcal{N}'\xi]}$, η can be so chosen that $[\mathcal{N}\eta]=[\mathcal{N}\xi]$.*

Proof. Set $\psi=\varphi+\varphi_\xi$. Then $\psi\geq\varphi_\xi$ and $s(\psi)=P_{[\mathcal{N}'\xi]}$. Thus, by 2.7.8, there is an element ξ_1 in $\mathcal{H}$ with $\psi=\varphi_{\xi_1}$ and $[\mathcal{N}\xi_1]=[\mathcal{N}\xi]$. But $\varphi\leq\psi$, so that there exists a positive element h in $\mathcal{N}$ with $0\leq h\leq 1$ such that for $x\in\mathcal{N}$, $\varphi(x)=\psi(hxh)$. Thus $\varphi(x)=\varphi_{\xi_1}(hxh)=\varphi_{h\xi_1}(x)$, that is $\varphi=\varphi_\eta$ where $\eta=h\xi_1$.

If e is the support projection of h, then $h(1-e)=0$ and so $\varphi(1-e)=\psi(h(1-e)h)=0$. Therefore $1-e\leq 1-s(\varphi)$, that is $s(\varphi)\leq e$. Let $h=\int_0^1\lambda\, de_\lambda$ be the spectral decomposition of h and set $p_n=\int_{1/n}^1 de_\lambda$,

$$t_n=\int_{1/n}^1\frac{1}{\lambda}\,de_\lambda.$$ Then $t_n\eta=t_nh\xi_1=p_n\xi_1\to e\xi_1$. Now $[\mathcal{N}'\xi_1]=s(\psi)=[\mathcal{N}'\xi]$. When $s(\varphi)=P_{[\mathcal{N}'\xi]}$, we have $e\xi_1=e(s(\psi)\xi_1)=e(s(\varphi)\xi_1)=s(\varphi)\xi_1=\xi_1$. Therefore $t_n\eta\to\xi_1$, $\xi_1\in[\mathcal{N}\eta]$, and so $[\mathcal{N}\xi_1]\subset[\mathcal{N}\eta]$. Thus, when $s(\varphi)=P_{[\mathcal{N}'\xi]}$, $[\mathcal{N}\eta]=[\mathcal{N}\xi]$. q.e.d.

2.7.10. Corollary. *Let $\mathcal{N}$ be a W^*-algebra, containing $1_{\mathcal{H}}$, on a Hilbert space $\mathcal{H}$ and let ξ,η be elements in $\mathcal{H}$. Then $P_{[\mathcal{N}'\eta]}\lesssim P_{[\mathcal{N}'\xi]}$ (in $\mathcal{N}$) if and only if $P_{[\mathcal{N}\eta]}\lesssim P_{[\mathcal{N}\xi]}$ (in $\mathcal{N}'$).*

Proof. It suffices to prove that $P_{[\mathcal{N}'\eta]}\prec P_{[\mathcal{N}'\xi]}$ implies $P_{[\mathcal{N}\eta]}\prec P_{[\mathcal{N}\xi]}$. There are a ξ_1 in $[\mathcal{N}'\xi]$ and a partial isometry v in $\mathcal{N}$ such that $\eta=v\xi_1$. Hence $[\mathcal{N}\eta]\subseteq[\mathcal{N}\xi_1]$. Since $\xi_1\in[\mathcal{N}'\xi]$, $[\mathcal{N}'\xi_1]\subseteq[\mathcal{N}'\xi]$. Hence $s(\varphi_{\xi_1})\leq s(\varphi_\xi)$, and so by 2.7.9 there exists an η_1 in $[\mathcal{N}\xi]$ with $\varphi_{\eta_1}(x)=(x\eta_1,\eta_1)=\varphi_{\xi_1}(x)=(x\xi_1,\xi_1)$ $(x\in\mathcal{N})$. Hence there is a partial isometry v' in $\mathcal{N}'$ such that $\xi_1=v'\eta_1$, $v'^*v'=P_{[\mathcal{N}\eta_1]}$ and $v'v'^*=P_{[\mathcal{N}\xi_1]}$. Therefore $P_{[\mathcal{N}\eta]}\leq P_{[\mathcal{N}\xi_1]}\sim P_{[\mathcal{N}\eta_1]}\leq P_{[\mathcal{N}\xi]}$. q.e.d.

2.7.11. Proposition. *Let $\{\pi,\mathcal{H}\}$ be a *-representation of a C^*-algebra $\mathcal{A}$. $\{\pi,\mathcal{H}\}$ has a joint cyclic and separating vector if it has cyclic and separating vectors.*

Proof. Let ξ_1 (resp. ξ_2) be a cyclic (resp. separating) vector in $\mathcal{H}$. Since $[\pi(\mathcal{A})\xi_1]\supseteq[\pi(\mathcal{A})\xi_2]$, $[\pi(\mathcal{A})'\xi_1]\succ[\pi(\mathcal{A})'\xi_2]$. Hence

$$[\pi(\mathcal{A})'\xi_1]\sim[\pi(\mathcal{A})'\xi_2],$$

since $[\pi(\mathcal{A})'\xi_2]=\mathcal{H}$. Let u be a partial isometry in $\pi(\mathcal{A})''$ with $u^*u=P_{[\pi(\mathcal{A})'\xi_1]}$, $uu^*=P_{[\pi(\mathcal{A})'\xi_2]}$. Put $\xi=u\xi_1$. Then

$$[\pi(\mathcal{A})\xi]\supseteq[\pi(\mathcal{A})u^*u\xi_1]=[\pi(\mathcal{A})\xi_1]=\mathcal{H}$$

and $[\pi(\mathcal{A})'\xi]=[u\pi(\mathcal{A})'\xi_1]=[\pi(\mathcal{A})'\xi_2]=\mathcal{H}$. q.e.d.

2.7.12. Theorem. *Let $\{\pi_1^w,\mathcal{H}_1\}$, $\{\pi_2^w,\mathcal{H}_2\}$ be two W^*-representations of a W^*-algebra $\mathcal{M}$, with joint cyclic and separating vectors ξ_1,ξ_2 respectively, such that $k(\pi_1)=k(\pi_2)$. Then they are equivalent.*

Proof. Consider the representation $\{\pi_1+\pi_2, \mathscr{H}_1\oplus\mathscr{H}_2\}$ of $\mathscr{M}$. $P_{[(\pi_1+\pi_2)(\mathscr{M})'\xi_i]}\geq P_{\mathscr{H}_i}(i=1,2)$. Since $P_{[(\pi_1+\pi_2)(\mathscr{M})'\xi_i]}\in(\pi_1+\pi_2)(\mathscr{M})$, for some projection e_i in $\mathscr{M}$, $P_{[(\pi_1+\pi_2)(\mathscr{M})'\xi_i]}=(\pi_1+\pi_2)(e_i)$. $(\pi_1+\pi_2)(1-e_i)P_{\mathscr{H}_i}=0$, and so $\pi_i(1-e_i)=0$. Hence $1-e_i\in k(\pi_i)$, and so $P_{[(\pi_1+\pi_2)(\mathscr{M})'\xi_i]}=1_{\mathscr{H}_1\oplus\mathscr{H}_2}$. Hence $P_{[(\pi_1+\pi_2)(\mathscr{M})\xi_1]}=P_{[\pi_1(\mathscr{M})\xi_1]}\sim P_{[(\pi_1+\pi_2)(\mathscr{M})\xi_2]}=P_{[\pi_2(\mathscr{M})\xi_2]}$, and so $P_{\mathscr{H}_1}\sim P_{\mathscr{H}_2}$ in $(\pi_1+\pi_2)(\mathscr{M})'$. q.e.d.

2.7.13. Definition. *Let $\mathscr{N}$ be a W^*-algebra, containing $1_{\mathscr{H}}$, on a Hilbert space $\mathscr{H}$, and let t be a closed linear operator in $\mathscr{H}$. If $u'^*tu'=t$ for all unitary u' in $\mathscr{N}'$, then t is said to be associated to $\mathscr{N}$ and this situation is denoted by $t\hat{\in}\mathscr{N}$.*

2.7.14. Theorem (BT theorem). *Let $\mathscr{N}$ be a W^*-algebra, containing $1_{\mathscr{H}}$, on a Hilbert space $\mathscr{H}$, ξ an element in $\mathscr{H}$ and let $[\mathscr{N}\xi]$ be the closed linear subspace of $\mathscr{H}$ generated by $\mathscr{N}\xi$. Then for each $\eta\in[\mathscr{N}\xi]$ there exist a bounded linear operator b and a closed linear operator t with a dense domain in $\mathscr{H}$ such that $bt\xi=\eta$, $b\in\mathscr{N}$ and $t\hat{\in}\mathscr{N}$.*

Proof. We can find for every $n=1,2,\ldots$ an $a_n\in\mathscr{N}$ with $\|a_n\xi-\eta\|<1/2^n$. Then $\|a_{n+1}\xi-a_n\xi\|<3/2^{n+1}$ and so

$$(2^n(a_{n+1}-a_n)^*(a_{n+1}-a_n)\xi,\xi)=2^n\|(a_{n+1}-a_n)\xi\|^2<9/2^{n+2}.$$

Set $b_0=1_{\mathscr{H}}$ and $b_n=2^n(a_{n+1}-a_n)^*(a_{n+1}-a_n)$ $(n=1,2,\ldots)$ and let $\mathscr{H}_0$ be the set of $\eta\in\mathscr{H}$ for which $\sum_{n=0}^{\infty}(b_n\eta,\eta)<+\infty$. For $\eta_1,\eta_2\in\mathscr{H}_0$, set $Q(\eta_1,\eta_2)=\sum_{n=0}^{\infty}(b_n\eta_1,\eta_2)$. Then one can easily see that Q defines an inner product in $\mathscr{H}_0$. Moreover $\mathscr{H}_0$ is a Hilbert space with this inner product. In fact, let (η_i) be a Q-Cauchy sequence. Then

$$Q(\eta_i-\eta_j,\eta_i-\eta_j)=\sum_{n=1}^{\infty}\|b_n^{\frac{1}{2}}(\eta_i-\eta_j)\|^2\to 0 \quad (i,j\to\infty).$$

Since $\|\eta_i-\eta_j\|^2\leq Q(\eta_i-\eta_j,\eta_i-\eta_j)$, (η_i) is a Cauchy sequence in $\mathscr{H}$. Let η_0 be the limit of (η_i) in $\mathscr{H}$. Then for arbitrary positive number ε there exists an integer i_0 such that $\sum_{n=0}^{\infty}\|b_n^{\frac{1}{2}}(\eta_i-\eta_j)\|^2<\varepsilon$ $(i,j\geq i_0)$. Therefore for an arbitrary positive integer m,

$$\sum_{n=1}^{m}\|b_n^{\frac{1}{2}}(\eta_i-\eta_j)\|^2<\varepsilon.$$

Hence $\sum_{n=1}^{m}\|b_n^{\frac{1}{2}}(\eta_0-\eta_j)\|^2\leq\varepsilon$ and so $\sum_{n=1}^{\infty}\|b_n^{\frac{1}{2}}(\eta_0-\eta_j)\|^2\leq\varepsilon$—i.e. (η_j) converges to η_0 in the Q-topology. For $\zeta\in\mathscr{H}$ and $\eta\in\mathscr{H}_0$,

$$|(\zeta,\eta)|\leq\|\zeta\|\,\|\eta\|\leq\|\zeta\|\,Q(\eta,\eta)^{\frac{1}{2}}.$$

Hence there exists a unique element ζ^* in $\mathscr{H}^0$ with $(\zeta,\eta)=Q(\zeta^*,\eta)$, and $Q(\zeta^*,\zeta^*)\leq\|\zeta\|^2$.

Put $c\zeta=\zeta^*$ $(\zeta\in\mathscr{H})$. Then c is a bounded positive linear operator on $\mathscr{H}$. In fact, $(c\zeta,\zeta)=(\zeta^*,\zeta)=\overline{(\zeta,\zeta^*)}=\overline{Q(\zeta^*,\zeta^*)}=Q(\zeta^*,\zeta^*)\geq 0$, and $(c\zeta,\zeta)=Q(\zeta^*,\zeta^*)\leq\|\zeta\|^2$.

If $\zeta\in\mathscr{H}\ominus[\mathscr{H}_0]$, $(\zeta,\eta)=Q(\zeta^*,\eta)=0$ for $\eta\in\mathscr{H}_0$; hence $c\zeta=0$ and so $c^{\frac{1}{2}}\zeta=0$. Moreover $s(c)=s(c^{\frac{1}{2}})=P_{[\mathscr{H}_0]}$. Let $\zeta\in[\mathscr{H}_0]$. Then there exists a sequence $c^{\frac{1}{2}}\zeta_n$ such that $\|c^{\frac{1}{2}}\zeta_n-\zeta\|\to 0$ $(n\to\infty)$.

$$\begin{aligned}Q(c\zeta_m-c\zeta_n, c\zeta_m-c\zeta_n)&=(\zeta_m-\zeta_n, c\zeta_m-c\zeta_n)\\&=(c^{\frac{1}{2}}(\zeta_m-\zeta_n), c^{\frac{1}{2}}(\zeta_m-\zeta_n))\to 0,\end{aligned}$$

$(m,n\to\infty)$. Hence there exists an element ζ_0 such that

$$Q(c\zeta_n-\zeta_0, c\zeta_n-\zeta_0)\to 0 \qquad (n\to\infty),$$

and so $\|c\zeta_n-\zeta_0\|\to 0$ $(n\to\infty)$. Since $\|c\zeta_n-c^{\frac{1}{2}}\zeta\|\to 0$ $(n\to\infty)$, $c^{\frac{1}{2}}\zeta=\zeta_0$. Hence $c^{\frac{1}{2}}[\mathscr{H}_0]\subset\mathscr{H}_0$. Moreover for $\zeta^1,\zeta^2\in[\mathscr{H}_0]$, by forming the sequences $(c^{\frac{1}{2}}\zeta_n^i)$ $(i=1,2)$ with $\|c^{\frac{1}{2}}\zeta_n^i-\zeta^i\|\to 0$ $(n\to\infty)$, we have

$$\begin{aligned}Q(c^{\frac{1}{2}}\zeta^1, c^{\frac{1}{2}}\zeta^2)&=\lim_n Q(c\zeta_n^1, c\zeta_n^2)=\lim_n(\zeta_n^1, c\zeta_n^2)\\&=\lim_n(c^{\frac{1}{2}}\zeta_n^1, c^{\frac{1}{2}}\zeta_n^2)=(\zeta^1,\zeta^2).\end{aligned}$$

Hence $c^{\frac{1}{2}}$ is an isometric mapping of $[\mathscr{H}_0]$ into $\mathscr{H}_0$.

Let $\eta\in\mathscr{H}_0\ominus c^{\frac{1}{2}}[\mathscr{H}_0]$; then $Q(\eta,c^{\frac{1}{2}}\zeta)=0$ for $\zeta\in[\mathscr{H}_0]$. For every $\zeta\in\mathscr{H}$, $c^{\frac{1}{2}}\zeta\in[\mathscr{H}_0]$. Hence $Q(\eta,c\zeta)=(\eta,\zeta)=0$ for $\zeta\in\mathscr{H}$ and so $\eta=0$, so that $c^{\frac{1}{2}}[\mathscr{H}_0]=\mathscr{H}_0$. Since $b_n\in\mathscr{N}$, $u'^*b_nu'=b_n$ for every unitary $u'\in\mathscr{N}'$. Therefore $\mathscr{H}_0$ and $Q(\zeta_1,\zeta_2)$ are invariant under these elements. Hence $c\in\mathscr{N}$ and so $c^{\frac{1}{2}}\in\mathscr{N}$.

Let t_0 be the inverse mapping of $c^{\frac{1}{2}}$. As $c^{\frac{1}{2}}\in\mathscr{N}$, $t_0\hat{\in}\mathscr{N}$. Let P be the orthogonal projection of $\mathscr{H}$ onto $[\mathscr{H}_0]$ and put $t=t_0P$. Then t is a closed linear operator with a dense domain in $\mathscr{H}$ and $t\hat{\in}\mathscr{N}$.

Consider $\zeta\in\mathscr{H}$. Then $c^{\frac{1}{2}}\zeta=c^{\frac{1}{2}}P\zeta\in\mathscr{H}_0$ and

$$Q(c^{\frac{1}{2}}\zeta,c^{\frac{1}{2}}\zeta)=(P\zeta,P\zeta)\leq\|\zeta\|^2.$$

But

$$\begin{aligned}Q(c^{\frac{1}{2}}\zeta,c^{\frac{1}{2}}\zeta)&=(c^{\frac{1}{2}}\zeta,c^{\frac{1}{2}}\zeta)+\sum_{n=1}^{\infty}(2^n(a_{n+1}-a_n)^*(a_{n+1}-a_n)c^{\frac{1}{2}}\zeta,c^{\frac{1}{2}}\zeta)\\&=\|c^{\frac{1}{2}}\zeta\|^2+\sum_{n=1}^{\infty}2^n\|(a_{n+1}-a_n)c^{\frac{1}{2}}\zeta\|^2\geq 2^n\|(a_{n+1}-a_n)c^{\frac{1}{2}}\zeta\|^2\end{aligned}$$

for $n=1,2,\ldots$. Hence $\|(a_{n+1}-a_n)c^{\frac{1}{2}}\zeta\|\leq\dfrac{1}{\sqrt{2^n}}\|\zeta\|$, and so

$$\|(a_{n+p}-a_n)c^{\frac{1}{2}}\zeta\|\leq\sum_{i=0}^{p-1}\frac{1}{\sqrt{2^{n+i}}}\|\zeta\|\leq\frac{1}{\sqrt{2^n}}\,\frac{1}{1-\dfrac{1}{\sqrt{2}}}\|\zeta\|=\frac{2+\sqrt{2}}{\sqrt{2^n}}\|\zeta\|.$$

Hence $\|(a_m - a_n)c^{\frac{1}{2}}\zeta\| \leq (2+\sqrt{2})/\sqrt{2^{\min(m,n)}}\|\zeta\|$. Thus the sequence $(a_n c^{\frac{1}{2}})$ converges uniformly to a bounded operator b belonging to $\mathcal{N}$. Now since $\xi \in \mathcal{H}_0$, $t\xi = t_0\xi \in [\mathcal{H}_0]$ and $c^{\frac{1}{2}} t\xi = c^{\frac{1}{2}} t_0 \xi = \xi$. Hence $\eta = \lim_n a_n \xi = \lim_n a_n c^{\frac{1}{2}} t\xi = bt\xi_0$. q.e.d.

Concluding remark on 2.7.

The proof of Theorem 2.7.9 is due to Vowden [216].

References. [9], [37], [44], [63], [64], [116].

2.8. The Commutation Theorem of Tensor Products

Let $\mathcal{N}_1$ (resp. $\mathcal{N}_2$) be a W^*-algebra, containing $1_{\mathcal{H}_1}$, (resp. $1_{\mathcal{H}_2}$) on a Hilbert space $\mathcal{H}_1$ (resp. $\mathcal{H}_2$) and let $R(\mathcal{N}_1 \otimes 1_{\mathcal{H}_2}, 1_{\mathcal{H}_1} \otimes \mathcal{N}_2)$ (resp. $R(\mathcal{N}_1' \otimes 1_{\mathcal{H}_2}, 1_{\mathcal{H}_1} \otimes \mathcal{N}_2'))$ be the W^*-algebra on $\mathcal{H}_1 \otimes \mathcal{H}_2$ generated by $\mathcal{N}_1 \otimes 1_{\mathcal{H}_2}$ and $1_{\mathcal{H}_1} \otimes \mathcal{N}_2$ (resp. $\mathcal{N}_1' \otimes 1_{\mathcal{H}_2}$ and $1_{\mathcal{H}_1} \otimes \mathcal{N}_2'$). Then we shall show

2.8.1. Theorem. $R(\mathcal{N}_1 \otimes 1_{\mathcal{H}_2}, 1_{\mathcal{H}_1} \otimes \mathcal{N}_2)' = R(\mathcal{N}_1' \otimes 1_{\mathcal{H}_2}, 1_{\mathcal{H}_1} \otimes \mathcal{N}_2')$.

To prove the theorem, we shall provide some considerations.

By 1.22.14 and 2.2.5, we can write as follows:

$$\mathcal{N}_i = \sum_{\alpha \in \mathbb{I}_i} \oplus \{\mathcal{L}_{i,\alpha} \bar{\otimes} B(\mathcal{H}_{i,\alpha})\} \quad (i = 1,2)$$

($\mathcal{L}_{i,\alpha}$ is a countably decomposable W^*-algebra). Therefore the proof can be reduced to the case that $\mathcal{N}_i = \mathcal{L}_i \bar{\otimes} B(K_i)$ $(i=1,2)$, ($\mathcal{L}_i$ is a countably decomposable W^*-algebra). Then by using the results of 1.20, we can reduce the proof to the case where $\mathcal{N}_i = \mathcal{L}_i$ $(i=1,2)$. Therefore we may assume that $\mathcal{N}_i$ $(i=1,2)$ is countably decomposable. Then $\mathcal{N}_i$ has a faithful normal state φ_i. Consider the W^*-representation $\{\pi^w_{\varphi_i}, \mathcal{H}_{\varphi_i}\}$ of $\mathcal{N}_i$. Then 1_{φ_i} is a cyclic and separating vector for $\mathcal{N}_i$. Then by using 2.7.4 and 2.7.6, we can reduce the proof to the case where $\mathcal{N}_i$ on $\mathcal{H}_i$ has a cyclic and separating vector ξ_i.

Now let $\mathcal{N}$ be a W^*-algebra, containing $1_{\mathcal{H}}$, on a Hilbert space $\mathcal{H}$ and suppose that $\mathcal{N}$ has a cyclic and separating vector ξ_0. Then mappings: $a \to a\xi_0$ and $a' \to a'\xi_0$ $(a \in \mathcal{N}, a' \in \mathcal{N}')$ are one-to-one.

We shall define a conjugate linear mapping Φ (resp. Ψ) on $\mathcal{N}\xi_0$ (resp. $\mathcal{N}'\xi_0$) as follows: $\Phi(a\xi_0) = a^*\xi_0$ (resp. $\Psi(a'\xi_0) = a'^*\xi_0$). Then $(\Phi(a\xi_0), a'\xi_0) = (a^*\xi_0, a'\xi_0) = (a'^*\xi_0, a\xi_0) = \overline{(a\xi_0, a'^*\xi_0)}$. Hence

$$\Phi^*(a'\xi_0) = a'^*\xi_0 = \Psi(a'\xi_0).$$

Since Φ^* has a dense domain, Φ has the least closed extension $\tilde{\Phi}(=\Phi^{**})$. Analogously $\Psi^* \supset \Phi$ and so Ψ has the least closed extension $\tilde{\Psi}(=\Psi^{**})$.

2.8.2. Lemma. $\tilde{\Psi}=\Phi^*$.

Proof. Let $\mathscr{D}(\Phi^*)$ be the domain of Φ^*. For $\xi\in\mathscr{D}(\Phi^*)$,

$$(\Phi(a\xi_0),\xi)=\overline{(a\xi_0,\Phi^*(\xi))} \qquad (a\in\mathcal{N}).$$

Since $\xi\in[\mathcal{N}'\xi_0]$, by 2.7.14. ($BT$ theorem), there exist a bounded linear operator b' and a closed linear operator t' such that $b't'\xi_0=\xi, b'\in\mathcal{N}'$ and $t'\hat{\in}\mathcal{N}'$. Thus

$$(\Phi(a\xi_0),\xi)=(a^*\xi_0,b't'\xi_0)=(b'^*\xi_0,t'a\xi_0).$$

Hence $(t'a\xi_0,b'^*\xi_0)=(a\xi_0,\Phi^*(\xi))\,(a\in\mathcal{N})$.

Let s' be the restriction of t' to $\mathcal{N}\xi_0$ and let $\tilde{s}'$ be its least closed extension. Then for every unitary $u\in\mathcal{N}$, $u^*s'u=s'$ and so $u^*\tilde{s}'u=\tilde{s}'$. Hence $\tilde{s}'\hat{\in}\mathcal{N}'$. Moreover $\tilde{s}'^*b'^*\xi_0=\Phi^*(\xi)$. Let $\tilde{s}'=v'h'$ be the polar decomposition of $\tilde{s}'$ (cf. [124]). Then by the uniqueness of polar decomposition, $v'\in\mathcal{N}'$ and $h'\hat{\in}\mathcal{N}'$. Let $h'=\int_0^\infty\lambda\,de'_\lambda$ be the spectral decomposition of h' and set $h'_n=\int_0^n\lambda\,de'_\lambda$. Then

$$\tilde{s}'^*b'^*\xi_0=h'v'^*b'^*\xi_0=\lim_n h'_n v'^*b'^*\xi_0$$

and $b'\tilde{s}'\xi_0=b'v'h'\xi_0=\lim_n b'v'h'_n\xi_0$. Since

$$\Psi(b'v'h'_n\xi_0)=h'_nv'^*b'^*\xi_0\to\Phi^*(\xi),\ \tilde{\Psi}=\Phi^*. \qquad \text{q.e.d.}$$

Let $\mathscr{H}_1$ (resp. $\mathscr{H}_2$) be a Hilbert space and let H_1 (resp. H_2) be a closed linear operator, with a dense domain, in $\mathscr{H}_1$ (resp. $\mathscr{H}_2$). Then one can define canonically a linear operator $H_1\odot H_2$ on

$$\mathscr{D}(H_1)\odot\mathscr{D}(H_2)\,(\subset\mathscr{H}_1\otimes\mathscr{H}_2)$$

such that $H_1\odot H_2(\eta_1\otimes\eta_2)=H_1\eta_1\otimes H_2\eta_2(\eta_1\in\mathscr{D}(H_1),\eta_2\in\mathscr{D}(H_2))$. Since $H_1^*\odot H_2^*\subset(H_1\odot H_2)^*$, $H_1\odot H_2$ has the least closed extension (denoted by $H_1\otimes H_2$).

2.8.3. Lemma. *If H_1 and H_2 are self-adjoint, then $H_1\otimes H_2$ is again self-adjoint.*

Proof. Let $H_1=\int_{-\infty}^{\infty}\lambda\,dE_\lambda$ and $H_2=\int_{-\infty}^{\infty}\mu\,dF_\mu$ be the spectral decomposition of H_1 and H_2 respectively. Consider a self-adjoint operator $K=\int_{-\infty}^{\infty}\int_{-\infty}^{\infty}\lambda\mu\,dE_\lambda\otimes F_\mu$ in $\mathscr{H}_1\otimes\mathscr{H}_2$, and put $P_n=\int_{-n}^{n}dE_\lambda$

and $Q_n = \int_{-n}^{n} dF_\mu$. Then $K(P_n \otimes Q_n) = (H_1 \otimes H_2)(P_n \otimes Q_n)$, and so $K \subset H_1 \otimes H_2$. Since $K^* = K \supset (H_1 \otimes H_2)^* \supset H_1 \otimes H_2$, $K = H_1 \otimes H_2$.

q.e.d.

Proof of Theorem 2.8.1. Set $\Phi_1(a_1 \xi_1) = a_1^* \xi_1$ and

$$\Phi_2(a_2 \xi_2) = a_2^* \xi_2 \qquad (a_1 \in \mathcal{N}_1, a_2 \in \mathcal{N}_2),$$

and let $\tilde{\Phi}_1$ (resp. $\tilde{\Phi}_2$) be the least closed extension of Φ_1 (resp. Φ_2) in $\mathcal{H}_1$ (resp. $\mathcal{H}_2$). Then by applying the polar decomposition to closed conjugate linear operators $\tilde{\Phi}_1, \tilde{\Phi}_2$ and using 2.8.3, we have $(\tilde{\Phi}_1 \odot \tilde{\Phi}_2)^* =$ the least closed extension of $\Phi_1^* \odot \Phi_2^*$.

Now consider the W^*-algebra $R(\mathcal{N}_1 \otimes 1_{\mathcal{H}_2}, 1_{\mathcal{H}_1} \otimes \mathcal{N}_2)$ (say $\mathcal{N}$). Then $\xi_1 \otimes \xi_2$ is a cyclic and separating vector of $\mathcal{N}$. Define

$$\Phi(a \xi_1 \otimes \xi_2) = a^* \xi_1 \otimes \xi_2 \quad (a \in \mathcal{N}).$$

Then clearly $\Phi_1 \odot \Phi_2 \subset \Phi$ and so $(\Phi_1 \odot \Phi_2)^* \supset \Phi^*$.

On the other hand, $\Phi^*(a' \xi_1 \otimes \xi_2) = a'^* \xi_1 \otimes \xi_2 \, (a' \in \mathcal{N}')$. Since $(\Phi_1 \odot \Phi_2)^*$ is the least closed extension of $\Phi_1^* \odot \Phi_2^*$ and Φ_1^* (resp. Φ_2^*) is the least closed extension of Ψ_1 (resp. Ψ_2)

$$(\Psi_1(a_1' \xi_1) = a_1'^* \xi_1, \Psi_2(a_2' \xi_2) = a_2'^* \xi_2 \text{ for } a_1' \in \mathcal{N}_1' \text{ and } a_2' \in \mathcal{N}_2')$$

by 2.8.2, $(\Phi_1 \odot \Phi_2)^*$ is the least closed extension of $\Psi_1 \odot \Psi_2$. Hence there exists a sequence (ζ_n) of elements in $\mathcal{H}_1 \otimes \mathcal{H}_2$, with the form

$$\zeta_n = \sum_{i=1}^{p_n} a'_{1,n,i} \xi_1 \otimes a'_{2,n,i} \xi_2 \qquad (a'_{1,n,i} \in \mathcal{N}_1', a'_{2,n,i} \in \mathcal{N}_2'),$$

such that $\|\zeta_n - a' \xi_1 \otimes \xi_2\| \to 0$ and $\|\Phi^*(\zeta_n) - \Phi^*(a' \xi_1 \otimes \xi_2)\| \to 0$. Since $\Phi^*(\zeta_n) = \sum_{i=1}^{p_n} a'^*_{1,n,i} \xi_1 \otimes a'^*_{2,n,i} \xi_2$ and $\Phi^*(a' \xi_1 \otimes \xi_2) = a'^* \xi_1 \otimes \xi_2$, there exists a sequence (b_n') of elements in $R(\mathcal{N}_1' \otimes 1_{\mathcal{H}_2}, 1_{\mathcal{H}_1} \otimes \mathcal{N}_2')$ such that $\|b_n' \xi_1 \otimes \xi_2 - a' \xi_1 \otimes \xi_2\| \to 0$ and $\|b_n'^* \xi_1 \otimes \xi_2 - a'^* \xi_1 \otimes \xi_2\| \to 0$. Let

$$b \in R(B(\mathcal{H}_1) \otimes 1_{\mathcal{H}_2}, 1_{\mathcal{H}_1} \otimes \mathcal{N}_2) \cap R(\mathcal{N}_1 \otimes 1_{\mathcal{H}_2}, 1_{\mathcal{H}_1} \otimes B(\mathcal{H}_2)) \text{ (say } \mathcal{L})$$

and $c, d \in R(\mathcal{N}_1 \otimes 1_{\mathcal{H}_2}, 1_{\mathcal{H}_1} \otimes \mathcal{N}_2)$. Then

$$\begin{aligned}(b a' c \xi_1 \otimes \xi_2, d \xi_1 \otimes \xi_2) &= (b c a' \xi_1 \otimes \xi_2, d \xi_1 \otimes \xi_2) \\ &= \lim_n (b c b_n' \xi_1 \otimes \xi_2, d \xi_1 \otimes \xi_2).\end{aligned}$$

Since $\mathscr{L}'=R\ (1_{\mathscr{H}_1}\otimes\mathscr{N}_2', \mathscr{N}_1'\otimes 1_{\mathscr{H}_2})$, we have

$$\begin{aligned}\lim_n (bcb_n'\xi_1\otimes\xi_2, d\xi_1\otimes\xi_2) &= \lim_n (b_n'bc\xi_1\otimes\xi_2, d\xi_1\otimes\xi_2)\\ &= \lim_n (bc\xi_1\otimes\xi_2, b_n'^* d\xi_1\otimes\xi_2)\\ &= \lim_n (bc\xi_1\otimes\xi_2, db_n'^*\xi_1\otimes\xi_2)\\ &= (bc\xi_1\otimes\xi_2, da'^*\xi_1\otimes\xi_2)\\ &= (bc\xi_1\otimes\xi_2, a'^* d\xi_1\otimes\xi_2)\\ &= (a'bc\xi_1\otimes\xi_2, d\xi_1\otimes\xi_2).\end{aligned}$$

Hence $(ba'c\xi_1\otimes\xi_2, d\xi_1\otimes\xi_2)=(a'bc\xi_1\otimes\xi_2, d\xi_1\otimes\xi_2)$. Since

$$[\mathscr{N}_1\xi_1\otimes\mathscr{N}_2\xi_2]=\mathscr{H}_1\otimes\mathscr{H}_2,$$

this implies $ba'=a'b$, so that $a'\in\mathscr{L}'=R\ (1_{\mathscr{H}_1}\otimes\mathscr{N}_2', \mathscr{N}_1'\otimes 1_{\mathscr{H}_2})$. Hence $R(\mathscr{N}_1\otimes 1_{\mathscr{H}_2}, 1_{\mathscr{H}_1}\otimes\mathscr{N}_2)'=R(\mathscr{N}_1'\otimes 1_{\mathscr{H}_2}, 1_{\mathscr{H}_1}\otimes\mathscr{N}_2')$. q.e.d.

Concluding remark on 2.8.

Theorem 2.8.1 was first proved by Tomita [210], using a different method. He proved also that a W^*-algebra $\mathscr{N}$ with $1_{\mathscr{H}}$ on a Hilbert space $\mathscr{H}$ is conjugate *-isomorphic to the commutant $\mathscr{N}'$ if $\mathscr{N}$ has a cyclic and separating vector.

References. [37], [117], [166], [206], [210].

2.9. Spatial Isomorphisms of W^*-Algebras

2.9.1. Definition. *Let $\mathscr{M}$ be a countably decomposable finite W^*-algebra, and let τ be a faithful normal tracial positive linear functional on $\mathscr{M}$. Then the W^*-representation $\{\pi_\tau^w, \mathscr{H}_\tau\}$ is called a standard representation of $\mathscr{M}$.*

$$(a_\tau, b_\tau)=\tau(b^*a)=\tau(ab^*)=(b_\tau^*, a_\tau^*) \qquad (a,b\in\mathscr{M}).$$

Hence a conjugate linear mapping $a_\tau\to a_\tau^*$ can be extended uniquely to a conjugate linear isometry J on $\mathscr{H}_\tau$ *(called the involution)*. Then

$$J\pi_\tau(x)J a_\tau = J\pi_\tau(x)a_\tau^*=(xa^*)_\tau^*=(ax^*)_\tau \qquad (x,a\in\mathscr{M}).$$

Set $R_x a_\tau=(ax)_\tau$ $(x,a\in\mathscr{M})$. Then R_x is extended to a bounded linear operator on $\mathscr{H}_\tau$ and $R_x\in\pi_\tau(\mathscr{M})'$.

Conversely let h' be a positive element in $\pi_\tau(\mathscr{M})'$ with $0\le h'\le 1_{\mathscr{H}_\tau}$ and set $\varphi(x)=(\pi_\tau(x)h'1_\tau, 1_\tau)$ $(x\in\mathscr{M})$. Then $0\le\varphi\le\tau$. Hence there exists an element h_0 in $\mathscr{M}$, with $0\le h_0\le 1$, such that $\varphi(x)=\tau(h_0xh_0)$ $(x\in\mathscr{M})$ by 1.24.3.

$$\begin{aligned}\tau(h_0xh_0)&=\tau(xh_0^2)=(\pi_\tau(xh_0^2)1_\tau, 1_\tau)\\ &=(\pi_\tau(x)R_{h_0^2}1_\tau, 1_\tau)=(\pi_\tau(x)h'1_\tau, 1_\tau) \qquad (x\in\mathscr{M}).\end{aligned}$$

Hence $h'1_\tau = R_{h_0^2} 1_\tau$. Since $[\pi_\tau(\mathcal{M})1_\tau] = \mathcal{H}_\tau$, $h' = R_{h_0^2}$. Therefore

$$\{R_x \mid x \in \mathcal{M}\} = \pi_\tau(\mathcal{M})'.$$

Hence we have

2.9.2. Proposition. *Let $\mathcal{H}$ be a countably decomposable finite W^*-algebra, and let τ be a faithful normal tracial state on $\mathcal{M}$. Then the standard representation $\{\pi_\tau, \mathcal{H}_\tau\}$ of $\mathcal{M}$ has a joint cyclic and separating vector. Moreover, the mapping $\pi_\tau(x) \to J\pi_\tau(x)J$ $(x \in \mathcal{M})$ is a conjugate linear $*$-isomorphism of $\pi_\tau(\mathcal{M})$* onto *$\pi_\tau(\mathcal{M})'$.*

2.9.3. Corollary. *Let $L^\infty(\Omega, \mu)$ be the commutative W^*-algebra of all essentially bounded μ-locally measurable functions on a localizable measure space (Ω, μ), and let $L^2(\Omega, \mu)$ be the Hilbert space of all μ-locally measurable square integrable functions on Ω. For $f \in L^\infty(\Omega, \mu)$, $g \in L^2(\Omega, \mu)$, set $(\pi(f)g)(t) = f(t)g(t)$ $(t \in \Omega)$. Then the mapping $f \to \pi(f)$ of $L^\infty(\Omega, \mu)$ into $B(L^2(\Omega, \mu))$ is a faithful W^*-representation, and $\pi(L^\infty(\Omega, \mu))$ is maximal commutative.*

Proof. The first part of the corollary is clear. We shall show the second part.

Since (Ω, μ) is localizable, we can write

$$(\Omega, \mu) = \sum_{\alpha \in \mathbb{I}} (\Omega_\alpha, \mu_\alpha)(\mu_\alpha(\Omega_\alpha) < +\infty).$$

Let χ_α be the characteristic function of Ω_α. Then

$$L^2(\Omega, \mu) = \sum_{\alpha \in \mathbb{I}} \oplus \pi(\chi_\alpha) L^2(\Omega, \mu) = \sum_{\alpha \in \mathbb{I}} \oplus L^2(\Omega_\alpha, \mu_\alpha),$$

and $\pi(L^\infty(\Omega, \mu)) = \sum_{\alpha \in \mathbb{I}} \oplus \pi(\chi_\alpha) L^\infty(\Omega, \mu) = \sum_{\alpha \in \mathbb{I}} \oplus L^\infty(\Omega_\alpha, \mu_\alpha)$. Since

$$\pi(L^\infty(\Omega, \mu))' = \Big(\sum_{\alpha \in \mathbb{I}} \oplus L^\infty(\Omega_\alpha, \mu_\alpha)\Big)' = \sum_{\alpha \in \mathbb{I}} \oplus L^\infty(\Omega_\alpha, \mu_\alpha)'$$

$(L^\infty(\Omega_\alpha, \mu_\alpha)'$ is the commutant of $L^\infty(\Omega_\alpha, \mu_\alpha)$ in $L^2(\Omega_\alpha, \mu_\alpha))$, we may assume that $\mu(\Omega) < +\infty$. Set $\mu(f) = \tau(f)$ $(f \in L^\infty(\Omega, \mu))$; then by the above result, $J\pi(L^\infty(\Omega, \mu))J = \{R_x \mid x \in L^\infty(\Omega, \mu)\} = \pi(L^\infty(\Omega, \mu))'$. But, since $L^\infty(\Omega, \mu)$ is commutative, $\{R_x \mid x \in L^\infty(\Omega, \mu)\} = \pi(L^\infty(\Omega, \mu))$. q.e.d.

2.9.4. Corollary. *Let $\mathcal{N}$ be a commutative W^*-algebra, with a cyclic vector ξ, on a Hilbert space $\mathcal{H}$. Then $\mathcal{N}$ is maximal commutative in $B(\mathcal{H})$.*

Proof. Let $\mathcal{N} = C(K)$, be the function representation of $\mathcal{N}$. For $f \in C(K)$, set $\mu(f) = (f\xi, \xi)$. Then μ is a positive Radon measure on K. It is easy to see that $\mathcal{N}$ on $\mathcal{H}$ is equivalent to the $*$-representation $\{\pi, L^2(\Omega, \mu)\}$ of $L^\infty(\Omega, \mu) = \mathcal{N}$. q.e.d.

2.9.5. Proposition. *Let $\mathcal{N}$ be a W^*-algebra, containing $1_{\mathcal{H}}$, on a Hilbert space $\mathcal{H}$, and let ξ be an element in $\mathcal{H}$. $P_{[\mathcal{N}\xi]}$ is abelian (resp. finite) if and only if $P_{[\mathcal{N}'\xi]}$ is abelian (resp. finite).*

Proof. Put $e=P_{[\mathcal{N}'\xi]}$ and $f'=P_{[\mathcal{N}\xi]}$. Consider the W^*-algebras $(ef')\mathcal{N}(ef')$ and $(ef')\mathcal{N}'(ef')$. Then

$$\{(ef')\mathcal{N}(ef')\}'=\{f'(e\mathcal{N}e)f'\}'=\{(e\mathcal{N}e)f'\}'=f'(e\mathcal{N}e)'f'$$
$$=f'(e\mathcal{N}'e)f'=(ef')\mathcal{N}'(ef').$$

Moreover, $P_{[(ef')\mathcal{N}'(ef')\xi]}=ef'$ and $P_{[(ef')\mathcal{N}(ef')\xi]}=ef'$. Therefore $(ef')\mathcal{N}(ef')$ has a cyclic and separating vector.

If e is abelian, $e\mathcal{N}e$ is commutative. Hence $(ef')\mathcal{N}(ef')$ is commutative, and so $(ef')\mathcal{N}(ef')$ is maximal commutative (2.9.4). Hence

$$(ef')\mathcal{N}'(ef')=(ef')\mathcal{N}(ef').\ (ef')\mathcal{N}'(ef')=e(f'\mathcal{N}'f')e=(f'\mathcal{N}'f')e.$$

The mapping $\Phi: f'x'f'\to f'x'f'e$ of $f'\mathcal{N}'f'$ onto $(f'\mathcal{N}'f')e$ $(x'\in\mathcal{N}')$ is a σ-continuous *-homomorphism. If $f'x'f'e=0$, $f'x'f'\xi=0$, and so $f'x'f'=0$. Hence Φ is a *-isomorphism. Therefore $f'\mathcal{N}'f'$ is commutative, and so f' is abelian.

If e is finite, $(ef')\mathcal{N}(ef')$ is finite and is countably decomposable, because it has a separating vector. Since $(ef')\mathcal{N}(ef')$ has a cyclic and separating vector, $(ef')\mathcal{N}'(ef')$ is anti-*-isomorphic to $(ef')\mathcal{N}(ef')$ (by 2.7.12 and 2.9.2). Hence $(ef')\mathcal{N}'(ef')$ is finite. By the above result, the mapping $f'x'f\to f'x'f'e\,(x'\in\mathcal{N}')$ is a *-isomorphism. Hence $f'\mathcal{N}'f'$ is finite, and so f' is finite. q.e.d.

2.9.6. Corollary. *Let $\mathcal{N}$ be a W^*-algebra, containing $1_{\mathcal{H}}$, on a Hilbert space $\mathcal{H}$. Then the following conditions are equivalent:*

1. *$\mathcal{N}$ is of type* I (*resp. type* II, *type* III);
2. *$\mathcal{N}'$ is of type* I (*resp. type* II, *type* III).

2.9.7. Theorem. *Let $\mathcal{N}$ be a finite W^*-algebra, containing $1_{\mathcal{H}}$, on a Hilbert space $\mathcal{H}$ such that $\mathcal{N}'$ is finite. Let $Z=L^\infty(\Omega,\mu)$ be the center of $\mathcal{N}$ and let $\natural$ (resp. $\natural'$) be the $\natural$-operation on $\mathcal{N}$ (resp. $\mathcal{N}'$). Then there exists a unique locally μ-measurable positive function C on Ω such that $P^\natural_{[\mathcal{N}'\xi]}=CP^{\natural'}_{[\mathcal{N}\xi]}$ $(\xi\in\mathcal{H})$ and $0<C(t)<+\infty$ $(t\in\Omega)$ locally almost everywhere.*

Proof. Since arbitrary finite W^*-algebra is a direct sum of countably decomposable finite W^*-algebras, we may assume that $\mathcal{N}$ is countably decomposable. Let $\{\xi_n\}$ be a maximal family of elements in $\mathcal{H}$ such that a family $\{[\mathcal{N}\xi_n]\}$ (resp. $\{[\mathcal{N}'\xi_n]\}$) of subspaces of $\mathcal{H}$ are mutually orthogonal. Put $e=\sum_{n=1}^{\infty} P_{[\mathcal{N}'\xi_n]}$ and $e'=\sum_{n=1}^{\infty} P_{[\mathcal{N}\xi_n]}$. Then

$$(1_{\mathcal{H}}-e)(1_{\mathcal{H}}-e')=0.$$

Let $z=c(1_{\mathscr{H}}-e)$ be the central support of $1_{\mathscr{H}}-e$ in $\mathscr{N}$. Then $1_{\mathscr{H}}-e\le z$ and $1_{\mathscr{H}}-e'\le 1_{\mathscr{H}}-z$. We may assume that $\sum_{n=1}^{\infty}\|\xi_n\|^2<+\infty$. Let $\xi_0=\sum_{n=1}^{\infty}\xi_n$. Then $[\mathscr{N}'\xi_0]\supseteq[\mathscr{N}'P_{[\mathscr{N}\xi_n]}\xi_0]=[\mathscr{N}'\xi_n]$ and

$$[\mathscr{N}\xi_0]\supseteq[\mathscr{N}P_{[\mathscr{N}'\xi_n]}\xi_0]=[\mathscr{N}\xi_n].$$

Thus $P_{[\mathscr{N}\xi_0]}\ge e'$ and $P_{[\mathscr{N}'\xi_0]}\ge e$. The reverse inequalities are clear. Hence $e'=P_{[\mathscr{N}\xi_0]}$ and $e=P_{[\mathscr{N}'\xi_0]}$.

Set $\zeta_1=z\xi_0$ and $\zeta_2=(1_{\mathscr{H}}-z)\xi_0$. Then

$$z\mathscr{H}=ze'\mathscr{H}=z[\mathscr{N}\xi_0]=[\mathscr{N}z\xi_0]=[\mathscr{N}\zeta_1]$$

and $(1_{\mathscr{H}}-z)\mathscr{H}=[\mathscr{N}'\zeta_2]$. Therefore the proof may reduce to the case of $[\mathscr{N}'\zeta_0]=\mathscr{H}$ or $[\mathscr{N}\zeta_0]=\mathscr{H}$ (some $\zeta_0\in\mathscr{H}$).

Suppose that $[\mathscr{N}'\zeta_0]=\mathscr{H}$ and let $B=P^{\natural\prime}_{[\mathscr{N}\zeta_0]}$. For

$$\xi\in\mathscr{H},\qquad[\mathscr{N}'\xi]\subset[\mathscr{N}'\zeta_0]=\mathscr{H},$$

and so $P_{[\mathscr{N}\xi]}\prec P_{[\mathscr{N}\zeta_0]}$. Since equivalent projections have the same value for the $\natural$-operation, we may assume that $[\mathscr{N}\xi]\subseteq[\mathscr{N}\zeta_0]$. Let

$$p'=P_{[\mathscr{N}\zeta_0]},\qquad q'=P_{[\mathscr{N}\xi]}$$

and $q=P_{[\mathscr{N}'\xi]}$. The mapping $ap'\to a^{\natural}p'$ $(a\in\mathscr{N})$ is the $\natural$-operation of $\mathscr{N}p'$. Therefore if we use $(ap')^0$ as the $\natural$-operation on $\mathscr{N}p'$, then $(ap')^0=a^{\natural}p'$. $[\mathscr{N}p'\zeta_0]=[\mathscr{N}\zeta_0]=[p'\mathscr{N}'p'\zeta_0]$. Hence $\mathscr{N}p'$ on $p'\mathscr{H}$ has a cyclic and separating vector ζ_0, and so it has a faithful normal tracial state. Therefore it is equivalent to a standard representation. Let J be the involution on $p'\mathscr{H}$ with $J\mathscr{N}p'J=p'\mathscr{N}'p'$. Then

$$Jqp'J=JP_{p'[\mathscr{N}'\xi]}J=JP_{[p'\mathscr{N}'p'\xi]}J=P_{[Jp'\mathscr{N}'p'\xi]}=P_{[Jp'\mathscr{N}'p'JJ\xi]}=P_{[\mathscr{N}p'J\xi]}.$$

We shall show now that $P_{[\mathscr{N}p'J\xi]}\sim P_{[\mathscr{N}p'\xi]}$. Since J commutes with central projections in $\mathscr{N}p'$, we may assume that $P_{[\mathscr{N}p'J\xi]}\lesssim P_{[\mathscr{N}p'\xi]}$. It suffices to assume $P_{[\mathscr{N}p'J\xi]}\prec P_{[\mathscr{N}p'\xi]}$. Then $JP_{[\mathscr{N}p'J\xi]}J\prec JP_{[\mathscr{N}p'\xi]}J$, and so $P_{[p'\mathscr{N}'p'\xi]}\prec P_{[p'\mathscr{N}'p'J\xi]}$. Hence by 2.7.10, $P_{[\mathscr{N}p'\xi]}\prec P_{[\mathscr{N}p'J\xi]}$, and so $P_{[\mathscr{N}p'J\xi]}\sim P_{[\mathscr{N}p'\xi]}$. Therefore $Jqp'J\sim P_{[\mathscr{N}p'\xi]}=P_{[\mathscr{N}\xi]}=q'$. Hence

$$J(qp')^0J=Jq^{\natural}p'J=q^{\natural}p'=(q')^0.$$

To find $(q')^0$, we invoke the definition of the $\natural$-operation and get a collection $u'_{k,n}$ in $\mathscr{N}'$, positive numbers $\lambda_{k,n}(k=1,2,\ldots,m_n;n=1,2,\ldots)$ such that $\sum_{k=1}^{m_n}\lambda_{k,n}=1$, $u'^*_{k,n}u'_{k,n}=p'=u'_{k,n}u'^*_{k,n}$ and $\left\{\sum_{k=1}^{m_n}\lambda_{k,n}u'^*_{k,n}q'u'_{k,n}\right\}$ converges to $(q')^0$ in the uniform topology. By the uniform continuity of the

$\natural$-operation, $\lim_{n\to\infty}\sum_{k=1}^{m_n}\lambda_{k,n}(q')^{\natural'}=(q')^{\natural'}=((q')^0)^{\natural'}=(q^\natural p')^{\natural'}=B\cdot q^\natural$. Hence $P^{\natural'}_{[\mathcal{N}\xi]}=B\cdot P^{\natural}_{[\mathcal{N}'\xi]}$ $(\xi\in\mathcal{H})$. If $B\xi=0$ (some $\xi\in\mathcal{H}$), then

$$(B\cdot P_{[\mathcal{N}'\xi]})^\natural=B\cdot P^\natural_{[\mathcal{N}'\xi]}=P^{\natural'}_{[\mathcal{N}\xi]}=0.$$

Hence $\xi=0$. Therefore $0<B(t)<+\infty$ locally almost everywhere, and so we have $P^\natural_{[\mathcal{N}'\xi]}=B^{-1}P^{\natural'}_{[\mathcal{N}\xi]}$ for $\xi\in\mathcal{H}$. Similarly, if $[\mathcal{N}\zeta_0]=\mathcal{H}$, we have $P^\natural_{[\mathcal{N}'\xi]}=P^{\natural'}_{[\mathcal{N}\zeta_0]}P^{\natural'}_{[\mathcal{N}\xi]}$ $(\xi\in\mathcal{H})$. One can easily see the uniqueness of C. q.e.d.

2.9.8. Definition. *Let $\mathcal{N}$ be a finite W^*-algebra, containing $1_{\mathcal{H}}$, on a Hilbert space $\mathcal{H}$ such that $\mathcal{N}'$ is finite. Then the above locally measurable function C is called the coupling function of $\mathcal{N}$ and $\mathcal{N}'$.*

2.9.9. Proposition. *Let $\mathcal{M}$ be a semi-finite W^*-algebra without a direct summand of finite type, and let $\{e_\alpha|\alpha\in\mathbb{I}\}$, $\{f_\beta|\beta\in\mathbb{J}\}$ be two families of mutually orthogonal finite equivalent projections in $\mathcal{M}$ (e_α is not necessarily equivalent to $f_\beta(\alpha\in\mathbb{I},\beta\in\mathbb{J})$). If $\sum_{\alpha\in\mathbb{I}}e_\alpha=1=\sum_{\beta\in\mathbb{J}}f_\beta$, then* $\mathrm{Card}(\mathbb{I})=\mathrm{Card}(\mathbb{J})$.

Proof. Let τ be a non-zero normal semi-finite trace on $\mathcal{M}$ with $\tau(e_{\alpha_0})<+\infty$ for some $\alpha_0\in\mathbb{I}$ (cf. 2.5.3). Then $e_\alpha s(\tau)\neq 0$ and $f_\beta s(\tau)\neq 0$. Let $\mathbb{J}_\alpha=\{\beta\in\mathbb{J}\,|\,e_\alpha f_\beta\, s(\tau)\neq 0\}$. Then $\mathbb{J}=\bigcup_{\alpha\in\mathbb{I}}\mathbb{J}_\alpha$. *Since*

$$\tau\Big(e_\alpha\Big(\sum_{\beta\in\mathbb{J}_\alpha}f_\beta\Big)e_\alpha\Big)=\sum_{\beta\in\mathbb{J}_\alpha}\tau(e_\alpha f_\beta e_\alpha)=\tau(e_\alpha)=\tau(e_0)<+\infty,$$

$\mathbb{J}_\alpha$ is countable. Hence $\mathrm{Card}(\mathbb{J})\leq\aleph_0\mathrm{Card}(\mathbb{I})=\mathrm{Card}(\mathbb{I})$. The reverse inequality is analogously obtained. q.e.d.

2.9.10. Definition. *A semi-finite W^*-algebra $\mathcal{M}$ is said to be of uniform type if it either satisfies the hypothesis of 2.9.9 or is finite. If the cardinal of 2.9.9 is γ, then the corresponding algebra is said to be of type S_γ. If it is finite, the algebra is said to be of type S_1.*

Then one can easily obtain.

2.9.11. Proposition. *Let $\mathcal{M}$ be a semi-finite W^*-algebra and let Δ be the collection of cardinals containing 1 plus all infinite cardinals $\alpha\leq$ the cardinal of $\mathcal{M}$. Then for $\alpha\in\Delta$ there exists a central projection z_α such that $\{z_\alpha\}$ are mutually orthogonal, $\sum_{\alpha\in\Delta}z_\alpha=1$, and $\mathcal{M}z_\alpha$ is of type S_α if $z_\alpha\neq 0$.*

2.9.12. Definition. *Let $\mathcal{N}$ be a semi-finite W^*-algebra, containing $1_{\mathcal{H}}$ on a Hilbert space $\mathcal{H}$. By 2.9.11 there exists a collection of mutually orthogonal central projections $\{z_\alpha\}_{\alpha\in\Delta_1}$ such that $\mathcal{M}z_\alpha$ is of type S_α ($\Delta_1=\{\alpha|z_\alpha\neq 0,\alpha\in\Delta\}$). For $\mathcal{N}'$, also there exists a collection $\{z'_\alpha|\alpha\in\Delta'\}$ of central projections such that $\mathcal{N}'z'_\alpha$ is of type S_α if $z'_\alpha\neq 0$, and $\sum_{\alpha\in\Delta'}z'_\alpha=1_{\mathcal{H}}$. Set $z_{\alpha,\beta}=z_\alpha z'_\beta$ ($\alpha\in\Delta,\beta\in\Delta'$). Then $\sum_{\alpha\in\Delta,\beta\in\Delta'}z_{\alpha,\beta}=1_{\mathcal{H}}$ with*

$\mathcal{N} z_{\alpha,\beta}$ (resp. $\mathcal{N}' z_{\alpha,\beta}$) of type S_α (resp. S_β) if $z_{\alpha,\beta} \neq 0$. Now we define a function C on the center of $\mathcal{N}$ as follows: $C = C_{(1,1)} + \sum_{(\alpha,\beta)\neq(1,1)} (\alpha,\beta) z_{\alpha,\beta}$, where $C_{(1,1)}$ is the function C in 2.9.7. This function is called the coupling function of the algebras $\mathcal{N}$ and $\mathcal{N}'$.

2.9.13. Definition. *Let $\mathcal{N}$ be a W^*-algebra, containing $1_{\mathcal{H}}$, on a Hilbert space $\mathcal{H}$, and let p be a projection in $\mathcal{N}$. p is said to be cyclic (relative to $\mathcal{N}$) if there exists an element ξ in $\mathcal{H}$ with $P_{[\mathcal{N}'\xi]} = p$.*

2.9.14. Proposition. *Let $(p_\alpha | \alpha \in \mathbb{I})$, $(q_\beta | \beta \in \mathbb{J})$ be two families of mutually orthogonal cyclic projections, with* $\mathrm{Card}(\mathbb{I})$, $\mathrm{Card}(\mathbb{J}) \geq \aleph_0$, *in a W^*-algebra $\mathcal{N}$ on a Hilbert space. If $\sum_{\alpha\in\mathbb{I}} p_\alpha = 1_{\mathcal{H}} = \sum_{\beta\in\mathbb{J}} q_\beta$, then* $\mathrm{Card}(\mathbb{I}) = \mathrm{Card}(\mathbb{J})$.

Proof. Let $\mathbb{J}_\alpha = \{\beta \in \mathbb{J} \,|\, p_\alpha q_\beta \neq 0\}$. Then $\mathbb{J} = \bigcup_{\alpha\in\mathbb{I}} \mathbb{J}_\alpha$. Since p_α is cyclic, there is a family $(\xi_\alpha | \alpha \in \mathbb{I})$ of elements in $\mathcal{H}$ with $[\mathcal{N}'\xi_\alpha] = p_\alpha \mathcal{H}$ $(\alpha \in \mathbb{I})$. If $(q_\beta \xi_\alpha, \xi_\alpha) = 0$, $q_\beta \xi_\alpha = 0$, and so $q_\beta \mathcal{N}' \xi_\alpha = 0$. Hence $q_\beta p_\alpha = 0$. Therefore $\mathbb{J}_\alpha$ is countable, and so $\mathrm{Card}(\mathbb{J}) \leq \mathrm{Card}(\mathbb{I})$. The reverse is analogously obtained. q.e.d.

2.9.15. Proposition. *Let $\mathcal{N}$ be a type* III *W^*-algebra, containing $1_{\mathcal{H}}$, on a Hilbert space $\mathcal{H}$, and let p be a projection in $\mathcal{N}$. p is cyclic relative to $\mathcal{N}$ if and only if p is countably decomposable.*

Proof. We may assume that $p = 1_{\mathcal{H}}$. Suppose that p is cyclic, and let $p = P_{[\mathcal{N}'\xi]}$ (some $\xi \in \mathcal{H}$). Let $(p_\beta | \beta \in \mathbb{J}\}$ be a family of mutually orthogonal projections in $\mathcal{N}$ with $\sum_{\beta\in\mathbb{J}} p_\beta = 1_{\mathcal{H}}$. Then

$$\Big\| \sum_{\beta\in\mathbb{J}} p_\beta \xi \Big\|^2 = \sum_{\beta\in\mathbb{J}} \|p_\beta \xi\|^2 < +\infty .$$

Hence $p_\beta \xi = 0$, except for a countable number of indices β. But $\mathcal{N}' p_\beta \xi = p_\beta \mathcal{N}' \xi$. Hence $p_\beta = 0$, except for a countable number of indices β, and so p is countably decomposable. Conversely suppose that $1_{\mathcal{H}}$ is countably decomposable. For an element $\xi (\neq 0) \in \mathcal{H}$, set $q = P_{[\mathcal{N}'\xi]}$. Let $\{q_\alpha | \alpha \in \mathbb{I}\}$ be a maximal family of mutually orthogonal equivalent projections with $q_\alpha \sim q$ $(\alpha \in \mathbb{I})$. Then $\mathrm{Card}(\mathbb{I}) \leq \aleph_0$. By the comparability theorem, there exists a central projection z such that $\left(1_{\mathcal{H}} - \sum_{\alpha\in\mathbb{I}} q_\alpha\right) z \prec qz$ and $\left(1_{\mathcal{H}} - \sum_{\alpha\in\mathbb{I}} q_\alpha\right)(1_{\mathcal{H}} - z) \succ q(1_{\mathcal{H}} - z)$. By the maximality of $\{q_\alpha | \alpha \in \mathbb{I}\}$, $qz \neq 0$. Since $\mathcal{N}$ is of type III,

$$z = \sum_{\alpha\in\mathbb{I}} q_\alpha z + \left(1_{\mathcal{H}} - \sum_{\alpha\in\mathbb{I}} q_\alpha\right) z \sim qz .$$

Hence z is cyclic.

Now let $\{z_\beta | \beta \in \mathbb{J}\}$ be a maximal family of mutually orthogonal central cyclic projections. Then $\operatorname{Card}(\mathbb{J}) \leq \aleph_0$ and $\sum_{\beta \in \mathbb{J}} z_\beta = 1$ by the above result.

Let $P_{[\mathcal{N}'\xi_\beta]} = z_\beta$ $(\xi_\beta \in \mathcal{H})$ with $\sum_{\beta \in \mathbb{J}} \|\xi_\beta\|^2 < +\infty$. Then $P_{[\mathcal{N}'\xi]} = 1_{\mathcal{H}}$ $\left(\xi = \sum_{\beta \in \mathbb{J}} \xi_\beta\right)$. q.e.d.

2.9.16. Definition. *Let $\mathcal{M}$ be a type* III *W^*-algebra, and let p be a projection in $\mathcal{M}$. p is said to be locally countably decomposable if for an arbitrary non-zero central projection z, there exists a non-zero central projection z_1, with $z_1 \leq z$, such that pz_1 is countably decomposable.*

Then one can easily show

2.9.17. Proposition. *Let $\mathcal{M}$ be a type* III *W^*-algebra, and let $\{p_\alpha | \alpha \in \mathbb{I}\}$, $\{q_\beta | \beta \in \mathbb{J}\}$ be two families of mutually orthogonal, equivalent, locally countably decomposable projections with* $\operatorname{Card}(\mathbb{I})$, $\operatorname{Card}(\mathbb{J}) \geq \aleph_0$. *If $\sum_{\alpha \in \mathbb{I}} p_\alpha = \sum_{\beta \in \mathbb{J}} q_\beta = 1$, then* $\operatorname{Card}(\mathbb{I}) = \operatorname{Card}(\mathbb{J})$.

2.9.18. Definition. *Let $\mathcal{M}$ be a type* III *W^*-algebra. If $\mathcal{M}$ satisfies the hypothesis in* 2.9.17, *then $\mathcal{M}$ is said to be of uniform type. If the cardinal of* 2.9.17 *is γ, the corresponding algebra is said to be of type P_γ.*

Then, one can easily see

2.9.19. Proposition. *Let $\mathcal{M}$ be a type* III *W^*-algebra, and let Δ be the collection of infinite cardinals which are less than or equal to the cardinal of $\mathcal{M}$. For each $\alpha \in \Delta$, there exists a central projection z_α such that $\mathcal{M} z_\alpha$ is of type P_α if $z_\alpha \neq 0$. Further this decomposition is unique.*

2.9.20. Definition. *Let $\mathcal{N}$ be a type* III *W^*-algebra, containing $1_{\mathcal{H}}$, on a Hilbert space $\mathcal{H}$; let $\{z_\alpha | \alpha \in \Delta\}$ and $\{z'_\beta | \beta \in \Delta'\}$ be the decompositions of $1_{\mathcal{H}}$ in $\mathcal{N}$ and $\mathcal{N}'$, where the decompositions being those mentioned in the statement of* 2.9.19 *above. We introduce a function*

$$C = \sum_{(\alpha,\beta) \in \Delta \times \Delta'} (\alpha, \beta) z_{\alpha,\beta} \quad (z_{\alpha,\beta} = z_\alpha z'_\beta).$$

C is called the coupling function of $\mathcal{N}$ and $\mathcal{N}'$.

2.9.21. Definition. *Let $\mathcal{N}$ be a W^*-algebra, containing $1_{\mathcal{H}}$, on a Hilbert space $\mathcal{H}$. Then $\mathcal{N}$ is uniquely decomposed into a direct sum of a semi-finite algebra and a type* III *algebra. By using* 2.9.12, 2.9.20, *we shall define the coupling function for $\mathcal{N}$, $\mathcal{N}'$ canonically.*

2.9.22. Definition. *Let $\mathcal{N}_1$ (resp. $\mathcal{N}_2$) be a W^*-algebra, containing $1_{\mathcal{H}_1}$ (resp. $1_{\mathcal{H}_2}$) on a Hilbert space $\mathcal{H}_1$ (resp. $\mathcal{H}_2$), and let C_1 (resp. C_2) be the coupling function of $\mathcal{N}_1$ and $\mathcal{N}'_1$ (resp. $\mathcal{N}_2$ and $\mathcal{N}'_2$).*

*Suppose that Φ is a *-isomorphism of $\mathcal{N}_1$ onto $\mathcal{N}_2$. Then Φ will induce a *-isomorphism of the center Z_1 of $\mathcal{N}_1$ onto the center Z_2 of $\mathcal{N}_2$. Therefore by using Φ, we can identify Z_1 with Z_2 as follows: $Z_1 = Z_2 = L^\infty(\Omega, \mu)$. Then we have two coupling functions C_1 and C_2 on Ω. If $C_1 = C_2$, it is denoted by $\Phi(C_1) = C_2$.*

2.9.23. Definition. *Let $\mathcal{N}_1$ (resp. $\mathcal{N}_2$) be a W^*-algebra, containing $1_{\mathcal{H}_1}$ (resp. $1_{\mathcal{H}_2}$), on a Hilbert space $\mathcal{H}_1$ (resp. $\mathcal{H}_2$), and let Φ be a *-isomorphism of $\mathcal{N}_1$ onto $\mathcal{N}_2$. Φ is said to be spatial if there exists a unitary mapping u of $\mathcal{H}_1$ onto $\mathcal{H}_2$ such that $uxu^* = \Phi(x)$ $(x \in \mathcal{N}_1)$.*

2.9.24. Proposition. *Let $\mathcal{N}_1$ (resp. $\mathcal{N}_2$) be a finite W^*-algebra, containing $1_{\mathcal{H}_1}$ (resp. $1_{\mathcal{H}_2}$), on a Hilbert space $\mathcal{H}_1$ (resp. $\mathcal{H}_2$), and let C_1 (resp. C_2) be the coupling function of $\mathcal{N}_1$ and $\mathcal{N}_1'$ (resp. $\mathcal{N}_2$ and $\mathcal{N}_2'$). Suppose that Φ is a *-isomorphism of $\mathcal{N}_1$ into $\mathcal{N}_2$ with $\Phi(C_1) = C_2$. Then Φ is spatial.*

Proof. There exist a W^*-algebra $\mathcal{N}$, containing $1_{\mathcal{H}}$, on a Hilbert space $\mathcal{H}$, and projections p', q' in $\mathcal{N}'$ with the central support $1_{\mathcal{H}}$ such that $\mathcal{N}_1$ on $\mathcal{H}_1$ (resp. $\mathcal{N}_2$ on $\mathcal{H}_2$) are equivalent to $\mathcal{N}p'$ on $p'\mathcal{H}$ (resp. $\mathcal{N}q'$ on $q'\mathcal{H}$).

We may reduce the proof to the case where both of $\mathcal{N}_1'$ and $\mathcal{N}_2'$ are finite or properly infinite. Suppose $\mathcal{N}_1'$ and $\mathcal{N}_2'$ are finite. Then $p' \vee q'$ is finite. It suffices to assume that $p' \vee q' = 1_{\mathcal{H}}$. Let $\natural$ (resp. $\natural'$) be the $\natural$-operation on $\mathcal{N}$ (resp. $\mathcal{N}'$) and let ρ (resp. ρ') be the $\natural$-operation on $\mathcal{N}p'$ (resp. $p'\mathcal{N}'p'$). Then for each $\xi \in \mathcal{H}$, $(P_{[\mathcal{N}'p'\xi]})^\natural = C(P_{[\mathcal{N}p'\xi]})^{\natural'}$, where C is the coupling function of $\mathcal{N}$ and $\mathcal{N}'$. On the other hand, $\rho(P_{[p'\mathcal{N}'p'\xi]}) = \rho(p'P_{[\mathcal{N}'p'\xi]}) = (P_{[\mathcal{N}'p'\xi]})^\natural p' = CP^{\natural'}_{[\mathcal{N}p'\xi]}p' = C_1\rho'(P_{[\mathcal{N}p'\xi]})$. By taking a family of elements (ξ_α) in $\mathcal{H}$ such that $\{[\mathcal{N}\xi_\alpha]\}$ are mutually orthogonal and $\sum_\alpha \oplus [\mathcal{N}\xi_\alpha] = \mathcal{H}$, and using the linearity and normality of $\natural$-operations, we have $Cp'^{\natural'}p' = C_1 p'$. Analogously $Cq'^{\natural'}q' = C_2 q'$. Since $\Phi(C_1) = C_2$, we have $p'^{\natural'} = q'^{\natural'}$, so that $p' \sim q'$ in $\mathcal{N}'$.

Next suppose that $\mathcal{N}_1'$ and $\mathcal{N}_2'$ are properly infinite. Since $\Phi(C_1) = C_2$, it suffices to assume that $\mathcal{N}_1'$ and $\mathcal{N}_2'$ are of type S_γ. Namely there exist families of mutually orthogonal equivalent finite projections (p_α') and (q_α') $(\alpha \in \mathbb{I}$ with $\mathrm{Card}(\mathbb{I}) = \gamma)$ in $\mathcal{N}'$ such that $\sum_\alpha p_\alpha' = p'$ and $\sum_\alpha q_\alpha' = q'$. Since $c(p') = c(q') = 1_{\mathcal{H}}$, $c(p_\alpha') = c(q_\alpha') = 1_{\mathcal{H}}$. If $\mathbb{I}_1$ is a denumberable subset of $\mathbb{I}$, $\sum_{\beta \in \mathbb{I}_1} p_\beta' \gtrsim q_\alpha'$. Hence $p' \gtrsim q'$ and analogously $p' \lesssim q'$, so that $p' \sim q'$. q.e.d.

2.9.25. Proposition. *Let $\mathcal{N}_1$ (resp. $\mathcal{N}_2$) be a semi-finite W^*-algebra, containing $1_{\mathcal{H}_1}$, (resp. $1_{\mathcal{H}_2}$) on a Hilbert space $\mathcal{H}_1$ (resp. $\mathcal{H}_2$) such that $\mathcal{N}_1'$ (resp. $\mathcal{N}_2'$) is properly infinite. Let C_1 (resp. C_2) be the coupling* *function of* $\mathcal{N}_1$ *and* $\mathcal{N}_1'$ *(resp.* $\mathcal{N}_2$ *and* $\mathcal{N}_2'$*), and let* Φ *be a *-isomorphism of* $\mathcal{N}_1$ *onto* $\mathcal{N}_2$ *with* $\Phi(C_1)=C_2$. *Then* Φ *is spatial.*

The proof is quite similar to that of 2.9.24.

Remark. In this proposition, we can not eliminate the assumptions that $\mathcal{N}_1'$ and $\mathcal{N}_2'$ are properly infinite (cf. [83]).

2.9.26. Proposition. *Let* $\mathcal{N}_1$ *(resp.* $\mathcal{N}_2$*) be a type* III W^**-algebra, containing* $1_{\mathcal{H}_1}$ *(resp.* $1_{\mathcal{H}_2}$*), on a Hilbert space* $\mathcal{H}_1$ *(resp.* $\mathcal{H}_2$*), and let* C_1 *(resp.* C_2*) be the coupling function of* $\mathcal{N}_1$ *and* $\mathcal{N}_1'$ *(resp.* $\mathcal{N}_2$ *and* $\mathcal{N}_2'$*). If* Φ *is a *-isomorphism of* $\mathcal{N}_1$ *onto* $\mathcal{N}_2$ *with* $\Phi(C_1)=C_2$, *then* Φ *is spatial.*

Proof. It suffices to assume that $\mathcal{N}_1'$ and $\mathcal{N}_2'$ are of type P_γ. Let $\mathcal{N}$ be a W^*-algebra, containing $1_{\mathcal{H}}$, on a Hilbert space $\mathcal{H}$ such that there exist two projections p', q' in $\mathcal{N}'$ such that $\mathcal{N}_1$ on $\mathcal{H}_1$ (resp. $\mathcal{N}_2$ on $\mathcal{H}_2$) is equivalent to $\mathcal{N}p'$ on $p'\mathcal{H}$ (resp. $\mathcal{N}q'$ on $q'\mathcal{H}$) and $c(p')=c(q')=1_{\mathcal{H}}$. Let $(p'_\alpha)_{\alpha\in\mathbb{I}}$, $(q'_\alpha)_{\alpha\in\mathbb{I}}$ be families of mutually orthogonal, locally countably decomposable projections of $\mathcal{N}'$ with $c(p'_\alpha)=c(q'_\alpha)=1_{\mathcal{H}}$, $\sum_{\alpha\in\mathbb{I}} p'_\alpha=p'$ and $\sum_{\alpha\in\mathbb{I}} q'_\alpha=q'$. Since $\mathcal{N}'$ is of type III, $p'_\alpha\sim q'_\alpha$, so that $p'\sim q'$. q.e.d.

2.9.27. Corollary. *Let* $\mathcal{N}_1$ *(resp.* $\mathcal{N}_2$*) be a type* III W^**-algebra, containing* $1_{\mathcal{H}_1}$ *(resp.* $1_{\mathcal{H}_2}$*) on a separable Hilbert space* $\mathcal{H}_1$ *(resp.* $\mathcal{H}_2$*). If* Φ *is a *-isomorphism of* $\mathcal{N}_1$ *onto* $\mathcal{N}_2$, *then* Φ *is spatial.*

2.9.28. Corollary. *Let* $\mathcal{N}$ *be a type* III W^**-algebra, containing* $1_{\mathcal{H}}$, *on a separable Hilbert space* $\mathcal{H}$. *Then* $\mathcal{N}$ *has a cyclic and separating vector.*

Proof. Since $\mathcal{H}$ is separable, $\mathcal{N}$ has a faithful normal state φ. Thus the *-representation $\{\pi_\varphi, \mathcal{H}_\varphi\}$ has a cyclic and separating vector. Hence by 2.9.27, $\mathcal{N}$ on $\mathcal{H}$ has a cyclic and separating vector. q.e.d.

2.9.29. Corollary. *Let* $\mathcal{N}$ *be a* W^**-algebra, containing* $1_{\mathcal{H}}$, *on a Hilbert space. Suppose that there is no central projection* z *in* $\mathcal{N}$ *such that* $\mathcal{N}z$ *is of type* S_γ $(\gamma\geq\aleph_0)$ *and* $\mathcal{N}'z$ *is of type* S_1. *If* Φ *is a *-automorphism on* $\mathcal{N}$ *leaving the center pointwise fixed, then there exists a unitary operator on* $\mathcal{H}$ *such that* $\Phi(x)=uxu^*$ $(x\in\mathcal{N})$.

2.9.30. Corollary. *Let* $\mathcal{N}$ *be a* W^**-algebra, containing* $1_{\mathcal{H}}$, *on a Hilbert space* $\mathcal{H}$ *with cyclic and separating vectors. If* Φ *is a *-automorphism on* $\mathcal{N}$ *then there exists a unitary operator* u *on* $\mathcal{H}$ *such that* $\Phi(x)=uxu^*$ $(x\in\mathcal{N})$.

The proof is obtained from 2.7.11 and 2.7.12.

2.9.31. Proposition. *Let $\mathcal{N}_1$ (resp. $\mathcal{N}_2$) be a W^*-algebra, containing $1_{\mathcal{H}_1}$ (resp. $1_{\mathcal{H}_2}$), on a Hilbert space $\mathcal{H}_1$ (resp. $\mathcal{H}_2$). If $\mathcal{N}_1'$ (resp. $\mathcal{N}_2'$) is a homogeneous type I_n algebra, then every *-isomorphism Φ of $\mathcal{N}_1$ onto $\mathcal{N}_2$ is spatial.*

This is almost clear.

2.9.32. Corollary. *Let $\mathcal{M}$ be a type* I *W^*-algebra. Then every *-automorphism Φ on $\mathcal{M}$ leaving the center pointwise fixed is inner.*

Proof. It suffices to assume that $\mathcal{M}$ is homogeneous. Then one can represent as follows: $\mathcal{M}=B(\mathcal{H})\overline{\otimes} Z$ on $\mathcal{H}\otimes\mathcal{K}$, where $1_{\mathcal{H}}\otimes Z$ is the center of $\mathcal{M}$ and Z is a maximal commutative W^*-algebra of $B(\mathcal{K})$ ($\mathcal{H}$, $\mathcal{K}$ Hilbert spaces).

By 2.9.31, there exists a unitary operator u on $\mathcal{H}\otimes\mathcal{K}$ such that $\Phi(x)=uxu^*$ $(x\in B(\mathcal{H})\overline{\otimes} Z)$. Since $u^*(1_{\mathcal{H}}\otimes z)u=1\otimes z$ $(z\in Z)$, $u\in(1_{\mathcal{H}}\otimes Z)'=B(\mathcal{H})\overline{\otimes} Z=\mathcal{M}$. q.e.d.

References. [9], [37], [44], [63], [64], [93], [98], [116], [183].

3. Decomposition Theory

3.1. Decompositions of States (Non-Separable Cases)

Let $\mathscr{A}$ be a C^*-algebra with identity, (in this section, we shall always assume that a C^*-algebra $\mathscr{A}$ has an identity), $\mathscr{A}^*$ the dual of $\mathscr{A}$, and let $\mathscr{S}$ be the state space, with the topology $\sigma(\mathscr{S},\mathscr{A})$, of $\mathscr{A}$. Let $C(\mathscr{S})$ be the Banach algebra of all complex valued continuous functions on the compact space $\mathscr{S}$ with the usual supremum norm. For $a\in\mathscr{A}$, a function $\hat{a}$ on $\mathscr{S}$ is defined by $\hat{a}(\varphi)=\varphi(a)$ $(\varphi\in\mathscr{S})$. By the mapping $a\to\hat{a}$ $(a\in\mathscr{A})$ the $\mathscr{A}$ may be topologically embedded into $C(\mathscr{S})$; moreover the self-adjoint portion $\mathscr{A}^s$ of $\mathscr{A}$ may be order-isomorphically embedded into the real Banach space $C_r(\mathscr{S})$ of all real valued continuous functions on $\mathscr{S}$.

Any state ψ on $\mathscr{A}$ extends to a state on $C(\mathscr{S})$ and thus, by the Riesz representation theorem, there is some probability Radon measure μ on $\mathscr{S}$ so that

1. $$\psi(a)=\int_{\mathscr{S}}\hat{a}(\varphi)\,d\mu(\varphi). \quad (a\in\mathscr{A}).$$

Let $\Omega(\psi)$ be the set of all probability Radon measures on $\mathscr{S}$ satisfying the equality 1. $\Omega(\psi)$ consists, in general, of many distinct measures. The Choquet-Bishop-de Leeuw theorem [10] in the theory of locally convex spaces assures us that $\Omega(\psi)$ contains at least one measure which concentrates on pure states in some sense.

But in the theory of operator algebras, the measure μ should be usually defined through a commutative W^*-algebra C, with $1_{\mathscr{H}_\psi}$, in the commutant $\pi_\psi(\mathscr{A})'$, because the corresponding decomposition of ψ should be related to the reduction theory (see later sections) ($\{\pi_\psi,\mathscr{H}_\psi\}$ is the *-representation of $\mathscr{A}$ constructed via ψ).

Let e be the orthogonal projection of $\mathscr{H}_\psi$ onto $[C1_\psi]$. Then $e\in C'$. Since Ce is a commutative W^*-algebra with a cyclic vector 1_ψ, on $e\mathscr{H}_\psi$, Ce is maximal commutative by 2.9.4, and so $(Ce)'=eC'e=Ce$. Therefore e is an abelian projection in C'. Let $c(e)$ be the central support of e in C'. Then $[C'c(e)1_\psi]=[C'1_\psi]\supseteq[\pi_\psi(\mathscr{A})1_\psi]=\mathscr{H}_\psi$. Hence $c(e)=1_{\mathscr{H}_\psi}$, and so e is maximal abelian in C'.

Let $Ce=C(K)$ be the function representation of Ce, and let $P(x_1,x_2,\ldots,x_n)$ be an arbitrary complex polynomial in n-variables. Then for $a_1,a_2,\ldots,a_n\in\mathscr{A}$,

$$\begin{aligned}\|P(e\pi_\psi(a_1)e,e\pi_\psi(a_2)e,\ldots,e\pi_\psi(a_n)e)\| &= \sup_{t\in K}|P((e\pi_\psi(a_1)e)(t),(e\pi_\psi(a_2)e)(t),\ldots,(e\pi_\psi(a_n)e)(t))| \\ &\leq \sup_{\varphi\in\mathscr{S}}|P(\hat{a}_1(\varphi),\hat{a}_2(\varphi),\ldots,\hat{a}_n(\varphi))|,\end{aligned}$$

since $a\to(e\pi_\psi(a)e)(t)$ $(a\in\mathscr{A})$ is a state on $\mathscr{A}$ $(t\in K)$. Therefore the mapping $P(\hat{a}_1,\hat{a}_2,\ldots,\hat{a}_n)\to P(e\pi_\psi(a_1)e,e\pi_\psi(a_2)e,\ldots,e\pi_\psi(a_n)e)$ can be extended uniquely to a *-homomorphism Λ of the C^*-algebra $C(\mathscr{S})$ into Ce with $\Lambda(\hat{a})=e\pi_\psi(a)e$ $(a\in\mathscr{A})$. Hence there exists a probability Radon measure ν on $\mathscr{S}$ such that

2. $$(\Lambda(f)1_\psi,1_\psi)=\int f(\varphi)d\nu(\varphi)\qquad (f\in C(\mathscr{S})).$$

Since $[e\pi_\psi(\mathscr{A})e1_\psi]=e[\pi_\psi(\mathscr{A})1_\psi]=e\mathscr{H}_\varphi$, $\{e\pi_\psi(\mathscr{A})e\}''=eC'e$. Hence Λ can be uniquely extended to a *-isomorphism (denoted again by Λ) of $L^\infty(\mathscr{S},\nu)$ onto $eC'e$, and $\Lambda(C(\mathscr{S}))$ is σ-dense in $eC'e$.

The measure ν satisfies

3. $$(e\pi_\psi(a_1)ee\pi_\psi(a_2)e\ldots e\pi_\psi(a_n)e1_\psi,1_\psi)=\int_{\mathscr{S}}\hat{a}_1(\varphi)\hat{a}_2(\varphi)\ldots\hat{a}_n(\varphi)d\nu(\varphi)$$
$$(a_1,a_2,\ldots,a_n\in\mathscr{A}).$$

Since $(e\pi_\psi(a)e1_\psi,1_\psi)=(\pi_\psi(a)1_\psi,1_\psi)=\psi(a)$ $(a\in\mathscr{A})$, ν belongs to $\Omega(\psi)$.

3.1.1. Proposition. *$\Omega(\psi)$ contains one and only one probability Radon measure ν which satisfies*

$$\begin{aligned}&(e\pi_\psi(a_1)ee\pi_\psi(a_2)e\ldots e\pi_\psi(a_n)e1_\psi,1_\psi)\\ &=\int_{\mathscr{S}}\hat{a}_1(\varphi)\hat{a}_2(\varphi)\ldots\hat{a}_n(\varphi)d\nu(\varphi)\qquad (a_1,a_2,\ldots,a_n\in\mathscr{A}).\end{aligned}$$

The uniqueness is clear from the Stone-Weierstrass theorem.

3.1.2. Definition. *The above unique measure ν satisfying the condition* 2. (*equivalently* 3.) *is called the C-measure of ψ.*

Since $c(e)=1_{\mathscr{H}_\psi}$, the mapping $x\to xe$ $(x\in C)$ is a *-isomorphism of C onto Ce. Put $\Gamma(x)=(\Lambda)^{-1}(xe)$ $(x\in C)$. Then Γ is a *-isomorphism of C onto $L^\infty(\mathscr{S},\nu)$, and it satisfies

4. $$\begin{aligned}(x\pi_\psi(a)1_\psi,1_\psi)&=(x\pi_\psi(a)e1_\psi,e1_\psi)\\ &=(xe\pi_\psi(a)e1_\psi,1_\psi)\\ &=\int(\Lambda)^{-1}(xe)(\varphi)\hat{a}(\varphi)d\nu(\varphi)\\ &=\int_{\mathscr{S}}\Gamma(x)(\varphi)\hat{a}(\varphi)d\nu(\varphi)\end{aligned}$$

$(x\in C,\ a\in\mathscr{A})$.

3.1.3. Proposition. *Let μ be another measure belonging to $\Omega(\psi)$ with a *-isomorphism Γ_1 of C onto $L^\infty(\mathcal{S},\mu)$ such that*

$$(x\pi_\psi(a)1_\psi, 1_\psi) = \int_{\mathcal{S}} \Gamma_1(x)(\varphi)\hat{a}(\varphi)\,d\mu(\varphi) \quad (x\in C, a\in\mathcal{A}).$$

Then $\Gamma=\Gamma_1$ and $\mu=\nu$.

Proof. Let $\Gamma_1(c_1^a)=\hat{a}\,\mu$-a.e. $(a\in\mathcal{A}; c_1^a\in C)$. Then

$$\begin{aligned}(c_1^a x 1_\psi, 1_\psi) &= \int \Gamma_1(c_1^a)(\varphi)\Gamma_1(x)(\varphi)\,d\mu(\varphi)\\ &= \int \hat{a}(\varphi)\Gamma_1(x)(\varphi)\,d\mu(\varphi)\\ &= (x\pi_\psi(a)1_\psi, 1_\psi)\\ &= (\pi_\psi(a)x1_\psi, 1_\psi) \qquad (x\in C).\end{aligned}$$

Hence $c_1^a e = e\pi_\psi(a)e$, and so $\Gamma(c_1^a)=\Lambda^{-1}(c_1^a e)=\Gamma_1(c_1^a)$ $(a\in\mathcal{A})$. Hence $\Gamma=\Gamma_1$. Then $\mu=\nu$ by 3.1.1. q.e.d.

3.1.4. Definition. *The above unique *-isomorphism of C onto $L^\infty(\mathcal{S},\nu)$ satisfying the equality* 4. *is called the C-isomorphism of ψ.*

Now according to Bishop-de Leeuw [10], we define an order $\prec$ in $\Omega(\psi)$ as follows: $\mu_1\prec\mu_2$ if $\int_{\mathcal{S}}\hat{a}(\varphi)^2\,d\mu_1(\varphi)\le\int_{\mathcal{S}}\hat{a}(\varphi)^2\,d\mu_2(\varphi)$ $(a\in\mathcal{A}^s)$.

3.1.5. Proposition. *Let C_1 and C_2 be two commutative W^*-algebras, containing $1_{\mathcal{H}_\psi}$, in the commutant $\pi_\psi(\mathcal{A})'$ with $C_1\subset C_2$. Let ν_1 (resp. ν_2) be the C_1 (resp. C_2)-measure of ψ. Then $\nu_1\prec\nu_2$.*

Proof. Let e_1 (resp. e_2) be the orthogonal projection of $\mathcal{H}_\psi$ onto $[C_1 1_\psi]$ (resp. $[C_2 1_\psi]$). Then $e_1\le e_2$. Therefore for $a\in\mathcal{A}^s$,

$$\begin{aligned}\int\hat{a}(\varphi)^2\,d\nu_1(\varphi) &= (e_1\pi_\psi(a)e_1 e_1\pi_\psi(a)e_1 1_\psi, 1_\psi)\\ &= (\pi(a)e_1\pi_\psi(a)1_\psi, 1_\psi)\\ &\le (\pi_\psi(a)e_2\pi_\psi(a)1_\psi, 1_\psi)\\ &= \int\hat{a}(\varphi)^2\,d\nu_2(\varphi).\end{aligned}$$

Hence $\nu_1\prec\nu_2$. q.e.d.

3.1.6. Theorem (Extremal decomposition). *Let A be a maximal commutative W^*-subalgebra in $\pi_\psi(\mathcal{A})'$. Then the A-measure ν of ψ is maximal in $\Omega(\psi)$ with respect to the order $\prec$. Hence by the Bishop-de Leeuw theorem* [10], *for any Baire subset Δ of $\mathcal{S}$ with $\Delta\cap\mathcal{E}=(\emptyset)$ ($\mathcal{E}$, the set of all pure states on $\mathcal{A}$), $\int_\Delta d\nu(\varphi)=0$.*

Proof. Let $\mathfrak{B}$ be the C^*-algebra generated by $\pi_\psi(\mathcal{A})$ and A, and set $\tilde{\psi}(x)=(x1_\psi, 1_\psi)$ $(x\in\mathfrak{B})$. Then $\mathfrak{B}'=A$, and $\tilde{\psi}$ is a state of $\mathfrak{B}$. Let $\tilde{\mathcal{S}}$ be the state space of $\mathfrak{B}$, and let $\tilde{\nu}$ be the A-measure of $\tilde{\psi}$ on $\tilde{\mathscr{S}}$. Let $\tilde{\mu}$ be a measure in $\Omega(\tilde{\psi})$ with $\tilde{\nu} \prec \tilde{\mu}$. For $0 \le a \le 1$ $(a \in A)$, put $\tilde{\psi}_{\hat{a}}(x) = \int_{\tilde{\mathscr{S}}} \hat{a}(\tilde{\varphi})\hat{x}(\tilde{\varphi})\,d\tilde{\mu}(\tilde{\varphi})$ $(x \in \mathfrak{B})$. Then $0 \le \tilde{\psi}_{\hat{a}} \le \tilde{\psi}$, and so there exists a unique element $h_{\hat{a}}$ in $\mathfrak{B}'$ $(=A)$ with $\tilde{\psi}_{\hat{a}}(x) = (x h_{\hat{a}} 1_\psi, 1_\psi) = \int \hat{a}(\tilde{\varphi})\hat{x}(\tilde{\varphi})\,d\tilde{\mu}(\tilde{\varphi})$. On the other hand, let e be the orthogonal projection of $\mathscr{H}_\psi$ onto $[A 1_\psi]$; then

$$\int \hat{y}(\tilde{\varphi})^2\,d\tilde{\nu}(\tilde{\varphi}) = (eyeeye 1_\psi, 1_\psi) = (y^2 1_\psi, 1_\psi) = \int \widehat{y^2}(\tilde{\varphi})\,d\tilde{\nu}(\tilde{\varphi})$$

$$= \int \widehat{y^2}(\tilde{\varphi})\,d\tilde{\mu}(\tilde{\varphi}) \ge \int \hat{y}(\tilde{\varphi})^2\,d\tilde{\mu}(\varphi) \qquad (y \in A^s).$$

Hence $\int \hat{y}(\tilde{\varphi})^2\,d\tilde{\nu}(\tilde{\varphi}) = \int \hat{y}(\tilde{\varphi})^2\,d\tilde{\mu}(\tilde{\varphi})$ $(y \in A^s)$. Since

$$\int \hat{y}(\tilde{\varphi})^2\,d\tilde{\nu}(\tilde{\varphi}) = \int \widehat{y^2}(\tilde{\varphi})\,d\tilde{\nu}(\tilde{\varphi})$$

and $\int \widehat{y^2}(\tilde{\varphi})\,d\tilde{\nu}(\tilde{\varphi}) = \int \widehat{y^2}(\tilde{\varphi})\,d\tilde{\mu}(\tilde{\varphi})$, $\int \widehat{y^2}(\tilde{\varphi})\,d\tilde{\mu}(\tilde{\varphi}) = \int \hat{y}(\tilde{\varphi})^2\,d\tilde{\mu}(\tilde{\varphi})$. Since

$$\widehat{y^2}(\tilde{\varphi}) \ge \hat{y}(\tilde{\varphi})^2,$$

we have $\hat{y}(\tilde{\varphi})^2 = \widehat{y^2}(\tilde{\varphi})$ $\tilde{\mu}$-a.e. Therefore,

$$\int \hat{y}_1(\tilde{\varphi})\hat{y}_2(\tilde{\varphi})\,d\tilde{\nu}(\tilde{\varphi}) = \int \hat{y}_1(\tilde{\varphi})\hat{y}_2(\tilde{\varphi})\,d\tilde{\mu}(\varphi)$$

$(y_1, y_2 \in A)$. Hence

$$(y h_{\hat{a}} 1_\psi, 1_\psi) = \int \hat{a}(\tilde{\varphi})\hat{y}(\tilde{\varphi})\,d\tilde{\mu}(\tilde{\varphi}) = \int \hat{a}(\tilde{\varphi})\hat{y}(\tilde{\varphi})\,d\tilde{\nu}(\varphi)$$
$$= (eyeeae 1_\psi, 1_\psi) \qquad (y \in A),$$

and so $h_{\hat{a}} e = ae$. Since the central support $c(e)$ of e in $A'(=\mathfrak{B}'')$ is $1_{\mathscr{H}_\psi}$, $h_{\hat{a}} = a$. Therefore

$$\int \hat{a}(\varphi)\hat{x}(\varphi)\,d\tilde{\mu}(\tilde{\varphi}) = (x a 1_\psi, 1_\psi) \qquad (a \in A,\ x \in \mathfrak{B}).$$

Similarly for $b \in \mathfrak{B}$ there exists a unique element h_b in A, satisfying $(x h_b 1_\psi, 1_\psi) = \int \hat{x}(\tilde{\varphi})\hat{b}(\tilde{\varphi})\,d\tilde{\mu}(\varphi)$ $(x \in \mathfrak{B})$. Then

$$(x h_b 1_\psi, 1_\psi) = \int \hat{x}(\tilde{\varphi})\hat{b}(\tilde{\varphi})\,d\tilde{\mu}(\tilde{\varphi}) = \int \hat{x}(\tilde{\varphi})\hat{h}_b(\tilde{\varphi})\,d\tilde{\mu}(\tilde{\varphi}).$$

Hence $\int (\hat{b}(\tilde{\varphi}) - \hat{h}_b(\tilde{\varphi}))\hat{x}(\tilde{\varphi})\,d\tilde{\mu}(\tilde{\varphi}) = 0$, and so

$$\int (\hat{b}(\tilde{\varphi}) - \hat{h}_b(\tilde{\varphi}))\,\widehat{(b^* - h_b^*)}(\tilde{\varphi})\,d\tilde{\mu}(\varphi) = \int |\hat{b}(\tilde{\varphi}) - \hat{h}_b(\tilde{\varphi})|^2\,d\tilde{\mu}(\varphi) = 0.$$

Hence $\hat{b}(\tilde{\varphi}) = \hat{h}_b(\tilde{\varphi})$ $\tilde{\mu}$-a.e. Therefore $L^\infty(\tilde{\mathscr{S}}, \tilde{\mu}) = \{\hat{y} \mid y \in A\}$, and so there exists a *-isomorphism Γ' of A onto $L^\infty(\tilde{\mathscr{S}}, \tilde{\mu})$ with

$$(x y 1_\psi, 1_\psi) = \int \hat{x}(\tilde{\varphi})\Gamma'(y)(\tilde{\varphi})\,d\tilde{\mu}(\tilde{\varphi}) \qquad (x \in \mathfrak{B} \text{ and } y \in A).$$

Hence $\tilde{\nu} = \tilde{\mu}$ by 3.1.3—i.e., $\tilde{\nu}$ is maximal in $\Omega(\tilde{\psi})$.

Let $\mathscr{S}'$ be the state space of the C^*-algebra $\pi_\psi(\mathscr{A})$, and for $\tilde{\varphi} \in \tilde{\mathscr{S}}$, let $\tilde{\varphi}|\pi_\psi(\mathscr{A})$ be the restriction of $\tilde{\varphi}$ to $\pi_\psi(\mathscr{A})$. Put $\gamma(\tilde{\varphi}) = \tilde{\varphi}|\pi_\psi(\mathscr{A})$. Then γ is a continuous mapping of $\tilde{\mathscr{S}}$ onto $\mathscr{S}'$. Let $d\nu'(\gamma(\tilde{\varphi})) = d\tilde{\nu}(\tilde{\varphi})$; then

ν' is a probability Radon measure on $\mathcal{S}'$. Clearly ν' is the A-measure of $\tilde{\psi}|\pi_\psi(\mathcal{A})$. Let $\mathcal{E}'$ be the set of all pure states on $\pi_\psi(\mathcal{A})$, and let Δ' be a Baire set of $\mathcal{S}'$ with $\mathcal{E}' \cap \Delta' = (\emptyset)$. Then $\gamma^{-1}(\Delta')$ is a Baire set in $\tilde{\mathcal{S}}$, and $\gamma^{-1}(\Delta') \cap \tilde{\mathcal{E}} = (\emptyset)$, where $\tilde{\mathcal{E}}$ is the set of all pure states on $\mathfrak{B}$. In fact, suppose $\tilde{\varphi}$ is a pure state on $\mathfrak{B}$ belonging to $\gamma^{-1}(\Delta')$. Then $\tilde{\varphi}|\pi_\psi(\mathcal{A})$ is pure, since $\mathfrak{B}$ is the C^*-algebra generated by $\pi_\psi(\mathcal{A})$ and A, and A is the center of $\mathfrak{B}$. Hence ν' is maximal in $\Omega(\tilde{\psi}|\pi(\mathcal{A}))$.

We shall consider a state on $\pi_\psi(\mathcal{A})$ as a state on $\mathcal{A}$ in the canonical way. Then $\mathcal{S}' \subset \mathcal{S}$, and so ν' can be considered as a probability Radon measure on $\mathcal{S}$. Then it is the A-measure of ψ and is maximal in $\Omega(\psi)$.

q.e.d.

3.1.7. Definition. *A state φ on a C^*-algebra $\mathcal{A}$ is said to be factorial if the weak closure $\overline{\pi_\varphi(\mathcal{A})}$ of $\pi_\varphi(\mathcal{A})$ is a factor.*

3.1.8. Theorem (Central decomposition). *Let Z be the center of $\pi_\psi(\mathcal{A})''$. The Z-measure (called the central measure) ν of ψ is concentrated on factorial states—namely, if Δ is a Baire set in $\mathcal{S}$ with $\Delta \cap \mathfrak{F} = (\emptyset)$ ($\mathfrak{F}$, the set of all factorial states on $\mathcal{A}$), then $\int_\Delta d\nu(\varphi) = 0$.*

Proof. Let $\mathfrak{B}$ be the C^*-algebra generated by $\pi_\psi(\mathcal{A})$ and $\pi_\psi(\mathcal{A})'$. Then $\mathfrak{B}' = Z$ and $Z \subset \mathfrak{B}$. Let $\tilde{\nu}$ be the Z-measure of $\mathfrak{B}$. By the same reasoning with the proof of 3.1.6, $\tilde{\nu}$ is maximal in $\Omega(\tilde{\psi})$ ($\tilde{\psi}(x) = (x 1_\psi, 1_\psi)$ for $x \in \mathfrak{B}$). For $\tilde{\varphi} \in \tilde{\mathcal{S}}$ ($\tilde{\mathcal{S}}$, the state space of $\mathfrak{B}$), set $\gamma(\tilde{\varphi}) = \tilde{\varphi}|\pi_\psi(\mathcal{A})$ and $d\nu'(\gamma(\tilde{\varphi})) = d\tilde{\nu}(\tilde{\varphi})$. Let $\mathfrak{F}'$ be the set of all factorial states on $\pi_\psi(\mathcal{A})$, and let Δ' be a Baire set in $\mathcal{S}'$ ($\mathcal{S}'$, the state space of $\pi_\psi(\mathcal{A})$) with $\Delta' \cap \mathfrak{F}' = (\emptyset)$. Then $\tilde{\mathcal{E}} \cap \gamma^{-1}(\Delta') = (\emptyset)$, where $\tilde{\mathcal{E}}$ is the set of pure states of $\mathfrak{B}$—in fact, suppose $\tilde{\varphi} \in \gamma^{-1}(\Delta') \cap \tilde{\mathcal{E}}$; then $\pi_{\tilde{\varphi}}(\pi_\psi(\mathcal{A}))''$ must be a factor. Hence we have the theorem. q.e.d.

In mathematical physics, a system $\{\mathcal{A}, G\}$ of a C^*-algebra $\mathcal{A}$ and a group G of $*$-automorphisms on $\mathcal{A}$ is very often considered. A state φ on $\mathcal{A}$ is said to be G-invariant if $\varphi(a^g) = \varphi(a)$ $(a \in \mathcal{A}, g \in G)$, where a^g is the image of a by a $*$-automorphism g.

Let $\mathcal{S}_G$ be the set of all G-invariant states on $\mathcal{A}$. Then $\mathcal{S}_G$ is a closed subset of the state space $\mathcal{S}$ of $\mathcal{A}$. For $\varphi \in \mathcal{S}_G$, consider the $*$-representation $\{\pi_\varphi, \mathcal{H}_\varphi\}$ of $\mathcal{A}$. Set $u_\varphi(g^{-1}) a_\varphi = (a^g)_\varphi$. Then

$$\|u_\varphi(g^{-1}) a_\varphi\|^2 = \|(a^g)_\varphi\|^2 = \varphi(a^{g*} a^g) = \varphi((a^* a)^g) = \varphi(a^* a) = \|a_\varphi\|^2 .$$

Hence $u_\varphi(g^{-1})$ can be extended uniquely to a unitary operator (denoted by $u_\varphi(g^{-1})$) on $\mathcal{H}_\varphi$. Then $g \to u_\varphi(g)$ $(g \in G)$ is a unitary representation of G—i.e., $u_\varphi(g_1 g_2) = u_\varphi(g_1) u_\varphi(g_2)$ $(g_1, g_2 \in G)$.

$$\begin{aligned} u_\varphi(g^{-1}) \pi_\varphi(a) u_\varphi(g) b_\varphi &= u_\varphi(g^{-1}) \pi_\varphi(a) (b^{g^{-1}})_\varphi = u_\varphi(g^{-1}) (a b^{g^{-1}})_\varphi \\ &= (a^g b)_\varphi \qquad (a, b \in \mathcal{A}). \end{aligned}$$

Hence $u_\varphi(g^{-1})\pi_\varphi(a)u_\varphi(g)=\pi_\varphi(a^g)$. Let

$$E_\varphi=\{\xi\,|\,u_\varphi(g)\xi=\xi \text{ for all } g\in G, \xi\in\mathscr{H}_\varphi\}.$$

Then E_φ is a closed subspace of $\mathscr{H}_\varphi$ containing 1_φ.

3.1.9. Definition. *An extreme point in $\mathscr{S}_G$ is said to be ergodic.*

3.1.10. Proposition. *Consider the following three conditions:*

I $\varphi\in\mathscr{S}_G$ *is ergodic;*

II $\{\pi_\varphi(\mathscr{A}),u_\varphi(G)\}'=\mathbb{C}1_{\mathscr{H}_\varphi}$;

III $\dim E_\varphi=1$.

Then I $\Leftrightarrow$ II $\Leftarrow$ III.

Proof. For ξ ($\|\xi\|=1$) in E_φ, put $\varphi_\xi(a)=(\pi_\varphi(a)\xi,\xi)$ $(a\in\mathscr{A})$. Then $\varphi_\xi(a^g)=(u_\varphi(g^{-1})\pi_\varphi(a)u_\varphi(g)\xi,\xi)=(\pi_\varphi(a)\xi,\xi)=\varphi_\xi(a)$. Hence $\varphi_\xi\in\mathscr{S}_G$.

I$\Rightarrow$II. Suppose φ is ergodic. For $h\in\{\pi_\varphi(\mathscr{A}),u_\varphi(G)\}'$ with $0<h\leq 1_{\mathscr{H}\varphi}$, put $\varphi_h(a)=(\pi_\psi(a)h1_\psi,1_\psi)(a\in\mathscr{A})$. Then $\varphi_h(a^g)=\varphi_h(a)$ and $\varphi_h\leq\varphi$. Hence $\varphi_h=\lambda\varphi$ $(\lambda>0)$, and so $h=\lambda 1_{\mathscr{H}_\varphi}$.

II$\Rightarrow$I. Suppose φ is not extreme in $\mathscr{S}_G$. Then there exists an invariant state ψ with $\psi(a)=(\pi_\varphi(a)h'1_\varphi,1_\varphi)$ and $h'\notin\mathbb{C}1_{\mathscr{H}}$ $(a\in\mathscr{A}$, some $h'\in\pi_\varphi(\mathscr{A})')$. Then

$$\begin{aligned}\psi(a^g)&=(\pi_\varphi(g^{-1})\pi_\varphi(a)u_\varphi(g)h'1_\varphi,1_\varphi)=(\pi_\varphi(a)u_\varphi(g)h'1_\varphi,1_\varphi)\\&=(\pi_\varphi(a)u_\varphi(g)h'u(g^{-1})1_\varphi,1_\varphi)=(\pi_\varphi(a)h'1_\varphi,1_\varphi)\qquad(a\in\mathscr{A},g\in G).\end{aligned}$$

Hence $u_\varphi(g)h'u_\varphi(g^{-1})1_\varphi=h'1_\varphi$. Since $u_\varphi(g)\pi_\varphi(\mathscr{A})u_\varphi(g^{-1})=\pi_\varphi(\mathscr{A})$, $u_\varphi(g)\pi_\varphi(\mathscr{A})'u_\varphi(g^{-1})=\pi_\varphi(\mathscr{A})'$. Hence $u_\varphi(g)h'u_\varphi(g^{-1})=h'$, and so $h'\in\{\pi_\varphi(\mathscr{A}),u_\varphi(G)\}'$.

III$\Rightarrow$II. Let e' be a projection in $\{\pi_\varphi(\mathscr{A}),u_\varphi(G)\}'$. Then

$$(u_\varphi(g^{-1})\pi_\varphi(a)u_\varphi(g)e'1_\varphi,1_\varphi)=(\pi_\varphi(a)e'1_\varphi,e'1_\varphi)\qquad(a\in\mathscr{A},g\in G).$$

Hence $e'1_\varphi=\lambda 1_\varphi$ for some complex number λ, and so

$$\pi_\varphi(a)e'1_\varphi=e'\pi_\varphi(a)1_\varphi=\lambda\pi_\varphi(a)1_\varphi\qquad(a\in\mathscr{A}).$$

Hence $e'=\lambda 1_{\mathscr{H}\varphi}$. q.e.d.

3.1.11. Definition. *A system $\{\mathscr{A},G\}$ of a C^*-algebra $\mathscr{A}$ and a group G of $*$-automorphisms on $\mathscr{A}$ is said to be G-abelian if for every $\varphi\in\mathscr{S}_G$, $P_\varphi\pi_\varphi(\mathscr{A})P_\varphi$ is a family of mutually commutative operators, where P_φ is the orthogonal projection of $\mathscr{H}_\varphi$ onto E_φ.*

3.1.12. Proposition. *Suppose a system $\{\mathscr{A},G\}$ is G-abelian. Then the three conditions in* 3.1.10 *are mutually equivalent.*

Proof. I$\Rightarrow$III. Suppose φ is ergodic. Let C be the commutative C^*-algebra generated by $P_\varphi\pi_\varphi(\mathscr{A})P_\varphi$, and let $C=C(K)$ be the function representation of C. Let μ be a probability Radon measure on K satisfying

$$(P_\varphi \pi_\varphi(a_1) P_\varphi P_\varphi \pi_\varphi(a_2) P_\varphi \dots P_\varphi \pi_\varphi(a_n) P_\varphi 1_\varphi, 1_\varphi)$$
$$= \int_K (P_\varphi \pi_\varphi(a_1) P_\varphi)(t)(P_\varphi \pi_\varphi(a_2) P_\varphi)(t) \dots (P_\varphi \pi_\varphi(a_n) P_\varphi)(t) d\mu(t)$$

$(a_1, a_2, \dots, a_n \in \mathscr{A})$. Since $P_\varphi \pi_\varphi(a^g) P_\varphi = P_\varphi u_\varphi(g^{-1}) \pi_\varphi(a) u_\varphi(g) P_\varphi = P_\varphi \pi_\varphi(a) P_\varphi$, $a \to (P_\varphi \pi_\varphi(a) P_\varphi)(t)$ $(a \in \mathscr{A})$ is a G-invariant state. Since φ is ergodic, μ must be a point measure.

On the other hand, $[C 1_\varphi] \supseteq [P_\varphi \pi_\varphi(\mathscr{A}) P_\varphi 1_\varphi] = [P_\varphi \pi_\varphi(\mathscr{A}) 1_\varphi] = E_\varphi$; hence if $(x^* x 1_\varphi, 1_\varphi) = 0$ for some $x \in C$, $x = 0$. Therefore K consists of a single point, and so $C = \mathbb{C} P_\varphi$. Hence $E_\varphi = [P_\varphi \pi_\varphi(\mathscr{A}) P_\varphi 1_\varphi] = [C 1_\varphi] = \mathbb{C} 1_\varphi$. q.e.d.

3.1.13. Proposition. *If $\{\mathscr{A}, G\}$ is a G-abelian system, then for $\varphi \in \mathscr{S}_G$, $\{\pi_\varphi(\mathscr{A}), u_\varphi(G)\}'$ is commutative.*

Proof. By the considerations in 3.1.10, $[\{\pi_\varphi(\mathscr{A}), u_\varphi(G)\}' E_\varphi] \subset E_\varphi$. Hence $P_\varphi \in \{\pi_\varphi(\mathscr{A}), u_\varphi(G)\}''$. $P_\varphi \pi_\varphi(a) u_\varphi(g) P_\varphi = P_\varphi \pi_\varphi(a) P_\varphi$, and so $P_\varphi \{\pi_\varphi(\mathscr{A}), u_\varphi(G)\}'' P_\varphi$ is commutative. Hence, P_φ is an abelian projection. Since the central support of P_φ in $\{\pi_\varphi(\mathscr{A}), u_\varphi(G)\}''$ is $1_{\mathscr{H}_\varphi}$, P_φ is a maximal abelian projection. Hence $\{\pi_\varphi(\mathscr{A}), u_\varphi(G)\}''$ is of type I (cf. 2.3). $P_\varphi \{\pi_\varphi(\mathscr{A}), u_\varphi(G)\}' P_\varphi = \{P_\varphi \{\pi_\varphi(\mathscr{A}), u_\varphi(G)\}'' P_\varphi\}'$ and

$$[P_\varphi \{\pi_\varphi(\mathscr{A}), u_\varphi(G)\}'' P_\varphi 1_\varphi] = P_\varphi \mathscr{H}_\varphi.$$

Hence $P_\varphi \{\pi_\varphi(\mathscr{A}), u_\varphi(G)\}'' P_\varphi$ is maximal commutative (cf. 2.9.4). Therefore $P_\varphi \{\pi_\varphi(\mathscr{A}), u_\varphi(G)\}' P_\varphi = \{\pi_\varphi(\mathscr{A}), u_\varphi(G)\}' P_\varphi$ is commutative. Since a mapping $y \to y P_\varphi$ $(y \in \{\pi_\varphi(\mathscr{A}), u_\varphi(G)\}')$ is a *-isomorphism, $\{\pi_\varphi(\mathscr{A}), u_\varphi(G)\}'$ is commutative. q.e.d.

3.1.14. Theorem (Ergodic decomposition). *If $\{\mathscr{A}, G\}$ is a G-abelian system, then $\mathscr{S}_G$ is a simplex in the sense of Choquet [23]. Therefore, for every $\psi \in \mathscr{S}_G$, there exists a unique probability Radon measure v on $\mathscr{S}_G$ such that $\psi(a) = \int_{\mathscr{S}_G} \hat{a}(\varphi) dv(\varphi)$ $(a \in \mathscr{A})$, and for any Baire set Δ in $\mathscr{S}_G$ with $\mathscr{E}_G \cap \Delta = (\emptyset)$, $\int_\Delta dv(\varphi) = 0$ ($\mathscr{E}_G$ is the set of all extreme points in $\mathscr{S}_G$).*

Proof. To prove that $\mathscr{S}_G$ is a simplex, it suffices to show the following: if E is the Banach space of all G-invariant self-adjoint bounded linear functionals on $\mathscr{A}$, then E is a lattice with the order $\leq$.

Let $f \in E$ and let $f = f^+ - f^-$ be the orthogonal decomposition of f. By the uniqueness of the orthogonal decomposition, $f^+, f^- \in E$. Now let $f_1, f_2 \in E$ and set $\psi = \dfrac{|f_1| + |f_2|}{\|f_1\| + \|f_2\|}$. Then $\psi \in \mathscr{S}_G$, and so $\{\pi_\psi(\mathscr{A}), u_\psi(G)\}'$ is commutative.

Since $|f_i| \leq (\|f_1\| + \|f_2\|)\psi$ $(i=1,2)$, there is a unique self-adjoint element h_i' in $\pi_\psi(\mathscr{A})'$ with $f_i(a) = (\pi_\psi(a) h_i' 1_\psi, 1_\psi)$ $(a \in \mathscr{A})$.

Since $f_i(a^g) = f_i(a)$, by a similar reasoning in the proof of 3.1.10, $u_\psi(g) h_i' u_\psi(g^{-1}) = h_i'$ $(g \in G)$. Hence $h_i' \in \{\pi_\psi(\mathscr{A}), u_\psi(G)\}'$. Since $\{\pi_\psi(\mathscr{A}), u_\psi(G)\}'$ is commutative, its self-adjoint portion is a lattice. Set $f_3(a) = (\pi_\psi(a)(h_1' \vee h_2') 1_\psi, 1_\psi)$ $(a \in \mathscr{A})$. Then $f_1, f_2 \leq f_3$. Suppose $f_1, f_2 \leq f_4$ with some $f_4 \in E$ and put

$$\psi_1 = \frac{\psi + |f_4|}{1 + \|f_4\|}.$$

Then $f_i(a) = (\pi_{\psi_1}(a) k_i' 1_{\psi_1}, 1_{\psi_1})$ and $\psi(a) = (\pi_{\psi_1}(a) k' 1_{\psi_1}, 1_{\psi_1})$ $(a \in \mathscr{A})$ with $k', k_i' \in \{\pi_{\psi_1}(\mathscr{A}), u_{\psi_1}(G)\}'$ $(i=1,2,3,4)$. Clearly $k_1' \vee k_2' \leq k_3'$. Put $f_5(a) = (\pi_{\psi_1}(a) k_1' \vee k_2' 1_{\psi_1}, 1_{\psi_1})$ $(a \in \mathscr{A})$. Since $|f_1|, |f_2| \leq (\|f_1\| + \|f_2\|)\psi$, $-(\|f_1\| + \|f_2\|)\psi \leq f_1, f_2 \leq (\|f_1\| + \|f_2\|)\psi$. Hence

$$-(\|f_1\| + \|f_2\|)k' \leq k_1', k_2' \leq (\|f_1\| + \|f_2\|)k',$$

and so

$$-(\|f_1\| + \|f_2\|)k' \leq k_1' \vee k_2' \leq (\|f_1\| + \|f_2\|)k'.$$

Hence $|k_1' \vee k_2'| \leq (\|f_1\| + \|f_2\|)k'$ and so there is an h_5' in $\{\pi_\psi(\mathscr{A}), u_\psi(G)\}'$ with $f_5(a) = (\pi_\psi(a) h_5' 1_\psi, 1_\psi)$ $(a \in \mathscr{A})$.

Since $f_1, f_2 \leq f_5$, $h_1', h_2' \leq h_5'$. Hence $h_1' \vee h_2' \leq h_5'$, and so $f_3 \leq f_5$. Therefore $f_3 = f_5 \leq f_4$ and so $f_1 \vee f_2 = f_3$. Similarly one can prove the existence of $f_1 \wedge f_2$. q.e.d.

3.1.15. Definition. *Let $\{\mathscr{A}, G\}$ be a system of a C*-algebra $\mathscr{A}$ and a group of *-automorphisms on $\mathscr{A}$. $\{\mathscr{A}, G\}$ is said to be asymptotically abelian if there exists a sequence $\{g_n\}$ in G satisfying $\|[a^{g_n}, b]\| \to 0$ $(n \to \infty)$ $(a, b \in \mathscr{A})$, where $[\,,\,]$ is the Lie product.*

3.1.16. Proposition. *If a system $\{\mathscr{A}, G\}$ is asymptotically abelian, then $\{\mathscr{A}, G\}$ is G-abelian.*

Proof. Let $\varphi \in \mathscr{S}_G$. For arbitrary $\eta_1, \eta_2 \in \mathscr{H}_\varphi$, $P_\varphi \eta_i$ $(i=1,2)$ is invariant under $u_\varphi(g)$ $(g \in G)$. Consider the space $\mathscr{H}_\varphi \oplus \mathscr{H}_\varphi$, and let Γ be the closed convex hull of $\{\{u_\varphi(g)\eta_1, u_\varphi(g)\eta_2\} \mid g \in G\}$ in $\mathscr{H}_\varphi \oplus \mathscr{H}_\varphi$. Then Γ is invariant under $u_\varphi(g) \oplus u_\varphi(g)$. Let $\{\eta', \eta''\}$ be a unique element in Γ with $\|\{\eta', \eta''\}\| = \inf_{\{\xi_1, \xi_2\} \in \Gamma} \|\{\xi_1, \xi_2\}\|$. Then $\{\eta', \eta''\}$ is invariant under $u_\varphi(g) \oplus u_\varphi(g)$ $(g \in G)$. Hence $\{\eta', \eta''\} \in P_\varphi \mathscr{H}_\varphi \oplus P_\varphi \mathscr{H}_\varphi$. Since $(P_\varphi \oplus P_\varphi)(u_\varphi(g) \oplus u_\varphi(g)) = P_\varphi \oplus P_\varphi$, $(P_\varphi \oplus P_\varphi)\Gamma = \{P_\varphi \eta_1, P_\varphi \eta_2\}$. Hence $\eta' = P_\varphi \eta_1$ and $\eta'' = P_\varphi \eta_2$. Therefore, for an arbitrary $\varepsilon > 0$, there exists a $\sum_{i=1}^{m} \lambda_i u_\varphi(h_i^{-1})$ with $\lambda_i \geq 0$, $\sum_{i=1}^{m} \lambda_i = 1$, $h_i \in G$ $(i=1,2,\ldots,m)$ such that for $\eta \in E_\varphi$

$$\left\| \left(\sum_{i=1}^{m} \lambda_i u_\varphi(h_i^{-1}) - P_\varphi \right) \pi_\varphi(a^*) \eta \right\| < \varepsilon$$

$$\left\| \left(\sum_{i=1}^{m} \lambda_i u_\varphi(h_i^{-1}) - P_\varphi \right) \pi_\varphi(a) \eta \right\| < \varepsilon .$$

Take an n_0 such that

$$\left| \left(\eta, \left[\pi_\varphi \left(\left(\sum_{i=1}^{m} \lambda_i a^{h_i} \right)^{g_n} \right), \pi_\varphi(b) \right] \eta \right) \right| < \varepsilon \qquad (n \geq n_0).$$

$$\begin{aligned}
&\left| \left(\eta, \left[\pi_\varphi \left(\left(\sum_{i=1}^{m} \lambda_i a^{h_i} \right)^{g_n} \right), \pi_\varphi(b) \right] \eta \right) \right| \\
&= \left| \left(\eta, \sum_{i=1}^{m} \lambda_i u_\varphi(g_n^{-1}) u_\varphi(h_i^{-1}) \pi_\varphi(a) u_\varphi(h_i) u_\varphi(g_n) \pi_\varphi(b) \eta \right) \right. \\
&\qquad \left. - \left(\eta, \sum_{i=1}^{m} \lambda_i \pi_\varphi(b) u_\varphi(g_n^{-1}) u_\varphi(h_i^{-1}) \pi_\varphi(a) u_\varphi(h_i) u_\varphi(g_n) \eta \right) \right| \\
&= \left| \left(\sum_{i=1}^{m} \lambda_i u_\varphi(h_i^{-1}) \pi_\varphi(a^*) \eta, u_\varphi(g_n) \pi_\varphi(b) \eta \right) \right. \\
&\qquad \left. - \left(\eta, \pi_\varphi(b) u_\varphi(g_n^{-1}) \sum_{i=1}^{m} \lambda_i u_\varphi(h_i^{-1}) \pi_\varphi(a) \eta \right) \right| \\
&\geq \left| (P_\varphi \pi_\varphi(a^*) \eta, u_\varphi(g_n) \pi_\varphi(b) \eta) - (\eta, \pi_\varphi(b) u_\varphi(g_n^{-1}) P_\varphi \pi_\varphi(a) \eta) \right| \\
&\qquad - \| \pi_\varphi(b) \eta \| \varepsilon - \| \pi_\varphi(b^*) \eta \| \varepsilon \\
&= \left| (P_\varphi \pi_\varphi(a^*) \eta, \pi_\varphi(b) \eta) - (\eta, \pi_\varphi(b) P_\varphi \pi_\varphi(a) \eta) \right| \\
&\qquad - \| \pi_\varphi(b) \eta \| \varepsilon - \| \pi_\varphi(b^*) \eta \| \varepsilon .
\end{aligned}$$

Hence $(\eta, \pi_\varphi(b) P_\varphi \pi_\varphi(a) \eta) = (P_\varphi \pi_\varphi(a^*) \eta, \pi_\varphi(b) \eta)$ $(\eta \in E_\varphi)$. Therefore

$$P_\varphi \pi_\varphi(b) P_\varphi \pi_\varphi(a) P_\varphi = P_\varphi \pi_\varphi(a) P_\varphi \pi_\varphi(b) P_\varphi \qquad (a, b \in \mathscr{A}).$$ q.e.d.

3.1.17. Example. *Now let* $\mathfrak{B}$ *be a* C^**-algebra with identity, and let* $\mathfrak{B}_m = \mathfrak{B}$ $(m=1,2,\ldots)$. *Let* $\mathscr{D} = \bigotimes_{m=1}^{\infty} \mathfrak{B}_m$ *be the infinite tensor product of* C^**-algebras* $\{\mathfrak{B}_m | m=1,2,\ldots\}$. *Let* P *be the group of all finite permutations of positive integers* N*—i.e., an element* $g \in P$ *is a one-to-one mapping of* N *onto itself which leaves all but a finite number of positive integers fixed. Then* g *will define a *-automorphism, also denoted by* g*, of* $\mathscr{D}$ *by* $(\sum \otimes a_m)^g = \sum \otimes a_{g(m)}$*, where* $a_m = 1$ *for all but a finite number of indices.*

For each integer n, we denote by g_n the permutation

$$g_n(k)=\begin{cases} 2^{n-1}+k & \text{if} \quad 1\le k\le 2^{n-1} \\ k-2^{n-1} & \text{if} \quad 2^{n-1}<k\le 2^n \\ k & \text{if} \quad 2^n<k. \end{cases}$$

Then one can easily see that $\|[a^{g_n},b]\|\to 0\,(n\to\infty)\,(a,b\in\mathscr{A})$. Therefore the system $\{\mathscr{D},P\}$ is asymptotically abelian.

Let $\mathscr{S}_{\mathfrak{B}}$ be the state space of $\mathfrak{B}$. For $\psi\in\mathscr{S}_{\mathfrak{B}}$, set $\varphi=\bigotimes_{m=1}^{\infty}\psi_m$ with $\psi_m=\psi$ $(m=1,2,\dots)$. Then φ is an infinite product state on $\bigotimes_{m=1}^{\infty}\mathfrak{B}_m$. Clearly φ is P-invariant. It is known that a P-invariant state on $\bigotimes_{m=1}^{\infty}\mathfrak{B}_m$ is ergodic if and only if it is an infinite product state (cf. [196]).

Let $\mathfrak{T}$ be the set of all tracial states on a C^*-algebra $\mathscr{A}$. Then $\mathfrak{T}$ is a compact subset in the state space $\mathscr{S}$ of $\mathscr{A}$. Let $\mathscr{A}^u$ be the set of all unitary elements in $\mathscr{A}$, and define $a^{i(v)}=v^{-1}av\,(a\in\mathscr{A},v\in\mathscr{A}^u)$. Then

$$G=\{i(v)\,|\,v\in\mathscr{A}^u\}$$

is a group of *-automorphisms on $\mathscr{A}$. For $\tau\in\mathfrak{T}$, $\tau(a^g)=\tau(a)\,(g\in G)$. Hence τ is a G-invariant state. Let $\overline{\pi_\tau(\mathscr{A})}$ be the weak closure of $\pi_\tau(\mathscr{A})$ on $\mathscr{H}_\tau$. Then it is a finite W^*-algebra. Set $\tilde{\tau}(x)=(x1_\tau,1_\tau)\,(x\in\overline{\pi_\tau(\mathscr{A})})$. Then $\tilde{\tau}$ is a faithful normal tracial state on $\overline{\pi_\tau(\mathscr{A})}$. Hence $\{\overline{\pi_\tau(\mathscr{A})},\mathscr{H}_\tau\}$ is a standard representation of $\overline{\pi_\tau(\mathscr{A})}$ (cf. 2.9.2).

For $a\in\mathscr{A}$, let Γ be the closed convex hull generated by $\{u_\tau(g)a_\tau|g\in G\}$. Then by the consideration in the proof of 3.1.16, $\Gamma\cap E_\tau=P_\tau a_\tau$. On the other hand, $u_\tau(g^{-1})a_\tau=(a^g)_\tau=(v^{-1}av)_\tau=\pi_\tau(v)^{-1}\pi_\tau(a)\pi_\tau(v)1_\tau\,(v^{-1}av=a^g)$. Hence there exists an element x in the unit sphere of $\overline{\pi_\tau(\mathscr{A})}$ with $x1_\tau=P_\tau a_\tau$.

Since $u_\tau(g^{-1})a_\tau=u_\tau(g^{-1})\pi_\tau(a)1_\tau$, $u_\tau(g^{-1})x1_\tau=\pi_\tau(v)^{-1}x\pi_\tau(v)1_\tau$. Hence $u_\tau(g^{-1})x1_\tau=x1_\tau$ implies $\pi_\tau(v)^{-1}x\pi_\tau(v)1_\tau=x1_\tau$. Since 1_τ is a separating vector for $\pi_\tau(\mathscr{A})''$, $x\in\pi_\tau(\mathscr{A})'\cap\pi_\tau(\mathscr{A})''$. Hence $E_\tau=[Z1_\tau]$, where Z is the center of $\overline{\pi_\tau(\mathscr{A})}$.

Since $Z'=R(\overline{\pi_\tau(\mathscr{A})},\pi_\tau(\mathscr{A})')$, $\{P_\tau R(\overline{\pi_\tau(\mathscr{A})},\pi_\tau(\mathscr{A})')P_\tau\}'=ZP_\tau$. Since ZP_τ is maximal commutative, $P_\tau R(\overline{\pi_\tau(\mathscr{A})},\pi_\tau(\mathscr{A})')P_\tau\subset ZP_\tau$. Hence $P_\tau\pi_\tau(\mathscr{A})P_\tau$ is commutative. Hence we have

3.1.18. Theorem (Tracial decomposition). *Let $\mathscr{A}$ be a C^*-algebra and let $\mathfrak{T}$ be the compact space of all tracial states on $\mathscr{A}$. Then $\mathfrak{T}$ is a simplex. Therefore for any $\tau\in\mathfrak{T}$, there exists a unique probability Radon measure ν on $\mathfrak{T}$, with $\tau(a)=\int\hat{a}(\varphi)d\nu(\varphi)\,(a\in\mathscr{A})$, such that for any Baire set Δ in $\mathscr{S}$ with $\Delta\cap\mathfrak{F}\cap\mathfrak{T}=(\emptyset)$, $\int_\Delta d\nu(\varphi)=0$ ($\mathfrak{F}$ is the set of factorial states on $\mathscr{A}$). Furthermore, this unique measure ν is the central measure of τ.*

Proof. If $\tau \in \mathfrak{J}$ is ergodic, $\dim(E_\tau)=1$. Hence $Z=\mathbb{C}1_{\mathscr{H}_\tau}$, and so τ is factorial. Conversely if $\tau \in \mathfrak{J}$ is factorial, then by the above result, $\dim(E_\tau)=1$. Hence it is ergodic.

Now let μ be the central measure of a tracial state τ, and let Z be the center of $\overline{\pi_\tau(\mathscr{A})}$. Then by 3.1.3, there exists a unique *-isomorphism Γ of Z onto $L^\infty(\mathscr{S},\mu)$ satisfying:

$$(x\pi_\tau(a)1_\tau, 1_\tau)=\int_{\mathscr{S}} \Gamma(x)(\varphi)\hat{a}(\varphi)d\mu(\varphi) \qquad (x\in Z, a\in\mathscr{A}).$$

$$\begin{aligned}(x\pi_\tau(v^* a v)1_\tau, 1_\tau) &= (\pi_\tau(v)^* x\pi_\tau(a)\pi_\tau(v)1_\tau, 1_\tau)\\ &= (x\pi_\tau(a)1_\tau, 1_\tau) \qquad (v\in\mathscr{A}^u).\end{aligned}$$

Hence $\int_{\mathscr{S}} \Gamma(x)(\varphi)\hat{a}(\varphi)d\mu(\varphi)=\int_{\mathscr{S}} \Gamma(x)(\varphi)\widehat{v^* a v}(\varphi)d\mu(\varphi)$ $(x\in Z)$. Hence $\hat{a}(\varphi)=\widehat{v^* a v}(\varphi)$ μ-a.e., and so $\hat{a}(\varphi)=\widehat{v^* a v}(\varphi)$ $(\varphi\in s(\mu))$ ($s(\mu)$ is the support of μ), because $\hat{a}$, $\widehat{v^* a v}$ are continuous. Hence $s(\mu)\subset\mathfrak{J}$, and so one can easily conclude that $\mu=\nu$. q.e.d.

Concluding remarks on 3.1.

Theorem 3.1.8 in non separable cases is due to Wils [218]. Wils [219] further extended the notion of central decomposition to an arbitrary convex compact set in a locally convex space, and defined the notions of central measures and factorial points. This result supplies interesting problems as follows: 1. Can we extend the notion of types I, II and III to these factorial points?; 2. Apply this central decomposition to some concrete cases which are important in the functional analysis, and obtain some detailed results.

Theorem 3.1.14 is due to Lanford and Ruelle [110]. Proposition 3.1.1 is due to Ruelle [146].

References. [10], [18], [22], [23], [40], [41], [103], [133], [144], [146], [159], [183], [207], [208], [218], [219].

3.2. Reduction Theory (Space-Free)

Let $L^1(\Omega,\mu)$ be the Banach space of all complex valued μ-integrable functions, with a positive measure μ, on a localizable measure space (Ω,μ), and let E be a Banach space. Then $L^1(\Omega,\mu)\otimes_\gamma E=L^1(\Omega,\mu,E)$ by Grothendieck [65], where $L^1(\Omega,\mu,E)$ is the Banach space of all E-valued strongly μ-integrable functions on the measure space (Ω,μ) and γ is the greatest cross norm. If E is separable, for any $x\in(L^1(\Omega,\mu)\otimes_\gamma E)^*$, there exists a unique E^*-valued essentially bounded weakly * locally μ-measurable function $f^x(t)$ on Ω such that

$$x(\xi\otimes\eta)=\int_\Omega \langle f^x(t),\xi\rangle \eta(t)\,d\mu(t)$$

and $\operatorname{ess.sup}_{t\in\Omega} \|f^x(t)\| = \|x\|$ by Dunford-Pettis theorem [43]

$$(\xi\in E, \eta\in L^1(\Omega,\mu)).$$

Under the mapping $x\to f^x$, $(L^1(\Omega,\mu)\otimes_\gamma E)^*$ is isometrically isomorphic to $L^\infty(\Omega,\mu,E^*)$, where $L^\infty(\Omega,\mu,E^*)$ is the Banach space of all E^*-valued essentially bounded weakly * locally μ-measurable functions on Ω, and so we can identify the dual of $L^1(\Omega,\mu,E)$ with $L^\infty(\Omega,\mu,E^*)$ as Banach spaces.

3.2.1. Proposition. *If E is separable and the dual E^* is a Banach algebra such that the multiplication in E^* is separately $\sigma(E^*,E)$-continuous, then $L^\infty(\Omega,\mu,E^*)$ is again a Banach algebra by the pointwise multiplication.*

Proof. It suffices to assume that $\mu(\Omega)=1$. Take $x,y\in L^\infty(\Omega,\mu,E^*)$. For $f\in E$, set $(R_b f)(a)=f(ab)$ $(a,b\in E^*)$. Since the multiplication is separately $\sigma(E^*,E)$-continuous, $R_{y(t)}f$ belongs to $E(t\in\Omega)$.

$$(R_{y(t)}f)(a)=f(ay(t))=(L_a f)(y(t)) \quad ((L_a f)(b)=f(ab)).$$

Put $g(t)=L_a f(t\in\Omega)$. Then $g\in L^1(\Omega,\mu,E)$, and so $(R_{y(t)}f)(a)$ is measurable for $a\in E^*$. Hence a function $t\to R_{y(t)}f$ is weakly measurable.

Since E is separable, $t\to R_{y(t)}f$ is strongly measurable. The set of all finite linear combinations of E-valued functions of form $\eta(t)f$ is norm-dense in $L^1(\Omega,\mu,E)$, and a function $\eta(t)R_{y(t)}f$ belongs to $L^1(\Omega,\mu,E)$. Hence there is a sequence (h_n) in $L^1(\Omega,\mu,E)$ such that $h_n\to h$ in $L^1(\Omega,\mu,E)$ and $R_{y(t)}h_n(t)$ is an E-valued function belonging to $L^1(\Omega,\mu,E)$ $(n=1,2,\ldots)$. Then

$$\|R_{y(t)}h_n(t)-R_{y(t)}h(t)\|\leq\|y\|\ \|h_n(t)-h(t)\|.$$

Since there exists a subsequence (n_j) of (n) such that $\|h_{n_j}(t)-h(t)\|\to 0$ μ—a.e., the above inequality implies the strong measurability of the function $R_{y(t)}h(t)$. Since $\|R_{y(t)}h(t)\|\leq\|y\|\ \|h(t)\|$, it belongs to $L^1(\Omega,\mu,E)$, so that $\langle x(t), R_{y(t)}h(t)\rangle=\langle x(t)y(t), h(t)\rangle$ is measurable. Therefore $x(t)y(t)$ is weakly *-measurable and $\|x(t)y(t)\|\leq\|x(t)\|\ \|y(t)\|$. q.e.d.

3.2.2. Corollary. *If $\mathcal{M}$ is a W^*-algebra with a separable predual $\mathcal{M}_*$, then $L^\infty(\Omega,\mu,\mathcal{M})$ is a W^*-algebra by the pointwise multiplication and its predual is $L^1(\Omega,\mu,\mathcal{M}_*)$.*

3.2.3. Proposition. *If $\mathcal{M}$ is a type I_n-factor $(n\leq\aleph_0)$, then $L^\infty(\Omega,\mu,\mathcal{M})$ is a type I_n W^*-algebra with the center $L^\infty(\Omega,\mu)\cdot 1$.*

Proof. We shall prove the case where $n=\aleph_0$, since other cases are similarly proved. For $a\in\mathcal{M}$, set $\tilde{a}(t)=a$ $(t\in\Omega)$. Then $\tilde{a}\in L^\infty(\Omega,\mu,\mathcal{M})$. If $(e_i|i=1,2,\ldots)$ is a maximal family of mutually orthogonal minimal projections in $\mathcal{M}$, then $\tilde{e}_i(t)x(t)\tilde{e}_i(t)=\lambda_i(t)e_i$ $(x\in L^\infty(\Omega,\mu,\mathcal{M}),\ \lambda_i\in L^\infty(\Omega,\mu))$. Hence $\tilde{e}_i$ is an abelian projection and clearly $\tilde{e}_i\tilde{e}_j=0$ $(i\neq j)$.

For $\eta\in L^1(\Omega,\mu)$, $f\in\mathcal{M}_*$ with $f\geq 0$, put $\varphi(t)=\eta(t)f$ $(t\in\Omega)$; then $\varphi\in L^1(\Omega,\mu,\mathcal{M}_*)$. Moreover

$$\left\langle\sum_{i=1}^{\infty}\tilde{e}_i,\varphi\right\rangle=\sum_{i=1}^{\infty}\langle\tilde{e}_i,\varphi\rangle=\sum_{i=1}^{\infty}\int_\Omega\langle e_i,\varphi(t)\rangle\,d\mu(t)=\sum_{i=1}^{\infty}\int_\Omega\langle e_i,f\rangle\eta(t)\,d\mu(t).$$

Since $f(e_i)\geq 0$ and

$$\sum_{i=1}^{\infty}f(e_i)=f\left(\sum_{i=1}^{\infty}e_i\right)=f(1),\qquad\left\langle\sum_{i=1}^{\infty}\tilde{e}_i,\varphi\right\rangle=\int_\Omega\langle 1,f\rangle\eta(t)\,d\mu(t)=\langle\tilde{1},\varphi\rangle$$

by the dominated convergence theorem.

Now one can easily conclude that $\sum_{i=1}^{\infty}\tilde{e}_i=\tilde{1}$. Let v_i be a partial isometry in $\mathcal{M}$ such that $v_i^*v_i=e_1$ and $v_iv_i^*=e_i$ $(i=2,3,\ldots)$. Then $\tilde{v}_i^*\tilde{v}_i=\tilde{e}_1$ and $\tilde{v}_i\tilde{v}_i^*=\tilde{e}_i$ $(i=2,3,\ldots)$. Hence $(\tilde{e}_i\,|\,i=1,2,\ldots)$ is a maximal family of mutually orthogonal equivalent maximal abelian projections in $L^\infty(\Omega,\mu,\mathcal{M})$.

Let Z be the center of $L^\infty(\Omega,\mu,\mathcal{M})$ and $z\in Z$. Then

$$\tilde{a}(t)z(t)=a\,z(t)=z(t)\tilde{a}(t)=z(t)a,$$

locally μ-almost everywhere. Let (a_n) be a family of elements in $\mathcal{M}$ which is $\sigma(\mathcal{M},\mathcal{M}_*)$-dense in $\mathcal{M}$. Then $a_nz(t)=z(t)a_n$ $(t\in\Omega-\Delta,\ n=1,2,\ldots)$, where Δ is a locally μ-null set, so that $Z\subset L^\infty(\Omega,\mu)\cdot 1$. q.e.d.

Now let C be a commutative W^*-algebra, with $1_{\mathcal{H}}$, on a Hilbert space $\mathcal{H}$. Then the commutant C' is a type I W^*-algebra. Hence there exists a family of mutually orthogonal central projections $(z_\alpha\,|\,\alpha\in\mathbb{I})$ such that $\sum_{\alpha\in\mathbb{I}}z_\alpha=1_{\mathcal{H}}$, $C'z_\alpha$ is a type I_{n_α} W^*-algebra, and $n_\alpha\neq n_\beta$ if $\alpha\neq\beta$.

Since $C''=C$, C is the center of C'. Now suppose that $z_\alpha=1_{\mathcal{H}}$ and $C=L^\infty(\Omega_\alpha,\mu_\alpha)$. Then C' is $*$-isomorphic to $L^\infty(\Omega_\alpha,\mu_\alpha)\mathbin{\overline{\otimes}}B(\mathcal{H}_\alpha)$ with $\dim(\mathcal{H}_\alpha)=n_\alpha$. $L^\infty(\Omega_\alpha,\mu_\alpha)\mathbin{\overline{\otimes}}B(\mathcal{H}_\alpha)$ can be considered as a W^*-algebra on a Hilbert space $L^2(\Omega_\alpha,\mu_\alpha)\otimes\mathcal{H}_\alpha$. Then

$$(L^\infty(\Omega_\alpha,\mu_\alpha)\mathbin{\overline{\otimes}}1_{\mathcal{H}_\alpha})'=L^\infty(\Omega_\alpha,\mu_\alpha)\mathbin{\overline{\otimes}}B(\mathcal{H}_\alpha).$$

Therefore C on $\mathcal{H}$ and $L^\infty(\Omega_\alpha,\mu_\alpha)\otimes 1_{\mathcal{H}_\alpha}$ on $L^2(\Omega_\alpha,\mu_\alpha)\otimes\mathcal{H}_\alpha$ have the same coupling function. Hence they are equivalent (cf. 2.9.31).

If $n_\alpha\leq\aleph_0$, $L^\infty(\Omega_\alpha,\mu_\alpha)\mathbin{\overline{\otimes}}B(\mathcal{H}_\alpha)=L^\infty(\Omega_\alpha,\mu_\alpha,B(\mathcal{H}_\alpha))$. Since

$$L^2(\Omega_\alpha,\mu_\alpha)\otimes\mathcal{H}_\alpha=L^2(\Omega_\alpha,\mu_\alpha,\mathcal{H}_\alpha)$$

$(L^2(\Omega_\alpha,\mu_\alpha,\mathcal{H}_\alpha)$ is the Hilbert space of all $\mathcal{H}_\alpha$-valued strongly μ-square integrable functions), a $(a\in L^\infty(Q_\alpha,\mu_\alpha,B(\mathscr{H}_\alpha))$ operates on $L^2(\Omega_\alpha,\mu_\alpha,\mathscr{H}_\alpha)$ as follows: $(a\xi)(t)=a(t)\xi(t)$ $(\xi\in L^2(\Omega_\alpha,\mu_\alpha,\mathscr{H}_\alpha))$. Hence we have

3.2.4. Theorem. *Let C be a commutative W^*-algebra, with $1_{\mathscr{H}}$, on a Hilbert space $\mathscr{H}$. If its commutant C' is a direct sum of type I_{n_α} $(n_\alpha\leq\aleph_0)$ W^*-algebras $(\alpha\in\mathbb{I})$, then C' on $\mathscr{H}$ is equivalent to the W^*-algebra $\sum_{\alpha\in\mathbb{I}}\oplus L^\infty(\Omega_\alpha,\mu_\alpha,B_\alpha(\mathscr{H}_\alpha))$ on the Hilbert space $\sum_{\alpha\in\mathbb{I}}\oplus L^2(\Omega_\alpha,\mu_\alpha,\mathscr{H}_\alpha)$. Furthermore, C on $\mathscr{H}$ is equivalent to $\sum_{\alpha\in\mathbb{I}}\oplus L^\infty(\Omega_\alpha,\mu_\alpha)\cdot 1_{\mathscr{H}_\alpha}$ on $\sum_{\alpha\in\mathbb{I}}\oplus L^2(\Omega_\alpha,\mu_\alpha,\mathscr{H}_\alpha)$.*

3.2.5. Definition. *The W^*-algebra $\sum_{\alpha\in\mathbb{I}}\oplus L^\infty(\Omega_\alpha,\mu_\alpha)\cdot 1_{\mathscr{H}_\alpha}$ is called the algebra of all diagonalizable operators.*

3.2.6. Corollary. *Let $\mathscr{N}$ be a W^*-algebra, with $1_{\mathscr{H}}$, on a separable Hilbert space $\mathscr{H}$, and let $R(\mathscr{N},\mathscr{N}')$ be the W^*-algebra generated by $\mathscr{N}$ and its commutant $\mathscr{N}'$. Then $R(\mathscr{N},\mathscr{N}')$ on $\mathscr{H}$ is equivalent to a W^*-algebra of the form $\sum_{\alpha\in\mathbb{I}}\oplus L^\infty(\Omega_\alpha,\mu_\alpha,B(\mathscr{H}_\alpha))$ on a Hilbert space $\sum_{\alpha\in\mathbb{I}}\oplus L^2(\Omega_\alpha,\mu_\alpha,\mathscr{H}_\alpha)$.*

Now let $\mathscr{M}$ be a type I_n $(n\leq\aleph_0)$ W^*-algebra on a separable Hilbert space $\mathscr{H}$ such that $\mathscr{M}'$ is commutative. Then under the identification, $\mathscr{M}=Z\,\overline{\otimes}\,B(\mathscr{H}_0)=L^\infty(\Omega,\mu,B(\mathscr{H}_0))$ and $\mathscr{H}=L^2(\Omega,\mu,\mathscr{H}_0)$ (Z is the center of $\mathscr{M}$ and $Z=L^\infty(\Omega,\mu)$).

Since $Z\otimes 1_{\mathscr{H}_0}$ has a separable predual, it suffices to assume that $\mu(\Omega)=1$. If $a\in Z\,\overline{\otimes}\,B(\mathscr{H}_0)$, then a is considered a $B(\mathscr{H}_0)$-valued essentially bounded weakly * μ-measurable function on Ω. We express such a situation by $a=\int_\Omega a(t)\,d\mu(t)$ (or $\int a(t)\,d\mu(t)$ or $\int a(t)$). Then $\|a\|=\operatorname{ess\,sup}_{t\in\Omega}\|a(t)\|$; for $a_1,a_2\in Z\,\overline{\otimes}\,B(\mathscr{H}_0)$, $a_1+a_2=\int(a_1(t)+a_2(t))\,d\mu(t)$, $a_1a_2=\int a_1(t)a_2(t)\,d\mu(t)$, $\lambda a=\int\lambda a(t)\,d\mu(t)$ (λ a complex number) and $a^*=\int a(t)^*d\mu(t)$.

3.2.7. Proposition. *Let $a_i=\int a_i(t)$ $(i=1,2,\ldots)$ and $a=\int a(t)$.*

I. *If (a_i) is s-convergent to a, there exists a subsequence (a_{i_j}) such that $(a_{i_j}(t))$ is s-convergent to $a(t)$ for almost all $t\in\Omega$.*

II. *If $(a_i(t))$ is s-convergent to $a(t)$ for almost all $t\in\Omega$ and if $\sup_i\|a_i\|<+\infty$, then (a_i) is s-convergent to a.*

Proof. I. Since (a_i) is s-convergent to a, $\sup_i\|a_i\|\leq k$ with a fixed positive number k. Hence $\|a_i(t)\|\leq k$ a.e. $(i=1,2,\ldots)$. Let $(\xi_n\,|\,n=1,2,\ldots)$ be a dense subset in the positive portion of $B(\mathscr{H}_0)_*$ and set $f_n=1\otimes\xi_n$ (1 is the constant one function on Ω). Then

$$\begin{aligned}&\langle(a_i-a)^*(a_i-a),f_n\rangle\\&=\int\langle\{a_i(t)-a(t)\}^*\{a_i(t)-a(t)\},\xi_n\rangle\,d\mu(t)\to 0\qquad(i\to\infty)\end{aligned}$$

for $n=1,2,\ldots$; hence there exists a subsequence (a_{i_j}) of (a_i) with $\sum_{j=1}^{\infty}\langle (a_{i_j}-a)^*(a_{i_j}-a), f_n\rangle < +\infty$. Therefore there is a null set Ω_n in Ω such that $\sum_{j=1}^{\infty}\langle\{a_{i_j}(t)-a(t)\}^*\{a_{i_j}(t)-a(t)\}, \xi_n\rangle < +\infty$ $(t\notin\Omega_n)$, and so $\lim_j\langle\{a_{i_j}(t)-a(t)\}^*\{a_{i_j}(t)-a(t)\}, \xi_n\rangle = 0$ $(t\notin\Omega_n)$.

By applying a diagonal process, we may assume that such a subsequence (i_j) does not depend on n. Then

$$\lim_j\langle\{a_{i_j}(t)-a(t)\}^*\{a_{i_j}(t)-a(t)\}, \xi_n\rangle = 0$$

$\left(t\notin\bigcup_{n=1}^{\infty}\Omega_n; n=1,2,\ldots\right)$. Since $\|a_{i_j}(t)\|\leq k$ a.e., one can conclude that $a_{i_j}(t)\to a(t)$ in the s-topology (a.e.).

II. Conversely, suppose that $a_i(t)\to a(t)$ in the s-topology (a.e.) and $\sup_i\|a_i\| < +\infty$. Then $\|a_i(t)\|\leq k$ a.e., so that $\|a_i(t)\|\leq k$ a.e.. For each n, $\langle\{a_i(t)-a(t)\}^*\{a_i(t)-a(t)\}, \xi_n\rangle\to 0$ a.e. $(i\to\infty)$ and

$$\langle\{a_i(t)-a(t)\}^*\{a_i(t)-a(t)\}, \xi_n\rangle \leq 4k^2\|f_n\|.$$

Hence for each n and $g\in L^1(\Omega,\mu)$,

$$\begin{aligned}&\lim_{i\to\infty}\langle (a_i-a)^*(a_i-a), g\otimes\xi_n\rangle\\ &=\lim_{i\to\infty}\int_\Omega\langle\{a_i(t)-a(t)\}^*\{a_i(t)-a(t)\}, \xi_n\rangle g(t)\,d\mu(t)=0.\end{aligned}$$

Since $\sup_i\|a_i\| < +\infty$, this implies that $a_i\to a$ in the s-topology $(i\to\infty)$.

q.e.d.

3.2.8. Proposition. *Let $(a_\beta)_{\beta\in\mathbb{I}}$ be a family of elements of $L^\infty(\Omega,\mu,B(\mathscr{H}_0))$ containing the identity, $\mathscr{A}$ the W^*-subalgebra of $B(\mathscr{H}_0)$ generated by $\{a_\beta\}_{\beta\in\mathbb{I}}$ and $Z\otimes 1_{\mathscr{H}_0}$, $\mathscr{A}(t)$ the W^*-subalgebra of $B(\mathscr{H}_0)$ generated by $\{a_\beta(t)\}_{\beta\in\mathbb{I}}$, and let $b\in L^\infty(\Omega,\mu,B(\mathscr{H}_0))$. Then*

I *if* $b\in\mathscr{A}$, $b(t)\in\mathscr{A}(t)$ a.e.;

II *if* $b(t)\in\mathscr{A}(t)$ a.e. *and if* $\mathbb{I}$ *is countable, then* $b\in\mathscr{A}$.

Proof. I. Let A be the *-algebra generated algebraically by $(a_\beta)_{\beta\in\mathbb{I}}$ and $Z\otimes 1_{\mathscr{H}_0}$. Then $b\in A$ implies $b(t)\in\mathscr{A}(t)$ a.e.. If $b\in\mathscr{A}$, there exists a sequence (b_n) in A with $b_n\to b$ in the s-topology. Hence $b(t)\in\mathscr{A}(t)$ a.e. by 3.2.7.

II. Conversely, suppose that $b(t)\in\mathscr{A}(t)$ a.e. and $\mathbb{I}$ is countable. Since $\mathscr{A}'\subset(Z\otimes 1_{\mathscr{H}_0})'=Z\,\bar{\otimes}\,B(\mathscr{H}_0)$, for any $a'\in\mathscr{A}'$, $a'=\int a'(t)\,d\mu(t)$. Since $a'a_\beta=a_\beta a'$ and $a'a_\beta^*=a_\beta^*a'$, $a'(t)a_\beta(t)=a_\beta(t)a'(t)$ and

$$a'(t)a_\beta^*(t)=a_\beta^*(t)a'(t) \qquad (t\notin\Omega_\beta \text{ with } \mu(\Omega_\beta)=0).$$

Hence $a'(t) \in \mathscr{A}(t)'$ a.e., so that $a'(t)b(t) = b(t)a'(t)$ and

$$a'(t)^* b(t) = b(t) a'(t)^* \qquad \text{a.e..}$$

Hence $b \in \mathscr{A}'' = \mathscr{A}$. q.e.d.

3.2.9. Definition. *A* family $\{\mathscr{A}(t) \mid t \in \Omega\}$ *of* W^**-subalgebras of* $B(\mathscr{H}_0)$ *is said to be measurable if there exists a sequence* (a_n) *of elements, containing the identity, in* $L^\infty(\Omega, \mu, B(\mathscr{H}_0))$ *such that* $\mathscr{A}(t)$ *is generated by* $\{a_n(t) \mid n = 1, 2, \dots\}$ *for almost all* t *in* Ω.

3.2.10. Proposition. *Let* $\{\mathscr{A}(t) \mid t \in \Omega\}$ *be a measurable family of* W^**-subalgebras of* $B(\mathscr{H}_0)$.

I. *the set* $\mathscr{A}$ *of all elements* a *in* $L^\infty(\Omega, \mu, B(\mathscr{H}_0))$ *with* $a(t) \in \mathscr{A}(t)$ *a.e. is a* W^**-subalgebra (called the* W^**-subalgebra defined by the measurable family* $\{\mathscr{A}(t) \mid t \in \Omega\}$*), containing* $Z \otimes 1_{\mathscr{H}_0}$, *of* $L^\infty(\Omega, \mu, B(\mathscr{H}_0))$.

II. *if a family* $\{\mathscr{B}(t) \mid t \in \Omega\}$ *is measurable and if it defines the same* W^**-subalgebra* $\mathscr{A}$, *then* $\mathscr{B}(t) = \mathscr{A}(t)$ *a.e.*

Proof. I. is clear by 3.2.8.

II. Let (a_n) be a sequence of elements in $L^\infty(\Omega, \mu, B(\mathscr{H}_0))$ which generates $\mathscr{A}(t)$ a.e.. Since $a_n \in \mathscr{A}$, $a_n(t) \in \mathscr{B}(t)$ a.e.. Hence $\mathscr{A}(t) \subset \mathscr{B}(t)$ a.e.. Similarly $\mathscr{A}(t) \supset \mathscr{B}(t)$ a.e.. q.e.d.

3.2.11. Definition. *By* 3.2.8 *and* 3.2.10, *there exists a one-to-one correspondence between a* W^**-subalgebra* $\mathscr{A}$, *containing* $Z \otimes 1_{\mathscr{H}_0}$, *of* $L^\infty(\Omega, \mu, B(\mathscr{H}_0))$ *and a measurable family* $\{\mathscr{A}(t) \mid t \in \Omega\}$ *of* W^**-subalgebras of* $B(\mathscr{H}_0)$. *We express such a situation by* $\mathscr{A} = \int \mathscr{A}(t) \, d\mu(t)$.

3.2.12. Theorem. *Suppose that* Ω *is a locally compact space satisfying the second countability axiom and* μ *is a positive Radon measure on* Ω. *If* $t \to \mathscr{A}(t)$ $(t \in \Omega)$ *is a measurable family, then* $t \to \mathscr{A}(t)'$ *is again measurable and* $\mathscr{A}' = \int \mathscr{A}(t)' \, d\mu(t)$ $(\mathscr{A} = \int \mathscr{A}(t) \, d\mu(t))$.

Proof. Let $(f_n \mid n = 1, 2, \dots)$ be a dense subset in the predual $B(\mathscr{H}_0)_*$. Set $d(x, y) = \sum_{n=1}^{\infty} \frac{1}{2^n \|f_n\|} |f_n(x - y)|$ $(x, y \in B(\mathscr{H}_0))$. Then the σ-topology on the bounded spheres of $B(\mathscr{H}_0)$ is equivalent to the metric topology $d(x, y)$. It suffices to assume that Ω is a compact space. Let $\{a_i(t) \mid i = 1, 2, \dots\}$ be a countable family of elements in $L^\infty(\Omega, \mu, B(\mathscr{H}_0))$ with

$$\{a_i(t) \mid i = 1, 2, \dots\}'' = \mathscr{A}(t) \qquad \text{a.e.}$$

Since $t \to f_n(a_i(t))$ is bounded measurable, there exists a compact set C_m in Ω such that $\mu(\Omega - C_m) < 1/m$ and $f_n(a_i(t))$ is continuous on C_m $(i, n = 1, 2, \dots)$ by Lusin's theorem. Hence $t \to a_i(t)$ is σ-continuous on

C_m $(i=1,2,\ldots)$. Since $\mu\left(\Omega - \bigcup_{m=1}^{\infty} C_m\right)=0$, it suffices to assume that $C_m=\Omega$—i.e. $t\to a_i(t)$ is σ-continuous on Ω $(i=1,2,\ldots)$.

Consider the product space, $\Omega\times S$ (S the unit sphere of $B(\mathscr{H}_0)$) and let $F=\{(t,a')\,|\,a_i(t)a'=a'a_i(t),\ a_i(t)^*a'=a'a_i(t)^*,\ a'\in S\,(i=1,2,\ldots)\}$. Then F is closed. Let $\{F_j|j=1,2,\ldots\}$ be a basis of closed neighborhoods of F, and let Ω_j be the projection of F_j into Ω. Then F_j is an analytic set. Hence there is a μ-measurable mapping $t\to a_j'(t)$ of Ω_j into S with $(t,a_j'(t))\in F_j$ (cf. [14]).

Put $a_j'(t)=0$ $(t\in\Omega-\Omega_j)$. Then $\int a_j'(t)\,d\mu(t)\in L^\infty(\Omega,\mu,B(\mathscr{H}_0))$. Now we shall show that $\{a_j'(t)|j=1,2,\ldots\}''=\mathscr{A}(t)'(t\in\Omega)$. If $a'\in\mathscr{A}(t)'$ with $\|a'\|\leq 1$, then $(t,a')\in F$. Let $(F_{n_1},F_{n_2},\ldots)$ be a fundamental system of neighborhoods of (t,a') in F. Then (t,a') is a limit of a sequence $\{(t,a_{n_p}(t))|p=1,2,\ldots\}$. Hence $a'\in\{a_j'(t)|j=1,2,\ldots\}''$, and so $\{\mathscr{A}(t)'|t\in\Omega\}$ is a measurable family.

Put $\mathscr{B}=\int\mathscr{A}(t)'\,d\mu(t)$. Then $\mathscr{A}'\supset\mathscr{B}$. On the other hand,

$\mathscr{A}'=\int\mathscr{A}'(t)\,d\mu(t)\subset\int\mathscr{A}(t)'\,d\mu(t)=\mathscr{B}$. Hence $\mathscr{A}'=\mathscr{B}$. q.e.d.

3.2.13. Corollary. *Suppose that the measure space* (Ω,μ) *satisfies the conditions in* 3.2.12. *If* $\mathscr{A}_i=\int\mathscr{A}_i(t)\,d\mu(t)$ $(i=1,2,\ldots)$ *and* $\mathscr{A}=\bigcap_{i=1}^{\infty}\mathscr{A}_i$, *then* $\mathscr{A}=\int\bigcap_{i=1}^{\infty}\mathscr{A}_i(t)\,d\mu(t)$.

Proof. $\mathscr{A}'=\left(\bigcap_{i=1}^{\infty}\mathscr{A}_i\right)'=R(\mathscr{A}_i'\,|\,i=1,2,\ldots)$. Hence

$$\mathscr{A}'=\int R(\mathscr{A}_i'(t)\,|\,i=1,2,\ldots)\,d\mu(t),$$

and so

$$\mathscr{A}=\int R(\mathscr{A}_i'(t)\,|\,i=1,2,\ldots)'\,d\mu(t)$$

$$=\int\bigcap_{i=1}^{\infty}\mathscr{A}_i(t)''\,d\mu(t)=\int\bigcap_{i=1}^{\infty}\mathscr{A}_i(t)\,d\mu(t). \qquad \text{q.e.d.}$$

References. [37], [47], [132], [154], [156], [179], [205].

3.3. Direct Integral of Hilbert Spaces

Let (Ω,μ) be a localizable measure space. For $t\in\Omega$, there corresponds a Hilbert space $\mathscr{H}(t)$. Let $\mathfrak{F}$ be a linear space of all vector valued functions $\xi(t)$ on Ω with $\xi(t)\in\mathscr{H}(t)$ $(t\in\Omega)$.

3.3.1. Definition. *$\{\mathscr{H}(t)|t\in\Omega\}$ is called a measurable family if there exists a countable family $\{\xi_n|n=1,2,\ldots\}$ in $\mathfrak{F}$ satisfying the following conditions:*

1. *$t\to(\xi_n(t),\xi_m(t))$ is locally μ-measurable $(m,n=1,2,\ldots)$;*
2. *$\{\xi_n(t)|n=1,2,\ldots\}$ is dense in $\mathscr{H}(t)$ for locally μ-almost all t in Ω.*

Now let $\{\mathscr{H}(t)|t\in\Omega\}$ be a measurable family, and let V be a linear subspace of $\mathfrak{F}$. V is said to be measurable if it satisfies

1. for $\xi,\eta\in V$, $t\to(\xi(t),\eta(t))$ is locally μ-measurable;
2. there exists a countable family $\{\eta_i|i=1,2,\ldots\}$ in V such that $\{\eta_i(t)|i=1,2,\ldots\}$ is dense in $\mathscr{H}(t)$ for locally μ-almost all t in Ω.

Such a family $\{\eta_i|i=1,2,\ldots\}$ is called fundamental in V. Let $\{V_\alpha\}_{\alpha\in\mathbb{I}}$ be the set of all measurable linear subspaces in $\mathfrak{F}$. We define an order in $\{V_\alpha\}_{\alpha\in\mathbb{I}}$ by set-inclusion. $\{V_\alpha\}_{\alpha\in\mathbb{I}}$ is not empty, because a linear subspace generated by $\{\xi_i|i=1,2,\ldots\}$ is measurable. Therefore one can easily conclude the existence of a maximal linear subspace V_0 by Zorn's lemma.

Let $M(\Omega,\mu)$ be the linear space of all locally μ-measurable complex valued functions on Ω. If V is measurable, the linear space generated by $\{f\cdot\xi|f\in M(\Omega,\mu),\xi\in V\}$ is again measurable. Hence V_0 is invariant under the multiplication of $M(\Omega,\mu)$.

Let $(\eta_i|i=1,2,\ldots)$ be a fundamental family in V_0. Define

$$\begin{aligned} e_1(t) &= \frac{\eta_1(t)}{\|\eta_1(t)\|} \quad \text{if } \eta_1(t)\neq 0,\\ &= \frac{\eta_2(t)}{\|\eta_2(t)\|} \quad \text{if } \eta_1(t)=0, \quad \eta_2\neq 0,\\ &\ \ \vdots\\ &= \frac{\eta_i(t)}{\|\eta_i(t)\|} \quad \text{if } \eta_i(t),\ldots,\eta_{i-1}(t)=0, \quad \eta_i(t)\neq 0 \end{aligned}$$

$(i=1,2,\ldots)$. Then $\|e_1(t)\|=1$ l.a.e. (locally almost everywhere). Define

$$e_2(t)=\frac{\eta_2(t)-(\eta_2(t),e_1(t))e_1(t)}{\|\eta_2(t)-(\eta_2(t),e_1(t))e_1(t)\|}$$

if

$$\|\eta_2(t)-(\eta_2(t),e_1(t))e_1(t)\|\neq 0$$

$$e_2(t)=\frac{\eta_3(t)-(\eta_3(t),e_1(t))e_1(t)}{\|\eta_3(t)-(\eta_3(t),e_1(t))e_1(t)\|}$$

if

$$\|\eta_2(t)-(\eta_2(t),e_1(t))e_1(t)\|=0$$

and

$$\|\eta_3(t)-(\eta_3(t),e_1(t))e_1(t)\|\neq 0,$$

and so on. Then one can define a countable family $(e_n|n=1,2,\ldots)$ in V_0 such that $(e_n(t)|n=1,2,\ldots)$ is a complete orthonormal system of $\mathscr{H}(t)$ l.a.e., and if $\dim(\mathscr{H}(t))=d(t)<+\infty$, then $e_n(t)=0$ $(n>d(t))$. Hence $\Omega_n=\{t|\dim(\mathscr{H}(t))=n\}$ $(n=1,2,\ldots)$ is locally μ-measurable and $\Omega_\infty=\{t|\dim(\mathscr{H}(t))=\aleph_0\}$ is also locally measurable.

Now let $\mathscr{K}_0$ be the set of all elements ξ in V_0 with $\int \|\xi(t)\|^2 d\mu(t)<+\infty$. If $\xi_1\,\xi_2\in\mathscr{K}_0$, then

$$(\int \|(\xi_1+\xi_2)(t)\|^2 d\mu(t))^{\frac{1}{2}} \leq (\int \|\xi_1(t)\|^2 d\mu(t))^{\frac{1}{2}}+(\int \|\xi_2(t)\|^2 d\mu(t))^{\frac{1}{2}}<+\infty.$$

Therefore $\mathscr{K}_0$ is a linear space. For $\xi,\eta\in\mathscr{K}_0$, set $(\xi,\eta)=\int(\xi(t),\eta(t))d\mu(t)$. Then one can easily see that $\mathscr{K}_0$ is a Hilbert space. $\mathscr{K}_0$ is called the associated Hilbert space of V_0. Clearly $\{L^2(\Omega,\mu)\cdot e_i|i=1,2,\ldots\}\subset\mathscr{K}_0$.

3.3.2. Proposition. *The linear subspace generated by*

$$\{L^2(\Omega,\mu)\cdot e_i\,|\,i=1,2,\ldots\}$$

is dense in $\mathscr{K}_0$.

Proof. Let $\mathscr{K}$ be the closed linear subspace of $\mathscr{K}_0$ generated by $\{L^2(\Omega,\mu)\cdot e_i|i=1,2,\ldots\}$. For $\xi\in\mathscr{K}_0$ with $(\xi,\mathscr{K})=0$,

$$\int f(t)\,(e_i(t),\,\xi(t))\,d\mu(t)=0 \qquad (f\in L^2(\Omega,\mu);\, i=1,2,\ldots).$$

Since $\int\|\xi(t)\|^2 d\mu(t)<+\infty$, there is a sequence $\{F_n\}$ of measurable subsets in Ω such that $F_n\subset F_{n+1}$, $\mu(F_n)<+\infty$ and $\xi(t)=0$ for all $t\in\Omega-\sum_{n=1}^{\infty}F_n$. Therefore the above equalities imply $\xi=0$ in $\mathscr{K}_0$. q.e.d.

Let $\mathscr{H}_n$ be an n-dimensional Hilbert space and let $(e_1^n,e_2^n,\ldots,e_n^n)$ be a complete orthonormal system of $\mathscr{H}_n$. Let $\mathscr{H}_\infty$ be an $\aleph_0$-dimensional Hilbert space and let $(e_1^\infty,e_2^\infty,\ldots)$ be a complete orthonormal system of $\mathscr{H}_\infty$. Define an isometry $U(t)$ of $\mathscr{H}(t)$ $(t\in\Omega_n)$ onto $\mathscr{H}_n$ by $U(t)e_i(t)=e_i^n$ $(i=1,2,\ldots,n)$, and an isometry $U(t)$ of $\mathscr{H}(t)$ $(t\in\Omega_\infty)$ onto $\mathscr{H}_\infty$ by $U(t)e_i(t)=e_i^\infty$ $(i=1,2,\ldots)$. Then for $f\in L^2(\Omega,\mu)$, $U(t)f(t)e_i(t)=f(t)e_i^n$, $(i=1,2,\ldots,n$ if $n<\infty$; $i=1,2,\ldots$ if $n=\infty)$. Hence the family $\{U(t)|t\in\Omega\}$ of isometries defines an isometry of $\mathscr{K}_0$ onto

$$\sum_{n=1}^{\infty}\oplus(L^2(\Omega_n,\mu)\otimes\mathscr{H}_n)\oplus(L^2(\Omega_\infty,\mu)\otimes\mathscr{H}_\infty).$$

Hence we have

3.3.3. Theorem. *Let* $\{\mathscr{H}(t)|t\in\Omega\}$ *be a measurable family of Hilbert spaces, and let* V_0, V_1 *be two maximal measurable subspaces of* $\mathfrak{F}$ *and let* $\mathscr{K}_0$, $\mathscr{K}_1$ *be the associated Hilbert spaces of* V_0, V_1 *respectively. Then there exists a unitary operator* $V(t)$ *on* $\mathscr{H}(t)$ $(t\in\Omega)$ *as follows: for* $\xi\in\mathscr{K}_0$, *set* $\eta(t)=V(t)\xi(t)$, *then* $\eta\in\mathscr{K}_1$ *and the mapping* $\xi\to\eta$ *gives a unitary* *mapping of $\mathcal{K}_0$ onto $\mathcal{K}_1$. Therefore we can obtain an essentially unique Hilbert space $\mathcal{K}_0$ from a given measurable family $\{\mathcal{H}(t)|t\in\Omega\}$ of Hilbert spaces.*

This Hilbert space $\mathcal{K}_0$ is called the direct integral of $\{\mathcal{H}(t)|t\in\Omega\}$ and is denoted by $\int \mathcal{H}(t)d\mu(t)$. Since $\int \mathcal{H}(t)d\mu(t)$ is isometrically isomorphic to $L^2(\Omega_\infty,\mu_\infty)\otimes\mathcal{H}_\infty\oplus\sum_{n=1}^{\infty}L^2(\Omega_n,\mu_n)\otimes\mathcal{H}_n$, the reduction theory $L^\infty(\Omega_\infty,\mu_\infty)\,\overline{\otimes}\,B(\mathcal{H}_\infty)\oplus\sum_{n=1}^{\infty}L^\infty(\Omega_n,\mu_n)\otimes B(\mathcal{H}_n)$ can be carried on $\int\mathcal{H}(t)d\mu(t)$ under this isomorphism, where μ_∞ (resp. μ_n) is the measure obtained by restricting μ to Ω_∞ (resp. Ω_n).

References. [37], [132], [179].

3.4. Decomposition of States (Separable Cases)

Let $\mathcal{A}$ be a C^*-algebra with identity. Throughout this section, we shall assume that $\mathcal{A}$ is uniformly separable. Let $\mathcal{S}$ be the state space of $\mathcal{A}$; then $\mathcal{S}$ is $\sigma(\mathcal{A}^*,\mathcal{A})$-compact. Let $\{a_n|n=1,2,\ldots\}$ be a sequence of non-zero elements in $\mathcal{A}$ which is uniformly dense in the self-adjoint portion $\mathcal{A}^s$ of $\mathcal{A}$.

For $\varphi,\psi\in\mathcal{S}$, set $d(\varphi,\psi)=\sum_{n=1}^{\infty}\frac{|(\varphi-\psi)(a_n)|}{2^n\|a_n\|}$. Then d is a metric on $\mathcal{S}$ which is equivalent to the topology $\sigma(\mathcal{S},\mathcal{A})$. Hence $\mathcal{S}$ is considered as a compact metric space, and so it satisfies the second countability axiom, since a compact metric space is separable. Throughout this section we shall deal with the topology $\sigma(\mathcal{S},\mathcal{A})$ on $\mathcal{S}$.

3.4.1. (General lemma of Choquet). *Let E be a locally convex space, C a metrizable compact convex subset of E and let ∂C be the set of all extreme points of C. Then ∂C is a G_δ-set.*

Proof. For $(\varphi,\psi)\in C\times C$, set $\rho((\varphi,\psi))=\frac{(\varphi+\psi)}{2}$. Then ρ is a continuous mapping of $C\times C$ onto C. Let $\Delta=\{(\varphi,\varphi)|\varphi\in C\}$. Then Δ is closed and so $C\times C-\Delta$ is open. Furthermore $\rho(C\times C-\Delta)=C-\partial C$. Let d be the metric on C and $U_n(\varphi)=\{\psi|d(\varphi,\psi)<1/n\}$. Then $\bigcup_{\varphi\in C}(U_n(\varphi),U_n(\varphi))$ is open, and so $C\times C-\bigcup_{\varphi\in C}(U_n(\varphi),U_n(\varphi))$ is closed. Clearly $\bigcup_{n=1}^{\infty}\Big(C\times C-\bigcup_{\varphi\in C}(U_n(\varphi),U_n(\varphi))\Big)\subset C\times C-\Delta$. If $(\varphi_0,\psi_0)\in C\times C-\Delta$ with $(\varphi_0,\psi_0)\notin\bigcup_{n=1}^{\infty}\Big(C\times C-\bigcup_{\varphi\in C}(U_n(\varphi),U_n(\varphi))\Big)$, then

$$(\varphi_0,\psi_0)\in\bigcap_{n=1}^{\infty}\bigcup_{\varphi\in C}(U_n(\varphi),U_n(\varphi)).$$

Since $\varphi_0 \neq \psi_0$, there exists an n_0 with $\psi_0 \notin U_{n_0}(\varphi_0)$. Hence $(\varphi_0, \psi_0) \notin (U_{n_0}(\varphi_0), U_{n_0}(\varphi_0))$. On the other hand, $(\varphi_0, \psi_0) \in (U_{2^{n_0}}(\varphi), U_{2^{n_0}}(\varphi))$ (some $\varphi \in C$). Hence $d(\varphi_0, \psi_0) \leq d(\varphi_0, \varphi) + d(\varphi, \psi_0) \leq 1/2^{n_0} + 1/2^{n_0} \leq 1/n_0$, a contradiction. Hence $\bigcup_{n=1}^{\infty} \left(C \times C - \bigcup_{\varphi \in C} (U_n(\varphi), U_n(\varphi))\right) = C \times C - \Delta$, and so $\rho(C \times C - \Delta) = \bigcup_{n=1}^{\infty} \rho\left(C \times C - \bigcup_{\varphi \in C} (U_n(\varphi), U_n(\varphi))\right)$. Therefore $\rho(C \times C - \Delta)$ is F_σ and so ∂C is G_δ. q.e.d.

3.4.2. Corollary. *Let $\mathscr{E}$ be the set of all pure states on $\mathscr{A}$. Then $\mathscr{E}$ is a G_δ-set.*

Now let $\mathscr{H}_\infty$ be an $\aleph_0$-dimensional Hilbert space and let $\mathbb{R}$ be the set of all nowhere trivial *-representations $\{\pi, \mathscr{H}_\infty\}$ of $\mathscr{A}$ on $\mathscr{H}_\infty$. We do not identify equivalent *-representations. Give $\mathbb{R}$ the weakest topology such that $\pi \to \pi_a \xi$ $(a \in \mathscr{A}, \xi \in \mathscr{H}_\infty)$ is continuous (the topology on $\mathscr{H}_\infty$ is the norm-topology). Then $\mathbb{R}$ is a complete space. In fact, if $\{\pi^\alpha\}_{\alpha \in \mathbb{I}}$ is a Cauchy directed set in $\mathbb{R}$, then $\{\pi_a^\alpha \xi\}$ $(a \in \mathscr{A}, \xi \in \mathscr{H}_\infty)$ is Cauchy in $\mathscr{H}_\infty$. Hence there exists a bounded operator T_a in $B(\mathscr{H}_\infty)$ with $\pi_a^\alpha \to T_a$ (strongly). $T_1 = 1_{\mathscr{H}_\infty}$, $T_{ab} = T_a T_b$ $(a, b \in \mathscr{A})$ and $T_a^* = T_{a^*}$, since $\pi_a^\alpha \to T_a$ (weakly) implies $\pi_a^{\alpha *} \to T_a^*$ (weakly).

Let (a_n) (resp. (ξ_n)) be a sequence of elements in $\mathscr{A}$ (resp. $\mathscr{H}_\infty$) which is norm-dense in the unit sphere of $\mathscr{A}$ (resp. $\mathscr{H}_\infty$). Set

$$d(\pi^1, \pi^2) = \sum_{m,n=1}^{\infty} \frac{1}{2^{m+n}} \|(\pi_{a_m}^1 - \pi_{a_m}^2)\xi_n\|.$$

Then d will define a metric on $\mathbb{R}$ which is equivalent to the topology on $\mathbb{R}$. Therefore $\mathbb{R}$ is a complete metric space.

Let $\mathbb{J} = \{(m,n) \mid m, n = 1, 2, \ldots\}$ and let E be the Banach space of all $\mathscr{H}_\infty$-valued functions $f(m,n)$ on $\mathbb{I}$ with $\sum_{m,n=1}^{\infty} \frac{1}{2^{m+n}} \|f(m,n)\| < +\infty$. Then E is separable, and the mapping $\pi \to \pi_{a_m} \xi_n = f(m,n)$ of $\mathbb{R}$ into E is isometric. Hence $\mathbb{R}$ is a separable complete metric space (called a polish space).

Let $B_n(\mathscr{H}_\infty)$ be the closed sphere, with radius n, of $B(\mathscr{H}_\infty)$.

3.4.3. Lemma. *Let δ be a metric on $B_1(\mathscr{H}_\infty)$ compatible with the weak operator topology, b an element in $B_1(\mathscr{H}_\infty)$, λ and μ non-negative real numbers, and $a_1, a_2, \ldots, a_n$ elements of $\mathscr{A}$. Let Y be the set of all π in $\mathbb{R}$ having the following properties: there exist $t_1, t_2, \ldots, t_n$ in $B_\mu(\mathscr{H}_\infty) \cap \pi(\mathscr{A})'$ such that $\pi_{a_1} t_1 + \pi_{a_2} t_2 + \cdots + \pi_{a_n} t_n \in B_1(\mathscr{H}_\infty)$ and*

$$\delta(\pi_{a_1} t_1 + \pi_{a_2} t_2 + \cdots + \pi_{a_n} t_n, b) \leq \lambda.$$

Then Y is closed in $\mathbb{R}$.

Proof. Let $\{\pi^\alpha\}_{\alpha\in\mathbb{I}}$ be a directed set of elements in Y which converges to π. For each α, there exist $t_1^\alpha,\ldots,t_n^\alpha$ in $B_\mu(\mathscr{H}_\infty)\cap\pi(\mathscr{A})'$ with $\delta(\pi_{a_1}^\alpha t_1^\alpha+\cdots+\pi_{a_n}^\alpha t_n^\alpha,b)\le\lambda$. By taking a subset of $\{\pi^\alpha\}_{\alpha\in\mathbb{I}}$, we may assume that $t_1^\alpha,\ldots,t_n^\alpha$ converges weakly to some elements $t_1,\ldots,t_n$ in $B_\mu(\mathscr{H}_\infty)$. For $a\in\mathscr{A}$ and $\xi,\eta\in\mathscr{H}_\infty$, $(t_1^\alpha\pi_a^\alpha\xi,\eta)=(\pi_a^\alpha\xi,t_1^{\alpha *}\eta)\to(\pi_a\xi,t_1^*\eta)=(t_1\pi_a\xi,\eta)$, and $(\pi_a^\alpha t_1^\alpha\xi,\eta)=(t_1^\alpha\xi,\pi_a^{\alpha *}\eta)\to(t_1\xi,\pi_a^*\eta)=(\pi_a t_1\xi,\eta)$. Hence $(t_1\pi_a\xi,\eta)=(\pi_a t_1\xi,\eta)$ and so $t_1\in\pi(\mathscr{A})'$. Similarly $t_2,\ldots,t_n\in\pi(\mathscr{A})'$. Furthermore $\delta(\pi_{a_1}t_1+\cdots+\pi_{a_n}t_n,b)\le\lambda$ and so $\pi\in Y$. q.e.d.

3.4.4. Theorem. *Let (a_n) be a sequence of elements in $\mathscr{A}$ which is uniformly dense in $\mathscr{A}$ and (b_n) a sequence of elements in $B_1(\mathscr{H}_\infty)$ which is weakly dense in $B_1(\mathscr{H}_\infty)$. For positive integers k,n,p, let $Y_{k,n,p}$ denote the set of all π in $\mathbb{R}$ posessing the following property: there exists $t_1,t_2,\ldots,t_n$ in $B_n(\mathscr{H}_\infty)\cap\pi(\mathscr{A})'$ with $\pi_{a_1}t_1+\cdots+\pi_{a_n}t_n\in B_1(\mathscr{H}_\infty)$ and $\delta(\pi_{a_1}t_1+\cdots+\pi_{a_n}t_n,b_k)\le 1/p$. Then $Y_{k,n,p}$ is closed in $\mathbb{R}$ and $\mathbb{P}=\bigcap_{k,p=1}^{\infty}\bigcup_{n=1}^{\infty}Y_{k,n,p}$ is the set of those π in $\mathbb{R}$ which are factorial (i.e. $\pi(\mathscr{A})''$ is a factor).*

Proof. $Y_{k,n,p}$ is closed by 3.4.3. If $\pi\in\mathbb{P}$, $b_k\in R(\pi(\mathscr{A})'',\pi(\mathscr{A})')$ $(k=1,2,\ldots)$. Hence $R(\pi(\mathscr{A})'',\pi(\mathscr{A})')=B(\mathscr{H}_\infty)$, and so π is factorial. The reverse is almost clear. q.e.d.

Let $L=\{\xi\mid\|\xi\|=1,\xi\in\mathscr{H}_\infty\}$ and consider a polish space $\mathbb{R}\times L$. Define a mapping l of $\mathbb{R}\times L$ into the state space $\mathscr{S}$ of $\mathscr{A}$ as follows: $l(\pi,\xi)(a)=(\pi_a\xi,\xi)$ $(a\in\mathscr{A})$. Then l is onto. In fact, for $\varphi\in\mathscr{S}$, $\mathscr{H}_\varphi$ is separable. If $\dim(\mathscr{H}_\varphi)=\aleph_0$, there is a unitary mapping U_φ of $\mathscr{H}_\varphi$ onto $\mathscr{H}_\infty$. By using U_φ, we can construct a *-representation $\{\pi,\mathscr{H}_\infty\}$ which is equivalent to $\{\pi_\varphi,\mathscr{H}_\varphi\}$. Hence $\varphi(a)=(\pi_a\xi,\xi)$ ($a\in\mathscr{A}$ and some $\xi\in\mathscr{H}_\infty$).

If $\dim(\mathscr{H}_\varphi)<\aleph_0$, take an amplification $\{\pi_\varphi\otimes 1_{\mathscr{H}},\mathscr{H}_\varphi\otimes\mathscr{H}\}$ with $\dim(\mathscr{H})=\aleph_0$. Then $\{\pi_\varphi\otimes 1_{\mathscr{H}},\mathscr{H}_\varphi\otimes\mathscr{H}\}$ is equivalent to a $\{\pi,\mathscr{H}_\infty\}$ $(\pi\in\mathbb{R})$. Hence l is onto Moreover,

$$|(\pi_a^1\xi,\xi)-(\pi_a^2\eta,\eta)|\le|(\pi_a^1\xi,\xi-\eta)|+|((\pi_a^1-\pi_a^2)\xi,\eta)|+|(\pi_a^2(\xi-\eta),\eta)|$$
$$\le\|a\|\,\|\xi-\eta\|+\|(\pi_a^1-\pi_a^2)\xi\|+\|a\|\,\|\xi-\eta\|$$
$$(a\in\mathscr{A},\ \xi,\eta\in\mathscr{H}_\infty).$$

Hence l is continuous.

3.4.5. Corollary. *Let $\mathfrak{F}$ be the set of all factorial states on $\mathscr{A}$. Then $\mathfrak{F}$ is a Borel subset in $\mathscr{S}$.*

Proof. $\mathbb{R}\times L$ is a polish space and the map l is continuous. Hence $l(\mathbb{P}\times L)=\mathfrak{F}$ is an analytic set. $(\mathbb{R}-\mathbb{P})\times L$ is also Borel and $l((\mathbb{R}-\mathbb{P})\times L)=\mathscr{S}-\mathfrak{F}$ is analytic. Hence $\mathfrak{F}$ is Borel. q.e.d.

The unit sphere $B_1(\mathscr{H}_\infty)$ of $B(\mathscr{H}_\infty)$ is a polish space in the s^*-topology.

3.4.6. Lemma. *The set of all pairs* (π, b) *in a polish space* $\mathbb{R} \times B_1(\mathcal{H}_\infty)$ *with* $b \in \pi(\mathcal{A})''$ *and* $\|b\| \leq 1$ *is a Borel set.*

Proof. Let (a_n) (resp. ξ_n) be a sequence of elements in the unit sphere of $\mathcal{A}$ (resp. $\mathcal{H}_\infty$) which is dense in $\mathcal{A}$ (resp. $\mathcal{H}_\infty$). $\{b$ is in the unit sphere of $\pi(\mathcal{A})''\}$ is equivalent to

$$\{\text{for each } n \text{ there is an } m \text{ such that } \|(\pi(a_m)-b)\xi_j\| < 1/n (j \leq n)\},$$

by the Kaplansky density theorem. Hence

$$\bigcap_{n=1}^{\infty} \bigcup_{m=1}^{\infty} \bigcap_{j=1}^{n} \{\|(\pi(a_m)-b)\xi_j\| < 1/n\}$$

$=$ the set of all pairs (π, b) in $\mathbb{R} \times B_1(\mathcal{H}_\infty)$ with $b \in \pi(\mathcal{A})''$ and $\|b\| \leq 1$.
q.e.d.

3.4.7. Proposition. *Let* $\mathcal{S}^{s,f}$ *be the set of all states* φ *on* $\mathcal{A}$ *such that* $\pi_\varphi(\mathcal{A})''$ *is semi-finite. Then* $\mathcal{S}^{s,f} \cap \mathfrak{F}$ *(denoted by* $\mathfrak{F}^{s,f}$*) is an analytic set, and so it is measurable for every Radon measure on* $\mathcal{S}$.

Proof. For $\pi \in \mathbb{P}$, $\pi(\mathcal{A})''$ is semi-finite if and only if there exists a non-zero projection $e \in \pi(\mathcal{A})''$ and a unit vector ξ in $\mathcal{H}_\infty$ with $(e\pi(a_m)e\pi(a_n)e\xi, \xi) = (e\pi(a_n)e\pi(a_m)e\xi, \xi)$ and $\xi \in e\mathcal{H}_\infty$ $(m, n = 1, 2, \ldots)$. In fact, if $\pi(\mathcal{A})''$ is semi-finite, there exists a non-zero finite projection p in $\pi(\mathcal{A})''$. For $\eta(\neq 0) \in p\mathcal{H}_\infty$, $P_{[\pi(\mathcal{A})'\eta]}$ is a finite non-zero projection in $\pi(\mathcal{A})''$.

Set $e = P_{[\pi(\mathcal{A})'\eta]}$; then $e\pi(\mathcal{A})''e$ is finite. Since $e\pi(\mathcal{A})''e$ on $e\mathcal{H}_\infty$ has a separating vector, any normal state φ on $e\pi(\mathcal{A})''e$ can be written as follows: $\varphi = \varphi_\xi$ (some $\xi \in e\mathcal{H}_\infty$) by 2.7.9. Conversely if

$$(e\pi(a_m)e\pi(a_n)e\xi, \xi) = (e\pi(a_n)e\pi(a_m)e\xi, \xi) \qquad (m, n = 1, 2, \ldots),$$

$(exeye\xi, \xi) = (eyexe\xi, \xi)$ $(x, y \in \pi(\mathcal{A})'')$. Hence $e\pi(\mathcal{A})''e$ is finite.

Let

$T = \{(\pi, e, \xi) \mid$ I. $e = e^*$, $e^2 = e$, $e \neq 0$, $e \in \pi(\mathcal{A})''$;

II. $(e\pi(a_m)e\pi(a_n)e\xi, \xi) = (e\pi(a_n)e\pi(a_m)e\xi, \xi)$ $(m, n = 1, 2, \ldots)$

and $e\xi = \xi\}$. Then the condition I. defines a Borel set in $\mathbb{R} \times B_1(\mathcal{H}) \times L$ by 3.4.6 and the s^*-continuity of the $*$-operation and the multiplication on $B_1(\mathcal{H}_\infty)$. $(e\pi(a_m)e\pi(a_n)e\xi, \xi)$ is a continuous function on $\mathbb{R} \times B_1(\mathcal{H}) \times L$. Hence T *is a Borel set, and so* $l\{(\pi, \xi) \mid (\pi, e, \xi) \in T\} \cap \mathfrak{F} = \mathcal{S}^{s,f} \cap \mathfrak{F}$ is analytic. q.e.d.

3.4.8. Proposition. *Let* $\mathcal{S}^f$ *be the set of all states* φ *on* $\mathcal{A}$ *such that* $\pi_\varphi(\mathcal{A})''$ *is finite. Then* $\mathcal{S}^f \cap \mathfrak{F}$ *is a Borel set.*

Proof. Suppose that $\pi(\mathcal{A})''$ $(\pi \in \mathbb{R})$ is a finite factor. Then its tracial state τ may be written as a finite linear combination of vector states. In fact, for a sufficiently large positive integer n, there exists a cyclic projection e in $\pi(\mathcal{A})''$ with $\tau(e) = 1/n$. Let $(e_1, e_2, \ldots, e_n)$ be a family of mutually orthogonal equivalent projections in $\pi(\mathscr{A})''$ with $e_i \sim e$ $(i=1,2,\ldots,n)$. Let τ_0 be the unique tracial state on $e\pi(\mathscr{A})''e$, and let v_i be a partial isometry in $\pi(\mathscr{A})''$ with $v_i^* v_i = e$ and $v_i v_i^* = e_i (i=1,2,\ldots,n)$. By 2.7.9 there exists a vector ξ in $e\mathscr{H}_\infty$ with $\tau_0(exe)=(x\xi,\xi)$ $(x\in\pi(\mathscr{A})'')$. Set $\varphi(x) = \frac{1}{n}\sum_{i=1}^{n} \tau_0(v_i^* x v_i)$ $(x\in\pi(\mathscr{A})'')$. Then φ is a tracial state on $\pi(\mathscr{A})''$. Hence by the uniqueness of the tracial state, $\varphi=\tau$. Moreover, $\frac{1}{n}\sum_{i=1}^{n} \tau_0(v_i^* x v_i) = \frac{1}{n}\sum_{i=1}^{n} (v_i^* x v_i \xi,\xi) = \frac{1}{n}\sum_{i=1}^{n} (x v_i \xi, v_i \xi)$. Thus for $\pi\in\mathbb{P}$, $\pi(\mathscr{A})''$ is finite if and only if there exist $\xi_1,\ldots,\xi_k\in\mathscr{H}_\infty$ with $\sum_{i=1}^{k} \|\xi_i\|^2=1$ and $\sum_{i=1}^{k} (\pi(a_m)\pi(a_n)\xi_i,\xi_i) = \sum_{i=1}^{k} (\pi(a_n)\pi(a_m)\xi_i,\xi_i)$ $(m,n=1,2,\ldots)$. Hence $\mathbb{P}\cap\mathbb{R}^f$ ($\mathbb{R}^f$ is the set of all π in $\mathbb{R}$ such that $\pi(\mathscr{A})''$ is finite) is the union over $k=1,2,\ldots$ of the projections (on $\mathbb{R}$) of the following subsets in

$$\mathbb{P}\times\underbrace{\mathscr{H}_\infty\times\cdots\times\mathscr{H}_\infty}_{k\,\text{times}}:\{(\pi,\xi_1,\ldots,\xi_k)\Big|\sum_{i=1}^{k}\|\xi_i\|^2=1,$$

$$\sum_{i=1}^{k}(\pi(a_m a_n)\xi_i,\xi_i)=\sum_{i=1}^{k}(\pi(a_n a_m)\xi_i,\xi_i)\ (m,n=1,2,\ldots)\}$$

which is clearly a Borel subset. Hence $\mathbb{P}\cap\mathbb{R}^f$ is analytic.

As for $\mathbb{P}-\mathbb{R}^f$, this is defined by the presence of an infinite projection in $\pi(\mathscr{A})''$. If $\pi\in\mathbb{P}-\mathbb{R}^f$ and $\{\xi_n\}$ is a dense sequence in $\mathscr{H}_\infty$, then there exists a partial isometry U in $\pi(\mathscr{A})''$ with $UU^*\leq U^*U$ and $UU^*\neq U^*U$. Thus $\pi(\mathscr{A})''$ is not finite if and only if it contains an element U such that

I $(U^*U)^2=(U^*U)$,

II $(U^*U\xi_i,\xi_i)\geq(UU^*\xi_i,\xi_i)$ $(i=1,2,\ldots)$ and the inequality is strict for at least one i. Then $\{(\pi,U)\,|\,U\in\pi(\mathscr{A})''$ and the above conditions I, II$\}$ is a Borel set in $\mathbb{R}\times B_1(\mathscr{H}_\infty)$. Hence $\mathbb{P}-\mathbb{R}^f$ is also analytic and so $\mathbb{P}\cap\mathbb{R}^f$ is a Borel set. q.e.d.

3.4.9. Theorem. *Let $\mathscr{S}^{\mathrm{I}_n}$ $(n=\aleph_0,1,2,\ldots)$ (resp. $\mathscr{S}^{\mathrm{II}_1}$, $\mathscr{S}^{\mathrm{II}_\infty}$, $\mathscr{S}^{\mathrm{III}}$) be the set of all states φ on $\mathscr{A}$ such that $\pi_\varphi(\mathscr{A})''$ is of type I_n (resp. II_1, II_∞, III), and let $\mathfrak{F}$ be the set of all factorial states on $\mathscr{A}$. Then, $\mathfrak{F}$, $\mathfrak{F}\cap\mathscr{S}^{\mathrm{I}_n}$, $\mathfrak{F}\cap\mathscr{S}^{\mathrm{II}_1}$, $\mathfrak{F}\cap\mathscr{S}^{\mathrm{II}_\infty}$ and $\mathfrak{F}\cap\mathscr{S}^{\mathrm{III}}$ are measurable subsets for every Radon measure on $\mathscr{S}$.*

Proof. For $\pi\in\mathbb{R}$, $e\in B_1(\mathscr{H}_\infty)$ let $G=\{(\pi,e)|$ I $e=e^*$, $e^2=e$, $e\neq0$, $e\in\pi(\mathscr{A})''$; II $e\pi(a_m)e\pi(a_n)e=e\pi(a_n)e\pi(a_m)e$ $(m,n=1,2,\ldots)\}$. Then G is a

Borel set in $\mathbb{R}\times B_1(\mathscr{H}_\infty)$ and so $\mathfrak{F}\cap\left(\bigcup_{n=\aleph_0,1,2,\ldots}\mathscr{S}^{\mathrm{I}_n}\right)$ is an analytic set in $\mathscr{S}$. For $p<+\infty$, $\pi\in\mathbb{R}$ and $e_1,\ldots,e_p\in B_1(\mathscr{H}_\infty)$, let

$$G_p=\Big\{(\pi,e_1,\ldots,e_p)\,|\,\mathrm{I}\ e_i=e_i^*,\ e_i^2=e_i,\ e_i\neq 0,\ e_i\in\pi(\mathscr{A})''\ \ (i=1,2,\ldots,p);$$

$$\mathrm{II}\ e_i\pi(a_m)e_i\pi(a_n)e_i=e_i\pi(a_n)e_i\pi(a_m)e_i\ \ (m,n=1,2,\ldots;\ i=1,2,\ldots,p);$$

$$\mathrm{III}\ \sum_{i=1}^{p}e_i=1\Big\}.$$

Then G_p is a Borel set in $\mathbb{R}\times\underbrace{B_1(\mathscr{H}_\infty)\times\cdots\times B_1(\mathscr{H}_\infty)}_{p\,\text{times}}$.

Hence $\mathfrak{F}\cap\mathscr{S}^{\mathrm{I}_p}$ is analytic ($p=1,2,\ldots$). Therefore, $\mathfrak{F}$, $\mathfrak{F}\cap\mathscr{S}^{\mathrm{I}_n}$, $\mathfrak{F}\cap\mathscr{S}^{\mathrm{II}_1}$, $\mathfrak{F}\cap\mathscr{S}^{\mathrm{II}_\infty}$ and $\mathfrak{F}\cap\mathscr{S}^{\mathrm{III}}$ are measurable for all Radon measures on $\mathscr{S}$ by 3.4.5, 3.4.7 and 3.4.8. q.e.d.

Now let ψ be a state of $\mathscr{A}$, C a commutative W^*-algebra, with $1_{\mathscr{H}_\psi}$, in $\pi_\psi(\mathscr{A})'$, μ the C-measure of ψ on $\mathscr{S}$ and let Γ be the C-isomorphism of ψ in 3.1. Then

$$(\pi_\psi(a)c\,1_\psi,1_\psi)=\int_{\mathscr{S}}\Gamma(c)(\varphi)\hat{a}(\varphi)\,d\mu(\varphi)$$

($c\in C$, $a\in\mathscr{A}$). For each $\varphi\in\mathscr{S}$, there corresponds the $*$-representation $\{\pi_\varphi,\mathscr{H}_\varphi\}$ of $\mathscr{A}$. Let (a_n) be a sequence of elements in $\mathscr{A}$ which is uniformly dense in $\mathscr{A}$. Then $[a_{n\varphi}\,|\,n=1,2,\ldots]=\mathscr{H}_\varphi$ ($\varphi\in\mathscr{S}$) and

$$(a_{n\varphi},a_{m\varphi})=\varphi(a_m^*a_n)=\widehat{a_m^*a_n}(\varphi)$$

is continuous on $\mathscr{S}$ ($m,n=1,2,\ldots$). Hence $\{\mathscr{H}_\varphi\,|\,\varphi\in\mathscr{S}\}$ is μ-measurable by 3.3, and so we have $\int\mathscr{H}_\varphi\,d\mu(\varphi)$.

Define a linear mapping U of $\mathscr{H}_\psi$ into $\mathscr{H}_\varphi$-valued functions on $(\mathscr{S},\mu)$ in the following way: For $\xi=\sum_{i=1}^{n}c_i\pi_\psi(a_i)1_\psi$ in $\mathscr{H}_\psi(c_i\in C,\ a_i\in\mathscr{A})$, set $(U\xi)(\varphi)=\sum_{i=1}^{n}\Gamma(c_i)(\varphi)\pi_\varphi(a_i)1_\varphi$. Then

$$\begin{aligned}\|\xi\|^2&=\left(\sum_{i=1}^{n}c_i\pi_\psi(a_i)1_\psi,\ \sum_{i=1}^{n}c_i\pi_\psi(a_i)1_\psi\right)\\&=\sum_{i,j=1}^{n}(c_j^*c_i\pi_\psi(a_j^*)\pi_\psi(a_i)1_\psi,1_\psi)\\&=\sum_{i,j=1}^{n}\int\overline{\Gamma(c_j)(\varphi)}\,\Gamma(c_i)(\varphi)\,\widehat{a_j^*a_i}(\varphi)\,d\mu(\varphi)\\&=\int\left\|\sum_{i=1}^{n}\Gamma(c_i)(\varphi)\pi_\varphi(a_i)1_\varphi\right\|^2d\mu(\varphi).\end{aligned}$$

Hence U can be extended uniquely to a unitary mapping (denoted by the same notation U) of $\mathscr{H}_\psi$ onto $\int \mathscr{H}_\varphi d\mu(\varphi)$, since elements of the form ξ are dense in $\mathscr{H}_\psi$.

For $c \in C$

$$(U c \xi)(\varphi) = \left(U c \sum_{i=1}^{n} c_i \pi_\psi(a_i) 1_\psi\right)(\varphi)$$
$$= \sum_{i=1} \Gamma(c)(\varphi) \Gamma(c_i)(\varphi) \pi_\varphi(a_i) 1_\varphi = \Gamma(c)(\varphi)(U\xi)(\varphi).$$

Hence $UCU^* = \Gamma(C)$.

Since $\Gamma(C) = L^\infty(\mathscr{S}, \mu)$, C is realized as the algebra of all diagonalizable operators in this reduction, and so we have the reduction:

$$\{\pi_\psi(\mathscr{A}), C\}'' = \int \pi_\varphi(\mathscr{A})'' \, d\mu(\varphi)$$

and $C = L^\infty(\Omega, \mu) 1_{\mathscr{H}_\psi}$.

Concluding remarks on 3.4.

Theorem 3.4.4 is due to Dixmier [35]. Theorem 3.4.9 is due to Schwartz [178]. The proof of 3.4.5 is due to Feldman [50].

References. [22], [35], [47], [50], [146], [159], [178].

3.5. Central Decomposition of States (Separable Cases)

Let $\mathscr{A}$ be a separable C^*-algebra with identity. For $\psi \in \mathscr{S}$, let $\Omega(\psi)$ be the set of all probability Radon measure μ with $\psi(a) = \int \hat{a}(\varphi) d\mu(\varphi)$ $(a \in \mathscr{A})$. $\Omega(\psi)$ consists, in general, of many different measures. In 3.1, we showed that for each commutative W^*-subalgebra C, with $1_{\mathscr{H}_\psi}$, in $\pi_\psi(\mathscr{A})'$, there is a unique C-measure of ψ in $\Omega(\psi)$. An important family of measures belonging to $\Omega(\psi)$ is the one which concentrates on the set $\mathscr{E}$ of all pure states on $\mathscr{A}$.

We can get such a measure by choosing a maximal commutative W^*-subalgebra in $\pi_\psi(\mathscr{A})'$—more generally, a maximal measure in $\Omega(\psi)$ in the sense of Bishop-de Leeuw. Thus the *-representation $\{\pi_\psi, \mathscr{H}_\psi\}$ of $\mathscr{A}$ can be, in a sense, expressed as a direct integral of irreducible *-representations $\{\pi_\varphi, \mathscr{H}_\varphi\}$ $(\varphi \in \mathscr{E})$. If $\mathscr{A}$ is commutative, the μ with $\mu \in \Omega(\psi)$ and $\mu(\mathscr{E}) = \mu(\mathscr{S})$ is unique. Hence we can reduce the study of the state ψ (therefore, the cyclic *-representation $\{\pi_\psi, \mathscr{H}_\psi\}$) to the study of the Radon measure μ and pure states (therefore irreducible *-representations). But if $\mathscr{A}$ is not commutative, the μ is, in general, not unique even for a finite-dimensional $\mathscr{A}$. Therefore inspite of their importance, those measures are not suitable for the decomposition theory of states.

On the other hand, the center C_z of $\pi_\psi(\mathscr{A})''$ is unique; therefore the central measure of ψ is unique. Furthermore the central measure ν will give the reduction: $\pi_\psi(\mathscr{A})''=\int \pi_\varphi(\mathscr{A})''\,d\nu(\varphi)$. Hence the W^*-algebra $\pi_\psi(\mathscr{A})''$ can be very well analysed by the set of $\pi_\varphi(\mathscr{A})''$ ($\varphi\in\mathfrak{F}$) and the central measure ν.

On the other hand, if $C\supsetneqq C_z$, $\pi_\psi(\mathscr{A})''\subsetneqq(\pi_\psi(\mathscr{A}),C)''=\int \pi_\varphi(\mathscr{A})''\,d\mu(\varphi)$ (μ is the C-measure of ψ). Hence $\pi_\psi(\mathscr{A})''$ may not be so closely related to $\int \pi_\varphi(\mathscr{A})''\,d\mu(\varphi)$. Also if $C\subsetneqq C_z$, we have again $\pi_\psi(\mathscr{A})''=\int \pi_\varphi(\mathscr{A})''\,d\mu(\varphi)$. But, $\pi_\varphi(\mathscr{A})''$ may not be so primitive—namely there may not be a big difference between $\pi_\psi(\mathscr{A})''$ and $\pi_\varphi(\mathscr{A})''$. In fact, in this case, we can not expect that $\pi_\varphi(\mathscr{A})''$ is a factor. However, still there are important applications of these two cases in mathematical physics (cf. [70], [146]). In any case, there is no doubt that the central decomposition is the nicest decomposition among various decompositions of ψ. In this section, we shall develop the central decomposition of states.

Let $\mathscr{A}^{**}$ be the second dual of $\mathscr{A}$. Each f in $\mathscr{A}^*$ can be uniquely extended to a σ-continuous linear functional (denoted by the same notation f) on the W^*-algebra $\mathscr{A}^{**}$. Let Z be the center of $\mathscr{A}^{**}$. For convenience, we shall give a new definition of central measures which is equivalent to the previous one.

3.5.1. Definition. *A measure $\mu\in\Omega(\psi)$ is said to be a central measure of ψ if it satisfies the following condition:* (*) *there exists a σ-continuous homomorphism Ψ (called the central homomorphism of ψ) of Z onto $L^\infty(\mathscr{S},\mu)$ such that $\psi(za)=\int_{\mathscr{S}} \Psi(z)(\varphi)\hat{a}(\varphi)\,d\mu(\varphi)$ $(z\in Z, a\in\mathscr{A})$.*

The existence and uniqueness of the central measure μ and the central homomorphism Ψ are seen as follows: take μ the central measure of ψ and the *-homomorphism Γ in the sense of 3.1. Then

$$(\pi_\psi(a)c\,1_\psi,1_\psi)=\int_{\mathscr{S}} \Gamma(c)(\varphi)\hat{a}(\varphi)\,d\mu(\varphi)$$

($a\in\mathscr{A}, c\in C_z$), where C_z is the center of $\pi_\psi(\mathscr{A})''$. The *-representation $\{\pi_\psi,\mathscr{H}_\psi\}$ of $\mathscr{A}$ can be extended uniquely to the W^*-representation $\{\pi_\psi^w,\mathscr{H}_\psi\}$ of $\mathscr{A}^{**}$ (cf. 1.21.13). Set $\Psi(z)=\Gamma(\pi_\psi^w(z))$ $(z\in Z)$. Then Ψ is a σ-continuous *-homomorphism of Z onto $L^\infty(\mathscr{S},\mu)$. Moreover

$$\begin{aligned}\psi(za)&=(\pi_\psi^w(z)\pi_\psi(a)1_\psi,1_\psi)=\int \Gamma(\pi_\psi^w(z))(\varphi)\hat{a}(\varphi)\,d\mu(\varphi)\\&=\int \Psi(z)(\varphi)\hat{a}(\varphi)\,d\mu(\varphi)\qquad(z\in Z, a\in\mathscr{A}).\end{aligned}$$

Hence $\{\mu,\Psi\}$ satisfies the above condition (*). Conversely suppose $\{\mu',\Psi'\}$ satisfies the condition $\psi(za)=\int \Psi'(z)(\varphi)\hat{a}(\varphi)\,d\mu'(\varphi)$ $(z\in Z, a\in\mathscr{A})$.

Then $(\pi_\psi^w(z)\pi_\psi(a)1_\psi, 1_\psi) = \int \Psi'(z)(\varphi)\hat{a}(\varphi)\,d\mu'(\varphi)$. Put $\Gamma'(\pi_\psi^w(z)) = \Psi'(z)$ $(z\in Z)$; then $\pi_\psi^w(z) \to \Gamma'(\pi_\psi^w(z))$ is well-defined and

$$(c\pi_\psi(a)1_\psi, 1_\psi) = \int \Gamma'(c)(\varphi)\hat{a}(\varphi)\,d\mu(\varphi) \qquad (c\in C_z, a\in\mathscr{A}).$$

Hence $\Gamma' = \Gamma$ and $\mu = \mu'$ by 3.1.3, and so $\Phi = \Phi'$.

Hence we have

3.5.2. Theorem. *Let $\mathscr{A}$ be a separable C*-algebra with identity, and let $\mathscr{S}$ be the state space of $\mathscr{A}$. For $\psi\in\mathscr{S}$, let μ (resp. Ψ) be the central measure (resp. the central homomorphism) of ψ. Then we have the reduction:*

$$\pi_\psi(\mathscr{A}^{**}) = \int_{\mathscr{S}} \pi_\varphi(\mathscr{A}^{**})\,d\mu(\varphi) \quad \text{and} \quad \Psi(Z) = L^\infty(\mathscr{S},\mu).$$

Moreover $\pi_\varphi(\mathscr{A}^{**})$ is a factor for μ-almost all φ in $\mathscr{S}$.

Next we shall consider a geometrical characterization of the central measure in $\Omega(\psi)$. Let μ (resp. Ψ) be the central measure (resp. homomorphism) of ψ, and let F be an arbitrary Borel set in $\mathscr{S}$. Set $\mu_F(f) = \int_F f(\varphi)\,d\mu(\varphi)$ $(f\in C(\mathscr{S}))$. Then μ_F is a Radon measure on $C(\mathscr{S})$. Put $\psi_F(a) = \int_F \hat{a}(\varphi)\,d\mu(\varphi)$ $(a\in\mathscr{A})$. Since $\Psi(Z) = L^\infty(\mathscr{S},\mu)$, there exist two mutually orthogonal central projections z_1, z_2 in Z with $\Psi(z_1) = \chi_F$ and $\Psi(z_2) = \chi_{\mathscr{S}-F}$, where χ_F (resp. $\chi_{\mathscr{S}-F}$) is the characteristic function of F (resp. $\mathscr{S}-F$). Then $\psi_F(a) = \psi(z_1 a)$ and $\psi_{\mathscr{S}-F}(a) = \psi(z_2 a)$. Hence $s(\pi_{\psi_F}^w)\cdot s(\pi_{\psi_{\mathscr{S}-F}}^w) = 0$.

3.5.3. Definition. *Let ψ_1, ψ_2 be two positive linear functionals on $\mathscr{A}$. ψ_1 and ψ_2 are said to be mutually centrally orthogonal (or disjoint) if $s(\pi_{\psi_1}^w)\cdot s(\pi_{\psi_2}^w) = 0$.*

3.5.4. Definition. *Let ν be a measure belonging to $\Omega(\psi)$. ν is said to be semi-central if ψ_F and $\psi_{\mathscr{S}-F}$ are mutually disjoint for every Borel set F in $\mathscr{S}$.*

Let $\Omega_c(\psi)$ be the set of all semi-central measures belonging to $\Omega(\psi)$.

3.5.5. Theorem. *For $\nu\in\Omega_c(\psi)$, there exists a unique commutative W*-subalgebra C, with $1_{\mathscr{H}_\psi}$, of the center C_z of $\pi_\psi(\mathscr{A})''$ such that ν is the C-measure of ψ—namely there is a one-to-one correspondence between $\Omega_c(\psi)$ and commutative W*-algebras C, with $1_{\mathscr{H}_\psi}$, in C_z, and this correspondence is given by $\nu\in\Omega_c(\psi)\to C$ such that ν is the C-measure of ψ.*

Proof. Let $\nu\in\Omega_c(\psi)$. For any $f\in L^\infty(\mathscr{S},\mu)$ with $0\le f\le 1$, put $\psi_f(a) = \int f(\varphi)\hat{a}(\varphi)\,d\nu(\varphi)$ $(a\in\mathscr{A})$. Then $\psi_f \le \psi$. Hence there exists a unique positive element h'_f in $\pi_\psi(\mathscr{A})'$ with

$$\psi_f(a) = (\pi_\psi(a)h'_f 1_\psi, 1_\psi) = \int f(\varphi)\hat{a}(\varphi)\,d\nu(\varphi).$$

The mapping $f \to h'_f$ can be extended uniquely to a linear mapping of $L^\infty(\mathcal{S}, \mu)$ into $\pi_\psi(\mathcal{A})'$. Since ψ_F is centrally orthogonal to $\psi_{\mathcal{S}-F}$, there exists a central projection z in Z such that $\psi_F(z) = \|\psi_F\|$ and $\psi_{\mathcal{S}-F}(z) = 0$. Hence $(\pi_\psi(z) h'_{\chi_F} 1_\psi, 1_\psi) = (h'_{\chi_F} 1_\psi, 1_\psi)$ and $(\pi_\psi(z) h'_{\chi_{\mathcal{S}-F}} 1_\psi, 1_\psi) = 0$. Therefore $h'_{\chi_F} \leq \pi_\psi(z)$ and $h'_{\chi_{\mathcal{S}-F}} \leq 1_{\mathcal{H}_\psi} - \pi_\psi(z)$.

On the other hand, $h'_{\chi_F} + h'_{\chi_{\mathcal{S}-F}} = 1_{\mathcal{H}_\psi}$ and so $h'_{\chi_F} = \pi_\psi(z)$. Hence we have

$$\begin{aligned} \int \chi_F(\varphi) \hat{a}(\varphi) d\nu(\varphi) &= (\pi_\psi(a) h'_{\chi_F} 1_\psi, 1_\psi) \\ &= (\pi_\psi(a) \pi_\psi(z) 1_\psi, 1_\psi) \\ &= \psi(az). \end{aligned}$$

Now one can easily see that $f \to h'_f$ is a *-isomorphism of $L^\infty(\mathcal{S}, \nu)$ onto a commutative W^*-subalgebra $C = \{h'_f \mid f \in L^\infty(\mathcal{S}, \mu)\}$.

Put $\Lambda(h'_f) = f$; then it is easily seen that ν is the C-measure of ψ. The uniqueness of C is clear. Conversely if C is a W^*-subalgebra, with $1_{\mathcal{H}_\psi}$, in C_z and ν is the C-measure of ψ, then ν is clearly semi-central.

q.e.d.

In 3.1, we showed that $C_1 \subset C_2 \subset C_z$ implies $\mu_1 \prec \mu_2$, where μ_1 (resp. μ_2) is the C_1 (resp. C_2) measure of ψ_1 (resp. ψ_2). Hence we have

3.5.6. Proposition. *Let $\mu \in \Omega(\psi)$. μ is the central measure of ψ if and only if μ is the greatest semi-central measure with respect to the order $\prec$.*

3.5.7. Theorem. *Let $\psi(a) = \int \hat{a}(\varphi) d\mu(\varphi)$ $(a \in \mathcal{A})$ be the central decomposition of ψ. Then there exists a Borel set Δ_μ in $\mathfrak{F}$ such that for two distinct $\varphi_1, \varphi_2 \in \Delta_\mu$, $s(\pi^w_{\varphi_1}) \cdot (\pi^w_{\varphi_2}) = 0$ and $\mu(\Delta_\mu) = \mu(\mathcal{S})$.*

Proof. Let $\{O_n \mid n = 1, 2, \ldots\}$ be a basis of open neighborhoods in $\mathcal{S}$, and let χ_n be the characteristic function of O_n. For each n, there exists a sequence $(a_{n,j})$ in $\mathcal{A}$, with $\|a_{n,j}\| \leq 1$, such that $\pi_\varphi(a_{n,j}) \to \chi_n(\varphi) 1_{\mathcal{H}_\varphi}$ in the s-topology $(j \to \infty)$ for all $\varphi \in \mathcal{S} - N_n$, with $\mu(N_n) = 0$. For two distinct $\varphi_1, \varphi_2 \in \mathcal{S} - \bigcup_{n=1}^{\infty} N_n$, suppose that $s(\pi^w_{\varphi_1}) = s(\pi^w_{\varphi_2})$; then $\{\pi_{\varphi_1}, \mathcal{H}_{\varphi_1}\}$ and $\{\pi_{\varphi_2}, \mathcal{H}_{\varphi_2}\}$ are quasi-equivalent. Hence there exists a *-isomorphism Φ of $\pi_{\varphi_1}(\mathcal{A})''$ onto $\pi_{\varphi_2}(\mathcal{A})''$ such that $\Phi(\pi_{\varphi_1}(a)) = \pi_{\varphi_2}(a)$ $(a \in \mathcal{A})$ and so $\chi_n(\varphi_1) = \chi_n(\varphi_2)$ $(n = 1, 2, \ldots)$. This is a contradiction.

q.e.d.

Now let $\{\mathcal{A}, G\}$ be a system of a separable C^*-algebra $\mathcal{A}$ with identity and a group of *-automorphisms on $\mathcal{A}$. For $\psi \in \mathcal{S}_G$ ($\mathcal{S}_G$ is the set of all G-invariant states), let μ (resp. Ψ) be the central measure (resp. homomorphism) of ψ. For $g \in G$, let g^* be the dual mapping on $\mathcal{A}^*$. Then it is an isometric linear mapping of $\mathscr{A}^*$ onto $\mathscr{A}^*$. Let g^{**} be the second dual of g. Then it is a $*$-automorphism of $\mathscr{A}^{**}$.

$$\begin{aligned}\psi(az) &= \psi((az)^{g^{**}}) = \psi(a^g z^{g^{**}})\\ &= \int \Psi(z^{g^{**}})(\varphi)\widehat{a^g}(\varphi)\,d\mu(\varphi)\\ &= \int \Psi(z^{g^{**}})(\varphi)\hat{a}(\varphi^{g^*})\,d\mu(\varphi)\\ &= \int \Psi(z^{g^{**}})(\varphi^{(g^{-1})^*})\hat{a}(\varphi)\,d\mu(\varphi^{(g^{-1})^*})\end{aligned}$$

$(a\in\mathscr{A}, z\in Z)$. Put $d\mu^{g^{-1}}(\varphi)=d\mu(\varphi^{(g^{-1})^*})$; then

$$\psi(az) = \int \Psi(z^{g^{**}})(\varphi^{(g^{-1})^*})\hat{a}(\varphi)\,d\mu^{g^{-1}}(\varphi).$$

Hence by the uniqueness of central measure, we have

3.5.8. Proposition. *$\Psi(z^{g^{**}})(\varphi^{(g^{-1})^*})=\Psi(z)(\varphi)$ and*

$$d\mu^{g^{-1}}(\varphi) = d\mu(\varphi^{(g^{-1})^*}) = d\mu(\varphi)$$

—i.e. μ is G-invariant.

3.5.9. Proposition. *If ψ is ergodic, then μ is G-ergodic.*

Proof. Suppose ψ is ergodic; then $\{\pi_\psi(\mathscr{A}), u_\psi(G)\}'=\mathbb{C}1_{\mathscr{H}_\psi}$. Consider the direct integral: $\mathscr{H}_\psi=\int \mathscr{H}_\varphi\,d\mu(\varphi)$. Let $F\in L^\infty(\mathscr{S},\mu)$ with $F(\varphi)=F(\varphi^{g^*})$ in $L^\infty(\mathscr{S},\mu)$. Take a central element z in Z with $\Psi(z)=F$; then

$$\begin{aligned}(\pi_\psi^w(z^{g^{**}})\pi_\psi^w(c)1_\psi, \pi_\psi^w(c)1_\psi) &= \int \Psi(z^{g^{**}})(\varphi)|\Psi(c)(\varphi)|^2\,d\mu(\varphi)\\ &= \int \Psi(z)(\varphi^{g^*})|\Psi(c)(\varphi)|^2\,d\mu(\varphi)\\ &= \int F(\varphi)|\Psi(c)(\varphi)|^2\,d\mu(\varphi)\\ &= (\pi_\psi^w(z)\pi_\psi^w(c)1_\psi, \pi_\psi^w(c)1_\psi)\end{aligned}$$

$(c\in Z)$. Hence $\pi_\psi^w(z^{g^{**}})=\pi_\psi^w(z)$. Since $u_\psi(g^{-1})\pi_\psi^w(z)u_\psi(g)=\pi_\psi^w(z^{g^{**}})=\pi_\psi^w(z)$, $\pi_\psi^w(z)\in\{\pi_\psi(\mathscr{A}), u_\psi(G)\}'=\mathbb{C}1_{\mathscr{H}_\psi}$. q.e.d.

3.5.10. Theorem. *Suppose that the system $\{\mathscr{A}, G\}$ is asymptotically abelian and G is amenable. Then the following two conditions are equivalent:*

1. *ψ is ergodic;*
2. *the central measure μ of ψ is G-ergodic.*

Proof. Suppose that $\|[a^{g_n}, b]\|\to 0$ $(n\to 0)$ $(a, b\in\mathscr{A})$ $(g_n\in G)$, and the central measure μ of ψ is G-ergodic. Then

$$\|[u_\psi(g_n^{-1})\pi_\psi(a)u_\psi(g_n), \pi_\psi(b)]\|\to 0 \quad (n\to\infty).$$

By the σ-compactness of the unit sphere of $\pi_\psi(\mathscr{A})''$, there exists a subsequence $(n_j)\subset(n)$ with $u_\psi(g_{n_j}^{-1})\pi_\psi(a)u_\psi(g_{n_j})\to T$ in the σ-topology, where T is an element in $\pi_\psi(\mathscr{A})''$.

$\|[T,\pi_\psi(b)]\|=0$ $(b\in\mathscr{A})$ and so $T\in\pi_\psi(\mathscr{A})''\cap\pi_\psi(\mathscr{A})'$. On the other hand, $P_\psi u_\psi(g_n^{-1})\pi_\psi(a)u_\psi(g_n)P_\psi=P_\psi\pi_\psi(a)P_\psi$ and so $P_\psi TP_\psi=P_\psi\pi_\psi(a)P_\psi$.

Since G is amenable, the weakly closed convex subset generated by $\{u_\psi(g^{-1})Tu_\psi(g)|g\in G\}$ contains an invariant point T_0—i.e., $u(g^{-1})T_0u(g)=T_0$ $(g\in G)$. Since $T_0\in\pi_\psi(\mathscr{A})''\cap\pi_\psi(\mathscr{A})'$, $T_0\in L^\infty(\mathscr{S},\mu)$. Since μ is G-ergodic, $T_0=\lambda 1_{\mathscr{H}_\psi}$ (λ a complex number). Since $P_\psi u_\psi(g^{-1})Tu_\psi(g)P_\psi=P_\psi TP_\psi, P_\psi TP_\psi=P_\psi T_0P_\psi$. Hence $P_\psi\pi_\psi(\mathscr{A})P_\psi=\mathbb{C}P_\psi$. On the other hand, the weakly closed linear subspace of $B(\mathscr{H}_\psi)$ generated by $\{\pi_\psi(a)u_\psi(g)|a\in\mathscr{A},g\in G\}$ is $\{\pi_\psi(\mathscr{A}),u_\psi(G)\}''$, and $\{\pi_\psi(\mathscr{A}),u_\psi(G)\}'$ is commutative, because $\{\mathscr{A},G\}$ is G-abelian. Since $P_\psi\pi_\psi(a)u_\psi(g)P_\psi=P_\psi\pi_\psi(a)P_\psi$, $P_\psi\{\pi_\psi(\mathscr{A}),u_\psi(G)\}''P_\psi=\mathbb{C}P_\psi$. Since $P_\psi\in\{\pi_\psi(\mathscr{A}),u_\psi(G)\}''$, $\{\pi_\psi(\mathscr{A}),u_\psi(G)\}''=B(\mathscr{H}_\psi)$. q.e.d.

If $\mathscr{A}$ is a commutative C^*-algebra, then $\mathscr{E}$ = the spectrum space of $\mathscr{A}$ and $\mathscr{A}=C(\mathscr{E})$. g^* $(g\in G)$ on $\mathscr{E}$ is a homeomorphism on the compact space $\mathscr{E}$. For a G-invariant measure μ on $\mathscr{E}$, put $\psi(a)=\int_{\mathscr{E}}a(t)d\mu(t)$ $(a\in\mathscr{A})$; then μ is the central measure of ψ.

$\pi_\psi(\mathscr{A})''$ is commutative and it has a cyclic vector. Hence it is maximal commutative. Therefore

$$\{\pi_\psi(\mathscr{A}),u_\psi(G)\}'=\pi_\psi(\mathscr{A})'\cap u_\psi(G)'=\pi_\psi(\mathscr{A})''\cap u_\psi(G)'.$$

If μ is G-ergodic, $\pi_\psi(\mathscr{A})''\cap u_\psi(G)'=\mathbb{C}1_{\mathscr{H}_\psi}$ and so ψ is ergodic. Therefore 3.5.9 and 3.5.10 are the exact non-commutative extension of Ergodic theory. Such a non-commutative Ergodic theory has been recently developed in Mathematical physics.

Concluding remarks on 3.5.

In mathematical physics, a G-invariant state φ in a system $\{\mathscr{A},G\}$ of a C^*-algebra $\mathscr{A}$ and a group G of $*$-automorphisms on $\mathscr{A}$ will be first decomposed into ergodic states, using the ergodic decomposition. Next to explain the broken symmetry, each ergodic state ψ will be decomposed into factorial states, using the central decomposition. Also the C-decomposition with $C\subsetneqq Z$ or $Z\subsetneqq C$ (Z, the center of $\pi_\psi(\mathscr{A})''$) is used (cf. [70], [146]).

In general, it is an important problem to find a nice characterization of central measures among probability Radon measures concentrating on the set of factorial states. There is a non-trivial example in which the central decomposition coincides with the ergodic decomposition (cf. [169]).

Theorem 3.5.7 is due to Guichardet [67].
Theorem 3.5.10 is due to Haag, Kastler and Michel [71].
References. [71], [146], [159].

4. Special Topics

4.1. Derivations and Automorphisms of C^*-Algebras and W^*-Algebras

4.1.1. Definition. *Let $\mathscr{A}$ be a C^*-algebra and let δ be a linear mapping of $\mathscr{A}$ into $\mathscr{A}$. δ is called a derivation if $\delta(xy)=\delta(x)y+x\delta(y)$ $(x,y\in\mathscr{A})$.*

For $a\in\mathscr{A}$, put $\delta_a(x)=[a,x]$ $(x\in\mathscr{A})$; then δ_a is a derivation. Such a derivation is called an inner derivation. If there is no element a in $\mathscr{A}$ with $\delta=\delta_a$, δ is called an outer derivation.

4.1.2. Lemma. *If C is a commutative C^*-algebra and δ is a derivation on C, then $\delta\equiv 0$.*

Proof. Put $\delta(x)=x'$ $(x\in C)$. It suffices to prove that $x'=0$ for every self-adjoint element x in C. Let $C=C(K)$ be the function space representation of C. For $t\in K$, $\{x-x(t)1\}'=x'-x(t)1'$. Since $1'=(1\cdot 1)'=1'\cdot 1+1\cdot 1'=21'$, $1'=0$. Hence $x'=\{x-x(t)1\}'$.

Write $x-x(t)1$ as the difference of two positive elements: $x-x(t)1=x_1-x_2$ with $x_1(t)=x_2(t)=0$. Let $x_1=h_1^2$. Then

$$x_1'=h_1'h_1+h_1h_1'=2h_1h_1';$$

hence $x_1'(t)=0$. Analogously $x_2'(t)=0$ and so $(x-x(t)1)'(t)=0$. Hence $x'=0$. q.e.d.

Remark. The existence of the identity in the above proof is not essential. If $\mathscr{A}$ has no identity, consider $\mathscr{A}_1=\mathscr{A}+\mathbb{C}1$ and set $\delta(1)=0$, then δ will be a derivation on $\mathscr{A}_1$.

4.1.3. Lemma. *Every derivation δ of a C^*-algebra $\mathscr{A}$ is norm-continuous.*

Proof. By the above remark, it suffices to assume that $\mathscr{A}$ has an identity.

Put $\delta(x)=x'$ $(x\in\mathscr{A})$. It is enough to show that δ is continuous on the self-adjoint portion $\mathscr{A}^s$. Suppose that δ is not continuous on $\mathscr{A}^s$. Then by the closed graph theorem, there is a sequence $\{x_n\}$ $(x_n\neq 0)$ in $\mathscr{A}^s$ such that $x_n\to 0$ and $x_n'\to a+ib$ $(\neq 0)$ $(n\to\infty)$ with $a,b\in\mathscr{A}^s$.

Suppose first that $a\neq 0$ and there is a positive number λ in the spectrum of a (otherwise, consider $\{-x_n\}$). It suffices to assume that

$\lambda=1$. Then there exists a positive element h ($\|h\|=1$) in $\mathscr{A}$ with $hah\geq\frac{1}{2}h^2$. Put $y_n=x_n+3\|x_n\|\cdot 1$. Then $y_n\to 0$, $y_n'=x_n'$ and

$$(hy_nh)'=h'y_nh+hy_n'h+hy_nh'\to h(a+ib)h.$$

Hence $\|(hy_{n_0}h)'-h(a+ib)h\|<\frac{1}{8}$ for some n_0 1.

On the other hand,

$$hy_nh\leq 4\|x_n\|h^2 \quad\text{and}\quad \frac{1}{2}\cdot\frac{hy_nh}{4\|x_n\|}\leq hah \qquad 2.$$

Since $\|x_n\|\cdot 1+x_n\geq 0$, $\dfrac{hy_nh}{4\|x_n\|}\geq\dfrac{1}{2}h^2$. Hence

$$\left\|\frac{hy_nh}{4\|x_n\|}\right\|\geq\frac{1}{2}\|h\|^2=\frac{1}{2} \qquad 3.$$

Let C be the C^*-subalgebra of $\mathscr{A}$ generated $hy_{n_0}h$ and 1. Then, by 3., there is a character φ on C with $\varphi\left(\dfrac{hy_{n_0}h}{4\|x_{n_0}\|}\right)\geq\dfrac{1}{2}$.

Let $\overline{\varphi}$ be an extended state of φ to $\mathscr{A}$, and let $\mathscr{L}=\{x|\overline{\varphi}(x^*x)=0, x\in\mathscr{A}\}$. Then $C\cap\mathscr{L}$ is a maximal ideal of C; it can be written

$$hy_{n_0}h-\varphi(hy_{n_0}h)1=u^2-v^2$$

with $u,v\in C\cap\mathscr{L}$ $(u,v\geq 0)$.

Hence $(hy_{n_0}h)'=u'u+uu'-v'v-vv'$, so that by Schwartz's inequality

$$\overline{\varphi}((hy_{n_0}h)')=0 \qquad 4.$$

Then by 1. and 4.,

$$|\overline{\varphi}(h(a+ib)h)|<\tfrac{1}{8} \qquad 5.$$

On the other hand, by 2.,

$$|\overline{\varphi}(h(a+ib)h)|\geq\overline{\varphi}(hah)\geq\frac{1}{2}\overline{\varphi}\left(\frac{hy_{n_0}h}{4\|x_{n_0}\|}\right)\geq\frac{1}{2}\cdot\frac{1}{2}=\frac{1}{4}.$$

This contradicts the inequality 5. Hence $a=0$.

Next suppose that $b\neq 0$ and there exists a positive number μ in the spectrum of b (otherwise, consider $\{-x_n\}$). It suffices to assume that $\mu=1$. Then there exists a positive element k ($\|k\|=1$) in $\mathscr{A}$ with $kbk\geq\frac{1}{2}k^2$.

$$\|(ky_{n_1}k)'-k(a+ib)k\|<\tfrac{1}{8} \quad\text{for some } n_1.$$

Let C_1 be the C^*-subalgebra of $\mathscr{A}$ generated by $k y_{n_1} k$ and 1. Then there is a character φ_1 of C_1 with $\varphi_1\left(\frac{k y_{n_1} k}{4\|x_{n_1}\|}\right) \geq \frac{1}{2}$. Let $\overline{\varphi}_1$ be an extended state of φ_1 to $\mathscr{A}$. Then

$$|\overline{\varphi}_1((k y_{n_1} k)')| = 0 \quad \text{and so } |\overline{\varphi}_1(k(a+ib)k)| < \tfrac{1}{8}.$$

On the other hand,

$$|\overline{\varphi}_1(k(a+ib)k)| \geq |\overline{\varphi}_1(k b k)| \geq \overline{\varphi}_1\left(\frac{1}{2} k^2\right) \geq \frac{1}{2} \overline{\varphi}_1\left(\frac{k y_{n_1} k}{4\|x_{n_1}\|}\right) \geq \frac{1}{4},$$

a contradiction. Hence $a+ib=0$ and we have a contradiction. q.e.d.

4.1.4. Lemma. *Let $\mathscr{A}$ be a C^*-algebra and let δ be a derivation on $\mathscr{A}$. Let $\{\pi, \mathscr{H}\}$ be a *-representation of $\mathscr{A}$. Put $\tilde{\delta}(\pi(a)) = \pi(\delta(a))$ $(a \in \mathscr{A})$. Then $\tilde{\delta}$ is a derivation on $\pi(\mathscr{A})$ and $\tilde{\delta}$ can be extended to a derivation $\tilde{\tilde{\delta}}$ on the weak closure $\overline{\pi(\mathscr{A})}$ of $\pi(\mathscr{A})$, and $\tilde{\tilde{\delta}}$ is σ-continuous on $\overline{\pi(\mathscr{A})}$.*

Proof. Let $\mathscr{A}^{**}$ be the second dual of $\mathscr{A}$, and let δ^{**} be the second dual of δ.

Then one can easily see that δ^{**} is a derivation on $\mathscr{A}^{**}$, since δ^{**} is σ-continuous on $\mathscr{A}^{**}$. Let $\{\pi^w, \mathscr{H}\}$ be the W^*-representation of $\mathscr{A}^{**}$ with $\pi(a) = \pi^w(a)$ $(a \in \mathscr{A})$.

Let $\mathscr{I}$ be the kernel of π^w. Then $\mathscr{I}$ is a σ-closed ideal of $\mathscr{A}^{**}$, and so there exists a central projection z in $\mathscr{A}^{**}$ with $\mathscr{I} = \mathscr{A}^{**} z$.

$$\delta^{**}(z) = \delta^{**}(z z) = \delta^{**}(z) z + z \delta^{**}(z) = 2\delta^{**}(z) z.$$

Hence $\delta^{**}(z) = 0$ and so $\delta^{**}(\mathscr{A}^{**} z) \subset \mathscr{A}^{**} z$.

Therefore δ^{**} defines a derivation on $\mathscr{A}^{**}(1-z)$. Since $\mathscr{A}^{**}(1-z)$ is *-isomorphic to $\pi^w(\mathscr{A}^{**}) = \overline{\pi(\mathscr{A})}$ by the mapping π^w, δ^{**} defines a derivation $\tilde{\tilde{\delta}}$ on $\overline{\pi(\mathscr{A})}$ with $\tilde{\delta} = \tilde{\tilde{\delta}}$ on $\pi(\mathscr{A})$. q.e.d.

4.1.5. Lemma. *Let $\mathscr{M}$ be a countably decomposable type III W^*-algebra and let a be a non-zero element in $\mathscr{M}$, and let $C(a)$ be the uniformly closed convex set generated by $\{u^* a u \mid u \in \mathscr{M}^u\}$ ($\mathscr{M}^u$ is the group of all unitary elements in $\mathscr{M}$). Then, $C(a) \cap Z$ contains a non-zero element (Z is the center of $\mathscr{M}$).*

Proof. $C(a) \cap Z \neq (\emptyset)$ by 2.1.16.

It suffices to assume that $\|a\| \leq 1$, $a^* = a$ and $a^+ \neq 0$ (otherwise consider $-a$). Then there is a non-zero projection e and a positive integer n_0 with $a \geq n_0^{-1} e - (1-e)$. If e majorizes a non-zero central projection z, then $a \geq n_0^{-1} z - (1-z)$. Hence $C(a) \cap Z \geq n_0^{-1} z - (1-z)$ and so $C(a) \cap Z$ contains a non-zero element.

Now suppose that e does not majorize any non-zero central element.

Let $c(e)$ be the central support of e in $\mathcal{M}$. Since

$$\mathcal{M} = \mathcal{M}c(e) \oplus \mathcal{M}(1-c(e)),$$

it suffices to assume that $c(e)=1$. Then $c(1-e)=1$, since e does not majorize any non-zero central projection. Hence $e \sim 1 \sim (1-e)$ by 2.2.14.

There exists a finite family $(e_1, e_2, \dots, e_{n_0+1})$ of mutually orthogonal equivalent projections in $\mathcal{M}$ with $\sum_{i=1}^{n_0+1} e_i = e$ and $e_i \sim e\,(i=1,2,\dots,n_0+1)$.

Put $1-e=e_0$. Then there is a finite family $\{u_0, u_1, \dots, u_{n_0+1}\}$ of unitary elements in $\mathcal{M}$ with $u_i e_j u_i^* = e_{\sigma_i(j)}$, where σ_i is the n_0+2 cyclic permutation of $0, 1, \dots, n_0+1$.

Then $\sum_{i=0}^{n_0+1} u_i e_j u_i^* = 1$ $(j=0, 1, 2, \dots, n_0+1)$.

Set $b=(n_0+2)^{-1} \sum_{i=0}^{n_0+1} u_i a u_i^*$. Then

$$\begin{aligned}(n_0+2)b &\geq \sum_{i=0}^{n_0+1} u_i(n_0^{-1}(e_1+e_2+\cdots+e_{n_0+1})-e_0)u_i^* \\ &= (n_0+1)n_0^{-1}1 - 1 = n_0^{-1}1.\end{aligned}$$

Hence $C(b) \cap Z \geq n_0^{-1}(n_0+2)^{-1}1$, but $C(a) \cap Z \supset C(b) \cap Z$. q.e.d.

4.1.6. Theorem. *Let δ be a derivation on a W^*-algebra $\mathcal{M}$. Then δ is inner—namely there exists an element a in $\mathcal{M}$ such that $\delta(x)=[a,x]\,(x \in \mathcal{M})$. Moreover we can choose such an element a as follows: $\|a\| \leq \|\delta\|$.*

Proof. Let $\mathcal{M}^u$ be the group of all unitary elements in $\mathcal{M}$. For $u \in \mathcal{M}^u$, put $T_u(x)=(ux+\delta(u))u^{-1}$ $(x \in \mathcal{M})$.

Then if $u, v \in \mathcal{M}^u$,

$$\begin{aligned}T_u T_v(x) &= \{u(vx+\delta(v))v^{-1}+\delta(u)\}u^{-1} = \{(uvx+u\delta(v))v^{-1}+\delta(u)\}u^{-1} \\ &= uvxv^{-1}u^{-1}+u\delta(v)v^{-1}u^{-1}+\delta(u)u^{-1} \\ &= (uvx+\delta(uv))(uv)^{-1} = T_{uv}(x).\end{aligned}$$

Hence $T_u T_v = T_{uv}$.

Let Δ be the set of all non-void σ-closed convex sets K in $\mathcal{M}$ satisfying the following conditions: 1. $T_u(K) \subset K$; 2. $\sup_{x \in K} \|x\| \leq \|\delta\|$.

Since $\|T_u(0)\| = \|\delta(u)u^{-1}\| \leq \|\delta\|$, Δ is not empty. Define an order in Δ by set inclusion. Let $(K_\alpha)_{\alpha \in \mathbb{I}}$ be a linearly ordered decreasing subset in Δ. Then $\bigcap_{\alpha \in \mathbb{I}} K_\alpha \in \Delta$, because K_α $(\alpha \in \mathbb{I})$ is compact. Hence there exists a minimal element K_0 in Δ by Zorn's lemma.

If $a,b\in K_0$, then for $u\in\mathscr{M}^u$,

$$u(a-b)u^{-1}=uau^{-1}-ubu^{-1}=uau^{-1}+\delta(u)u^{-1}-(ubu^{-1}+\delta(u)u^{-1})$$
$$=T_u(a)-T_u(b).$$

Hence K_0-K_0 is invariant under the mapping $\Phi^u: x\to uxu^{-1}$ $(x\in\mathscr{M})$.

Suppose now that $\mathscr{M}$ is a countably decomposable finite W^*-algebra; then $\mathscr{M}$ has a faithful normal tracial state τ. Put $\|x\|_2=\tau(x^*x)^{\frac{1}{2}}$. $(x\in\mathscr{M})$. Suppose that K_0-K_0 contains a non-zero element c with

$$c=a-b \qquad (a,b\in K_0).$$

Let $\lambda=\sup_{x\in K_0}\|x\|_2$. Then for an arbitrary positive number ε, there is a u in $\mathscr{M}^u$ with $\left\|T_u\left(\frac{a+b}{2}\right)\right\|_2>\lambda-\varepsilon$, since K_0 is minimal. Since $\|T_u(a)\|_2$, $\|T_u(b)\|_2\le\lambda$,

$$\|\tfrac{1}{2}(T_u(a)-T_u(b))\|_2^2=\tfrac{1}{2}(\|T_u(a)\|^2+\|T_u(b)\|^2)-\|\tfrac{1}{2}(T_u(a)+T_u(b))\|_2^2$$
$$\le\lambda^2-(\lambda-\varepsilon)^2=(2\lambda\varepsilon-\varepsilon^2),$$

since $\frac{1}{2}(T_u(a)+T_u(b))=T_u(\frac{1}{2}(a+b))$.

On the other hand,

$$\|T_u(a)-T_u(b)\|_2=\|u(a-b)u^{-1}\|_2=\|a-b\|_2.$$

Hence $\|a-b\|_2=0$ and so $c=0$, a contradiction. Therefore K_0 consists of a single element a_0.

$T_u(a_0)=ua_0u^{-1}+\delta(u)u^{-1}=a_0$ $(u\in\mathscr{M}^u)$ and so $\delta(u)=[a_0,u]$. Since any element of $\mathscr{M}$ is a finite linear combination of unitary elements in $\mathscr{M}$, we have $\delta(x)=[a_0,x]$ $(x\in\mathscr{M})$. Clearly $\|a_0\|\le\|\delta\|$.

Next suppose that $\mathscr{M}$ is an arbitrary semi-finite algebra. For a countably decomposable finite projection e in $\mathscr{M}$, put $\delta_e(exe)=e\delta(exe)e$ $(x\in\mathscr{M})$; then δ_e is a derivation on $e\mathscr{M}e$.

Hence there is an element a_e in $e\mathscr{M}e$ with $e\delta(exe)e=[a_e,exe]$ $(x\in\mathscr{M})$.

Let $(e_\alpha)_{\alpha\in\Pi}$ be an increasing directed set of countably decomposable finite projections in $\mathscr{M}$ with $\underset{\alpha}{\text{l.u.b.}}\, e_\alpha=1$. Then there is an a_{e_α} in $e_\alpha\mathscr{M}e_\alpha$ with $\delta_{e_\alpha}(e_\alpha xe_\alpha)=[a_{e_\alpha},e_\alpha xe_\alpha]$ $(x\in\mathscr{M})$, and $\|a_{e_\alpha}\|\le\|\delta\|$.

Let a_0 be an accumulation point of $(a_{e_\alpha}|\alpha\in\Pi)$ in the σ-topology.

If $\beta\le\alpha$, $\delta_{e_\alpha}(e_\beta xe_\beta)=e_\alpha\delta(e_\beta xe_\beta)e_\alpha=[a_{e_\alpha},e_\beta xe_\beta]$. Hence

$$\delta(e_\beta xe_\beta)=[a_0,e_\beta xe_\beta].$$

Since δ is σ-continuous by 4.1.4, $\delta(x)=[a_0,x]$ $(x\in\mathscr{M})$ and $\|a_0\|\le\|\delta\|$.

Next suppose $\mathcal{M}$ is a countable decomposable type III W^*-algebra. If $K_0 - K_0$ contains a non-zero element, then $(K_0 - K_0) \cap Z$ contains a non-zero element c by 4.1.5. Let $a - b = c$ with $a, b \in K_0$. For $f \in \mathcal{M}_*$, put $\lambda = \sup_{x \in K_0} |f(x)|$. Then for an arbitrary positive number ε, there exists a u in $\mathcal{M}^u$ such that $\left| f\left(T_u\left(\frac{a+b}{2} \right) \right) \right| > \lambda - \varepsilon$ and $|f(T_u(a))| \leq \lambda$, $|f(T_u(b))| \leq \lambda$.

Since $|f(T_u(a) - T_u(b))| = |f(u(a-b)u^{-1})| = |f(a-b)|$, $|f(a-b)|$ can be arbitrarily small. Hence $a = b$.

Therefore K_0 consists of a single point a_0, and so $\delta(x) = [a_0, x]$ $(x \in \mathcal{M})$ and $\|a_0\| \leq \|\delta\|$.

To move from a countably decomposable type III-algebra to an arbitrary type III algebra, we need only to do a similar discussion to the semi-finite case.

Finally let $\mathcal{M}$ be an arbitrary W^*-algebra. Then $\mathcal{M} = \mathcal{M}_1 \oplus \mathcal{M}_2$ with a semi-finite algebra $\mathcal{M}_1$ and a type III-algebra $\mathcal{M}_2$. Therefore we can reduce the proof to $\mathcal{M}_1$ and $\mathcal{M}_2$. q.e.d.

4.1.7. Corollary. *Let $\mathcal{A}$ be a C^*-algebra on a Hilbert space $\mathcal{H}$ and let δ be a derivation on $\mathcal{A}$. Then there exists an element a in the weak closure $\bar{\mathcal{A}}$ of $\mathcal{A}$ on $\mathcal{H}$ with $\delta(x) = [a, x]$ $(x \in \mathcal{A})$.*

Proof. By 4.1.4, δ can be extended to a derivation $\tilde{\delta}$ on $\bar{\mathcal{A}}$. Then $\tilde{\delta}$ is inner by 4.1.6. q.e.d.

A derivation on a C^*-algebra is, in general, not inner.

4.1.8. Example. *Let $C(\mathcal{H})$ be the C^*-algebra of all compact operators on a separable infinite dimensional Hilbert space $\mathcal{H}$. For $a \in B(\mathcal{H})$, put $\delta_a(x) = [a, x]$ $(x \in C(\mathcal{H}))$. Then δ_a is a derivation on $C(\mathcal{H})$. But it is not inner if a does not belong to $C(\mathcal{H}) + \mathbb{C}1_{\mathcal{H}}$.*

Remark. $C(\mathcal{H})$ is simple, but it does not have an identity. In the following, we shall show that every derivation of a simple C^*-algebra with identity is inner.

Let $\mathcal{A}$ be a simple C^*-algebra with identity, and let δ be a derivation on $\mathcal{A}$. Let $\{\pi, \mathcal{H}\}$ be a $*$-representation of $\mathcal{A}$ on a Hilbert space $\mathcal{H}$, $\overline{\pi(\mathcal{A})}$ the weak closure of $\pi(\mathcal{A})$ on $\mathcal{H}$.

Then there exists an element a in $\overline{\pi(\mathcal{A})}$ with $\pi(\delta(x)) = [a, \pi(x)]$ $(x \in \mathcal{A})$.

Put $\delta^*(x) = \pi^{-1}([a^*, \pi(x)])$ $(x \in \mathcal{A})$. Then δ^* is also a derivation on $\mathcal{A}$. Hence it suffices to assume that a is self-adjoint.

By considering $\|a\| \cdot 1 + a$, we may assume that a is positive. Let $\mathcal{B}$ be the C^*-subalgebra of $\overline{\pi(\mathcal{A})}$ generated by $\pi(\mathcal{A})$ and a. Take a maximal ideal $\mathcal{I}$ of $\mathcal{B}$ and consider the quotient algebra $\mathcal{D} = \mathcal{B}/\mathcal{I}$; then $\mathcal{D}$ is a simple C^*-algebra, and $\pi(\mathcal{A}) \cap \mathcal{I} = (0)$, since $\mathcal{A}$ is simple and has an identity.

Hence the image of $\pi(\mathscr{A})$ in $\mathscr{D}$ is *-isomorphic to $\pi(\mathscr{A})$ under the canonical mapping.

Let $\tilde{y}$ be the images of elements y of $\mathscr{B}$ in $\mathscr{D}$. Then

$$\widetilde{\pi(\delta(x))} = \widetilde{[a, \pi(x)]} = [\tilde{a}, \widetilde{\pi(x)}] \qquad (x \in \mathscr{A}).$$

By identifying $\mathscr{A}$ with the image of $\pi(\mathscr{A})$ in $\mathscr{D}$, we have the following situation: there exists a simple C^*-algebra $\mathscr{D}$, containing $\mathscr{A}$, such that $\mathscr{D}$ is generated by $\mathscr{A}$ and a positive element d, and $\delta(x) = [d, x]$ $(x \in \mathscr{A})$.

In the following reasoning, we shall show that $\mathscr{A} = \mathscr{D}$ and so δ is inner.

Suppose that $\mathscr{A} \subsetneqq \mathscr{D}$ and let S be the set of all self-adjoint linear functionals f on $\mathscr{D}$ with $f(\mathscr{A}) = 0$ and $\|f\| \leq 1$. S is a $\sigma(\mathscr{D}^*, \mathscr{D})$-compact convex set. Let g be an extreme point in S and let $g = g_1 - g_2$ be the orthogonal decomposition of g with $g_1, g_2 \geq 0$, $\|g\| = \|g_1\| + \|g_2\|$.

Put $\xi = g_1 + g_2$ and let $\{\pi_\xi, \mathscr{H}_\xi\}$ be the *-representation of $\mathscr{D}$ constructed via ξ.

4.1.9. Lemma. *Let $\overline{\pi_\xi(\mathscr{A})}$ be the weak closure of $\pi_\xi(\mathscr{A})$ on $\mathscr{H}_\xi$; then it is a factor.*

Proof. Let $\xi(y) = (\pi_\xi(y) 1_\xi, 1_\xi)$ $(y \in \mathscr{D})$. There are two elements η_1, η_2 in $\mathscr{H}_\xi$ with $g_i(y) = (\pi_\xi(y)\eta_i, \eta_i)$ for $y \in \mathscr{D}$ $(i = 1, 2)$. Since $g_1 = g_2$ on $\mathscr{A}$, a mapping $\pi_\xi(x)\eta_1 \to \pi_\xi(x)\eta_2$ $(x \in \mathscr{A})$ defines a partial isometry u' on $\mathscr{H}_\xi$ with $u' \in \pi_\xi(\mathscr{A})'$.

Suppose that $\overline{\pi_\xi(\mathscr{A})}$ contains a non-trivial central projection z. Set $h_i(y) = (\pi_\xi(y) z \eta_i, \eta_i)$ $(i = 1, 2)$ and $k_i(y) = (\pi_\xi(y)(1_{\mathscr{H}} - z)\eta_i, \eta_i)$ $(i = 1, 2)$ $(y \in \mathscr{D})$. Then

$$\begin{aligned} h_2(x) &= (\pi_\xi(x) z \eta_2, \eta_2) = (\pi_\xi(x) z u' \eta_1, u' \eta_1) \\ &= (\pi_\xi(x) z \eta_1, \eta_1) = h_1(x) \quad (x \in \mathscr{A}). \end{aligned}$$

Analogously $k_2 = k_1$ on $\mathscr{A}$. On the other hand, there is a positive element b in $\overline{\pi_\xi(\mathscr{A})}$ with $\pi_\xi(\delta(x)) = [b, \pi_\xi(x)]$ $(x \in \mathscr{A})$, and so

$$[b, \pi_\xi(x)] = \pi_\xi([d, x]) = [\pi_\xi(d), \pi_\xi(x)].$$

Hence $b - \pi_\xi(d) \in \pi_\xi(\mathscr{A})'$. This implies that z commutes with $\pi_\xi(d)$ and so z belongs to the center of $\overline{\pi_\xi(\mathscr{D})}$.

Hence $h_i, k_i \geq 0$ $(i = 1, 2)$, and

$$\begin{aligned} 1 = \|g\| &= \|h_1 - h_2 + k_1 - k_2\| \leq \|h_1 - h_2\| + \|k_1 - k_2\| \\ &\leq \|h_1\| + \|h_2\| + \|k_1\| + \|k_2\| \\ &= \|\eta_1\|^2 + \|\eta_2\|^2 = \|g_1\| + \|g_2\| = 1. \end{aligned}$$

Since z is a non-trivial central projection in $\overline{\pi_\xi(\mathscr{D})}$ and

$$[\pi_\xi(\mathscr{D}) 1_\xi] = \mathscr{H}_\xi, \quad 0 < \|h_1 - h_2\| < 1.$$

Hence $g=\|h_1-h_2\|\cdot\frac{h_1-h_2}{\|h_1-h_2\|}+\|k_1-k_2\|\frac{k_1-k_2}{\|k_1-k_2\|}$ and so $\frac{h_1-h_2}{\|h_1-h_2\|}=\frac{k_1-k_2}{\|k_1-k_2\|}=g_1-g_2$. But $s(h_1+h_2)\cdot s(k_1+k_2)=0$, a contradiction. q.e.d.

4.1.10. Lemma. *There exist two states φ_1, φ_2 on $\mathcal{D}$ such that $\varphi_1=\varphi_2$ on $\mathcal{A}$, $\varphi_1(d)\neq\varphi_2(d)$ and, moreover, the *-representation $\{\pi_{\varphi_i},\mathcal{H}_{\varphi_i}\}$ of $\mathcal{D}$ is factorial* $(i=1,2)$.

Proof. Take the *-representation $\{\pi_\xi,\mathcal{H}_\xi\}$ of $\mathcal{D}$ in the above lemma. Then $\overline{\pi_\xi(\mathcal{A})}$ is a factor. $\pi_\xi(d)=b+(\pi_\xi(d)-b)$ with $b\in\overline{\pi_\xi(\mathcal{A})}$ and $\pi_\xi(d)-b\in\pi_\xi(\mathcal{A})'$. Let C be the commutative C^*-algebra generated by $\pi_\xi(d)-b$ and $1_{\mathcal{H}_\xi}$, and let R be the C^*-algebra generated by $\overline{\pi_\xi(\mathcal{A})}$ and C. Then R can be canonically identified with the C^*-tensor product $\overline{\pi_\xi(\mathcal{A})}\otimes C$.

In fact, let R_0 be the *-subalgebra generated algebraically by $\overline{\pi_\xi(\mathcal{A})}$ and C, and suppose that $\sum_{i=1}^n x_i c_i=0$ $(x_i\in\overline{\pi_\xi(\mathcal{A})},\ c_i\in C\ (i=1,2,\ldots,n))$. It suffices to assume that $(x_i\,|\,i=1,2,\ldots,n)$ are linearly independent.

Since $\overline{\pi_\xi(\mathcal{A})}$ is a factor and $C\subset\pi_\xi(\mathcal{A})'$, there exists a family of complex numbers $(\lambda_{ij})_{i,j=1,2,\ldots,n}$ with $\sum_{k=1}^n x_k\lambda_{kj}=0$ $(j=1,2,\ldots,n)$ and $\sum_{k=1}^n \lambda_{ik}c_k=c_i$ $(i=1,2,\ldots,n)$ by 1.20.5. Hence $\lambda_{kj}=0$ $(k,j=1,2,\ldots,n)$ and so $c_i=0$ $(i=1,2,\ldots,n)$. Hence the canonical mapping Φ of R_0 onto the algebraic tensor product $\overline{\pi_\xi(\mathcal{A})}\odot C$ with $\Phi(xc)=x\otimes c$ $(x\in\overline{\pi_\xi(\mathcal{A})},\ c\in C)$ is a *-isomorphism, and so by 1.22.5, Φ can be uniquely extended to a *-isomorphism of R onto $\overline{\pi_\xi(\mathcal{A})}\otimes C$.

Now let ξ_1 be a state on $\overline{\pi_\xi(\mathcal{A})}$ with $\xi_1(w)=(w1_\xi,1_\xi)$ $(w\in\overline{\pi_\xi(\mathcal{A})})$, and let χ_1, χ_2 be two different characters on C with

$$\chi_1(\pi_\xi(d)-b)\neq\chi_2(\pi_\xi(d)-b).$$

Then $\xi_1\otimes\chi_1$ and $\xi_1\otimes\chi_2$ are two different states on R.

Put $\varphi_i(y)=\xi_1\otimes\chi_i(\pi_\xi(y))$ $(y\in\mathcal{D})$ $(i=1,2)$. Then

$$\varphi_i(d)=\xi_1\otimes\chi_i(\pi_\xi(d))=\xi_1\otimes\chi_i(b+\pi_\xi(d)-b)=\xi_1(b)+\chi_i(\pi_\xi(d)-b).$$

Hence $\varphi_1(d)\neq\varphi_2(d)$.

We shall show next that $\{\pi_{\varphi_1},\mathcal{H}_{\varphi_1}\}$ is factorial. The *-representation $\{\pi_{\varphi_1},\mathcal{H}_{\varphi_1}\}$ of $\mathcal{D}$ can be canonically considered as the *-representation of $\pi_\xi(\mathcal{D})$, denoted by $\{\tilde{\pi}_{\varphi_1},\mathcal{H}_{\varphi_1}\}$. Consider the *-representation

$$\{\pi_{\xi_1\otimes\chi_1},\mathcal{H}_{\xi_1\otimes\chi_1}\}$$

of R. Then $\pi_{\xi_1 \otimes \chi_1}(R) = \overline{\pi_\xi(\mathscr{A})} E \otimes 1$, where E is the orthogonal projection of $\mathscr{H}_\xi$ onto $[\pi_\xi(\mathscr{A}) 1_\xi]$ and 1 is the identity operator on a one-dimensional space.

On the other hand, $\pi_\xi(\mathscr{A}) \subset \pi_\xi(\mathscr{D}) \subset R$. Hence

$$\overline{\pi_{\xi_1 \otimes \chi_1}(\pi_\xi(\mathscr{A}))} = \overline{\pi_\xi(\mathscr{A})} E \otimes 1 \subset \overline{\pi_{\xi_1 \otimes \chi_1}(\pi_\xi(\mathscr{D}))} \subset \overline{\pi_{\xi_1 \otimes \chi_1}(R)} = \overline{\pi_\xi(\mathscr{A})} E \otimes 1 .$$

The representation $\{\tilde{\pi}_{\varphi_1}, \mathscr{H}_{\varphi_1}\}$ of $\pi_\xi(\mathscr{D})$ is considered as a restriction of the representation $\{\pi_{\xi_1 \otimes \chi_1}, \mathscr{H}_{\xi_1 \otimes \chi_1}\}$ of $\pi_\xi(\mathscr{D})$ to some invariant subspace. Hence $\{\tilde{\pi}_{\varphi_1}, \mathscr{H}_{\varphi_1}\}$ and so $\{\pi_{\varphi_1}, \mathscr{H}_{\varphi_1}\}$ are factorial. Analogously, one can show that $\{\pi_{\varphi_2}, \mathscr{H}_{\varphi_2}\}$ is factorial. q.e.d.

4.1.11. Theorem. *Every derivation of a simple C^*-algebra with identity is inner.*

Proof. It is enough to show that $\mathscr{D} = \mathscr{A}$. Suppose $\mathscr{D} \supsetneqq \mathscr{A}$; then we shall take two states φ_1, φ_2 on $\mathscr{D}$ as in the above lemma.

Let d_i be a positive element in $\overline{\pi_{\varphi_i}(\mathscr{A})}$ with

$$\pi_{\varphi_i}(\delta(x)) = [d_i, \pi_{\varphi_i}(x)] \qquad (x \in \mathscr{A}) \qquad (i = 1, 2).$$

Then $[d_i, \pi_{\varphi_i}(x)] = [\pi_{\varphi_i}(d), \pi_{\varphi_i}(x)]$. Hence $\pi_{\varphi_i}(d) - d_i \in \pi_{\varphi_i}(\mathscr{A})'$.

On the other hand, $\overline{\pi_{\varphi_i}(\mathscr{A})}$ and $\pi_{\varphi_i}(d) - d_i$ generate $\overline{\pi_{\varphi_i}(\mathscr{D})}$. Hence $\pi_{\varphi_i}(d) - d_i$ belongs to the center of $\overline{\pi_{\varphi_i}(\mathscr{D})}$ and so $\pi_{\varphi_i}(d) - d_i = \lambda_i 1_{\mathscr{H}_{\varphi_i}}$, where λ_i is a real number. Hence $\overline{\pi_{\varphi_i}(\mathscr{A})} = \overline{\pi_{\varphi_i}(\mathscr{D})}$. Since $\varphi_1 = \varphi_2$ on $\mathscr{A}$, there exists a *-isomorphism Φ of $\overline{\pi_{\varphi_1}(\mathscr{D})}$ onto $\overline{\pi_{\varphi_2}(\mathscr{D})}$ with $\Phi(\pi_{\varphi_1}(x)) = \pi_{\varphi_2}(x)$ and $(\Phi(\pi_{\varphi_1}(x)) 1_{\varphi_2}, 1_{\varphi_2}) = (\pi_{\varphi_1}(x) 1_{\varphi_1}, 1_{\varphi_1})$ $(x \in \mathscr{A})$.

Put $\Phi(\pi_{\varphi_1}(d)) = p$. Then

$$\begin{aligned}[p, \pi_{\varphi_2}(x)] &= [\Phi(\pi_{\varphi_1}(d)), \pi_{\varphi_2}(x)] = [\Phi(\pi_{\varphi_1}(d)), \Phi(\pi_{\varphi_1}(x))] = \Phi([\pi_{\varphi_1}(d), \pi_{\varphi_1}(x)]) \\ &= \pi_{\varphi_2}([d, x]) = [\pi_{\varphi_2}(d), \pi_{\varphi_2}(x)] \qquad (x \in \mathscr{A}).\end{aligned}$$

Hence $\Phi(\pi_{\varphi_1}(d)) - \pi_{\varphi_2}(d) \in \pi_{\varphi_2}(\mathscr{A})'$ and so $\Phi(\pi_{\varphi_1}(d)) = \pi_{\varphi_2}(d) + \lambda 1_{\mathscr{H}_{\varphi_2}}$, since $\overline{\pi_{\varphi_2}(\mathscr{A})} = \overline{\pi_{\varphi_2}(\mathscr{D})}$ (λ is a real number).

Suppose $\lambda > 0$; then $\|\Phi(\pi_{\varphi_1}(d))\| > \|\pi_{\varphi_2}(d)\|$. But $\|\pi_{\varphi_2}(d)\| = \|d\|$, since $\mathscr{D}$ is simple, a contradiction. Analogously if $\lambda < 0$, then $\|\pi_{\varphi_2}(d)\| > \|\Phi(\pi_{\varphi_1}(d))\| = \|d\|$, a contradiction. Hence $\lambda = 0$—namely $\Phi(\pi_{\varphi_1}(d)) = \pi_{\varphi_2}(d)$. Thus $\varphi_2(d) = (\pi_{\varphi_2}(d) 1_{\varphi_2}, 1_{\varphi_2}) = (\Phi(\pi_{\varphi_1}(d)) 1_{\varphi_2}, 1_{\varphi_2})$.

Take a directed set $\{\pi_{\varphi_1}(x_\alpha)\}$ $(x_\alpha \in \mathscr{A})$ with $\pi_{\varphi_1}(x_\alpha) \to \pi_{\varphi_1}(d)$ in the σ-topology. Then

$$\begin{aligned}(\Phi(\pi_{\varphi_1}(d)) 1_{\varphi_2}, 1_{\varphi_2}) &= \lim_\alpha (\Phi(\pi_{\varphi_1}(x_\alpha)) 1_{\varphi_2}, 1_{\varphi_2}) = \lim (\pi_{\varphi_1}(x_\alpha) 1_{\varphi_1}, 1_{\varphi_1}) \\ &= (\pi_{\varphi_1}(d) 1_{\varphi_1}, 1_{\varphi_1}) = \varphi_1(d).\end{aligned}$$

This is a contradiction. q.e.d.

Now let δ be a derivation on a C^*-algebra $\mathscr{A}$. Then δ is norm-continuous on $\mathscr{A}$, and so δ belongs to $B(\mathscr{A})$ ($B(\mathscr{A})$ is the algebra of all bounded linear operators on $\mathscr{A}$). Put $\Phi(t)=\exp t\delta$ $(-\infty<t<\infty)$. Then $\Phi(t)$ is a one-parameter subgroup of automorphisms on $\mathscr{A}$. Furthermore, $t\to\Phi(t)\in B(\mathscr{A})$ is norm-continuous.

Conversely suppose that $t\to\Phi(t)\in B(\mathscr{A})$ $(-\infty<t<\infty)$ is a norm-continuous one-parameter subgroup of automorphisms on $\mathscr{A}$. Then by the general theory of operators on Banach spaces ([75]) there exists a bounded linear operator δ on $\mathscr{A}$ with $\Phi(t)=\exp t\delta$. One can easily see that δ is a derivation on $\mathscr{A}$.

4.1.12. Lemma. *Let $\mathscr{A}$, $\mathscr{B}$ be two C^*-algebras and let Φ be a homomorphism of $\mathscr{A}$ onto $\mathscr{B}$. Then Φ is uniformly continuous.*

Proof. Suppose first that Φ is an isomorphism. Put

$$\|x\|_1=\|\Phi(x)\| \qquad (x\in\mathscr{A}).$$

By 1.2.4, $\|x^*x\|\leq\|x^*x\|_1\leq\|x^*\|_1\,\|x\|_1$ $(x\in\mathscr{A})$. Now we shall show that the $*$-operation on $\mathscr{A}$ is $\|\cdot\|_1$-continuous.

If $\|x_n\|_1\to 0$ and $\|x_n^*-y\|_1\to 0$, then $\|x_n^*x_n\|\leq\|x_n^*\|_1\,\|x_n\|_1\to 0$. Hence $\|x_n\|\to 0$.

On the other hand,

$$\|(x_n-y^*)(x_n^*-y)\|\leq\|x_n-y^*\|_1\,\|x_n^*-y\|_1\to 0.$$

Hence $\|x_n^*-y\|\to 0$ and so $y=0$.

By the closed graph theorem, the $*$-operation is $\|\cdot\|_1$-continuous. If $\|x_n\|_1\to 0$ and $\|x_n-z\|_1\to 0$, then $\|x_n-z\|\to 0$ and so $z=0$. This implies that Φ is $\|\cdot\|_1$-continuous.

Suppose next that Φ is a homomorphism of $\mathscr{A}$ onto $\mathscr{B}$. It suffices to assume that $\mathscr{A}$ and $\mathscr{B}$ have identities. Let $\Phi^{-1}(0)$ be the kernel of Φ. Then it is an ideal. Let $\{\mathscr{L}_\alpha\}_{\alpha\in\mathbb{I}}$ be the set of all maximal left ideals in $\mathscr{B}$. Then $\bigcap_{\alpha\in\mathbb{I}}\mathscr{L}_\alpha=(0)$. Since $\Phi^{-1}(\mathscr{L}_\alpha)$ is a maximal left ideal in $\mathscr{A}$, it is closed in $\mathscr{A}$, and $\bigcap_{\alpha\in\mathbb{I}}\Phi^{-1}(\mathscr{L}_\alpha)=\Phi^{-1}(0)$. Hence $\Phi^{-1}(0)$ is closed.

Consider the C^*-algebra $\mathscr{A}/\Phi^{-1}(0)$; then Φ induces an isomorphism of $\mathscr{A}/\Phi^{-1}(0)$ onto $\mathscr{B}$. Hence by the preceding result, Φ is continuous. q.e.d.

Let $A(\mathscr{A})$ be the group of all automorphisms on $\mathscr{A}$. Then by the above lemma, $A(\mathscr{A})\subset B(\mathscr{A})$.

4.1.13. Proposition. *$A(\mathscr{A})$ is a topological group in the uniform topology (i.e. the norm topology in $B(\mathscr{A})$), and it is closed in $B(\mathscr{A})$.*

Proof. We shall show that $A(\mathscr{A})$ is closed. The remainder of the proof is almost clear.

If $\|\Phi_n - T\| \to 0$ $(n \to \infty)$ with $\Phi_n \in A(\mathscr{A})$ and $T \in B(\mathscr{A})$, then one can easily see that T is a homomorphism of $\mathscr{A}$ into $\mathscr{A}$, and so $T^{-1}(0)$ is a closed ideal of $\mathscr{A}$.

If $h \in T^{-1}(0)$ with $h^* = h$, then $\|T(h)\| = \lim_{n\to\infty} \|\Phi_n(h)\| \geq \|h\|$. Hence $h = 0$ and so T is an isomorphism. Put $\|x\|_1 = \|T(x)\|$ $(x \in \mathscr{A})$. Then $\|x\|_1 \leq \|T\|\,\|x\|$ and

$$\|x^*\|\,\|x\| = \|x^*x\| \leq \|x^*x\|_1 \leq \|x^*\|_1\,\|x\|_1 \leq \|T\|\,\|x^*\|\,\|x\|_1 .$$

Hence $\|x\| \leq \|T\|\,\|x\|_1$ and so $\dfrac{\|x\|}{\|T\|} \leq \|x\|_1 \leq \|T\|\,\|x\|$. Therefore the image $T(\mathscr{A})$ is a closed set in $\mathscr{A}$.

Suppose $T(\mathscr{A}) \subsetneqq \mathscr{A}$; then there exists an element $f \in \mathscr{A}^*$ with $\|f\| = 1$ and $f(T(\mathscr{A})) = 0$. Since $\|\Phi_n^* - T^*\| = \|\Phi_n - T\| \to 0$ $(n \to \infty)$, $\|\Phi_n^*(f) - T^*(f)\| = \|\Phi_n^*(f)\| \to 0$ $(n \to \infty)$. On the other hand, Φ_n is onto, so that $\{\Phi_n(x) \mid \|\Phi_n(x)\| \leq 1, x \in \mathscr{A}\}$ is the unit sphere S of $\mathscr{A}$.

Put $S_n = \{x \mid \|\Phi_n(x)\| \leq 1, x \in \mathscr{A}\}$ and $\|x\|_n = \|\Phi_n(x)\|$ $(x \in \mathscr{A})$. By a similar reasoning with the previous one, we have

$$\frac{1}{\|\Phi_n\|}\,\|x\| \leq \|x\|_n \leq \|\Phi_n\|\,\|x\|$$

and so $\dfrac{1}{\|\Phi_n\|} S_n \subset S \subset \|\Phi_n\|\,S_n$. Hence $\dfrac{1}{\|\Phi_n\|} \Phi_n(S_n) \subset \Phi_n(S)$ and so $\dfrac{1}{\|\Phi_n\|} S \subset \Phi_n(S)$. Therefore $\|\Phi_n^*(f)\| = \sup_{\|x\| \leq 1} |f(\Phi_n(x))| \geq \dfrac{\|f\|}{\|\Phi_n\|}$, a contradiction. q.e.d.

4.1.14. Corollary. *Let $\mathscr{A}$ be a C^*-algebra on a Hilbert space $\mathscr{H}$ such that its weak closure $\bar{\mathscr{A}}$ contains $1_{\mathscr{H}}$, and let $t \to \Phi(t)$ $(-\infty < t < \infty)$ be a uniformly continuous one-parameter subgroup of automorphisms on $\mathscr{A}$. Then there exists an element b in $\bar{\mathscr{A}}$ such that $\Phi(t)(x) = (\exp tb)x(\exp - tb)$ for $x \in \mathscr{A}$ $(-\infty < t < \infty)$. Furthermore if $\Phi(t)$ is a *-automorphism for all t, b is skew-symmetric and so $\exp tb$ is unitary $(-\infty < t < \infty)$.*

In particular, if $\mathscr{A}$ is a W^*-algebra or a simple C^*-algebra with identity, b belongs to $\mathscr{A}$.

Proof. Let $\delta = \lim_{t\to 0} \dfrac{\Phi(t) - 1_{\mathscr{A}}}{t}$ in the uniform topology ($1_{\mathscr{A}}$ is the identity operator on $\mathscr{A}$). Then δ is a derivation on $\mathscr{A}$. Hence there exists an element b in $\bar{\mathscr{A}}$ with $\delta(x) = [b, x]$ $(x \in \mathscr{A})$. One can easily see that $(\exp t\delta)(x) = (\exp tb)x(\exp - tb)$ for $x \in \mathscr{A}$ $(-\infty < t < \infty)$.

If $\Phi(t)$ is a *-automorphism $(-\infty < t < \infty)$, then $\delta(a^*) = \delta(a)^*$ and so $b^* = -b$—i.e. it is skew-symmetric. q.e.d.

In mathematical physics, recently it has become usual to assume that the set of observables is the set of all self-adjoint elements in a C^*-algebra $\mathscr{A}$. The symmetries of the physical system under the consideration are expressed in terms of a representation of the physical symmetry group G by *-automorphisms of $\mathscr{A}$. In general, G is a Lie group. For example, in the case of the Haag-Araki description of relativistically invariant local quantum fields in terms of W^*-algebras of bounded local observables [3], the dynamics and relativistic invariance are expressed in terms of a strong-operator continuous unitary representation $g \to u_g$ of the inhomogeneous Lorentz group satisfying certain conditions.

The u_gs induce automorphisms of $\mathscr{A}$, the C^*-algebra of bounded global observables by $x \to u_g x u_g^{-1}$. The infinitesimal generators of the translation part of G correspond to the energy and momenta of the field.

Assuming the positive energy condition and by using 4.1.6 and the analytic function theory of several complex variables, Borchers [13] proved that the corresponding automorphism subgroup is inner—namely, observability of energy and momenta.

In the following we shall show a simple form of Borchers' result.

4.1.15. Theorem. *Let $\mathscr{M}$ be a W^*-algebra, with $1_{\mathscr{H}}$, on a Hilbert space $\mathscr{H}$, and let $t \to u(t)$ $(-\infty < t < \infty)$ be a strong-operator continuous one-parameter subgroup of unitary operators on $\mathscr{H}$. Suppose that*

$$\Phi(t): x \to u(t) x u(t)^{-1} \qquad (x \in \mathscr{M})$$

*defines a *-automorphism on $\mathscr{M}$. Let $u(t) = \int_{-\infty}^{\infty} e^{it\lambda} dE_\lambda$ be the Stone-representation of the one-parameter group $\{u(t)\}$, and let $h = \int_{-\infty}^{\infty} \lambda dE_\lambda$ be the infinitesimal generator of $u(t)$.*

If h is lower bounded, then there exists a strong operator-continuous one parameter subgroup $v(t)$ of unitary elements in $\mathscr{M}$ such that $u(t) x u(t)^{-1} = v(t) x v(t)^{-1}$ $(x \in \mathscr{M},\ t \in (-\infty, \infty))$.

To prove the theorem, we shall provide some considerations. Since $\exp it(h + \lambda 1) = \exp ith \exp it\lambda 1$ for a real number λ, it suffices to assume that $h \geq 0$. Put $Q_\mu = \int_0^\mu dE_\lambda$ $(\mu > 0)$ and $e_\mu = P_{[\mathscr{M}' Q_\mu \mathscr{H}]}$. Then $e_\mu \in \mathscr{M}$. Since $Q_\mu \mathscr{H}$ is invariant under $u(t)$ and

$$u(t) \mathscr{M}' Q_\mu \mathscr{H} = u(t) \mathscr{M}' u(t)^{-1} u(t) Q_\mu \mathscr{H} \subset \mathscr{M}' Q_\mu \mathscr{H},$$

e_μ commutes with $u(t)$ $(t \in (-\infty, \infty))$.

Hence a mapping: $e_\mu x e_\mu \to u(t) e_\mu x e_\mu u(t)^{-1}$ $(x \in \mathscr{M})$ is a *-automorphism on $e_\mu \mathscr{M} e_\mu$.

4.1.16. Lemma. $\|u(t)e_\mu x e_\mu u(t)^{-1} - e_\mu x e_\mu\| \leq 2|t| \, \|x\| \, e^{2\mu}$ $(x \in \mathcal{M})$.

Proof. For $x \in \mathcal{M}$, $x'_j, y'_k \in \mathcal{M}'$ and $\xi_j, \eta_k \in \mathcal{H}$ $(j=1,2,\ldots,p; k=1,2,\ldots,q)$, put

$$f(t) = \left(\sum_{j=1}^{p} x'_j u(t) x u(t)^{-1} Q_\mu \xi_j, \ \sum_{k=1}^{q} y'_k Q_\mu \eta_k \right).$$

Then $|f(t)| \leq \|x\| \left\| \sum_{j=1}^{p} x'_j Q_\mu \xi_j \right\| \left\| \sum_{k=1}^{q} y'_k Q_\mu \eta_k \right\|$ (i.e. f is bounded).

On the other hand, one can easily see that $f(t)$ is the boundary value of an analytic function holomorphic in $\operatorname{Im} z > 0$ as follows:

$$f(t+is) = \left(\sum_{j=1}^{p} x'_j u(t+is) x u(t+is)^{-1} Q_\mu \xi_j, \ \sum_{k=1}^{q} y'_j Q_\mu \eta_k \right)$$

with $z = t+is$ and $u(t+is) = \int_0^\infty e^{i\lambda(t+is)} dE_\lambda$. Moreover $|f(t+is)| \leq K e^{2\mu s}$ (K is a fixed number; $s>0$).

In the same way we have:

$$f(t) = \left(\sum_{j=1}^{p} x'_j Q_\mu \xi_j, \ \sum_{k=1}^{q} y'_k u(t) x^* u(t)^{-1} Q_\mu \eta_k \right)$$

is the boundary value of an analytic function holomorphic in $\operatorname{Im} z < 0$ and $|f(t+is)| \leq K' e^{2\mu|s|}$ (K' is a fixed number; $s<0$). Since both functions have the same boundary value, $f(z)$ is an entire function bounded by $|f(z)| \leq \max(K, K') e^{2\mu|\operatorname{Im} z|}$ by the "edge of the wedge" theorem (cf. [193]). Hence by the phragmen-Lindelöf theorem,

$$|f(z)| \leq \|x\| \left\| \sum_{j=1}^{p} x'_j Q_\mu \xi_j \right\| \left\| \sum_{k=1}^{q} y'_k Q_\mu \eta_k \right\| e^{2\mu|\operatorname{Im} z|}.$$

Finally by using Schwartz's lemma,

$$|(u(t)e_\mu x e_\mu u(t)^{-1} \xi, \eta) - (e_\mu x e_\mu \xi, \eta)| \leq 2|t| \, \|x\| \, \|\xi\| \, \|\eta\| \, e^{2\mu}.$$

q.e.d.

Proof of 4.1.15. Set $\rho_\mu(t)(e_\mu x e_\mu) = u(t) e_\mu x e_\mu u(t)^{-1}$ $(x \in \mathcal{M})$. Then $t \to \rho_\mu(t)$ is uniformly continuous. Hence there exists a derivation δ_μ on $e_\mu \mathcal{M} e_\mu$ with $\rho_\mu(t) = \exp(t\delta_\mu)$.

By 4.1.6, there exists a self-adjoint element h_μ in $e_\mu \mathcal{M} e_\mu$ with $\delta_\mu(e_\mu x e_\mu) = i[h_\mu, e_\mu x e_\mu]$ $(x \in \mathcal{M})$.

Put $v_\mu(t) = \exp ith_\mu$. Then $u(t)e_\mu x e_\mu u(t)^{-1} = v_\mu(t) e_\mu x e_\mu v_\mu(t)^{-1}$ $(x \in \mathcal{M})$. Hence $v_\mu(t)^{-1} u(t) e_\mu \in (e_\mu \mathcal{M} e_\mu)' = \mathcal{M}' e_\mu$. Since $t \to u(t)$ and $t \to v_\mu(t)$ are strongly continuous, $t \to v_\mu(t)^{-1} u(t)$ is again strongly continuous.

Let $c(e_\mu)$ be the central support of e_μ. Then a mapping

$$\Phi: x' e_\mu \to x' c(e_\mu) \qquad (x' \in \mathcal{M}')$$

is a *-isomorphism. Hence there exists a strong operator continuous one parameter subgroup $\{\overline{w}_\mu(t)\}$ of unitary elements in $\mathcal{M}' c(e_\mu)$ with

$$\overline{w}_\mu(t) e_\mu = v_\mu(t)^{-1} u(t) e_\mu.$$

Since $u(t) e_\mu u(t)^{-1} = e_\mu$, $c(e_\mu) = \bigvee_{\substack{e \sim e_\mu \\ e \in \mathcal{M}}} e = \bigvee_{\substack{e \sim e_\mu \\ e \in \mathcal{M}}} u(t) e u(t)^{-1}$. Hence

$$u(t) c(e_\mu) u(t)^{-1} = c(e_\mu).$$

Now set $\tilde{v}_\mu(t) = u(t) \overline{w}_\mu(t)^{-1} c(e_\mu)$. Then $\tilde{v}_\mu(t)$ is a unitary operator on $c(e_\mu)\mathcal{H}$.

Moreover,

$$\begin{aligned}
\tilde{v}_\mu(t) x' c(e_\mu) \tilde{v}_\mu(t)^{-1} &= u(t) \overline{w}_\mu(t)^{-1} x' c(e_\mu) \overline{w}_\mu(t) u(t)^{-1} \\
&= u(t) \Phi\big(\Phi^{-1}(\overline{w}_\mu(t)^{-1} x' c(e_\mu) \overline{w}_\mu(t))\big) u(t)^{-1} \\
&= u(t) \Phi(u(t)^{-1} v_\mu(t) x' e_\mu v_\mu(t)^{-1} u(t)) u(t)^{-1} \\
&= u(t) \Phi(u(t)^{-1} x' e_\mu u(t)) u(t)^{-1} = x' c(e_\mu).
\end{aligned}$$

Hence $\tilde{v}_\mu(t) \in \mathcal{M} c(e_\mu)$ and clearly

$$\tilde{v}_\mu(t) x c(e_\mu) \tilde{v}_\mu(t)^{-1} = u(t) x c(e_\mu) u(t)^{-1} \qquad (x \in \mathcal{M}).$$

Now put $v(t) = \sum_{n=0}^{\infty} \tilde{v}_{n+1}(t)(c(e_{n+1}) - c(e_n))$ $(t \in (-\infty, \infty))$. Since $Q_n \uparrow 1$ strongly $(n \to \infty)$, $e_n \uparrow 1$ strongly $(n \to \infty)$. Hence $\sum_{n=0}^{\infty} (c(e_{n+1}) - c(e_n)) = 1$. Therefore $t \to v(t)$ is a strong-operator continuous one parameter subgroup of unitary elements in $\mathcal{M}$ and $v(t) x v(t)^{-1} = u(t) x u(t)^{-1}$

$$(x \in \mathcal{M}; t \in (-\infty, \infty)). \qquad \text{q.e.d.}$$

Remark 1. It is an interesting problem how much can we relax the spectrum condition in Theorem 4.1.15. It is clear that it can not be unconditional, since Blattner [11] proved the following result: Let $\mathcal{M}$ be a hyperfinite II_1-factor (cf. 4.4) on a separable Hilbert space and let G be a separable locally compact group. Then there exists a faithful strong-operator continuous unitary representation $g \to u(g)$ of G on $\mathcal{H}$ such that

1. $u(g) \mathcal{M} u(g)^{-1} = \mathcal{M}$;
2. if $g \neq e$ (e, the identity in G), a mapping $x \to u(g) x u(g)^{-1}$ $(x \in \mathcal{M})$ is an outer *-automorphism $(g \in G)$.

Remark 2. Let G be a separable locally compact group and let $\mathcal{M}$ be a W^*-algebra.

Suppose that there exists a homomorphism ρ of G into the group of inner *-automorphisms on $\mathscr{M}$ such that for each $a \in \mathscr{M}$, a mapping $g \to \rho(g)a$ of G into $\mathscr{M}$, with the σ-topology, is continuous. Then it is clear that for each $g \in G$, there exists a unitary element $u(g)$ in $\mathscr{M}$ with $\rho(g)a = u(g)au(g)^{-1}$ $(a \in \mathscr{M})$. However it is not trivial whether or not we may choose $u(g)s$ such as $g \to u(g)$ is a strong-operator continuous unitary representation of G.

Kallman [91] proved that if G is the group of all real numbers, the solution to this problem is yes.

If $g \to \rho(g)$ is uniformly continuous and G is the group of real numbers, the affirmative solution is obtained immediately by 4.1.6.

Dixmier [38] proved that the solution is yes if $g \to \rho(g)$ is uniformly continuous and G is a Lie group.

It is an interesting problem to find a solution to this problem in a more general group under a suitable continuity condition of the homomorphism $g \to \rho(g)$.

Let $\mathscr{D}$ be a Banach algebra with identity, and let a be an element in $\mathscr{D}$. The spectrum $\mathrm{Sp}(a)$ of a is defined as the set of all complex numbers λ such that $a - \lambda 1$ has no inverse.

Suppose $\mathrm{Sp}(a) \subset \Omega = \{\lambda \in \mathbb{C} \mid \mathrm{Re}\,\lambda > 0\}$, where $\mathbb{C}$ is the field of all complex numbers.

Denote the principal determination of logarithms by Log. Then we can define $y = \mathrm{Log}\, a$ such that $a = \exp y$. Put $i_a(x) = a x a^{-1}$ and $(ad\, y)(x) = yx - xy$ $(x \in \mathscr{D})$. Then we have

4.1.17. Lemma. *If C is a closed subspace of $\mathscr{D}$ which is invariant under i_a, then it is again invariant under $ad\, y$.*

Proof. $(\exp t\, ad\, y)(x) = (\exp t\, y)x(\exp -t\, y)$ for real t, and

$$(\exp ad\, y)(x) = a x a^{-1}.$$

Put $R_y x = xy$ and $L_y x = yx$ $(x, y \in \mathscr{D})$. Then $ad\, y = L_y - R_y$. Since $y \to R_y$ (resp. L_y) is an isometric anti-isomorphism (resp. isomorphism) and $\mathscr{D}$ has an identity, $\mathrm{Sp}(y) = \mathrm{Sp}(R_y) = \mathrm{Sp}(L_y)$ and so

$$\mathrm{Sp}(ad\, y) \subset \{\lambda - \mu \mid \lambda, \mu \in \mathrm{Sp}(y)\},$$

where $\mathrm{Sp}(R_y)$, $\mathrm{Sp}(L_y)$ and $\mathrm{Sp}(ad\, y)$ are considered in the Banach algebra $B(\mathscr{D})$.

Since $\mathrm{Sp}(y) \subset \left\{\lambda \in \mathbb{C} \,\middle|\, |\mathrm{Im}\,\lambda| < \frac{\pi}{2}\right\}$, $\mathrm{Sp}(ad\, y) \subset \{\lambda \in \mathbb{C} \mid |\mathrm{Im}\,\lambda| < \pi\}$ (say Σ).

Since $\mathrm{Log}\exp\lambda = \lambda$ for $\lambda \in \Sigma$, $\mathrm{Log}\, i_a = \mathrm{Log}(\exp(ad\, y)) = ad\, y$. Finally by Rünge's theorem, $\mathrm{Log}\,\lambda$ is a uniform limit of polynomials in a variable λ on all compact subsets in Σ. Hence $ad\, y$ is a norm-limit of polynomials of i_a. q.e.d.

4.1.18. Lemma. *Let $\mathcal{D}$ be a Banach algebra and let Φ be a continuous automorphism of $\mathcal{D}$ with $\mathrm{Sp}(\Phi)\subset\Omega$. Then $\delta=\mathrm{Log}\,\Phi$ is a derivation on $\mathcal{D}$, so that Φ belongs to a one-parameter subgroup $\{t\to\exp t\delta\}(-\infty<t<\infty)$ of automorphisms on $\mathcal{D}$.*

Proof. Let $\tilde{\mathcal{D}}$ be the Banach algebra obtained by adjoining an identity to $\mathcal{D}$, and let $\tilde{\Phi}$ be a unique continuous automorphism of $\tilde{\mathcal{D}}$ with $\Phi=\tilde{\Phi}$ on $\mathcal{D}$. Then $\mathrm{Sp}(\tilde{\Phi})=\mathrm{Sp}(\Phi)\cup\{1\}$ and $\mathrm{Log}\,\tilde{\Phi}|\mathcal{D}=\mathrm{Log}\,\Phi$.

Hence it suffices to assume that $\mathcal{D}$ has an identity. Let L (resp. R) be the left (resp. right) regular representation of $\mathcal{D}$ in $B(\mathcal{D})$—i.e. $L_x y=xy$ and $R_x y=yx\,(x,y\in\mathcal{D})$. Put $C=\{L_x|x\in\mathcal{D}\}$. Then C is a closed subalgebra of $B(\mathcal{D})$ and $\Phi L_x\Phi^{-1}=L_{\Phi(x)}\,(x\in\mathcal{D})$. Hence C is invariant under i_Φ and so by 4.1.17, it is again invariant under $ad\,\mathrm{Log}\,\Phi$.

Set $L_{x'}=(ad\,\mathrm{Log}\,\Phi)(L_x)=[\mathrm{Log}\,\Phi, L_x]$ $(x\in\mathcal{D})$. Then $\delta_1: x\to x'$ is a bounded derivation on $\mathcal{D}$.

Since $\mathrm{Sp}(i_\Phi)\subset\{\lambda\mu^{-1}|\lambda,\mu\in\Omega\}\subset\{\lambda|\lambda\notin(-\infty,0]\}$, we can define $\mathrm{Log}\,i_\Phi$. Since $\exp\mathrm{Log}\,i_\Phi=\exp ad\,\mathrm{Log}\,\Phi$ and $\mathrm{Sp}(ad\,\mathrm{Log}\,\Phi)\subset\Sigma$, $\mathrm{Log}\,i_\Phi=ad\,\mathrm{Log}\,\Phi$. Since $\mathrm{Log}\,i_\Phi(L_x)=L_{\mathrm{Log}\,\Phi(x)}$ for $x\in\mathcal{D}$, $\delta_1=\mathrm{Log}\,\Phi$. q.e.d.

4.1.19. Theorem *Let $\mathcal{A}$ be a C^*-algebra on a Hilbert space $\mathcal{H}$ such that the weak closure $\bar{\mathcal{A}}$ contains $1_{\mathcal{H}}$. Let Φ be an automorphism on $\mathcal{A}$ with $\mathrm{Sp}(\Phi)\subset\Omega$ (in particular, $\|\Phi-1_{\mathcal{A}}\|<1$). Then there exists an invertible element a in $\bar{\mathcal{A}}$ such that $\Phi(x)=axa^{-1}$ for $x\in\mathcal{A}$. In particular, if $\mathcal{A}$ is a W^*-algebra or a simple C^*-algebra with identity, then Φ is inner. Also if Φ is a *-automorphism satisfying the above condition, then a is a unitary element in $\mathcal{A}$.*

Proof. By 4.1.18, $\delta=\mathrm{Log}\,\Phi$ is a derivation on $\mathcal{A}$. Hence there exists an element b in $\bar{\mathcal{A}}$ with $\delta(x)=[b,x]$ $(x\in\mathcal{A})$. Thus

$$\Phi(x)=(\exp\delta)(x)=(\exp b)x(\exp -b) \qquad (x\in\mathcal{A}).$$

The remainder of the proof is almost clear. q.e.d.

Remark. If Φ is a *-automorphism, the above condition can be replaced by $\|\Phi-1_{\mathcal{A}}\|<2$ (cf. [88]).

Now let $\mathcal{A}$ and $\mathfrak{B}$ be two C^*-algebras, and let Φ be an isomorphism of $\mathcal{A}$ onto $\mathfrak{B}$. Then by 4.1.12, Φ is continuous.

Let $\mathcal{E}_{\mathcal{A}}$ (resp. $\mathcal{E}_{\mathfrak{B}}$) be the set of all pure states on $\mathcal{A}$ (resp. $\mathfrak{B}$). We shall consider a *-representation $\left\{\sum_{\varphi\in\mathcal{E}_{\mathcal{A}}}\pi_\varphi, \sum_{\varphi\in\mathcal{E}_{\mathcal{A}}}\oplus\mathcal{H}_\varphi\right\}$ (resp. $\left\{\sum_{\psi\in\mathcal{E}_{\mathfrak{B}}}\pi_\psi, \sum_{\psi\in\mathcal{E}_{\mathfrak{B}}}\oplus\mathcal{H}_\psi\right\}$) of $\mathcal{A}$ (resp. $\mathfrak{B}$).

Let $\mathcal{L}_\varphi=\{x|\varphi(x^*x)=0, x\in\mathcal{A}\}$ $(\varphi\in\mathcal{E}_{\mathcal{A}})$ and

$$\mathcal{L}_\psi=\{y|\psi(y^*y)=0,\ y\in\mathfrak{B}\} \qquad (\psi\in\mathcal{E}_{\mathfrak{B}}).$$

Then $\mathcal{L}_\varphi$ (resp. $\mathcal{L}_\psi$) is a regular maximal left ideal of $\mathcal{A}$ (resp. $\mathfrak{B}$).

By 1.21.18, $\mathscr{H}_\varphi = \mathscr{A}/\mathscr{L}_\varphi$ and $\mathscr{H}_\psi = \mathfrak{B}/\mathscr{L}_\psi$.

Since $\Phi(\mathscr{L}_\varphi)$ is a regular maximal left ideal in $\mathfrak{B}$, there exists a unique pure state φ' on $\mathfrak{B}$ with $\Phi(\mathscr{L}_\varphi) = \mathscr{L}_{\varphi'}$ by 1.21.19. Thus Φ induces a one-to-one mapping $\varphi \to \varphi'$ of $\mathscr{E}_\mathscr{A}$ onto $\mathscr{E}_\mathfrak{B}$.

Moreover it induces a bounded linear mapping T_φ of $\mathscr{A}/\mathscr{L}_\varphi$ onto $\mathfrak{B}/\mathscr{L}_{\varphi'}$ by $T_\varphi(a+\mathscr{L}_\varphi) = \Phi(a) + \mathscr{L}_{\varphi'}$ $(a \in \mathscr{A})$. Clearly $\|T_\varphi\| \leq \|\Phi\|$. Put $T = \sum_{\varphi \in \mathscr{E}} \oplus T_\varphi$. Then T is a bounded linear mapping of $\sum_{\varphi \in \mathscr{E}_\mathscr{A}} \oplus \mathscr{H}_\varphi$ onto $\sum_{\varphi' \in \mathscr{E}_\mathfrak{B}} \oplus \mathscr{H}_{\varphi'}$ with $\|T\| \leq \|\Phi\|$.

For simplicity, we shall identify $\mathscr{A}$ (resp. $\mathfrak{B}$) with $\left(\sum_{\varphi \in \mathscr{E}_\mathscr{A}} \pi_\varphi\right)(\mathscr{A})$ (resp. $\left(\sum_{\varphi' \in \mathscr{E}_\mathfrak{B}} \pi_{\varphi'}\right)(\mathfrak{B})$). Then

$$Tax_\varphi = T(ax)_\varphi = (\Phi(ax))_{\varphi'} = \Phi(a)\Phi(x)_{\varphi'} = \Phi(a)Tx_\varphi \qquad (a, x \in \mathscr{A};\ \varphi \in \mathscr{E}_\mathscr{A}).$$

By symmetry, one can easily see that T^{-1} is again a bounded linear mapping of $\sum_{\varphi' \in \mathscr{E}_\mathfrak{B}} \oplus \mathscr{H}_{\varphi'}$ onto $\sum_{\varphi \in \mathscr{E}_\mathscr{A}} \oplus \mathscr{H}_\varphi$. Hence $TaT^{-1} = \Phi(a)\,(a \in \mathscr{A})$.

Let $T = u|T|$ be the polar decomposition of T. Then

$$TaT^{-1} = u|T|\,a|T|^{-1}u^*.$$

$|T|$ is a positive element in $B\left(\sum_{\varphi \in \mathscr{E}_\mathscr{A}} \oplus \mathscr{H}_\varphi\right)$, and u is a unitary operator of $\sum_{\varphi \in \mathscr{E}_\mathscr{A}} \oplus \mathscr{H}_\varphi$ onto $\sum_{\varphi' \in \mathscr{E}_\mathfrak{B}} \oplus \mathscr{H}_{\varphi'}$.

$$\begin{aligned} |T|^{-2}a|T|^2 &= (T^*T)^{-1}a(T^*T) = T^{-1}T^{*-1}aT^*T = T^{-1}(Ta^*T^{-1})^*T \\ &= T^{-1}(\Phi(a^*)^*)T = \Phi^{-1}(\Phi(a^*)^*) \in \mathscr{A}. \end{aligned}$$

Hence the mapping $a \to |T|^{-2}a|T|^2$ $(a \in \mathscr{A})$ is an automorphism on $\mathscr{A}$, since $\{\Phi^{-1}(\Phi(a^*)^*) \mid a \in \mathscr{A}\} = \mathscr{A}$.

Now consider $B\left(\sum_{\varphi \in \mathscr{E}_\mathscr{A}} \oplus \mathscr{H}_\varphi\right)$. Since $\mathscr{A}$ is invariant under the mapping $x \to |T|^{-2}x|T|^2 \left(x \in B\left(\sum_{\varphi \in \mathscr{E}_\mathscr{A}} \oplus \mathscr{H}_\varphi\right)\right)$, it is also invariant under the mapping $ad\,\mathrm{Log}|T|^{-2}$ by 4.1.17.

$$\begin{aligned} (\exp t\, ad\,\mathrm{Log}|T|^{-2})(a) &= (\exp t\,\mathrm{Log}|T|^{-2})a(\exp -t\,\mathrm{Log}|T|^{-2}) \\ &= (\exp 2t\,\mathrm{Log}|T|^{-1})a(\exp -2t\,\mathrm{Log}|T|^{-1}) \quad (a \in \mathscr{A}). \end{aligned}$$

Set $t = \frac{1}{2}$; then we have that $a \to |T|^{-1}a|T|$ is an automorphism on $\mathscr{A}$, because $ad\,\mathrm{Log}|T|^{-2}$ is a derivation on $\mathscr{A}$.

Thus, $u\mathscr{A}u^* = T|T|^{-1}\mathscr{A}|T|T^{-1} = \mathfrak{B}$. Hence the mapping $a \to uau^*$ of $\mathscr{A}$ onto $\mathfrak{B}$ is a $*$-isomorphism. Hence we have

4.1.20. Theorem. *Let $\mathscr{A}$ and $\mathfrak{B}$ be two C^*-algebras and suppose that $\mathscr{A}$ and $\mathfrak{B}$ are isomorphic. Then they are $*$-isomorphic.*

4.1.21. Corollary. *Let $\mathscr{A}$ and $\mathfrak{B}$ be two C^*-algebras and let Φ be an isomorphism of $\mathscr{A}$ onto $\mathfrak{B}$. Then Φ can be uniquely written as follows: $\Phi = \Phi_1 \Phi_2$, where Φ_1 is a *-isomorphism of $\mathscr{A}$ onto $\mathfrak{B}$ and Φ_2 is an automorphism on $\mathscr{A}$ with $\Phi_2 = \exp\delta$ (δ is a derivation on $\mathscr{A}$).*

Proof. By the previous consideration,

$$\Phi(a) = TaT^{-1} = u|T|\,a|T|^{-1}u^* \qquad (a \in \mathscr{A}).$$

Put $\Phi_2(a) = |T|\,a|T|^{-1}$ and $\Phi_1(a) = uau^*$ ($a \in \mathscr{A}$). Then Φ_2 is an automorphism on $\mathscr{A}$ and Φ_1 is a *-isomorphism of $\mathscr{A}$ onto $\mathfrak{B}$. Moreover, $|T|\,a|T|^{-1} = (\exp ad \operatorname{Log}|T|)a$. Since $ad\operatorname{Log}|T|^{-2} = -2ad\operatorname{Log}|T|$ is a derivation on $\mathscr{A}$, $ad\operatorname{Log}|T|$ is again a derivation on $\mathscr{A}$. Put $\delta = ad\operatorname{Log}|T|$; then we have $\exp\delta = \Phi_2$.

The uniqueness of such a decomposition is not so difficult, because the polar decomposition of T is unique. q.e.d.

4.1.22. Corollary. *Let $\mathscr{A}$ (resp. $\mathfrak{B}$) be a C^*-algebra on a Hilbert space $\mathscr{H}$ (resp. $\mathscr{K}$) such that its weak closure $\bar{\mathscr{A}}$ (resp. $\bar{\mathfrak{B}}$) contains $1_{\mathscr{H}}$ (resp. $1_{\mathscr{K}}$). Let Φ be an isomorphism of $\mathscr{A}$ onto $\mathfrak{B}$. Then there exists an invertible positive element h in $\bar{\mathscr{A}}$ such that $\Phi(a) = \Phi_1(hah^{-1})$ ($a \in \mathscr{A}$), where Φ_1 is a *-isomorphism of $\mathscr{A}$ onto $\mathfrak{B}$. In particular if $\mathscr{A}$ is a W^*-algebra or a simple C^*-algebra with identity, then h belongs to $\mathscr{A}$.*

4.1.23. Corollary. *Let $\mathscr{M}$ and $\mathscr{N}$ be two W^*-algebras and let Φ be an isomorphism of $\mathscr{M}$ onto $\mathscr{N}$. Then Φ is σ and s-continuous.*

Remark. A homomorphism (even a *-homomorphism) of a W^*-algebra onto another W^*-algebra is not, in general, σ-continuous. For example, let l^∞ be the W^*-algebra of all bounded sequences, and let $l^\infty = C(K)$. Then K is the Cech compactification of all positive integers. Let t be a limit point in K and let $\mathscr{I}_t$ be the corresponding maximal ideal of $C(K)$. Then $C(K)/\mathscr{I}_t$ is a W^*-algebra, but the canonical homomorphism of $C(K)$ onto $C(K)/\mathscr{I}_t$ is not σ-continuous, since $\mathscr{I}_t$ is not σ-closed.

Concluding remarks on 4.1.

It would be an interesting problem to study derivations on more general C^*-algebras (or Banach algebras). For example, let $\mathscr{A}$ be a simple C^*-algebra with identity, $C(K)$ the C^*-algebra of all continuous functions on a compact space K, and let $\mathscr{A} \otimes C(K)$ be the C^*-tensor product of $\mathscr{A}$ and $C(K)$.

Is there an outer derivation on $\mathscr{A} \otimes C(K)$?

Lance [107] proved that $B(\mathscr{H}) \otimes C(K)$ has only inner derivations, where $\dim(\mathscr{H}) = \aleph_0$.

Let E be a Banach space and let $B(E)$ be the algebra of all bounded linear operators on E.

Is there an outer derivation on $B(E)$?

Johnson [78] proved that every derivation on a semi-simple commutative Banach algebra is identically zero. Johnson and Sinclair [79] proved that every derivation on a semi-simple Banach algebra is automatically continuous.

Johnson and Ringrose [80] proved that every derivation on the group algebra $l^1(G)$ of a discrete group G is inner.

Kadison and Ringrose [90] developed a cohomology theory on $B(\mathscr{H})$. It would be an interesting problem to study a cohomology theory on more general W^*-algebras.

There are various studies concerning derivations and automorphisms of W^*-algebras and C^*-algebras (cf. [48], [49], [88], [89], [108], [109], [189], [190]).

Johnson [77] proved that every automorphism on a semi-simple Banach algebra is automatically continuous.

Theorem 4.1.20 is due to Gardner [53], Lemma 4.1.4 is due to Kadison [86], Lemma 4.1.18 is due to Zeller-Meier [223], and Corollary 4.1.21 is due to Okayasu [135]. The proof of Theorem 4.1.6 is due to Johnson and Ringrose [80]. Lemma 4.1.12 is due to Rickart [140].

References. [13], [53], [80], [86], [87], [98], [135], [140], [153], [160], [165].

4.2. Examples of Factors, 1 (General Construction)

We have classified W^*-algebras into ones of types I, II and III in chapter 2. To a large extent the study of W^*-algebras may be reduced to the case of a factor by a reduction theory devised by von Neumann, which was discussed in chapter 3 (at least if the W^*-algebra has a separable predual). Thus the problem of classifying W^*-algebras may be reduced to the one of classifying factors. In the following sections, we shall show that there are uncountably many examples of types II_1, II_∞ and III-factors on a separable Hilbert space.

We can easily construct a factor of type I_n (n, cardinal) as follows: Let $\mathscr{H}$ be a Hilbert space with dimension n. Then $B(\mathscr{H})$ is a factor of type I_n. Moreover all type I_n-factors are *-isomorphic to $B(\mathscr{H})$ (2.3.3).

Next let $\mathscr{M}$ be a type II_∞-factor, e a non-zero finite projection of $\mathscr{M}$ and let $(e_\alpha)_{\alpha\in\mathbb{I}}$ be a maximal family of mutually orthogonal equivalent projections in $\mathscr{M}$ with $e_\alpha \sim e$ $(\alpha\in\mathbb{I})$. Then $p = 1 - \sum_{\alpha\in\mathbb{I}} e_\alpha \precsim\!\!\!\!/\; e$. Take a sequence (α_i) in $\mathbb{I}$. Since $p + \sum_{i=1}^{\infty} e_{\alpha_{2i}} \sim \sum_{i=1}^{\infty} e_{\alpha_{2i}}$, we have $p + \sum_{i=1}^{\infty} e_{\alpha_{2i}} = \sum_{i=1}^{\infty} e_{\beta_i}$

with $e_{\beta_i} \sim e$ $(i=1,2,\ldots)$. Thus $1 = \sum_{\alpha \in \mathbb{I} - (\alpha_i)}^{\infty} e_\alpha + \sum_{i=1}^{\infty} e_{\alpha_{2i+1}} + \sum_{i=1}^{\infty} e_{\beta_i}$ and $e_\alpha \sim e$, $e_{\alpha_{2i+1}} \sim e$ and $e_{\beta_i} \sim e$ $(\alpha \in \mathbb{I} - (\alpha_i);\ i=1,2,\ldots)$. Hence there exists a family of mutually orthogonal equivalent projections $(p_\gamma)_{\gamma \in \mathbb{J}}$ with $p_\gamma \sim e$ and $\sum_{\gamma \in \mathbb{J}} p_\gamma = 1$. Now let q be a non-zero finite projection in $\mathcal{M}$ and let $(q_\alpha)_{\alpha \in \mathbb{H}}$ be a family of mutually orthogonal equivalent projections in $\mathcal{M}$ with $q_\alpha \sim q$ $(\alpha \in \mathbb{H})$ and $\sum_{\alpha \in \mathbb{H}} q_\alpha = 1$.

Let τ_α be a normal tracial state on $q_\alpha \mathcal{M} q_\alpha$. Since $q_\alpha \mathcal{M} q_\alpha$ is a factor, such a τ_α is unique and faithful by the uniqueness of the $\natural$-operation. Hence $\tau_\alpha(q_\alpha) = \tau_\alpha\left(q_\alpha\left(\sum_{\gamma \in \mathbb{J}} p_\gamma\right) q_\alpha\right) = \sum_{\gamma \in \mathbb{J}} \tau_\alpha(q_\alpha p_\gamma q_\alpha)$, so that there is a countable subset $\mathbb{J}_\alpha$ of $\mathbb{J}$ such that $\tau_\alpha(q_\alpha p_\gamma q_\alpha) = 0$ for $\gamma \in \mathbb{J} - \mathbb{J}_\alpha$. Hence $p_\gamma q_\alpha = 0$ for $\gamma \in \mathbb{J} - \mathbb{J}_\alpha$, and so if $\gamma \in \mathbb{J} - \bigcup_{\alpha \in \mathbb{H}} \mathbb{J}_\alpha$, $p_\gamma q_\alpha = 0$ for all $\alpha \in \mathbb{H}$, so that $p_\gamma = 0$. Hence $\mathrm{Card}(\mathbb{J}) \leq \aleph_0 \mathrm{Card}(\mathbb{H}) = \mathrm{Card}(\mathbb{H})$. Analogously we have $\mathrm{Card}(\mathbb{H}) \leq \mathrm{Card}(\mathbb{J})$ and so $\mathrm{Card}(\mathbb{J}) = \mathrm{Card}(\mathbb{H})$. Hence by 1.22.14, $\mathcal{M} = \mathcal{N} \bar{\otimes} B$, where B is a type I_n-factor with $n = \mathrm{Card}(\mathbb{J})$ (unique) and $\mathcal{N}$ is *-isomorphic to $e\mathcal{M}e$, so that it is a II_1-factor.

Remark. It is not known whether or not such a II_1-factor $\mathcal{N}$ is unique—namely the following problem is open: Let $\mathcal{L}$ be a II_1-factor and let p be a non-zero projection in $\mathcal{L}$. Then, is $\mathcal{L}$ *-isomorphic to $p\mathcal{L}p$?

Conversely let $\mathcal{N}$ be a type II_1-factor, B a type I_n-*factor* $(n \geq \aleph_0)$. Then the W^*-tensor product $\mathcal{N} \bar{\otimes} B$ is a type II_∞-factor. Thus the study of II_∞-factors can be, more or less, reduced to the one of II_1-factors. Therefore the problem of constructing examples of types II_1, II_∞ and III may be reduced to the one of constructing types II_1 and III-factors. The essential point of the problem is to construct factors of type II_1 (resp. III) which act on a separable Hilbert space. Otherwise we can easily construct infinitely many examples of type II_1 (resp. III)-factors, by using infinite tensor products (resp. tensor products). Therefore in this section, we shall always consider constructions under the restriction of separability.

Let $\mathcal{H}$ be a Hilbert space, $\mathcal{M}$ a W^*-subalgebra of $B(\mathcal{H})$ containing $1_{\mathcal{H}}$, $\mathcal{G}$ a discrete group and let $g \to U_g$ $(g \in \mathcal{G})$ be a unitary representation of $\mathcal{G}$ in $\mathcal{H}$.

We assume that $U_g^{-1} \mathcal{M} U_g = \mathcal{M}$ $(g \in \mathcal{G})$. Then a mapping $T \to U_g^{-1} T U_g$ (denoted T^g) is a *-automorphism of $\mathcal{M}$. For every $g \in \mathcal{G}$, let $\mathcal{H}_g$ be a Hilbert space which is isomorphic to $\mathcal{H}$ and let J_g be a unitary mapping of $\mathcal{H}$ onto $\mathcal{H}_g$. Let $\tilde{\mathcal{H}} = \sum_{g \in \mathcal{G}} \oplus \mathcal{H}_g$. Then by the results of 1.20, we may represent each element R in $B(\tilde{\mathcal{H}})$ by a matrix $(R_{g,h})_{g,h \in \mathcal{G}}$ with $R_{g,h} = J_g^* R J_h \in B(\mathcal{H})$. For $T \in \mathcal{M}$, let $\Phi(T)$ be an element in $B(\tilde{\mathcal{H}})$ with

$R_{g,h}=0$ if $g\neq h$; $R_{g,g}=T(g\in\mathscr{G})$. Then Φ is a *-isomorphism of $\mathscr{M}$ onto a W^*-subalgebra $\tilde{\mathscr{M}}$ of $B(\tilde{\mathscr{H}})$.

For $x\in\mathscr{G}$, let $\tilde{U}_x$ be an element in $B(\tilde{\mathscr{H}})$ defined by a matrix $(R_{g,h})$ with $R_{g,h}=0$ if $gh^{-1}\neq x$; $R_{xg,g}=U_x$ $(g\in\mathscr{G})$. Then

$$\tilde{U}_{x_1x_2}=\tilde{U}_{x_1}\tilde{U}_{x_2} \qquad (x_1,x_2\in\mathscr{G});$$

moreover

$$J_g^*(\tilde{U}_{x^{-1}}\Phi(T)\tilde{U}_x)J_g=(J_g^*\tilde{U}_{x^{-1}}J_{xg})(J_{xg}^*\Phi(T)J_{xg})(J_{xg}^*\tilde{U}_xJ_g)=U_{x^{-1}}TU_x=T^x.$$

Hence $\tilde{U}_{x^{-1}}\Phi(T)\tilde{U}_x=\Phi(T^x)$.

Elements of the form

$$\{\Phi(T_1)\tilde{U}_{x_1}+\cdots+\Phi(T_n)\tilde{U}_{x_n}\mid T_i\in\mathscr{M},\ x_i\in\mathscr{G}\ \ (i=1,2,\ldots,n;n=1,2,3,\ldots)\}$$

make a *-subalgebra $\mathscr{N}_0$ of $B(\tilde{\mathscr{H}})$—in fact,

$$(\Phi(T)\tilde{U}_x)^*=\tilde{U}_{x^{-1}}\Phi(T^*)=\Phi(T^{*x})\tilde{U}_{x^{-1}}$$

and $\Phi(T_1)\tilde{U}_{x_1}\Phi(T_2)\tilde{U}_{x_2}=\Phi(T_1)\Phi(T_2^{x_1^{-1}})\tilde{U}_{x_1}\tilde{U}_{x_2}=\Phi(T_1T_2^{x_1^{-1}})\tilde{U}_{x_1x_2}$. Let $\mathscr{N}$ be the W^*-subalgebra of $B(\tilde{\mathscr{H}})$ generated by $\mathscr{N}_0$. One can easily see that the matrix $(R_{g,h})$ of $\Phi(T)\tilde{U}_x$ satisfies: $R_{g,h}=0$ if $gh^{-1}\neq x$, and $R_{xg,g}=TU_x$. Therefore there is a family $(T_g)_{g\in\mathscr{G}}$ of $\mathscr{M}$ such that $R_{g,h}=T_{gh^{-1}}U_{gh^{-1}}$. This property is preserved for the elements of $\mathscr{N}_0$ and their weak operator limits. Hence every element S in $\mathscr{N}$ is represented by a matrix of the form $(T_{gh^{-1}}U_{gh^{-1}})$ with $T_g\in\mathscr{M}$ $(g\in\mathscr{G})$.

If $S=(T_{gh^{-1}}U_{gh^{-1}})$ and $S'=(T'_{gh^{-1}}U_{gh^{-1}})$, then

$$SS'=\left(\left(\sum_r T_{gr^{-1}}T'^{\,rg^{-1}}_{rh^{-1}}\right)U_{gh^{-1}}\right).$$

Hence $SS'=(T''_{gh^{-1}}U_{gh^{-1}})$ with $T''_g=\sum_r T_r T'^{\,r^{-1}}_{r^{-1}g}$.

4.2.1. Lemma. *Let φ be a faithful semi-finite normal trace on $\mathscr{M}$ which is invariant under $\mathscr{G}$ (i.e., $\varphi(T^g)=\varphi(T)$ $(g\in\mathscr{G})$). For every*

$$S=(T_{gh^{-1}}U_{gh^{-1}}) \qquad (S\geq 0)\in\mathscr{N},$$

put $\psi(S)=\varphi(T_e)$ (e is the unit of $\mathscr{G}$). Then ψ is a semi-finite faithful normal trace on $\mathscr{N}$. ψ is finite if and only if φ is finite. Finally for $T(\geq 0)\in\mathscr{M}$, $\psi(\Phi(T))=\varphi(T)$.

Proof. Since $T_e=R_{e,e}=J_e^*SJ_e$, a linear mapping $\Psi: S\to T_e$ of $\mathscr{N}$ onto $\mathscr{M}$ is σ-continuous. $T_e\geq 0$ if $S\geq 0$, and so ψ is normal.

Since $S^*=(U^*_{hg^{-1}}T^*_{hg^{-1}})$, $\Psi(SS^*)=\sum_{g\in\mathscr{G}}T_gT_g^*$ and

$$\Psi(S^*S)=\sum_{g\in\mathscr{G}}U_g^*T_g^*T_gU_g.$$

Hence $\psi(S^*S)=\psi(SS^*)$. If $\psi(S^*S)=0$, $\varphi(T_g^*T_g)=0$ for $g\in\mathscr{G}$. Hence $T_g=0$ for $g\in\mathscr{G}$, and so $S=0$.

Since the matrix of $\Phi(T)$ is: $R_{e,e}=T$, $\psi(\Phi(T))=\varphi(T)$. Since φ is semi-finite, there exists an increasing directed set (T_α) of positive elements in $\mathscr{M}$ with $\underset{\alpha}{\text{l.u.b.}}\, T_\alpha=1_{\mathscr{H}}$ and $\varphi(T_\alpha)<+\infty$. Hence $\underset{\alpha}{\text{l.u.b.}}\, \Phi(T_\alpha)=1_{\tilde{\mathscr{H}}}$ and $\psi(\Phi(T_\alpha))<+\infty$. Thus ψ is semi-finite. It is clear that ψ is finite if and only if φ is finite. q.e.d.

4.2.2. Lemma. I. *Suppose that $\tilde{\mathscr{M}}$ is a maximal commutative *-subalgebra of $\mathscr{N}$. Then $\mathscr{N}$ is semi-finite if and only if there is a faithful normal semi-finite trace on $\mathscr{M}$ which is invariant under $\mathscr{G}$.* II. *If $\mathscr{M}$ is purely infinite, $\mathscr{N}$ is again purely infinite.*

Proof. I. By 4.2.1, the condition is sufficient. Conversely suppose that $\mathscr{N}$ is semi-finite. Let ψ be a faithful normal semi-finite trace on $\mathscr{N}$ (cf. 2.5.7) and put $\varphi(T)=\psi(\Phi(T))$ ($T\in\mathscr{M}$ with $T\geq 0$). Then clearly φ is a faithful normal trace on $\mathscr{M}$. The problem is to show that φ is semi-finite.

Let $T_1=\Phi(T)$ be a unitary element in $\tilde{\mathscr{M}}$ and set $S=(T_{gh^{-1}}U_{gh^{-1}})$. Then $J_g^*(T_1ST_1^{-1})J_g=J_g^*T_1J_gJ_g^*SJ_gJ_g^*T_1^{-1}J_g=T\,T_eT^{-1}=T_e=J_g^*SJ_g$. Let K_S be the convex span of $\{T_1ST_1^{-1}\,|\,T_1\in\mathscr{M}^u\}$. Then $J_g^*K_SJ_g$ consists of a single point, so that $J_g^*\bar{K}_SJ_g$ is again one point, where $\bar{K}_S$ is the σ-closure of K_S. On the other hand, $\bar{K}_S$ is σ-compact and invariant under mappings $R\to T_\alpha RT_\alpha^{-1}$, $(T_\alpha\in\tilde{\mathscr{M}}^u)$. Since $\tilde{\mathscr{M}}$ is commutative, there exists a fixed point S_0 in $\bar{K}_S$ by the fixed point theorem of Kakutani-Markoff (cf. [16]). Then $S_0\in\tilde{\mathscr{M}}'\cap\mathscr{N}=\tilde{\mathscr{M}}$ and so $J_g^*SJ_g=J_g^*S_0J_g=T_e$ with $\Phi(T_e)=S_0$. If $S(\geq 0)\in\mathscr{N}$ and $\psi(S)<+\infty$, then $\psi(R)=\psi(S)<+\infty$ for all $R\in K_S$. Let $R_1\in\bar{K}_S$. Then there is a directed set (S_α) in K_S such that $S_\alpha\to R_1$ (σ-weakly).

Put $\mathscr{D}=\{V\,|\,0\leq V\leq 1, \psi(V)<+\infty, V\in\mathscr{N}\}$; then

$$\lim_\alpha \psi(VS_\alpha)=\psi(VR_1)\qquad (V\in\mathscr{D}).$$

Since $\psi(VS_\alpha)\leq\psi(S_\alpha)=\psi(S)$, $\psi(VR_1)\leq\psi(S)$. Since $\underset{V\in\mathscr{D}}{\text{l.u.b.}}\,\psi(VR_1)=\psi(R_1)$, we have $\psi(R_1)\leq\psi(S)$ and so $\psi(S_0)=\varphi(T_e)<+\infty$.

Since $1_{\tilde{\mathscr{H}}}$ is the strong limit of a directed set $\{V_\alpha\}$ of elements in $\mathscr{D}$, $\{J_g^*V_\alpha J_g\}$ converges strongly to $1_{\mathscr{H}}$. Hence φ is semi-finite.

II. Consider a σ-continuous linear mapping P of $\mathscr{N}$ onto $\tilde{\mathscr{M}}$ as follows: $P(S)=\Phi(T_e)$ with $S=(T_{gh^{-1}}U_{gh^{-1}})$. Then P satisfies the conditions 1. $P(1_{\tilde{\mathscr{H}}})=1_{\tilde{\mathscr{H}}}$; 2. $\|P(S)\|\leq\|S\|$; 3. $P(H)\geq 0$ $(H\geq 0)$; 4. $P(XSY)=XP(S)Y$ $(X, Y\in\tilde{\mathscr{M}})$; 5. $P(S)^*P(S)\leq P(S^*S)$; 6. $P(S^*S)=0$ implies $S=0$. Hence $\mathscr{N}$ is of type III by 2.6.5. q.e.d.

Remark. If $\mathscr{M}$ is a factor and $\mathscr{G}$ consists of outer *-automorphisms except for the identity, then $\mathscr{N}$ is again a factor (cf. [198]).

4.2.3. Lemma. *Let $\mathscr{M}$ be a maximal commutative *-subalgebra of $B(\mathscr{H})$. Suppose that $\mathscr{M} \cap \mathscr{M} U_g = (0)$ for $g \neq e$ $(g \in \mathscr{G})$. Then $\tilde{\mathscr{M}}$ is a maximal commutative *-subalgebra of $\mathscr{N}$.*

Proof. Let $S = (T_{gh^{-1}} U_{gh^{-1}})$ be an element in $\mathscr{N}$ which commutes with $\tilde{\mathscr{M}}$. For $T \in \mathscr{M}$, $S\Phi(T) = \Phi(T)S$ and so $TT_{gh^{-1}} U_{gh^{-1}} = T_{gh^{-1}} U_{gh^{-1}} T$. Since $\mathscr{M}$ is maximal commutative, $T_g U_g \in \mathscr{M} \cap \mathscr{M} U_g$ and so $T_g U_g = 0$ for $g \neq e$ $(g \in \mathscr{G})$, so that $S \in \tilde{\mathscr{M}}$. q.e.d.

4.2.4. Lemma. *Suppose that $\tilde{\mathscr{M}}$ is maximal commutative in $\mathscr{N}$, and elements in $\mathscr{M}$ which are invariant under $\mathscr{G}$ are scalar operators. Then $\mathscr{N}$ is a factor.*

Proof. Let S be an element in the center of $\mathscr{N}$. Then $S \in \tilde{\mathscr{M}}$. Put $S = \Phi(T)$ for some $T \in \mathscr{M}$. $\tilde{U}_{g^{-1}} \Phi(T) \tilde{U}_g = \Phi(T^g) = \Phi(T)$. Hence $T^g = T$, so that $T = \lambda 1_{\mathscr{H}}$ (λ a complex number). q.e.d.

Now let Ω be a locally compact Hausdorff space satisfying the second countability axiom. Let Δ be the set of all Borel subsets of Ω, and let μ be a σ-finite, countably additive, positive measure on Δ (μ is not necessarily a Radon measure).

Let $\mathscr{G}$ be a countable discrete group of homeomorphisms on Ω. For $\xi \in \Omega$ and $a \in \mathscr{G}$, we denote the effect of the mapping corresponding to a on ξ by ξa. The measure μ is said to be quasi-invariant under $\mathscr{G}$ if $\mu(E) = 0$ for a measurable set E in Ω implies $\mu(Ea) = 0$ $(a \in \mathscr{G})$. In this case, the translated measure μ_a defined for a measurable set F by $\mu_a(F) = \mu(Fa)$ is absolutely continuous with respect to μ, thus we can form the Radon-Nikodym's derivative $r_a(\xi)$. Since

$$r_{ab}(\xi) d\mu(\xi) = d\mu_{ab}(\xi) = d\mu(\xi ab) = r_b(\xi a) d\mu(\xi a) = r_b(\xi a) r_a(\xi) d\mu(\xi),$$

$r_{ab}(\xi) = r_b(\xi a) r_a(\xi)$ a.e. The group $\mathscr{G}$ is said to be: I. free if for $a \in \mathscr{G}$ $(a \neq e)$, the set of points satisfying the condition $\xi = \xi a$ $(\xi \in \Omega)$ is of a μ-measure 0; II. ergodic if $\mu((E \cup Ea) - (E \cap Ea)) = 0$ for a measurable set E and every $a \in \mathscr{G}$ implies either $\mu(E) = 0$ or $\mu(\Omega - E) = 0$; III. measurable if there is a σ-finite invariant positive measure ν (i.e., $\nu(Ea) = \nu(E)$ $(a \in \mathscr{G})$) which is equivalent to μ; IV. non-measurable if it is not measurable.

Now suppose that μ is quasi-invariant. We formulate the Hilbert space $L^2(\Omega, \mu)$ of complex valued μ-square integrable functions on Ω. For $f \in L^\infty(\Omega, \mu)$ and $k \in L^2(\Omega, \mu)$, set $(T_f k)(\xi) = f(\xi) k(\xi)$. Then $L^\infty(\Omega, \mu)$ is a weakly closed *-subalgebra of $B(\mathscr{H})$ with $\mathscr{H} = L^2(\Omega, \mu)$. $L^\infty(\Omega, \mu)$ is a maximal commutative *-subalgebra of $B(\mathscr{H})$ (2.9.3). For $a \in \mathscr{G}$, put $(U_a k)(\xi) = \sqrt{r_a(\xi)}\, k(\xi a)$ $(k \in L^2(\Omega, \mu))$. Then U_a is a unitary and $U_{ab} = U_a U_b$ $(a, b \in \mathscr{G})$, so that $a \to U_a$ is a unitary representation of $\mathscr{G}$ in $L^2(\Omega, \mu)$.

$(U_a^{-1} T_f U_a k)(\xi) = \sqrt{r_{a^{-1}}(\xi)}\, f(\xi a^{-1}) \sqrt{r_a(\xi a^{-1})}\, k(\xi) = f(\xi a^{-1}) k(\xi)$. Hence $U_a^{-1} T_f U_a = T_{f_a}$ with $f_a(\xi) = f(\xi a^{-1})$ $(f \in L^\infty(\Omega, \mu))$.

Now put $L^\infty(\Omega, \mu) = \mathscr{M}$ and we shall apply the previous result to $\mathscr{M}$ and $\mathscr{G}$, and construct a weakly closed self-adjoint subalgebra $\mathscr{N}$ of $B(\tilde{\mathscr{H}})$.

4.2.5. Lemma. *If $\mathscr{G}$ is free and ergodic, then $\mathscr{N}$ is a factor.*

Proof. We shall show first that $\mathscr{M} \cap \mathscr{M} U_a = (\dot{0})$ for $a\ (\neq e) \in \mathscr{G}$. Let $T_{f_1} = T_{f_2} U_a$ $(f_1, f_2 \in L^\infty(\Omega, \mu))$ and let $E = \{\xi \,|\, f_1(\xi) \neq 0\}$. Since $\mathscr{G}$ is free and a is a homeomorphism on Ω, $F_a = \{\xi \,|\, \xi a = \xi, \xi \in \Omega\}$ is a closed null set. By considering $\Omega - F_a$, we may assume that $F_a = (\emptyset)$.

For any $\xi \in E$, there is a compact neighborhood V_ξ of ξ with $V_\xi \cap V_\xi a = (\emptyset)$. Put $W_\xi = E \cap V_\xi$. Then $W_\xi \cap W_\xi a = (\emptyset)$ and $W_\xi \subset E$. If $\mu(E) \neq 0$, there exists a set W_{ξ_0} with $\mu(W_{\xi_0}) \neq 0$. Take a subset E_1 in W_{ξ_0} with $0 < \mu(E_1) < +\infty$. Then $\mu(E_1 \cap E_1 a) = 0$. Let χ be the characteristic function of E_1. Then $\chi \in L^2(\Omega, \mu)$ and

$$(T_{f_1} \chi)(\xi) = f_1(\xi) \chi(\xi) = (T_{f_2} U_a \chi)(\xi) = f_2(\xi) \sqrt{r_a(\xi)}\, \chi(\xi a),$$

a contradiction. Hence $f_1 = 0$. By 4.2.3, $\tilde{\mathscr{M}}$ is a maximal commutative *-subalgebra of $\mathscr{N}$. Next let T_f be a positive element in $\mathscr{M}$ which is invariant under $\mathscr{G}$. $U_a^{-1} T_f U_a = T_{f_a} = T_f$ and so $f(\xi a^{-1}) = f(\xi)$ a.e. $(a \in \mathscr{G})$. Let $E_{p,q} = \{\xi \,|\, 0 \le p \le f(\xi) \le q,\ p < q,\ \xi \in \Omega\}$. Then

$$\mu\{(E_{p,q} \cup E_{p,q} a) - (E_{p,q} \cap E_{p,q} a)\} = 0 \quad (a \in \mathscr{G})$$

and so $\mu(E_{p,q}) = 0$ or $\mu(\Omega - E_{p,q}) = 0$. This implies $f(\xi)$ is a constant function, so that $\mathscr{N}$ is a factor by 4.2.4. q.e.d.

4.2.6. Lemma. *Suppose that the group $\mathscr{G}$ is free, ergodic and measurable. Then if an invariant positive measure ν which is equivalent to μ has the following properties: $\nu(\{\xi\}) = 0$ $(\xi \in \Omega)$ and $0 < \nu(\Omega) < +\infty$ (resp. $\nu(\Omega) = +\infty$), then $\mathscr{N}$ is a type II_1 (resp. II_∞)-factor.*

Proof. For $T_f (\ge 0) \in \mathscr{M}$, put $\varphi(T_f) = \int_\Omega f(\xi) d\nu(\xi)$. Then φ is a faithful normal semi-finite trace on $\mathscr{M}$. Moreover

$$\varphi(U_a^* T_f U_a) = \varphi(T_{f_a}) = \int_\Omega f(\xi a^{-1}) d\nu(\xi) = \int_\Omega f(\xi) d\nu(\xi) = \varphi(T_f) \qquad (a \in \mathscr{G}).$$

Hence $\mathscr{N}$ is a semi-finite factor by 4.2.1. If $0 < \nu(\Omega) < +\infty$, $\mathscr{N}$ is a finite factor and if $\nu(\Omega) = +\infty$, $\mathscr{N}$ is a semi-finite, properly infinite factor. Since $\nu(\{\xi\}) = 0$, there is a decreasing sequence $\{E_n\}$ of measurable sets such that $\nu(E_n) > \nu(E_{n+1})$ $(n = 1, 2, \ldots)$ and $\lim_n \nu(E_n) = 0$. Thus

$$\psi(\Phi(T_{\chi_{E_n}})) = \varphi(T_{\chi_{E_n}}) \to 0 \qquad (n \to \infty),$$

where ψ is a normal trace on $\mathcal{N}$ and χ_{E_n} is the characteristic function of E_n. Hence $\mathcal{N}$ is continuous. q.e.d.

4.2.7. Lemma. *If $\mathcal{G}$ is free, ergodic and non-measurable, then $\mathcal{N}$ is a type* III*-factor.*

Proof. Suppose that $\mathcal{N}$ is semi-finite and let ψ be a faithful normal semi-finite trace on $\mathcal{N}$. Clearly ψ will define a σ-finite countably additive positive measure ν on Ω which is equivalent to μ. It is clear that ν is invariant under $\mathcal{G}$, so that $\mathcal{G}$ is measurable, a contradiction. q.e.d.

4.2.8. Lemma. *Let $\mathcal{G}_0=\{a \mid r_a(\xi)=1$ a.e., $a\in\mathcal{G}\}$. Then $\mathcal{G}_0$ is a subgroup of $\mathcal{G}$. If $\mathcal{G}_0$ is ergodic and $\mathcal{G}_0\subsetneqq\mathcal{G}$, then $\mathcal{G}$ is non-measurable.*

Proof. Suppose that $\mathcal{G}$ is measurable, and let ν be a positive invariant measure which is equivalent to μ. Then $d\mu(\xi)=\left(\frac{d\mu}{d\nu}\right)(\xi)d\nu(\xi)$ and

$$d\mu_a(\xi)=d\mu(\xi)=\left(\frac{d\mu}{d\nu}\right)(\xi a)d\nu_a(\xi)=\left(\frac{d\mu}{d\nu}\right)(\xi a)d\nu(\xi)$$

for $a\in\mathcal{G}_0$. Since $\mathcal{G}_0$ is ergodic, $\left(\frac{d\mu}{d\nu}\right)(\xi)=a$ constant a.e. Hence $\mathcal{G}_0=\mathcal{G}$, a contradiction. q.e.d.

4.2.9. Example (II$_1$-factors). *Let Ω be the one-dimensional torus group, μ the Haar measure on Ω with $\mu(\Omega)=1$, and let $\mathcal{G}$ be a countably infinite subgroup of Ω which is dense in Ω. For $a\in\mathcal{G}$ and $\xi\in\Omega$, define a homeomorphism $\xi\to\xi a$ by $\xi a=\xi+a$. $\mathcal{G}$ is free.*

Let E be a measurable set in Ω with $\mu\{(E\cup Ea)-(E\cap Ea)\}=0$ $(a\in\mathcal{G})$, and let χ_E be the characteristic function of E. Then $\chi_E\in L^2(\Omega,\mu)$. $\{\varphi_n(\xi)=e^{2\pi i n\xi}\}$ $(n=0,\pm1,\pm2,\ldots)$ is a complete orthonormal system of $L^2(\Omega,\mu)$. Hence $\chi_E(\xi)=\sum_{n=-\infty}^{\infty}\lambda_n e^{2\pi i n\xi}$ with $\sum_{n=-\infty}^{\infty}|\lambda_n|^2<+\infty$.

$$\chi_E(\xi a)=\sum_{n=-\infty}^{\infty}\lambda_n e^{2\pi i n(\xi+a)}=\sum_{n=-\infty}^{\infty}\lambda_n e^{2\pi i n a}e^{2\pi i n\xi}=\chi_E(\xi)$$

in $L^2(\Omega,\mu)$ and so $\lambda_n e^{2\pi i n a}=\lambda_n$ $(a\in\mathcal{G})$, so that $\lambda_n=0$ for $n\neq0$. Hence $\mu(E)=0$ or $\mu(\Omega-E)=0$, and so $\mathcal{G}$ is ergodic. Since μ is invariant under $\mathcal{G}$, $\mu(\{\xi\})=0$ and $\mu(\Omega)=1$, $\mathcal{N}$ is a type II$_1$*-factor. For instance, we may take:*

1. $\mathcal{G}=\{n\theta(\text{mod }1) \mid n=0,\pm1,\pm2,\ldots;\ \theta$ *an irrational number*$\}$;
2. $\mathcal{G}=\{\gamma(\text{mod }1) \mid \gamma$ *all rational numbers*$\}$;
3. $\mathcal{G}=\{m/p^n(\text{mod }1) \mid m=0,\pm1,\pm2,\ldots;\ n=0,1,2,\ldots;\ p=$ *any fixed number of* $2,3,4,\ldots\}$.

4.2.10. Example (II_∞-factors). *Let Ω be the locally compact group of all real numbers, μ the Haar measure on Ω and let $\mathcal{G}$ be a countable subgroup of Ω which is dense in Ω. Then by a similar reasoning to* 4.2.9 *we may show that $\mathcal{G}$ is free, ergodic and measurable. Since $\mu(\Omega)=+\infty$, $\mathcal{N}$ is a type* II_∞*-factor. For instance, we may take:*

1. $\mathcal{G}=\{m+n\theta \mid m,n=0,\pm 1,\pm 2,\ldots;\ \theta$ *an irrational number*$\}$,
2. $\mathcal{G}=\{\gamma \mid \gamma$ *all rational numbers*$\}$;
3. $\mathcal{G}=\{m/p^n \mid m=0,\pm 1,\pm 2,\ldots;\ p=$ *any fixed number of* $2,3,4,\ldots\}$.

4.2.11. Example (a III-factor). *Let Ω be the locally compact group of all real numbers, μ the Haar measure on Ω. Consider the following homeomorphisms: $\alpha_1(\rho,\sigma)$ ($\rho>0$, ρ,σ rationals): $\xi\to\rho\xi+\sigma$ ($\xi\in\Omega$). Put $\mathcal{G}=\{\alpha_1(\rho,\sigma)\}$. Then $\mathcal{G}$ is a countable group and μ is quasi-invariant under $\mathcal{G}$. Moreover $\mathcal{G}$ is free. $\mu(E\alpha_1(\rho,\sigma))=\rho\,\mu(E)$ for a measurable set E and so $\mathcal{G}_0=\{\alpha_1(\rho,\sigma)\mid \mu_{\alpha_1(\rho,\sigma)}=\mu\}=\{\alpha_1(1,\sigma)\}$. Since $\mathcal{G}_0$ is ergodic, $\mathcal{G}$ is non-measurable by* 4.2.8, *and so $\mathcal{N}$ is a type* III*-factor.*

4.2.12. Example (a III-factor). *Let Ω be the one-dimensional torus group and we shall consider Ω as the set of all complex numbers z with $|z|=1$. Let μ be the Haar measure of Ω with $\mu(\Omega)=1$.*

Consider the following homeomorphisms on Ω: $\alpha_2(\theta,u)$: $z\to\theta\dfrac{z+u}{1+uz}$ ($|\theta|=1, |u|<1$). Let $\mathcal{G}$ be a countable group generated by

$$\{a_2(\theta,u)\,|\,\theta=e^{2\pi i\rho}(\rho,\ \textit{all rationals}),\ u=0,\ \textit{and}\ \theta_1=1,\ u=1/2\}.$$

Then we may show that $\mathcal{G}$ is free, ergodic and non-measurable. Hence $\mathcal{N}$ is a type III*-factor*

4.2.13. Example (III-factors). *Let Ω_n ($n=1,2,\ldots$) be the additive groups of integers, reduced* mod 2; *therefore Ω_n is a compact group composed of two elements $\{0,1\}$ as follows: $0+0=0$, $0+1=1+0=1$ and $1+1=0$. Let $\Omega=\prod_{n=1}^{\infty}\Omega_n$ be the weakly infinite direct product of $\{\Omega_n \mid n=1,2,\ldots\}$. Then Ω is a compact group. Let μ_n be a Radon measure on Ω_n with $\mu_n\{(0)\}=\dfrac{1-\alpha_n}{2}$, $\mu_n\{(1)\}=\dfrac{1+\alpha_n}{2}$ with $0<\delta<\alpha_n<1-\delta$ for some fixed δ and $n=1,2,\ldots$. Let $\mu=\prod_{n=1}^{\infty}\mu_n$ be the infinite product measure of $\{\mu_n\}$ on Ω. Ω is the set of all elements $\xi=(\xi_n \mid n=1,2,\ldots)$ with $\xi_n=0$ or* 1. *Let $\mathcal{G}$ be the set of those $a=(a_n \mid n=1,2,\ldots)$ in Ω for which $a_n\neq 0$ occurs for a finite number of n only. Then $\mathcal{G}$ is a countable group.*

For $a\in\mathcal{G}$ and $\xi\in\Omega$, define a homeomorphism $\xi\to\xi a$ by $\xi a=\xi+a$. Then in the following consideration we shall show that $\mathcal{N}$ is a type III-factor.

4.2.14. Lemma. *The measure μ is quasi-invariant under $\mathscr{G}$.*

Proof. Put $p_n = \dfrac{1-\alpha_n}{2}$ and $q_n = \dfrac{1+\alpha_n}{2}$. For a fixed positive integer *k, set*

$$f_k(\xi) = \begin{cases} \dfrac{p_k}{q_k} & \text{if } \xi_k = 1 \\ \\ \dfrac{q_k}{p_k} & \text{if } \xi_k = 0 \end{cases}$$

and let $\gamma_k = (\gamma_{k,n} = 0 \text{ if } k \neq n,\ \gamma_{k,k} = 1;\ n = 1, 2, \ldots) \in \mathscr{G}$ and let

$$E = \{\xi \,|\, \xi_{m_i} = 0,\ \xi_{n_j} = 1;\ i = 1, 2, \ldots, u,\ j = 1, 2, \ldots, v\}.$$

Suppose that k occurs among the numbers m_i; then

$$\int_E f_k(\xi)\, d\mu(\xi) = \frac{q_k}{p_k}\, \mu(E) = \mu(E\gamma_k).$$

Analogously we have

$$\mu(E\gamma_k) = \int_E f_k(\xi)\, d\mu(\xi) \quad \text{for all } k.$$

Since the set of the forms E is a fundamental family of neighborhoods of Ω, μ_{γ_k} is absolutly continuous with respect to μ and

$$\left(\frac{d\mu_{\gamma_k}}{d\mu}\right)(\xi) = f_k(\xi).$$

Thus for $a = \sum_{i=1}^{n} \gamma_{k_i} \in \mathscr{G}$,

$$\left(\frac{d\mu_a}{d\mu}\right)(\xi) = f_{k_1}(\xi)\, f_{k_2}(\xi) \cdots f_{k_n}(\xi) = \prod_{n=1}^{\infty} \left(\frac{p_n}{q_n}\right)^{(2\xi_n - 1)a_n}$$

and so $\mathscr{G}$ is quasi-invariant under $\mathscr{G}$. q.e.d.

4.2.15. Lemma. *The system of functions*

$$W_a(\xi) = (-1)^{\sum_{n=1}^{\infty} a_n \xi_n} \prod_{n=1}^{\infty} \left(\frac{p_n}{q_n}\right)^{(\xi_n - \frac{1}{2})a_n} \qquad (a \in \mathscr{G})$$

forms a complete orthonormal system in $L^2(\Omega, \mu)$.

Proof. For $a, b \in \mathscr{G}$,

$$\begin{aligned} \int W_a(\xi)\, W_b(\xi)\, d\mu(\xi) &= \int W_{a+b}(\xi)\, W_{a\wedge b}^2(\xi)\, d\mu(\xi) \\ &= \int W_{a+b}(\xi)\, d\mu(\xi) \int W_{a\wedge b}^2(\xi)\, d\mu(\xi) \end{aligned}$$

$\{a \wedge b = (\beta_n \mid \beta_n = 1 \text{ if } a_n = b_n = 1; \text{otherwise } \beta_n = 0)\}$. If $a = \gamma_{k_1} + \gamma_{k_2} + \cdots + \gamma_{k_n}$, then

$$\int W_a(\xi)\, d\mu(\xi) = \prod_{i=1}^{n} \int W_{\gamma_{k_i}}(\xi)\, d\mu(\xi)$$

and

$$\int W_a^2(\xi)\, d\mu(\xi) = \prod_{i=1}^{n} \int W_{\gamma_{k_i}}^2(\xi)\, d\mu(\xi).$$

For every k, $\int W_{\gamma_k}(\xi)\, d\mu(\xi) = -q_k \sqrt{\frac{p_k}{q_k}} + p_k \sqrt{\frac{q_k}{p_k}} = 0$, and

$$\int W_{\gamma_k}^2(\xi)\, d\mu(\xi) = q_k \frac{p_k}{q_k} + p_k \frac{q_k}{p_k} = 1.$$

To prove the completeness, it suffices to show that the characteristic functions of the form $E = \{\xi \mid \xi_{m_i} = 0, \xi_{n_j} = 1; i = 1, 2, \ldots, u; j = 1, 2, \ldots, v\}$ are linear combinations of the system $\{W_a(\xi)\}$. Such a set is the intersection of a finite number of sets $E_j = \{\xi \mid \xi_j = 0\}$ and $F_k = \{\xi \mid \xi_k = 1\}$ $(j, k = 1, 2, \ldots)$. Let e_j and f_k be the characteristic functions of E_j and F_k respectively. Then $e_j(\xi) = \sqrt{p_j q_j}\, W_{\gamma_j}(\xi) + p_j W_0(\xi)$ and $f_k(\xi) = -\sqrt{p_k q_k}\, W_{\gamma_k}(\xi) + q_k W_0(\xi)$. For $a, b \in \mathscr{G}$ with $a \wedge b = 0$, $W_{a+b}(\xi) = W_a(\xi)\, W_b(\xi)$. Hence we have the completeness of $\{W_a\}$. q.e.d.

4.2.16. Lemma. *The group $\mathscr{G}$ is free, ergodic and non-measurable.*

Proof. Clearly, $\mathscr{G}$ is free. We shall show the ergodicity ot $\mathscr{G}$. Let $f(\xi)$ be a bounded measurable function on Ω with $f(\xi\gamma_k) = f(\xi)$ a.e. $(k = 1, 2, \ldots)$. Since $r_{\gamma_k}(\xi) = \left(\frac{p_k}{q_k}\right)^{(2\xi_k - 1)} = \frac{q_k - p_k}{\sqrt{p_k q_k}} W_{\gamma_k}(\xi) + 1$, if $a_k = 0$ for $a = (a_n)$,

$$C_a = \int f(\xi)\, W_a(\xi)\, d\mu(\xi) = \int f(\xi\gamma_k)\, W_a(\xi)\, d\mu(\xi) = \int f(\xi)\, W_a(\xi\gamma_k)\, r_{\gamma_k}(\xi)\, d\mu(\xi)$$

$$= \frac{q_k - p_k}{\sqrt{p_k q_k}} \int f(\xi)\, W_a(\xi\gamma_k)\, W_{\gamma_k}(\xi)\, d\mu(\xi) + \int f(\xi)\, W_a(\xi\gamma_k)\, d\mu(\xi)$$

$$= \frac{q_k - p_k}{\sqrt{p_k q_k}} \int f(\xi)\, W_{a+\gamma_k}(\xi)\, d\mu(\xi) + \int f(\xi)\, W_a(\xi)\, d\mu(\xi)$$

$$= \frac{q_k - p_k}{\sqrt{p_k q_k}} C_{a+\gamma_k} + C_a \quad \Big(f(\xi) = \sum_{a \in \mathscr{G}} C_a W_a(\xi)\Big).$$

Therefore $C_{a+\gamma_k} = 0$, so that $C_a = 0$ for $a (\neq e) \in \mathscr{G}$.

Finally we shall show the non-measurability of $\mathscr{G}$. Suppose that there exists a positive measure ν on Ω which is equivalent to μ and is invariant under $\mathscr{G}$. By Radon-Nikodym theorem, there exists a measurable positive function f on Ω such that $\nu(E)=\int_E f(\xi)d\mu(\xi)$ for $E\in\Delta$, where Δ is the set of all Borel sets in Ω.

Since

$$\int_E f(\xi)d\mu(\xi)=\nu(E)=\nu(E\gamma_k)=\int_{E\gamma_k} f(\xi)d\mu(\xi)=\int_E f(\xi\gamma_k)\frac{d\mu_{\gamma_k}}{d\mu}(\xi)d\mu(\xi),$$

$$f(\xi\gamma_k)\frac{d\mu_{\gamma_k}}{d\mu}(\xi)=f(\xi)$$

a.e. Take an $F\in\Delta$ such that $\mu(F)>0$ and $f(x)$ is bounded on F, and let $f'(\xi)=f(\xi)$ on $F\cup\left(\bigcup_{k=1}^{\infty}F\gamma_k\right)$ and $f'(\xi)=0$, otherwise.

Then

$$\begin{aligned}\int_\Omega |f'(\xi\gamma_k)-f'(\xi)|^2 d\mu(\xi) &\geq \int_F |f(\xi\gamma_k)-f(\xi)|^2 d\mu(\xi)\\ &=\int_F \left|\left\{\frac{d\mu_{\gamma_k}}{d\mu}(\xi)\right\}^{-1}-1\right|^2 |f(\xi)|^2 d\mu(\xi)\\ &\geq \min\left(\left|1-\frac{p_k}{q_k}\right|^2,\left|1-\frac{q_k}{p_k}\right|^2\right)\int_F |f(\xi)|^2 d\mu(\xi)\\ &\geq\left(\frac{2\delta}{1-\delta}\right)^2\int_F |f(\xi)|^2 d\mu(\xi).\end{aligned}$$

On the other hand, for $a\in\mathscr{G}$, if k is sufficiently large, $W_a(\xi\gamma_k)=W_a(\xi)$. Hence

$$\lim_{k\to+\infty}\int |W_a(\xi\gamma_k)-W_a(\xi)|^2 d\mu(\xi)=0.$$

Since $\{W_a\}$ is a complete orthonormal system in $L^2(\Omega,\mu)$, this implies $\lim_{k\to\infty}\int |g(\xi\gamma_k)-g(\xi)|^2 d\mu(\xi)=0$ $(g\in L^2(\Omega,\mu))$, and so $\int_F |f(\xi)|^2 d\mu(\xi)=0$. Hence $\nu(F)=\int_F f(\xi)d\mu(\xi)=0$, a contradiction. q.e.d.

Now let $\mathscr{H}=\mathscr{H}_0=\mathbb{C}$ be the one-dimensional Hilbert space, $\mathscr{M}=\mathbb{C}1_{\mathscr{H}_0}$, $\mathscr{G}=G$ a discrete group, and let $a\to U_a=1_{\mathscr{H}_0}$ $(a\in G)$ be the unitary representation of G. Then the corresponding W^*-algebra $\mathscr{N}$ is $*$-isomorphic to the W^*-algebra generated by the left regular representation of G.

We shall explain this special case in a different form.

Let G be a discrete group and let $l^2(G)$ be the set of all complex valued square summable functions on G. For $f_1, f_2 \in l^2(G)$, the convolution $f_1 * f_2$ is defined as follows: $(f_1 * f_2)(a) = \sum_{b \in G} f_1(b) f_2(b^{-1}a)$. $f_1 * f_2$ is a bounded function on G (does not necessarily belong to $l^2(G)$), and $\|f_1 * f_2\|_\infty \leq \|f_1\|_2 \|f_2\|_2$, where $\| \ \|_\infty$ is the norm of $l^\infty(G)$ and $\| \ \|_2$ is the norm of $l^2(G)$. For $a \in G$ and $h \in l^2(G)$, define $(U(a)h)(b) = h(a^{-1}b)$ and $(V(a)h)(b) = h(ba)$ $(b \in G)$. Then $U(a)$ and $V(a)$ are unitary on $l^2(G)$ and the mapping $a \to U(a)$ (resp. $V(a)$) is the left (resp. right) regular representation of G. Clearly $U(a_1)V(a_2) = V(a_2)U(a_1)$ $(a_1, a_2 \in G)$.

Let $U(G)$ be the W^*-subalgebra of $B(l^2(G))$ generated by $\{U(a) | a \in G\}$. Denote by ε_a the function on G with $\varepsilon_a(b) = 0$ if $b \neq a$ and $\varepsilon_a(a) = 1$.

Consider a mapping $\Phi: T \to T\varepsilon_e$ of $U(G)$ into $l^2(G)$ (e, the unit of G). If $T_1 \varepsilon_e = T_2 \varepsilon_e$ $(T_1, T_2 \in U(G))$,

$$V(a)T_1\varepsilon_e = T_1\varepsilon_{a^{-1}} = V(a)T_2\varepsilon_e = T_2\varepsilon_{a^{-1}} \qquad (a \in G).$$

Since the set $\mathscr{D}$ of all linear combinations of $\{\varepsilon_a | a \in G\}$ is dense in $l^2(G)$, $T_1 = T_2$. Hence Φ is one-to-one. Moreover $V(a^{-1})T\varepsilon_e = TV(a^{-1})\varepsilon_e = T\varepsilon_a$ and $V(a^{-1})T\varepsilon_e = (T\varepsilon_e) * \varepsilon_a$. *Hence* $(T\varepsilon_e) * \varepsilon_a = T\varepsilon_a$ $(a \in G)$, *and so* $Tk = (T\varepsilon_e) * k$ for $k \in \mathscr{D}$. For $h \in l^2(G)$, take a sequence $\{k_n\}$ in $\mathscr{D}$ with $\|k_n - h\|_2 \to 0$ $(n \to \infty)$. Then $Tk_n \to Th$ in $l^2(G)$ and $(T\varepsilon_e) * k_n \to (T\varepsilon_e) * h$ in $l^\infty(G)$. Since $l^2(G) \subset l^\infty(G)$ and $Tk_n \to Th$ in the topology $\sigma(l^\infty(G), l^1(G))$, $Th = (T\varepsilon_e) * h$. By using the mapping Φ, we shall identify $U(G)$ with a subspace of $l^2(G)$. Then by the above result, if $f \in U(G) \subset l^2(G)$ and $h \in l^2(G)$, the corresponding operator $\Phi^{-1}(f)$ is as follows: $\Phi^{-1}(f)h = f * h$. We shall denote $\Phi^{-1}(f)$ by U_f. Put $\tau(f) = f(e)$ $(f \in U(G))$. Then one can easily see that τ is a faithful normal tracial state on $U(G)$. Hence $U(G)$ is a finite algebra.

4.2.17. Definition. *A discrete group G is said to be an infinite conjugacy group if every conjugacy class $\{bab^{-1} | b \in G\}$ $(a \in G)$ is infinite except for $a = e$.*

4.2.18. Lemma. *If G is an infinite conjugacy group, $U(G)$ is a type* II_1-*factor.*

Proof. Let $f \in U(G)$ be a central element in $U(G)$. Then

$$(U(a)fU(a^{-1}))(b) = (\varepsilon_a f \varepsilon_{a^{-1}})(b) = f(a^{-1}ba) = f(b) \qquad (a, b \in G).$$

Hence f is a constant on each conjugacy class. On the other hand, f belongs to $l^2(G)$. Hence $f(b) = 0$ for $b \neq e$, and so $U(G)$ is a finite factor. Since $U(G)$ is infinite-dimensional, it must be a type II_1-factor. q.e.d.

References. [121], [131], [139].

4.3. Examples of Factors, 2. (Uncountable Families of Types II_1, II_∞ and III)

In this section, we shall show the existence of uncountably many examples of type II_1 (resp. II_∞, III)-factors on a separable Hilbert space.

Let G be a discrete group and let $U(G)$ be the W^*-algebra generated by the left regular representation of G.

4.3.1. Definition. *A uniformly bounded sequence (T_n) of elements in $U(G)$ is called a central sequence if for all $x \in U(G)$, $\|[x, T_n]\|_2 \to 0$ $(n \to \infty)$, where $\|x\|_2$ $(x \in U(G))$ is the $l^2(G)$-norm of x, when $U(G)$ is embedded into $l^2(G)$ canonically. Two central sequences (T_n), (T'_n) in $U(G)$ are said to be equivalent, if $\|T_n - T'_n\|_2 \to 0$ $(n \to \infty)$.*

4.3.2. Definition. *For a W^*-algebra $\mathcal{M}$, its unit sphere is denoted by $(\mathcal{M})_1$. If $\mathcal{M}$ and $\mathcal{N}$ are W^*-subalgebras of a W^*-algebra $U(G)$ and $\delta > 0$, then we shall write $\mathcal{N} \overset{\delta}{\subset} \mathcal{M}$ to mean that given any $T \in (\mathcal{N})_1$, there exists some $S \in (\mathcal{M})_1$ with $\|T - S\|_2 < \delta$.*

4.3.3. Lemma. *Let G be a discrete group and let E be a subset of G. Suppose that there exist a subset $F \subset E$ and two elements $a_1, a_2 \in G$ such that I. $F \cup a_1 F a_1^{-1} = E$; II. $a_2^{-1} F a_2, F, a_2 F a_2^{-1} \subset E$ and they are mutually disjoint. Let $f(a)$ be a complex valued function on G such that*

$$\sum_{a \in G} |f(a)|^2 < +\infty$$

and $\left(\sum_{a \in G} |f(a_i a a_i^{-1}) - f(a)|^2\right)^{\frac{1}{2}} < \varepsilon$ $(i = 1, 2)$. *Then* $\left(\sum_{a \in E} |f(a)|^2\right)^{\frac{1}{2}} < 14\varepsilon$.

Proof. Put $v(M) = \sum_{a \in M} |f(a)|^2$ for every subset M in G. Then

$$\varepsilon > \left(\sum_{a \in G} |f(a_1 a a_1^{-1}) - f(a)|^2\right)^{\frac{1}{2}} \geq |v(a_1 F a_1^{-1})^{\frac{1}{2}} - v(F)^{\frac{1}{2}}|.$$

Putting $v(E)^{\frac{1}{2}} = s$, then

$$|v(a_1 F a_1^{-1}) - v(F)| = |v(a_1 F a_1^{-1})^{\frac{1}{2}} + v(F)^{\frac{1}{2}}|\,|v(a_1 F a_1^{-1})^{\frac{1}{2}} - v(F)^{\frac{1}{2}}| < 2s\varepsilon$$

and so $v(a_1 F a_1^{-1}) < v(F) + 2s\varepsilon$; hence $s^2 \leq v(a_1 F a_1^{-1}) + v(F) < 2(v(F) + s\varepsilon)$, so that $v(F) > s^2/2 - s\varepsilon$. Since

$$\left(\sum_{a \in G} |f(a_2 a a_2^{-1}) - f(a)|^2\right)^{\frac{1}{2}} = \left(\sum_{a \in G} |f(a_2 a_2^{-1} a a_2 a_2^{-1}) - f(a_2^{-1} a a_2)|^2\right)^{\frac{1}{2}},$$

we have $|v(a_2 F a_2^{-1}) - v(F)| < 2s\varepsilon$ and $|v(a_2^{-1} F a_2) - v(F)| < 2s\varepsilon$. Hence, $v(a_2 F a_2^{-1}) > v(F) - 2s\varepsilon > \frac{s^2}{2} - 3s\varepsilon$ and $v(a_2^{-1} F a_2) > \frac{s^2}{2} - 3s\varepsilon$. Therefore, $s^2 = v(E) \geq v(F) + v(a_2^{-1} F a_2) + v(a_2 F a_2^{-1}) > \frac{3}{2} s^2 - 7s\varepsilon$, that is $s < 14\varepsilon$. q.e.d.

4.3.4. Definition. *Let G be a discrete group and let H be a subgroup of G. H is called residual in G if there exists a subset S of $G\backslash H$ (the complement of H in G) and elements a_1, a_2 of G such that $G\backslash H = S \cup a_1^{-1} S a_1$, and S, $a_2^{-1} S a$, $a_2 S a_2^{-1}$ are mutually disjoint subsets in $G\backslash H$.*

4.3.5. Definition. *A sequence (T_n) in $\{U(G)\}_1$ is called an ε-central sequence if $\limsup_n \|[T_n, X]\|_2 < \varepsilon$ for all $X \in \{U(G)\}_1$.*

4.3.6. Lemma. *Let H be a residual subgroup of G and let (T_n) be an ε-central sequence in $U(G)$. Then there exists a sequence (T_n') in $\{U(H)\}_1$ such that $\limsup_n \|T_n - T_n'\|_2 < 14\varepsilon$ where, $U(H)$ is canonically considered as a subalgebra of $U(G)$.*

Proof. Since H is residual in G, there exist two elements a_1, a_2 in G and a subset S of $G\backslash H$ such that $G\backslash H = S \cup a_1^{-1} S a_1$ and S, $a_2^{-1} S a_2$, $a_2 S a_2^{-1}$ are mutually disjoint subsets in $G\backslash H$. Put $T_n = U_{f_n}$ $(f_n \in l^2(G))$. Then

$$\begin{aligned}\|[T_n, U(a_i)]\|_2 &= \|T_n U(a_i) - U(a_i) T_n\|_2 = \|U(a_i^{-1}) T_n U(a_i) - T_n\|_2 \\ &= \Big(\sum_{a \in G} |f_n(a_i a a_i^{-1}) - f_n(a)|^2\Big)^{\frac{1}{2}} < \varepsilon \qquad (n \geq \text{some } n_i;\ i = 1, 2).\end{aligned}$$

Hence by 4.3.3, $\Big(\sum_{a \in G\backslash H} |f_n(a)|^2\Big)^{\frac{1}{2}} < 14\varepsilon$. Put $f_n'(a) = f_n(a)$ for $a \in H$ and $f_n'(a) = 0$ for $a \varepsilon G\backslash H$. Then one can easily see that $U_{f_n'} \in U(H)$ with

$$\|U_{f_n'}\| \leq \|U_{f_n}\|.$$

Furthermore, $\|U_{f_n} - U_{f_n'}\|_2 = \|f_n - f_n'\|_2^2 = \Big(\sum_{a \in G\backslash H} |f_n(a)|^2\Big)^{\frac{1}{2}} < 14\varepsilon$ $(n \geq n_0)$. q.e.d.

4.3.7. Definition. *Let H be a subgroup of a group G. H is said to be strongly residual if there exist elements a_1, a_2 in G and a subset S in $G\backslash H$ such that* I. $a_1 H a_1^{-1} = H$; II. $S \cup a_1 S a_1^{-1} = G\backslash H$; III. $\{a_2^n S a_2^{-n}\}_{n = 0, \pm 1, \pm 2, \ldots}$ *forms a family of mutually disjoint subsets in $G\backslash H$.*

One can easily see that only one strongly residual subgroup of a commutative group G is G itself. In this case, S is the empty set.

4.3.8. Lemma. *Let G_i $(i = 1, 2, \ldots, n)$ be a finite family of groups and let H_i $(i = 1, 2, \ldots, n)$ be a subgroup of G_i. If H_i is strongly residual in G_i for each i, then $\sum_{i=1}^{n} \oplus H_i$ is strongly residual in $\sum_{i=1}^{n} \oplus G_i$.*

Proof. It suffices to assume that $n = 2$. Let $\{S_i, a_{i1}, a_{i2}\}$ be a system which defines the strong residuality of H_i in G_i $(i = 1, 2)$. Put

$$S = (S_1 \oplus G_2) \cup (H_1 \oplus S_2),$$

$a_1 = a_{11} \oplus a_{21}$ and $a_2 = a_{12} \oplus a_{22}$. Then one can easily see that $\{S, a_1, a_2\}$ will define the strong residuality of $H_1 \oplus H_2$ in $G_1 \oplus G_2$. q.e.d.

Now let $G_1, G_2, \ldots; H_1, H_2, \ldots$ be two sequences of groups. We denote by $(G_1, G_2, \ldots; H_1, H_2, \ldots)$ the group generated by the G_i's and the H_i's with additional relations that H_i, H_j commute elementwise for $i \neq j$ and G_i, H_j commute elementwise for $i \leq j$.

Let $L_1 = (G_1, G_2, \ldots; H_1, H_2, \ldots)$ with $G_i = Z$ and $H_i = Z$ $(i = 1, 2, \ldots)$, where Z is the group of all integers. Define L_k inductively by

$$L_k = (G_1, G_2, \ldots; H_1, H_2, \ldots)$$

with $G_i = Z$, $H_i = L_{k-1}$ $(i = 1, 2, \ldots; k = 2, 3, \ldots)$.

Let $\mathbb{I}_1$ be a sequence of positive integers. Put $M_n(\mathbb{I}_1) = \sum_{i=1}^{n} \oplus L_{p_i}$ $(n = 1, 2, \ldots)$ if $\mathbb{I}_1 = (p_1, p_2, \ldots)$ is infinite, and $M_n(\mathbb{I}_1) = \sum_{i=1}^{n} \oplus L_{p_i}$ for $n \leq n_0$ and $M_n(\mathbb{I}_1) = M_{n_0}(\mathbb{I}_1)$ for $n > n_0$ if $\mathbb{I}_1 = (p_1, p_2, \ldots, p_{n_0})$ is finite. Define $G[\mathbb{I}_1] = (G_1, G_2, \ldots; H_1, H_2, \ldots)$ with $G_i = Z$ and $H_i = M_i(\mathbb{I}_1)$ $(i = 1, 2, \ldots)$. Then we shall show the following theorem.

4.3.9. Theorem. *Let $\mathbb{I}_1 = (p_i)$ and $\mathbb{I}_2 = (q_i)$ be two sequences of positive integers such that $\mathbb{I}_1$ and $\mathbb{I}_2$ are different as sets—i.e., there exists a p_{i_0} or q_{i_0} such that $p_{i_0} \neq q_i$ for all i or $q_{i_0} \neq p_i$ for all i. Then $U(G[\mathbb{I}_1])$ is not *-isomorphic to $U(G[\mathbb{I}_2])$.*

4.3.10. Corollary. *There exist uncountably many type II_1-factors on a separable Hilbert space.*

To prove Theorem 4.3.9, we shall provide some considerations. Clearly $U(G[\mathbb{I}_i])$ $(i = 1, 2)$ is an infinite conjugacy group, so that it is a type II_1-factor by 4.2.18.

If H is a strongly residual subgroup of G, then H must contain the center of G.

4.3.11. Definition. *Let $\{H_n\}$ be a sequence of subgroups of a group G. $\{H_n\}$ is called a strongly residual sequence in G if it satisfies the following conditions:* I. *H_n is strongly residual in G;* II. *$H_n = H_{n+1} \oplus K_n$ (K_n, a subgroup of G);* III. *$\bigcup_{n=1}^{\infty} H'_n = G$ (H'_n, the commutant of H_n in G) $(n = 1, 2, \ldots)$.*

Let G_i $(i = 1, 2, \ldots, m)$ be a finite family of groups and let $\{H_{i,n}\}$ $(i = 1, 2, \ldots, m)$ be a strongly residual sequence in G_i. Then one can easily see that $\left\{\sum_{i=1}^{m} \oplus H_{i,n}\right\}$ is a strongly residual sequence in $\sum_{i=1}^{m} \oplus G_i$.

4.3.12. Lemma. *If $\{H_n\}$ is a strongly residual sequence in G and $T_n \in \{U(H_n)\}_1$ $(n = 1, 2, \ldots)$, then (T_n) is a central sequence in $U(G)$.*

Proof. Since $\bigcup_{n=1}^{\infty} H_n' = G$ and $H_n \supset H_{n+1}$, $[T_n, X] = 0$ for $X \in U(H_{n_0}')$ with $n \geq n_0$. Since $\bigcup_{n=1}^{\infty} U(H_n')$ is σ-dense in $U(G)$, we have $\|[T_n, X]\|_2 \to 0$ $(n \to \infty)$ for $X \in U(G)$. Hence (T_n) is central. q.e.d.

Now let $G = (G_1, G_2, \ldots; H_1, H_2, \ldots)$ with $G_i = Z$ $(i = 1, 2, \ldots)$. Put

$$Q(G,n) = \sum_{j=n}^{\infty} \oplus H \qquad (n = 1, 2, \ldots)$$

and $Q(G,m,n) = \sum_{j=m}^{n} \oplus H \quad (m < n;\ m = 1, 2, \ldots;\ n = 2, 3, \ldots)$.

4.3.13. Lemma. *$\{Q(G,n)\}$ is a strongly residual sequence in G.*

Proof. Let δ_n be a generator of G_n and let S_n be the set of all those elements which, when powers of δ_n are taken as far to the right as possible, end in a non-zero power of δ_n, and put $a_{n,1} = \delta_n$ and $a_{n,2} = \delta_{n+1}$. Then

$$a_{n,1}\, Q(G,n)\, a_{n,1}^{-1} = Q(G,n); \quad S_n \cup a_{n,1}\; S_n a_{n,1}^{-1} = G \backslash Q(G,n);$$

$\{a_{n,2}^{r} S_n a_{n,2}^{-r}\}_{r = 0, \pm 1, \pm 2, \ldots}$ are mutually disjoint subsets in $G \backslash Q(G,n)$. Hence $Q(G,n)$ is strongly residual in G. Moreover

$$Q(G,n) = H_n \oplus Q(G, n+1) \qquad (n = 1, 2, \ldots)$$

and $Q(G,n)' \supset$ the subgroup of G generated by $G_1, G_2, \ldots, G_n$ and $H_1, H_2, \ldots H_{n-1}$ $(n = 2, 3, \ldots)$, so that $\bigcup_{n=1}^{\infty} Q(G,n)' = G$. q.e.d.

If Γ_k $(k = 1, 2, \ldots, r)$ is a finite family with the form

$$\Gamma_k = (G_1, G_2, \ldots; H_1, H_2, \ldots)$$

with $G_i = Z$ $(i = 1, 2, \ldots)$, then $Q\left(\sum_{k=1}^{r} \oplus \Gamma_k, n\right) = \sum_{k=1}^{r} \oplus Q(\Gamma_k, n)$ is a strongly residual sequence in $\sum_{k=1}^{r} \oplus \Gamma_k$. This strong residual sequence is called the canonical, strongly residual sequence. Denote

$$Q\left(\sum_{k=1}^{r} \oplus \Gamma_k, m, n\right) = \sum_{k=1}^{r} \oplus Q(\Gamma_k, m, n).$$

A group G is said to be of type 0 if it is commutative; type i if $G = \sum_{j=1}^{n} \oplus G_j$ with $G_j = L_i$; type i_∞ if $G = \sum_{j=1}^{\infty} \oplus G_j$ with $G_j = L_i$; type

$(i_1, i_2, \ldots, i_n)$ if $G = \sum_{j=1}^{n} \oplus G_j$, where G_j is of type i_j; type $(i_1, i_2, \ldots, i_n)_\infty$ if $G = \sum_{j=1}^{n} \oplus G_j$ and some of the G_j's are of type $i_{j\infty}$ and others are of type i_j.

Now suppose that $U(G[\mathbb{I}_1])$ is *-isomorphic to $U(G[\mathbb{I}_2])$. Then under the identification $U(G[\mathbb{I}_1]) = U(G[\mathbb{I}_2])$, we have two expressions $U(G[\mathbb{I}_1])$ and $U(G[\mathbb{I}_2])$. Henceforward we shall assume

$$U(G[\mathbb{I}_1]) = U(G[\mathbb{I}_2])$$

and conclude a contradiction.

Let $\mathbb{I}_1 = (p_i)$ and $\mathbb{I}_2 = (q_i)$ and assume that $q_{i_0} \neq p_i$ for all i.

4.3.14. Lemma. *For $\delta > 0$ and positive integers n_1 and r, there exists a positive integer n_2 such that $n_2 > r$ and*

$$U\{Q(G[\mathbb{I}_2], n_2)\} \overset{\delta}{\subset} U\{Q(G[\mathbb{I}_1], n_1)\}.$$

Proof. Suppose this is not true. Then for each n with $n > r$ there exists a $T_n \in (U\{Q(G[\mathbb{I}_2], n)\})_1$ such that $\|T_n - S\|_2 \geq \delta$ for all $S \in (U\{Q(G[\mathbb{I}_1], n_1)\})_1$. Since $\{Q(G[\mathbb{I}_2], n)\}$ is a strongly residual sequence, (T_n) is a central sequence in $U(G[\mathbb{I}_2]) = U(G[\mathbb{I}_1])$ by 4.3.12. Since $Q(G[\mathbb{I}_1], n_1)$ is residual in $G[\mathbb{I}_1]$, there exists a central sequence (T_n') in $(U\{Q(G[\mathbb{I}_1], n_1)\})_1$ such that $\|T_n - T_n'\|_2 \to 0$ $(n \to \infty)$ by 4.3.6. This is a contradiction. q.e.d.

Put $t = 10^{q_{i_0} + i_0}$. By applying the above lemma for $\mathbb{I}_1$ and its symetric form for $\mathbb{I}_2$ we can choose positive integers $n_1, n_2, \ldots, n_t$ such that $n_2 < n_4 < \cdots < n_t$ and $n_1 < n_3 < \cdots < n_{t-1}$, and

$$\begin{aligned} U\{Q(G[\mathbb{I}_2], n_t)\} &\overset{\delta}{\subset} U\{Q(G[\mathbb{I}_1], n_{t-1})\} \\ &\overset{\delta}{\subset} \cdots \\ &\overset{\delta}{\subset} U\{Q(G[\mathbb{I}_2], n_2)\} \\ &\overset{\delta}{\subset} U\{Q(G[\mathbb{I}_1], n_1)\}. \end{aligned}$$

q.e.d.

4.3.15. Lemma. *If $U\{Q(G[\mathbb{I}_1], h)\} \overset{\delta}{\subset} U\{Q(G[\mathbb{I}_2], i)\}$, then for arbitrary positive integers r, w with $h < r$ there exists a positive integer s such that $U\{Q(G[\mathbb{I}_1], h, r)\} \overset{5\delta}{\subset} U\{Q(G[\mathbb{I}_2], i, s)\}$ with $i < s$ and $w < s$.*

Proof. For $T \in (U\{Q(G[\mathrm{II}_1],h,r)\})_1 \subset (U\{Q(G[\mathrm{II}_1],h)\})_1$, take

$$T' \in (U\{Q(G[\mathrm{II}_2],i)\})_1$$

with $\|T-T'\|_2 < \delta$. Since $\{Q(G[\mathrm{II}_2],n) \mid n > i+1, n > w\}$ is a strongly residual sequence in $G[\mathrm{II}_2]$, there exists an n' such that $n' > i+1$, $n' > w$ and $U\{Q(G[\mathrm{II}_2],n')\} \overset{\delta}{\subset} U\{Q(G[\mathrm{II}_1],r+1)\}$ (4.3.14). For

$$X' \in (U\{Q(G[\mathrm{II}_2],n')\})_1,$$

take $X \in (U\{Q(G[\mathrm{II}_1],r+1)\})_1$ with $\|X'-X\|_2 < \delta$. Then

$$\|[T',X']\|_2 \leq \|[T'-T,X']\|_2 + \|[X'-X,T]\|_2 + \|[X,T]\|_2 < 2\delta + 2\delta,$$

since $[X,T]=0$.

Let U' be a unitary element in $U\{Q(G[\mathrm{II}_2],n')\}$. Then

$$\|[U',T']\|_2 = \|U'T'U'^* - T'\|_2.$$

Let C be the $\| \|_2$-closed convex subset of $U(G[\mathrm{II}_2])$ generated by $\{U'T'U'^* \mid U' \text{ (unitary)} \in U\{Q(G[\mathrm{II}_2],n')\}\}$ and let Y' be a unique element in C such that $\|Y'\|_2 = \inf_{W' \in C} \|W'\|_2$. Then $\|U'Y'U'^*\|_2 = \|Y'\|_2$ and so $U'Y'U'^* = Y'$, so that $Y' \in [U\{Q(G[\mathrm{II}_2],n')\}]'$, where $[()]'$ is the commutant of $()$. Since $U\{Q(G[\mathrm{II}_2],n')\} = U\left(\sum_{i=n'}^{\infty} \oplus M_i(\mathrm{II}_2)\right)$ is a factor,

$$\begin{aligned} U\{Q(G[\mathrm{II}_2],1)\} \cap [U\{Q(G[\mathrm{II}_2],n')\}]' &= U\left(\sum_{i=1}^{n'-1} \oplus M_i(\mathrm{II}_2)\right) \\ &= U\{Q(G[\mathrm{II}_2],1,n'-1)\}. \end{aligned}$$

Since

$Y' \in (U\{Q(G[\mathrm{II}_2],i)\})_1 \cap [U\{Q(G[\mathrm{II}_2],n'\}]'$, $Y' \in (U\{Q(G[\mathrm{II}_2],i,n'-1)\})_1$. Moreover $\|Y'-T'\|_2 \leq 4\delta$. Hence

$$\|T-Y'\|_2 \leq \|T-T'\|_2 + \|T'-Y'\|_2 < \delta + 4\delta = 5\delta. \qquad \text{q.e.d.}$$

By applying this lemma for II_1 and its symmetric form for II_2, one can choose positive integers $m_1, m_2, \ldots, m_t$ such that $m_2 > m_4 > \cdots > m_t$ and $m_1 > m_3 \ldots > m_{t-1}$, with $m_t > n_t$, and

$$\begin{aligned} U\{Q(G[\mathrm{II}_2],n_t,m_t)\} &\overset{5\delta}{\subset} U\{Q(G[\mathrm{II}_1],n_{t-1},m_{t-1})\} \\ &\overset{5\delta}{\subset} \cdots \\ &\overset{5\delta}{\subset} U\{Q(G[\mathrm{II}_1],n_1,m_1)\}. \end{aligned}$$

Since $Q(G[\mathrm{II}_i],h,k)$ $(i=1,2)$ is a finite sum of the form

$$(G_1, G_2, \ldots; H_1, H_2, \ldots)$$

with $G_j = Z$ $(j=1,2,\ldots)$, it has the canonical, strongly residual sequence $[Q\{Q(G[\mathrm{II}_i],h,k),n\}]$. For simplicity, we shall denote $Q(G[\mathrm{II}_i],h,k)$ (resp. $Q\{Q(G[\mathrm{II}_i],h,k),n\}$) by $Q_i(h,k)$ (resp. $Q_i^2[(h,k),n]$).

4.3.16. Lemma. *If* $U\{Q_1(h,k)\}\overset{5\delta}{\subset}U\{Q_2(i,j)\}\overset{5\delta}{\subset}U\{Q_1(p,q)\}$ *with* $h>p$ *and* $q>k$, *then for arbitrary positive integers* r *and* w *there exists a positive integer* s *such that* $U\{Q_1^2[(h,k),s]\}\overset{(10)^3\delta}{\subset}U\{Q_2^2[(i,j),r]\}$ *with* $s>w$.

Proof. Suppose this is not true. Then for each n with $n>w$ there is a $T_n\in(U\{Q_1^2[(h,k),n]\})_1$ such that $\|T_n-S\|_2\geq(10)^3\,\delta$ for all

$$S\in(U\{Q_2^2[(i,j),r]\})_1.$$

Since $(Q_1^2[(h,k),n])$ is a strongly residual sequence in $Q_1(h,k)$, (T_n) is a central sequence in $U[Q_1(h,k)]$. On the other hand, $Q_1(p,q)=Q_1(h,k)\oplus C$ (C, some subgroup of $Q_1(p,q)$), Hence (T_n) is a central sequence in $U[Q_1(p,q)]$. Take $T'_n\in\{U[Q_2(i,j)]\}_1$ with $\|T_n-T'_n\|<5\delta$, and for an arbitrary $X'\in\{U[Q_2(i,j)]\}_1$, take $X\in\{U[Q_1(p,q)]\}_1$ with

$$\|X-X'\|_2<5\delta.$$

Then

$$\begin{aligned}\|[X',T'_n]\|_2&=\|[X',T'_n-T_n]\|_2+\|[T_n,X]\|_2+\|[T_n,X-X']\|_2\\&\leq 2\|T'_n-T_n\|_2+\|[T_n,X]\|_2+2\|X-X'\|_2.\end{aligned}$$

Hence $\limsup_n\|[X',T'_n]\|_2<20\delta$. Therefore there exists a sequence (T''_n) in $\{U[Q_2^2(i,j),r]\}_1$ such that $\limsup_n\|T'_n-T''_n\|_2<14\cdot 20\delta$. Thus

$$\limsup_n\|T_n-T''_n\|\leq\limsup_n\|T_n-T'_n\|+\limsup_n\|T'_n-T''_n\|_2<(10)^3\,\delta$$

This is a contradiction. q.e.d.

Applying this lemma for $\mathbb{I}_1$ and its symmetric one for $\mathbb{I}_2$, there exist positive integers $r_2,r_3,\dots,r_t$ such that $r_2<r_4<\cdots<r_t$ and $r_3<r_5<\cdots<r_{t-1}$, and

$$\begin{aligned}U\{Q_2^2[(n_t,m_t),r_t]\}\overset{(10)^3\delta}{\subset}\cdots\overset{(10)^3\delta}{\subset}&\;U\{Q_1^2[(n_3,m_3),r_3]\}\\\overset{(10)^3\delta}{\subset}&\;U\{Q_2^2[(n_2,m_2),r_2]\}.\end{aligned}$$

Since $Q(G[\mathbb{I}_1],n,m)=\sum_{j=n}^{m}\oplus M_j(\mathbb{I}_1)$ is of type $(p_1,p_2,\dots,p_m)$, $Q_1^2[(n,m),r]$ is of type $(p_1-1,p_2-1,\dots,p_m-1)_\infty$. At this time, $Q_i^2[(h,k),r]$ $(i=1,2)$ may contain a type 0 group as a direct summand. For $r<s$, we define

$$\begin{aligned}R\,Q_i^2[(h,k),(r,s)]=&\{\text{the center of } Q_i^2[(h,k),r]\}\\&+\{Q_i^2[(h,k),r]\ominus Q_i^2[(h,k),s+1]\}.\end{aligned}$$

4.3.17. Lemma. *For an arbitrary positive integer* s_t *with* $s_t > r_t$, *there exist positive integers* $s_4, s_5, \ldots, s_t$ *such that* $s_4 > s_6 > \cdots > s_t$ *and* $s_5 > s_7 > \cdots > s_{t-1}$, *and*

$$U[RQ_2^2[(n_t,m_t),(r_t,s_t)]\} \overset{(10)^5\delta}{\subset} \cdots \overset{(10)^5\delta}{\subset} U\{RQ_1^2[(n_5,m_5),(r_5,s_5)]\} \overset{(10)^5\delta}{\subset} U\{RQ_2^2[(n_4,m_4),(r_4,s_4)]\}.$$

Proof. Write $Q_2(n_{t-2},m_{t-2}) = Q_2(n_t,m_t) \oplus H$, where H is a subgroup of $Q_2(n_{t-2},m_{t-2})$. Since $Q_2^2[(n_t,m_t),r]$ is strongly residual in $Q_2(n_t,m_t)$, $Q_2^2[(n_t,m_t),r] \oplus H$ is strongly residual in $Q_2(n_{t-2},m_{t-2})$ for each r with $r > s_t$. On the other hand,

$$U\{Q_1(n_{t-1},m_{t-1})\} \overset{5\delta}{\subset} U\{Q_2(n_{t-2},m_{t-2})\} \overset{5\delta}{\subset} U\{Q_1(n_{t-3},m_{t-3})\}.$$

Hence by the reasoning similar to the proof of 4.3.16, for each r there exists a k with $k-1 > r_{t-1}$ such that

$$U\{Q_1^2[(n_{t-1},m_{t-1}),k]\} \overset{(10)^3\delta}{\subset} U\{Q_2^2[(n_t,m_t),r] \oplus H\}.$$

For $T \in (U\{RQ_2^2[(n_t,m_t),(r_t,s_t)]\})_1 \subset (U\{Q_2^2[(n_t,m_t),r_t]\})_1$ there exists a $T' \in (U\{Q_1^2[(n_{t-1},m_{t-1}),r_{t-1}]\})_1$ such that $\|T-T'\|_2 < (10)^3\delta$. For $X' \in (U\{Q_1^2[(n_{t-1},m_{t-1}),k]\})_1$, take $X \in (U\{Q_2^2[(n_t,m_t),r] \oplus H\})_1$ with $\|X-X'\|_2 < (10)^3\delta$. Then

$$\begin{aligned}\|[T',X']\|_2 &\le \|[T'-T,X]\|_2 + \|[X'-X,T]\|_2 + \|[T,X]\|_2 \\ &< 2(10)^3\delta + 2(10)^3\delta,\end{aligned}$$

because $[T,X]=0$. Hence there exists a

$$T'' \in ((U\{Q_1^2[(n_{t-1},m_{t-1}),k]\})')_1 \cap U\{Q_1^2[(n_{t-1},m_{t-1}),r_{t-1}]\}$$

with $\|T'-T''\|_2 < 5\cdot(10)^3\delta$. Thus

$$\|T-T''\|_2 \le \|T-T'\|_2 + \|T'-T''\|_2 < (10)^5\delta.$$

Clearly

$$\begin{aligned}(U\{Q_1^2[(n_{t-1},m_{t-1}),k]\})' \cap U\{Q_1^2[(n_{t-1},m_{t-1}),r_{t-1}]\} \\ = U\{RQ_1^2[(n_{t-1},m_{t-1}),(r_{t-1},k-1)]\}.\end{aligned}$$

Take $k-1$ as s_{t-1}. The remainder of the proof is quite similar. q.e.d.

$RQ_1^2[(h,k),(i,j)]$ is of type $(p_1-1, p_2-1, \ldots, p_k-1)$ and $RQ_2^2[(h,k),(i,j)]$ is of type $(q_1-1, q_2-1, \ldots, q_k-1)$. They may contain a type 0 group as a direct summand. Write $RQ_1^2[(h,k),(i,j)] = D \oplus W$, where D is the center of $RQ_1^2[(h,k),(i,j)]$ and W is of type $(i_1, i_2, \ldots, i_n)$ with $i_u \ge 1$ for $u = 1,2,\ldots,n$. Define the canonical strongly residual sequence of

$RQ_1^2[(h,k),(i,j)]$ as follows: $Q_1 RQ_1^2[(h,k),(i,j),n]=D\oplus Q(W,n)$. Quite similarly, we can define the canonical, strongly residual sequence of RQ_2^2. Then from the results in 4.3.17 and the reasoning similar to 4.3.16, we obtain

$$\begin{aligned} U\{D_t\oplus Q(W_t,l_t)\} &\overset{(10)^8\delta}{\subset} U\{D_{t-1}\oplus Q(W_{t-1},l_{t-1})\} \\ &\overset{(10)^8\delta}{\subset} \cdots \\ &\overset{(10)^8\delta}{\subset} U\{D_5\oplus Q(W_5,l_5)\}, \end{aligned}$$

where D_i is the center of $RQ_1^2[(n_i,m_i),(r_i,s_i)]$ or $RQ_2^2[(n_i,m_i),(r_i,s_i)]$ $(i=5,6,\ldots,t)$ and $l_5<l_7<\cdots<l_{t-1}$, $l_6<l_8<\cdots<l_t$.

Next, by applying the reasoning similar to the proof of 4.3.17 for these relations we have

$$\begin{aligned} U\{D_t\oplus RQ[W_t,(l_t,p_t)]\} & \\ &\overset{(10)^{10}\delta}{\subset} \cdots \\ &\overset{(10)^{10}\delta}{\subset} U\{D_7\oplus RQ[W_7,(l_7,p_7)]\} \\ &\overset{(10)^{10}\delta}{\subset} U\{D_6\oplus RQ[W_6,(l_6,p_6)]\} \end{aligned}$$

with $l_t<p_t, p_t<p_{t-2}<\cdots<p_6$ and $p_{t-1}<p_{t-3}<\cdots<p_7$. Continuing this process q_{i_0}-times, we have the following situation,

$$U(\Omega_t)\overset{K\delta}{\subset}U(\Omega_{t-1})\overset{K\delta}{\subset}U(\Omega_{t-2})\overset{K\delta}{\subset}U(\Omega_{t-3}),$$

where Ω_t contains a type 1 group as a direct summand and Ω_{t-1} does not contain a type 1 group as a direct summand. Moreover $\Omega_{t-2}=\Omega_t\oplus R$, where R is a subgroup of Ω_{t-2} and K is a constant which does not depend on δ, and by the "$q_{i_0}+1$"-th process, we have

$$U(\Delta_t)\overset{K_1\delta}{\subset}U(\Delta_{t-1})\overset{K_1\delta}{\subset}U(\Delta_{t-2}),$$

where K_1 does not depend on δ. Moreover, put $\Omega_t=E\oplus H$ (E, the center of Ω_t). Then $\Delta_t=E\oplus E_1\oplus W$, where E_1 is contained in the center of Δ_t and $E_1=Q(L_1,n)$ for some n.

On the other hand, the center of Δ_{t-1} is the same as the center C of Ω_{t-1}, because Ω_{t-1} does not contain a type 1 group as a direct summand.

4.3.18. Lemma. *For* $X\in\{U(E_1)\}_1$, *there exists an element* $X'\in\{U(C)\}_1$ *with* $\|X-X'\|_2<(10)^2 K_1\delta$.

Proof. Put $X_n = X$. Then (X_n) is a central sequence in $U(E_1)$. Since $\Delta_{t-2} = \Delta_t \oplus \Gamma$ for some subgroup Γ of Δ_{t-2}, (X_n) is a central sequence in $U(\Delta_{t-2})$. Let $Y' \in \{U(\Delta_{t-1})\}_1$ with $\|X - Y'\| < K_1 \delta$. Then by the same reasoning with the proof of 4.3.15, $\|[Y', W']\|_2 < 5 K_1 \delta$ for all $W' \in \{U(\Delta_{t-1})\}_1$. Hence there exists a central element X' in $\{U(\Delta_{t-1})\}_1$ with $\|X' - Y'\|_2 \leq 2 \cdot 5 K_1 \delta$ and so

$$\|X - X'\|_2 \leq \|X - Y'\|_2 + \|Y' - X'\|_2 < (10)^2 K_1 \delta. \quad \text{q.e.d.}$$

Now we shall prove Theorem 4.3.9.

Proof of 4.3.9. $\{U(E_1)\}_1 \subset \{U(L_1)\}_1 \subset \{U(\Omega_t)\}_1 \overset{K\delta}{\subset} U(\Omega_{t-1})$. By 4.3.18, for $X \in \{U(E_1)\}_1$ there exists an $X' \in \{U(C)\}_1$ with

$$\|X - X'\|_2 < (10)^2 K_1 \delta.$$

For an arbitrary $Y \in \{U(L_1)\}_1$, take $Y' \in \{U(\Omega_{t-1})\}_1$ with $\|Y' - Y\|_2 < K \delta$. Then

$$\begin{aligned} \|[Y, X]\|_2 &\leq \|[Y - Y', X]\|_2 + \|[Y', X - X']\|_2 + \|[Y', X']\|_2 \\ &\leq 2K\delta + 2 \cdot 10^2 K_1 \delta. \end{aligned}$$

Hence there exists an element $X'' \in U(L_1) \cap U(L_1)' = \mathbb{C}1$ such that $\|X - X''\|_2 < 4(K + 10^2 K_1)\delta$. One can choose δ as an arbitrary small number and so $U(E_1)$ must be the center of $U(L_1)$. On the other hand, $U(E_1)$ is not the center, because $U(E_1) = U\left(\sum_{j=n}^{\infty} \oplus H_j\right)$ with $H_j = Z$ $(j = n, n+1, \ldots)$. This is a contradiction and completes the proof. q.e.d.

Next, by using the tensor product of the above examples of type II_1-factors and a special type III-factor, we shall construct an uncountable family of type III-factors and as its corollary, an uncountable family of type II_∞-factors.

We shall construct first a special type III-factor due to Pukanszky [139].

Let G be the free group of two generators, Ω_g $(g \in G)$ be the additive group of two elements $(0, 1)$. Define a measure μ_g on Ω_g such that $\mu_g(\{0\}) = p$, $\mu(\{1\}) = q$ $(q > p > 0$ and $p + q = 1)$. Let $\Omega = \prod_{g \in G} \Omega_g$ be the infinite direct product of $\{\Omega_g | g \in G\}$. Then Ω is a compact group in the weak topology. Let μ be the Radon measure on Ω defined by $\prod_{g \in G} \mu_g$. Every $\xi \in \Omega$ may be identified with a function (ξ_g) $(g \in G)$ defined on G taking the values 0 and 1 only.

Let $\mathscr{G}_1$ be the set of those $\alpha = (\alpha_g) \in \Omega$ for which $\alpha_g \neq 0$ occurs for a finite number of g only. We denote the set of pairs (α, a) $(\alpha \in \mathscr{G}_1, a \in G)$ by $\mathscr{G}$.

To an element $(\alpha, a)=d$ of $\mathscr{G}$ we associate a mapping $\xi \to \xi d$ of Ω onto itself defined by $\xi d=(\xi_{ag}+\alpha_g)$ $(g\in G)$. These mappings are one-to-one. Introducing the notation $\alpha^a=(\alpha_{ag})$ $(g\in G)$, we get

$$\{\xi(\alpha,a)\}(\beta,b)=(\xi_{abg}+\alpha_{bg}+\beta_g)$$

and so $(\alpha,a)(\beta,b)=(\alpha^b+\beta, ab)$. If $\xi(\alpha,a)=\xi(\alpha',a')$ for $\xi\in\Omega$, we have $\alpha=\alpha'$ and $a=a'$. Hence $\mathscr{G}$ is a semi-group. Observing

$$(\alpha,a)(0,e)=(\alpha^e+0,ae)=(\alpha,a),$$

where 0 (resp. e) is the unit of $\mathscr{G}_1$ (resp. G), and

$$(0,e)(\alpha,a)=(0^a+\alpha,ea)=(\alpha,a);$$

$$(\alpha,a)(\alpha^{a^{-1}},a^{-1})=(\alpha^{a^{-1}a}+\alpha,a^{-1}a)=(0,e).$$

We have that $(0,e)$ is the unit and the inverse $(\alpha,a)^{-1}=(\alpha^{a^{-1}},a^{-1})$, so that $\mathscr{G}$ is a group.

One can easily see that the μ is quasi-invariant under the mappings $\xi\to\xi(\alpha,a)$ $(\alpha\in\mathscr{G}_1, a\in G)$. We shall denote $\dfrac{d\mu(\alpha,a)}{d\mu}(\xi)$ by $r(\alpha,a)(\xi)$. It is easily seen that the correspondences $\alpha\to(\alpha,e)$ and $a\to(0,a)$ define isomorphisms of $\mathscr{G}_1$ and G with subgroups of $\mathscr{G}$ respectively. We denote these subgroups in the sequel again by $\mathscr{G}_1$ and G.

Now put $L^2(\Omega,\mu)=\mathscr{H}$, $L^\infty(\Omega,\mu)=\mathscr{M}$ and

$$U(\alpha,a)f(\xi)=\sqrt{r_{(\alpha,a)}(\xi)}\,f(\xi(\alpha,a))$$

for $f\in L^2(\Omega,\mu)$ and $(\alpha,a)\in\mathscr{G}$, and we shall construct a weakly closed *-subalgebra $\mathscr{N}$ of $B(\tilde{\mathscr{H}})$ according to the method of 4.2. Then by the same reasoning as 4.2.13, $\mathscr{N}$ is a type III-factor. We shall denote this special $\mathscr{N}$ by $\mathbb{P}$.

4.3.19. Proposition. *$\mathbb{P}$ has a cyclic and separating vector η_0 ($\|\eta_0\|=1$) in $\tilde{\mathscr{H}}$ and moreover there exist two elements T_1, T_2 in $\mathbb{P}$ and a fixed positive number K such that for $X\in\mathbb{P}$, $\|[X,T_i]\|_2^2<\varepsilon$ $(i=1,2)$ imply*

$$|(X\eta_0,\eta_0)|^2\geq\|X\|_2^2-K\varepsilon,$$

where $\|X\|_2=\|X\eta_0\|$ and ε is an arbitrary positive number.

To prove this proposition, we shall provide some lemmas.

4.3.20. Lemma. *Let G be the free group of two generators a_1, a_2. If a function $f(a)$ on G satisfies $\sum_{a\in G}|f(a)|^2<+\infty$ and $\left(\sum_{a\in G}|f(aa_i)-f(a)|^2\right)^{\frac{1}{2}}<\varepsilon$ $(i=1,2)$, then $\left(\sum_{a\in G}|f(a)|^2\right)^{\frac{1}{2}}<14\varepsilon$.*

Proof. Let F be the set of those $a \in G$, which when written as a power of a_1, a_2 of minimum length, end with a_1^n $(n = \pm 1, \pm 2, \ldots)$. Then it is clear that $F \cup F a_1 = G$. Moreover $F \cap F a_2 = (\emptyset)$, $F \cap F a_2^{-1} = (\emptyset)$ and $F a_2 \cap F a_2^{-1} = (\emptyset)$. Put $\nu(M) = \sum_{a \in M} |f(a)|^2$ for a subset M in G. Then

$$\varepsilon > \Big(\sum_{a \in G} |f(a a_1) - f(a)|^2 \Big)^{\frac{1}{2}} \geq |\nu(F a_1)^{\frac{1}{2}} - \nu(F)^{\frac{1}{2}}| \,.$$

Putting $\nu(G) = s^2$, then

$$|\nu(F a_1) - \nu(F)| = |\nu(F a_1)^{\frac{1}{2}} + \nu(F)^{\frac{1}{2}}| \, |\nu(F a_1)^{\frac{1}{2}} - \nu \mathrm{F})^{\frac{1}{2}}| < 2 s \varepsilon$$

and so $\nu(F a_1) < \nu(F) + 2 s \varepsilon$. Hence $s^2 < \nu(F a_1) + \nu(F) < 2(\nu(F) + s \varepsilon)$, so that $\varepsilon(F) > \frac{s^2}{2} - s \varepsilon$.

Analogously we have $|\nu(F a_2) - \nu(F)| < 2 s \varepsilon$, $|\nu(F a_2^{-1}) - \nu\,(F)| < 2 s \varepsilon$. Hence $\nu(F a_2) > \nu(F) - 2 s \varepsilon > \frac{s^2}{2} - 3 s \varepsilon$ and $\nu(F a_2^{-1}) > \frac{s^2}{2} - 3 s \varepsilon$. Thus

$$s^2 = \nu(G) \geq \nu(F) + \nu(F a_2) + \nu(F a_2^{-1}) > \tfrac{3}{2} s^2 - 7 s \varepsilon,$$

that is $s < 14 \varepsilon$. q.e.d.

Proof of 4.3.19. Now put $T_i = \tilde{U}_{a_i}$ and let f_0 be the function on Ω with $f_0(\xi) \equiv 1$. Set $\eta_0 = J_{(0,e)} f_0$ and suppose that

$$\| [\tilde{U}_{a_i}, X] \|_2^2 < \varepsilon (\| Y \|_2 = \| Y \eta_0 \| \text{ for } Y \in \mathbb{P}).$$

Let $X^* = (\varphi_{st^{-1}} U_{st^{-1}})$ with $\varphi_t \in L^\infty(\Omega, \mu)$ and $s, t \in \mathcal{G}$, and put $(X^*)_t(\xi) = \varphi_t(\xi)$. Then $(X)_t(\xi) = \overline{\varphi_{t^{-1}}(\xi t)}$. Therefore $(X \tilde{U}_{a_i})_t(\xi) = \overline{\varphi_{a_i t^{-1}}(\xi t a_i^{-1})}$,

$$(\tilde{U}_{a_i} X)_t(\xi) = \overline{\varphi_{t^{-1} a_i}(\xi t)}$$

and so $(X \tilde{U}_{a_i} - \tilde{U}_{a_i} X)_t(\xi) = \overline{\varphi_{a_i t^{-1}}(\xi t a_i^{-1})} - \overline{\varphi_{t^{-1} a_i}(\xi t)}$.

$$\begin{aligned} \| [X, \tilde{U}_{a_i}] \|_2^2 &= \sum_{t \in \mathcal{G}} \int |\varphi_{a_i t^{-1}}(\xi t a_i^{-1}) - \varphi_{t^{-1} a_i}(\xi t)|^2 \, |U_t f_0(\xi)|^2 \, d\mu(\xi) \\ &= \sum_{t \in \mathcal{G}} \int |\varphi_{a_i t^{-1}}(\xi t a_i^{-1}) - \varphi_{t^{-1} a_i}(\xi t)|^2 \, r_t(\xi) \, d\mu(\xi) \\ &= \sum_{t \in \mathcal{G}} \int |\varphi_{a_i t^{-1}}(\xi a_i^{-1}) - \varphi_{t^{-1} a_i}(\xi)|^2 \, d\mu(\xi) \\ &= \sum_{t \in \mathcal{G}} \int |\varphi_{a_i t}(\xi a_i^{-1}) - \varphi_{t a_i}(\xi)|^2 \, d\mu(\xi) \\ &= \sum_{t \in \mathcal{G}} \int |\varphi_{a_i t a_i^{-1}}(\xi a_i^{-1}) - \varphi_t(\xi)|^2 \, d\mu(\xi). \end{aligned}$$

Put $f(t)=\Big(\int_\Omega |\varphi_t(\xi)|^2\, d\mu(\xi)\Big)^{\frac{1}{2}}$. Since μ is invariant under G,

$$\Big(\sum_{t\in\mathscr{G}} |f(a_i t a_i^{-1})-f(t)|^2\Big)^{\frac{1}{2}}$$

$$=\Big(\sum_{t\in\mathscr{G}} |(\int |\varphi_{a_i^{-1}ta_i}(\xi a_i^{-1})|^2\, d\mu(\xi)\Big)^{\frac{1}{2}}-(\int |\varphi_t(\xi)|^2\, d\mu(\xi))^{\frac{1}{2}}|^2)^{\frac{1}{2}}$$

$$\leq\Big(\sum_{t\in\mathscr{G}} \int |\varphi_{a_i^{-1}ta_i}(\xi a_i^{-1})-\varphi_t(\xi)|^2\, d\mu(\xi)\Big)^{\frac{1}{2}}=\|[X,\tilde{U}_{a_i}]\|_2<\sqrt{\varepsilon}.$$

Put $\mathscr{E}=\{(\alpha,a)\,|\,a\neq e, \alpha\in\mathscr{G}_1, a\in G\}$ and $\mathfrak{F}=\{(\alpha,a)\,|\,\alpha\in\mathscr{G}_1, a\in F\}$, where F is the set in the proof of 4.3.20. Then $\mathfrak{F}\cup a_1\mathfrak{F}a_1^{-1}=\mathscr{E}$ and $\mathfrak{F}$, $a_2\mathfrak{F}a_2^{-1}$, $a_2^{-1}\mathfrak{F}a_2$ $(\subset\mathscr{E})$ are mutually disjoint. Hence by 4.3.3,

$$\Big(\sum_{t\in\mathscr{E}} |f(t)|^2\Big)^{\frac{1}{2}}<14\sqrt{\varepsilon}.$$

Since $a_i\alpha a_i^{-1}=\alpha^{a_i^{-1}}$, $\Big(\sum_{\alpha\in\mathscr{G}_1} |f(\alpha^{a_i^{-1}})-f(\alpha)|^2\Big)^{\frac{1}{2}}<\sqrt{\varepsilon}\ (i=1,2)$. For $\alpha,\beta\in\mathscr{G}_1$, we write $\alpha\sim\beta$ if there exists an $a\in G$ with $\alpha^a=\beta$. The relation $\sim$ will define an equivalence relation in $\mathscr{G}_1$. Let Λ be all equivalence classes under the relation $\sim$. If $\alpha_\lambda\in\mathscr{G}_1$ is an element of the class λ, then every element of λ can be written uniquely in the form α_λ^a $(a\in G)$. Put

$$h^\lambda(a)=f(\alpha_\lambda^a)\quad(\lambda\in\Lambda, a\in G)$$

and $a_\lambda=\Big(\sum_{\alpha\in G} |h^\lambda(a)|^2\Big)^{\frac{1}{2}}$, $b_\lambda=\Big(\sum_{i=1}^{2}\sum_{a\in G} |h^{(\lambda)}(a a_i)-h^{(\lambda)}(a)|^2\Big)^{\frac{1}{2}}$. Then by 4.3.20, $a_\lambda\leq 14 b_\lambda$. Hence

$$\sum_{\substack{\alpha\in\mathscr{G}_1\\ \alpha\neq 0}} |f(\alpha)|^2=\sum_{\substack{\lambda\in\Lambda\\ \lambda\neq(0)}}\sum_{a\in G} |h^\lambda(a)|^2=\sum_{\substack{\lambda\in\Lambda\\ \lambda\neq(0)}} a_\lambda^2\leq(14)^2\sum_{\substack{\lambda\in\Lambda\\ \lambda\neq(0)}} b_\lambda^2$$

$$=(14)^2\sum_{\substack{\lambda\in\Lambda\\ \lambda\neq(0)}}\Big(\sum_{a\in G} |f(\alpha_\lambda^{aa_1})-f(\alpha_\lambda^a)|^2+\sum_{\substack{a\in G\\ \lambda\neq(0)}} |f(\alpha_\lambda^{aa_2})-f(\alpha_\lambda^a)|^2\Big)$$

$$=(14)^2\sum_{\alpha\neq 0}\sum_{i=1}^{2} |f(\alpha^{a_i})-f(\alpha)|^2<2(14)^2\varepsilon.$$

Hence

$$\int |\varphi_{(0,e)}(\xi)|^2\, d\mu(\xi)=|f(0,e)|^2$$

$$\geq\sum_{t\in\mathscr{G}} |f(t)|^2-(14)^2\varepsilon-2(14)^2\varepsilon$$

$$=\sum_{t\in\mathscr{G}}\int |\varphi_t(\xi)|^2\, d\mu(\xi)-3(14)^2\varepsilon$$

$$=\|X\eta_0\|^2-3(14)^2\varepsilon.$$

On the other hand, by the results of 4.2.15, we may write

$$\varphi_{(0,e)}(\xi)=\sum_{\alpha\in\mathscr{G}_1} C_\alpha W_\alpha(\xi)$$

with $\sum_{\alpha\in\mathscr{G}_1}|C_\alpha|^2<+\infty$. Since the mapping $\xi\to\xi a$ $(\xi\in\Omega, a\in G)$ leaves the measure μ invariant and $W_\alpha(\xi a)=W_{\alpha^{a^{-1}}}(\xi)$,

$$\begin{aligned}\int\varphi_{(0,e)}(\xi a_i)\,W_\alpha(\xi)\,d\mu(\xi)&=\int\varphi_{(0,e)}(\xi)\,W_\alpha(\xi a_i^{-1})\,d\mu(\xi)\\&=\int\varphi_{(0,e)}(\xi)\,W_{\alpha^{a_i}}(\xi)\,d\mu(\xi)\\&=C_{\alpha^{a_i}}\end{aligned}$$

Hence

$$\int|\varphi_{(0,e)}(\xi a_i)-\varphi_{(0,e)}(\xi)|^2\,d\mu(\xi)=\sum_{\alpha\in\mathscr{G}_1}|C_{\alpha^{a_i}}-C_\alpha|^2.$$

Put $k^\lambda(a)=C_{\alpha_\lambda^a}$ $(\lambda\in\Lambda, a\in G)$, $d_\lambda=\left(\sum_{a\in G}|k^\lambda(a)|^2\right)^{\frac{1}{2}}$ and

$$e_\lambda=\left(\sum_{i=1}^{2}\sum_{a\in G}|k^{(\lambda)}(a a_i)-k^{(\lambda)}(a)|^2\right)^{\frac{1}{2}}.$$

Then

$$\begin{aligned}&\left|\left|\int\varphi_{(0,e)}(\xi)\,d\mu(\xi)\right|^2-\int\left|\varphi_{(0,e)}(\xi)\right|^2d\mu(\xi)\right|\\
&=\sum_{\substack{\alpha\in\mathscr{G}_1\\ \alpha\neq 0}}|C_\alpha|^2=\sum_{\substack{\lambda\in\Lambda\\ \lambda\neq(0)}}\sum_{\alpha\in G}|C_{\alpha_\lambda^a}|^2\\
&=\sum_{\substack{\lambda\in\Lambda\\ \lambda\neq(0)}}d_\lambda^2\le(14)^2\sum_{\substack{\lambda\in\Lambda\\ \lambda\neq(0)}}e_\lambda^2\\
&=(14)^2\sum_{\substack{\lambda\in\Lambda\\ \lambda\neq(0)}}\sum_{i=1}^{2}|C_{\alpha_\lambda^{aa_i}}-C_{\alpha_\lambda^a}|^2\\
&=(14)^2\sum_{i=1}^{2}\sum_{\alpha\in\mathscr{G}_1}|C_{\alpha^{a_i}}-C_\alpha|^2\\
&=(14)^2\sum_{i=1}^{2}\int|\varphi_{(0,e)}(\xi a_i)-\varphi_{(0,e)}(\xi)|^2\,d\mu(\xi)\\
&\le(14)^2\sum_{i=1}^{2}\sum_{t\in\mathscr{G}}\int|\varphi_{a_i t a_i^{-1}}(\xi a_i^{-1})-\varphi_t(\xi)|^2\,d\mu(\xi)\\
&<2(14)^2\,\varepsilon.\end{aligned}$$

Hence

$$\left|\int \varphi_{(0,e)}(\xi)\,d\mu(\xi)\right|^2 \geq \int |\varphi_{(0,e)}(\xi)|^2\,d\mu(\xi) - 2(14)^2\,\varepsilon$$
$$\geq \|X\eta_0\|^2 - 5\cdot(14)^2\,\varepsilon.$$

On the other hand,

$$\left|\int \varphi_{(0,e)}(\xi)\,d\mu(\xi)\right|^2 = |(X^*\eta_0,\eta_0)|^2 = |(X\eta_0,\eta_0)|^2.$$

This completes the proof.

Now let H be a countable discrete group such that $U(H)$ is a factor. Then the W^*-tensor product $\mathbb{P}\,\overline{\otimes}\,U(H)$ is a type III-factor (cf. 2.6). The tensor product $\mathbb{P}\,\overline{\otimes}\,U(H)$ is a special case of the construction in 4.2 such that $h\to U_h=$ the identity $(h\in H)$ and $\mathbb{P}=\mathscr{M}$. Therefore every element $S\in\mathbb{P}\,\overline{\otimes}\,U(H)$ can be represented by a matrix of the form $(T_{st^{-1}})$ with $T_h\in\mathbb{P}$ $(h,s,t\in H)$. Define a normal state φ_H on $\mathbb{P}\,\overline{\otimes}\,U(H)$ as follows: $\varphi_H((T_{st^{-1}}))=(T_{e'}\eta_0,\eta_0)$, where e' is the unit of H.

Put $A_{i,e'}=T_i$ and $A_{i,h}=0$ for $h\neq e'$ $(i=1,2)$, where T_i $(i=1,2)$ are the elements of $\mathbb{P}$ stated in 4.3.19. Then

$$[(A_{i,st^{-1}})(T_{st^{-1}})]=([T_i,T_{st^{-1}}]).$$

Now we shall generalize the definition of central sequences into an arbitrary W^*-algebra.

4.3.21. Definition. *Let $\mathscr{L}$ be a W^*-algebra and let (X_n) be a uniformly bounded sequence in $\mathscr{L}$. (X_n) is called a central sequence if $[X_n,X]\to 0$ in the s-topology $(X\in\mathscr{L})$. Let (X_n), (Y_n) be two central sequences in $\mathscr{L}$. (X_n) is said to be equivalent to (Y_n) if $X_n-Y_n\to 0$ in the s-topology and is denoted by $(X_n)\sim(Y_n)$.*

4.3.22. Lemma. *A central sequence in $\mathbb{P}\,\overline{\otimes}\,U(H)$ is equivalent to a central sequence in $1\otimes U(H)$.*

Proof. Put $|||(T_{st^{-1}})|||_2=\varphi_H((T_{st^{-1}})^*(T_{st^{-1}}))^{\frac{1}{2}}=\left(\sum_{h\in H}\|T_h\eta_0\|^2\right)^{\frac{1}{2}}$ for $X=(T_{st^{-1}})$. Suppose that

$$|||[(A_{i,st^{-1}}),(T_{st^{-1}})]|||_2^2=|||([T_i,T_{st^{-1}}])|||_2^2=\sum_{h\in H}\|[T_i,T_h]\|_2^2<\varepsilon.$$

Then

$$|(T_h\eta_0,\eta_0)|^2\geq\|T_h\|_2^2-K(\|[T_1,T_h]\|_2^2+\|[T_2,T_h]\|_2^2).$$

Hence

$$|||(T_{st^{-1}})-((T_{st^{-1}}\eta_0,\eta_0)1)|||_2^2=\sum_{h\in H}(\|T_h\eta_0\|^2-|(T_h\eta_0,\eta_0)|^2)$$
$$\leq\sum_{h\in H}K(\|[T_1,T_h]\|_2^2+\|[T_2,T_h]\|_2^2)<2K\varepsilon.$$

Now we shall show that $((T_{st^{-1}}\eta_0,\eta_0)1)$ is a bounded operator and belongs to $1\otimes U(H)$.

Put $\psi_0(X)=(X\eta_0,\eta_0)$ for $X\in\mathbb{P}$ and let $f\in U(H)_*$ ($U(H)_*$ is the predual of $U(H)$). Then $\psi_0\otimes f$ is a normal linear functional on $\mathbb{P}\otimes U(H)$. Moreover, $|\psi_0\otimes f((T_{st^{-1}}))|\leq\|f\|\,\|(T_{st^{-1}})\|$. Hence there exists an element d in $U(H)$ such that $\psi_0\otimes f((T_{st^{-1}}))=f(d)$ and $\|d\|\leq\|(T_{st^{-1}})\|$. Let $f_{a,b}(y)=(y\varepsilon_a,\varepsilon_b)$ for $y\in U(H)$. Then

$$\psi_0\otimes f_{a,b}((T_{st^{-1}}))=((T_{st^{-1}})\eta_0\otimes\varepsilon_a,\eta_0\otimes\varepsilon_b)=(T_{ba^{-1}}\eta_0,\eta_0).$$

On the other hand, $f_{a,b}(d)=(d\varepsilon_a,\varepsilon_b)=d(ba^{-1})$. Hence

$$(T_h\eta_0,\eta_0)=d(h)\qquad(h\in H)$$

and so $((T_{st^{-1}}\eta_0,\eta_0)1)$ belongs to $1\otimes U(H)$ and

$$\|((T_{st^{-1}}\eta_0,\eta_0)1)\|\leq\|(T_{st^{-1}})\|.\qquad\text{q.e.d.}$$

4.3.23. Theorem. *Let $\mathbb{I}_1$ and $\mathbb{I}_2$ be two sequences of positive integers which are mutually different as sets. Then $\mathbb{P}\mathbin{\bar\otimes} U(G[\mathbb{I}_1])$ is not *-isomorphic to $\mathbb{P}\mathbin{\bar\otimes} U(G[\mathbb{I}_2])$.*

Remark. These factors are not hyperfinite (cf. 4.4).

4.3.24. Corollary. *These exists an uncountable family of type* III-*factors on a separable Hilbert space.*

To prove Theorem 4.3.23, we shall provide some lemmas.

Let $\mathscr{A}=\mathbb{P}\mathbin{\bar\otimes} U(G[\mathbb{I}_1])$ and $\mathscr{B}=\mathbb{P}\mathbin{\bar\otimes} U(G[\mathbb{I}_2])$. Suppose that $\mathscr{A}$ is *-isomorphic to $\mathscr{B}$. Then under the identification $\mathscr{A}=\mathscr{B}$, we have two expressions $\mathbb{P}\mathbin{\bar\otimes} U(G[\mathbb{I}_1])$ and $\mathbb{P}\mathbin{\bar\otimes} U(G[\mathbb{I}_2])$. Pick the expression $\mathscr{A}$ and take the normal state $\varphi_{G[\mathbb{I}_1]}$ and define $|||X|||_2=\varphi_{G[\mathbb{I}_1]}(X^*X)^{\frac{1}{2}}$ for $X\in\mathscr{A}$. Then this metric is equivalent to the s-topology on bounded spheres.

We shall extend the notion $V_1\overset{\delta}{\subset}V_2$ to the algebra $\mathscr{A}$, using the norm $|||\;|||_2$ instead of the trace norm in finite algebras. One can easily see that $\varphi_{G[\mathbb{I}_1]}(XY)=\varphi_{G[\mathbb{I}_1]}(YX)$ for $X\in\mathbb{P}\mathbin{\bar\otimes} U(G[\mathbb{I}_1])$ and

$$Y\in 1\otimes U(G[\mathbb{I}_1]).$$

Let (U_n) be a central sequence in $U(G[\mathbb{I}_1])$ with $U_n^*U_n=1$. Then we shall show that $(1\otimes U_n)$ is a central sequence in $\mathbb{P}\mathbin{\bar\otimes} U(G[\mathbb{I}_1])$. For $X\in\mathbb{P}\mathbin{\bar\otimes} U(G[\mathbb{I}_1])$ and $\varepsilon>0$, there exists an element X_0 in the algebraic tensor product $\mathbb{P}\odot U(G[\mathbb{I}_1])$ with $|||X-X_0|||_2<\varepsilon$. Then

$$\begin{aligned}|||(1\otimes U_n)^*X(1\otimes U_n)-X|||_2&\leq|||(1\otimes U_n^*)X(1\otimes U_n)-(1\otimes U_n^*)X_0(1\otimes U_n)|||_2\\&\quad+|||(1\otimes U_n^*)X_0(1\otimes U_n)-X_0|||_2+|||X-X_0|||_2.\end{aligned}$$

Hence $\limsup_n |||(1\otimes U_n)^* X(1\otimes U_n)-X|||_2 \leq 2|||X-X_0|||_2 < 2\varepsilon$, and so $[X, 1\otimes U_n]\to 0$ in the s-topology.

Now, consider a sequence $\{1\otimes U(Q(G[\mathrm{II}_1], n)\}$ of W^*-subalgebras in $\mathbb{P}\,\bar{\otimes}\, U(G[\mathrm{II}_1])$, and take $Y_n \in 1\otimes U(Q(G[\mathrm{II}_1], n))$ with $\|Y_n\|\leq 1$, $Y_n^* = Y_n$, then $Y_n = \dfrac{(Y_n + i\sqrt{1-Y_n^2}) + (Y_n - i\sqrt{1-Y_n^2})}{2}$, and $Y_n \pm i\sqrt{1-Y_n^2}$ are unitary.

Since $(Y_n \pm i\sqrt{1-Y_n^2})$ is central in $1\otimes U(G[\mathrm{II}_1])$, they are central in $\mathbb{P}\,\bar{\otimes}\, U(G[\mathrm{II}_1])$. Now one can easily see that an arbitrary uniformly bounded sequence (X_n) with $X_n \in 1\otimes U\{Q(G[\mathrm{II}_1], n)\}$ is central. Analogously, (X_n) with $X_n \in 1\otimes U\{Q(G[\mathrm{II}_2], n)\}$ is also central in $\mathbb{P}\,\bar{\otimes}\, U(G[\mathrm{II}_2])$.

Conversely an arbitrary central sequence is equivalent to a central sequence in $1\otimes U(G[\mathrm{II}_i])$ $(i=1,2)$ by 4.3.22. For simplicity, we shall identify $1\otimes U(G[\mathrm{II}_i])$ with $U(G[\mathrm{II}_i])$.

4.3.25. Lemma. *For a positive number δ with $0<\delta<1$ and a positive integer n_1 there exists a positive integer n_2 such that*

$$U(Q(G[\mathrm{II}_2], n_2) \overset{\delta}{\subset} U(Q(G[\mathrm{II}_1], n_1)).$$

Proof. Suppose that this is not true. Then there exists

$$T_n \in [U\{Q(G[\mathrm{II}_2], n)\}]_1$$

for each n such that $|||T_n - S|||_2 \geq \delta$ for all $S\in [U\{Q(G[\mathrm{II}_1], n_1)\}]_1$. Since $\{Q(G[\mathrm{II}_2], n)\}$ is a strongly residual sequence in $G[\mathrm{II}_2]$, (T_n) is central in $\mathbb{P}\,\bar{\otimes}\, U(G[\mathrm{II}_1])$. Hence there exists a sequence (T'_n) in $[U\{Q(G[\mathrm{II}_1], n_1)\}]_1$ such that $|||T_n - T'_n|||_2 \to 0$. This is a contradiction. q.e.d.

4.3.26. Lemma. *For a positive integer m_2 with $m_2 > n_2$, there exists a positive integer m_1 such that $m_1 > n_1$ and*

$$U\{Q(G[\mathrm{II}_2], n_2, m_2)\} \overset{9\delta^{\frac{1}{2}}}{\subset} U\{Q(G[\mathrm{II}_1], n_1, m_1)\}.$$

Proof. For $T\in [U\{Q(G[\mathrm{II}_2], n_2, m_2)\}]_1$, take $T' \in [U\{Q(G[\mathrm{II}_1], n_1)\}]_1$ such that $|||T-T'|||_2 < \delta$. By 4.3.25, we can choose a positive integer r such that $r-1>n_1$ and $U\{Q(G[\mathrm{II}_1], r)\} \overset{\delta}{\subset} U\{Q(G[\mathrm{II}_2], m_2+1)\}$. For $X' \in [U\{Q(G[\mathrm{II}_1], r)\}]_1$, take $X\in [U\{Q(G[\mathrm{II}_2], m_2+1)\}]_1$ such that $|||X'-X|||_2 < \delta$. Then

$$\begin{aligned} |||[X', T']|||_2 &\leq |||[X', T'-T]|||_2 + |||[X, T]|||_2 + |||[X-X', T]|||_2 \\ &= |||[X', T'-T]|||_2 + |||[X-X', T]|||_2, \end{aligned}$$

because $[X, T]=0$.

Since $\varphi_{G[\mathrm{II}_1]}(XY)=\varphi_{G[\mathrm{II}_1]}(YX)$ for $Y\in U(G[\mathrm{II}_1])$ and

$$X\in \mathbb{P}\mathbin{\bar{\otimes}} U(G[\mathrm{II}_1]),$$

$$\begin{aligned}
|||[X',T'-T]|||_2 &\le |||X'(T'-T)|||_2+|||(T'-T)X'|||_2\\
&\le |||T'-T|||_2+\varphi_{G[\mathrm{II}_1]}(X'^*(T'-T)^*(T'-T)X')^{\frac{1}{2}}\\
&\le \delta+\varphi_{G[\mathrm{II}_1]}(X'X'^*(T'-T)^*(T'-T))^{\frac{1}{2}}\\
&\le \delta+\varphi_{G[\mathrm{II}_1]}(X'X'^*(T'-T)^*(T'-T)X'X'^*)^{\frac{1}{4}}\,\varphi_{G[\mathrm{II}_1]}((T'-T)^*(T'-T))^{\frac{1}{4}}\\
&< \delta+4^{\frac{1}{4}}\delta^{\frac{1}{2}}<3\delta^{\frac{1}{2}}.
\end{aligned}$$

Moreover

$$\begin{aligned}
|||[X'-X,T]|||_2 &\le |||(X'-X)T|||_2+|||T(X'-X)|||_2\\
&< \delta+|||(X'-X)T|||_2\\
&= \delta+\varphi_{G[\mathrm{II}_1]}(T^*(X'-X)^*(X'-X)T)^{\frac{1}{2}}.
\end{aligned}$$

Now we shall show that if a W^*-subalgebra C of $\mathbb{P}\mathbin{\bar{\otimes}} U(G[\mathrm{II}_1])$ satisfies $C\overset{\delta}{\subset} U(G[\mathrm{II}_1])$, then $|\varphi_{G[\mathrm{II}_1]}(XZ)-\varphi_{G[\mathrm{II}_1]}(ZX)|<4\delta$ for $X\in\{\mathbb{P}\mathbin{\bar{\otimes}} U(G[\mathrm{II}_1])\}_1$ and $Z\in(C)_1$.

Let $Z^*=Z$ and take $Y\in\{U(G[\mathrm{II}_1])\}_1$ with $|||Z-Y|||_2<\delta$. Then

$$\begin{aligned}
&|\varphi_{G[\mathrm{II}_1]}(XZ)-\varphi_{G[\mathrm{II}_1]}(ZX)|\\
&\le |\varphi_{G[\mathrm{II}_1]}(X(Z-Y))|+|\varphi_{G[\mathrm{II}_1]}((Z-Y)X)|+|\varphi_{G[\mathrm{II}_1]}(XY-YX)|\\
&\le \varphi_{G[\mathrm{II}_1]}(XX^*)^{\frac{1}{2}}\,\varphi_{G[\mathrm{II}_1]}((Z-Y)^*(Z-Y))^{\frac{1}{2}}\\
&\quad+\varphi_{G[\mathrm{II}_1]}((Z-Y)(Z-Y)^*)^{\frac{1}{2}}\,\varphi_{G[\mathrm{II}_1]}(X^*X)^{\frac{1}{2}}\\
&\le |||Z-Y|||_2+|||Z-Y^*|||_2.
\end{aligned}$$

On the other hand,

$$\begin{aligned}
|||Z-Y^*|||_2^2 &= \varphi_{G[\mathrm{II}_1]}((Z-Y)(Z-Y)^*)\\
&= \varphi_{G[\mathrm{II}_1]}(Z^2-YZ-ZY^*+YY^*)\\
&= \varphi_{G[\mathrm{II}_1]}(Z^2-ZY-Y^*Z+Y^*Y)=|||Z-Y|||_2^2.
\end{aligned}$$

Hence

$$|\varphi_{G[\mathrm{II}_1]}(XZ)-\varphi_{G[\mathrm{II}_1]}(ZX)|<2\delta$$

and so for an arbitrary $Z\in(C)_1$, we have

$$|\varphi_{G[\mathrm{II}_1]}(XZ)-\varphi_{G[\mathrm{II}_1]}(ZX)|<4\delta.$$

(This implies that $|\varphi_{G[\mathrm{II}_1]}(U^*XU)-\varphi_{G[\mathrm{II}_1]}(X)|<4\delta$ for every unitary $U\in U\{Q(G[\mathrm{II}_2],n_2)\}$ and $X\in[U\{Q(G[\mathrm{II}_2],n_2)\}]_1$, and so

$$|\varphi_{G[\mathrm{II}_1]}(\tau(X)1)-\varphi_{G[\mathrm{II}_1]}(X)|\le 4\delta,$$

where $\tau(X)$ is the unique tracial state on the factor $U\{Q(G[\mathrm{II}_2],n_2)\}$; hence $|\,|||X|||_2^2-\|X\|_2^2\,|\leq 4\delta$ and so $|\,|||X|||_2-\|X\|_2\,|\leq 2\delta^{\frac{1}{2}}$ for

$$X\in[U\{Q(G[\mathrm{II}_2],n_2)\}]_1,$$

where $\|X\|_2$ is the trace norm on $U\{Q(G[\mathrm{II}_2],n_2)\}$.)

Therefore

$$|\varphi_{G[\mathrm{II}_1]}(T^*(X'-X)^*(X'-X)\,T)-\varphi_{G[\mathrm{II}_1]}(T\,T^*(X'-X)^*(X'-X))|<4\delta.$$

On the other hand,

$$|\varphi_{G[\mathrm{II}_1]}(T\,T^*(X'-X)^*(X'-X))|\leq 2\,|||X'-X|||_2.$$

Hence $\varphi_{G[\mathrm{II}_1]}(T^*(X'-X)^*(X'-X)\,T)^{\frac{1}{2}}<3\delta^{\frac{1}{2}}$ and so $|||[X-X',T]|||_2<4\delta^{\frac{1}{2}}$ and $|||[X',T']|||_2<7\delta^{\frac{1}{2}}$. Therefore, for an arbitrary unitary U' in $U\{Q(G[\mathrm{II}_1],r)\}$, we have $|||U'^*\,T'\,U'-T'|||_2<7\delta^{\frac{1}{2}}$. Let K be the strongly closed convex set in $U\{Q(G[\mathrm{II}_1],r)\}$ generated by $\{U'^*\,T'\,U'\}$. Then it contains a unique element

$$T''\in U\{Q(G[\mathrm{II}_1],r)\}'\cap[U\{Q(G[\mathrm{II}_1],n_1)\}]_1=[U\{Q(G[\mathrm{II}_1],n_1,r-1)\}]_1.$$

Hence $|||T''-T'|||_2<8\delta^{\frac{1}{2}}$. Thus

$$|||T-T''|||_2<|||T-T'|||_2+|||T'-T''|||_2<\delta+8\delta^{\frac{1}{2}}<9\delta^{\frac{1}{2}}.\qquad \text{q.e.d.}$$

4.3.27. Lemma. *Suppose that* $U\{Q(G[\mathrm{II}_1],n_3)\}\overset{\delta}{\subset}U\{Q(G[\mathrm{II}_2],n_2)\}$. *Then, for a positive integer* m_3 *with* $m_3>n_3$, *there exists a positive integer* m_2 *with* $m_2>n_2$ *such that* $U\{Q(G[\mathrm{II}_1],n_3,m_3)\}\overset{9\delta^{\frac{1}{2}}}{\subset}U\{Q(G[\mathrm{II}_2],n_2,m_2)\}$.

Proof. By 4.3.25, we can choose a positive integer s with $s-1>n_2$ such that $U\{Q(G(\mathrm{II}_2],s)\}\overset{\delta}{\subset}U\{Q(G[\mathrm{II}_1],m_3+1)\}$. For

$$T'\in[U\{Q(G[\mathrm{II}_1],n_3,m_3)\}]_1,$$

take $T\in[U\{Q(G[\mathrm{II}_2],n_2)\}]_1$ such that $|||T'-T|||_2<\delta$. For $X\in[U\{Q(G[\mathrm{II}_2],s)\}]_1$, take $X'\in[U\{Q(G[\mathrm{II}_1],m_3+1)\}]_1$ such that $|||X-X'|||_2<\delta$. Then

$$|||[X,T]|||_2\leq|||[X,T-T']|||_2+|||[X',T']|||_2+|||[X-X',T']|||_2.$$

Since $X\in U\{Q(G[\mathrm{II}_2],s)\}\overset{\delta}{\subset}U(G[\mathrm{II}_1])$ and $T'\in U(G[\mathrm{II}_1])$, by the argument similar to that of the proof of 4.3.26, we have $|||[X,T]|||_2<7\delta^{\frac{1}{2}}$. The remainder of the proof is quite similar to that of 4.3.26. q.e.d.

The remainder of the proof of Theorem 4.3.23 is obtained by using $|\,|||X|||_2-\|X\|_2\,|\leq 2\delta^{\frac{1}{2}}$ for $X\in[U\{Q(G[\mathrm{II}_2],n_2)\}]_1$ and a slight modification of the corresponding part of that of Theorem 4.3.9—i.e.

finally we have the following situation: $D \overset{R\delta^{\alpha}}{\subset} \mathbb{C}1$ and $\mathbb{C}1 \subset D$, where D is an infinite-dimensional commutative W^*-algebra, and R and α are positive numbers which do not depend on δ. This is a contradiction and completes the proof.

4.3.28. Corollary. *Suppose that* $\mathbb{I}_1$, $\mathbb{I}_2$ *satisfy the conditions of Theorem* 4.3.23, *and let* $B(\mathscr{H})$ *be a type* I-*factor with* $\dim \mathscr{H} \leq \aleph_0$. *Then* $B(\mathscr{H}) \bar{\otimes} U(G[\mathbb{I}_2])$ *is not *-isomorphic to* $B(\mathscr{H}) \bar{\otimes} U(G[\mathbb{I}_1])$.

Proof. Since $\mathbb{P}$ is a type III-factor, $\mathbb{P} \bar{\otimes} B(\mathscr{H})$ is *-isomorphic to $\mathbb{P}$. Hence if $B(\mathscr{H}) \bar{\otimes} U(G[\mathbb{I}_1])$ is *-isomorphic to $B(\mathscr{H}) \bar{\otimes} U(G[\mathbb{I}_2])$, $\mathbb{P} \bar{\otimes} U(G[\mathbb{I}_1])$ is *-isomorphic to $\mathbb{P} \bar{\otimes} U(G[\mathbb{I}_2])$, a contradiction.

4.3.29. Corollary. *There exists a uncountable family of type* II_∞*-factors on a separable Hilbert space.*

Concluding remarks on 4.3.

The problem of finding non-isomorphic factors on a separable Hilbert space has been one of central problems in the theory of W^*-algebras. Powers [137] proved the existence of uncountably many type III-factors (cf. 4.4). The existence of uncountably many type II_1-factors was proved by McDuff [112].

References. [39], [111], [112], [170], [171].

4.4. Examples of Factors, 3 (Other Results and Problems)

We shall begin with general results about the infinite tensor product of W^*-algebras. Let $\{\mathfrak{B}_n\}$ be a sequence of W^*-algebras and let $\mathscr{A} = \bigotimes_{n=1}^{\infty} \mathfrak{B}_n$ be the C^*-infinite tensor product of $\{\mathfrak{B}_n\}$. Let φ_n be a normal state on $\mathfrak{B}_n$ $(n=1,2,\ldots)$ and let $\bigotimes_{n=1}^{\infty} \varphi_n$ (say φ) be the infinite product state on $\mathscr{A}$.

4.4.1. Definition. *The* W^*-*algebra* $\pi_\varphi(\mathscr{A})''$ *is called the* W^*-*infinite tensor product of* $\{\mathfrak{B}_n\}$ *by the infinite product state* φ. *We shall call simply a* W^*-*infinite tensor product of* $\{\mathfrak{B}_n\}$, *when an infinite product state is not specified.*

4.4.2. Proposition. *Let* $\mathscr{A} = \bigotimes_{n=1}^{\infty} \mathfrak{B}_n$ *be the* C^*-*infinite tensor product of* C^*-*algebras* $\{B_n\}$ *with identities, and let* φ_n *be a factorial state on* $\mathfrak{B}_n$ $(n=1,2,\ldots)$. *Then the infinite product state* $\varphi = \bigotimes_{n=1}^{\infty} \varphi_n$ *on* $\mathscr{A}$ *is again factorial.*

Proof. Let $\mathscr{A}_n = \bigotimes_{i=1}^{n} \mathfrak{B}_i \otimes 1 \otimes 1 \otimes \cdots$ and $\mathscr{D}_n = 1 \otimes 1 \otimes \cdots \bigotimes_{i=n+1}^{\infty} \mathfrak{B}_i$. Then $\pi_\varphi(\mathscr{A})'' = \pi_\varphi(\mathscr{A}_n)'' \bar{\otimes} \pi_\varphi(\mathscr{D}_n)''$ as abstract W^*-algebras. Set $\psi_n = \bigotimes_{i=n+1}^{\infty} \varphi_i$. Then there exists a σ-continuous conditional expectation P_n of $\pi_\varphi(\mathscr{A})''$ onto $\pi_\varphi(\mathscr{A}_n)''$ satisfying $\langle x, f \otimes \psi_n \rangle = \langle P_n(x), f \rangle$ $(f \in \pi_\varphi(\mathscr{A}_n)''_*)$ (cf. 2.6.4). Let z be a central element in $\pi_\varphi(\mathscr{A})''$. Then $P_n(zx) = P_n(z)x = P_n(xz) = xP_n(z)$ $(x \in \pi_\varphi(\mathscr{A}_n)'')$. Hence $P_n(z)$ is a central element in $\pi_\varphi(\mathscr{A}_n)''$.

Since $\pi_\varphi(\mathscr{A}_n)'' = \pi_{\varphi_1}(\mathfrak{B}_1)'' \bar{\otimes} \pi_{\varphi_2}(\mathfrak{B}_2)'' \bar{\otimes} \cdots \bar{\otimes} \pi_{\varphi_n}(\mathfrak{B}_n)''$ as abstract W^*-algebras, it is a factor and so $P_n(z) = \lambda_n 1_{\mathscr{H}}$ for some complex number λ_n.

Let $\mathscr{A}_0$ be the algebraic *-subalgebra of $\mathscr{A}$ generated by elements of the form $\bigotimes_{i=1}^{\infty} a_i$ $(a_i = 1$ for all but a finite number of indices), and let V be a linear subspace of $\pi_\varphi(\mathscr{A})''_*$ composed of all finite linear combinations $\{L_a R_b \varphi \mid a, b \in \mathscr{A}_0\}$. Then for an arbitrary $g \in V$, there exists a positive integer m such that $g = f_n \otimes \psi_n$ with $f_n \in \pi_\varphi(\mathscr{A}_n)''_*$ $(n \geq m)$. Hence for $x \in \pi_\varphi(\mathscr{A})''$, $\langle x, g \rangle = \langle P_n(x), f_n \rangle = \langle P_n(x), g \rangle$ $(n \geq m)$.

Since $\{P_n(x)\}$ is uniformly bounded and V is norm-dense in $\pi_\varphi(\mathscr{A})''_*$, this implies that $\lim_n P_n(x) = x$ in the σ-topology. Hence $P_n(x) = \lambda_n 1_{\mathscr{H}} \to z$ in the σ-topology, so that $\pi_\varphi(\mathscr{A})''$ is a factor. q.e.d.

4.4.3. Proposition. *Let $\mathscr{A} = \bigotimes_{n=1}^{\infty} \mathfrak{B}_n$ be the C^*-infinite tensor product of C^*-algebras $\{\mathfrak{B}_n\}$ with identities, and let φ_n be a pure state on $\mathfrak{B}_n$ $(n = 1, 2, \ldots)$. Then the infinite product state $\varphi = \bigotimes_{n=1}^{\infty} \varphi_n$ on $\mathscr{A}$ is again pure.*

Proof. One can easily see that $\bigotimes_{n=1}^{m} \varphi_n$ is pure on $\bigotimes_{n=1}^{m} \mathfrak{B}_n$. Suppose ψ is a positive linear functional on $\mathscr{A}$ with $\psi \leq \varphi$. Then $\psi = \psi(1)\varphi$ on $\bigotimes_{n=1}^{m} \mathfrak{B}_n \otimes 1 \otimes \cdots$ $(m = 1, 2, \ldots)$. Hence $\psi = \psi(1)\varphi$ on $\bigotimes_{n=1}^{\infty} \mathfrak{B}_n$. q.e.d.

Now let $\mathscr{A} = \bigotimes_{n=1}^{\infty} \mathfrak{B}_n$ be the C^*-infinite tensor product of W^*-algebras $\{\mathfrak{B}_n\}$, and let φ_n be a normal state on $\mathfrak{B}_n$ $(n = 1, 2, \ldots)$. Let $\varphi = \bigotimes_{n=1}^{\infty} \varphi_n$ be the infinite product state on $\mathscr{A}$. For each n, $1 \otimes \cdots \otimes \pi_{\varphi_n}(\mathscr{A})' \otimes 1 \otimes \cdots$ (say $\mathscr{N}_n$) is a W^*-subalgebra of $B(\mathscr{H}_\varphi)$ and $\mathscr{N}_n \subset \pi_\varphi(\mathscr{A})'$.

4.4.4. Proposition. $\pi_\varphi(\mathscr{A})' = R(\mathscr{N}_n | n=1,2,\ldots)$, *where* $R(\mathscr{N}_n | n=1,2,\ldots)$ *is the* W^**-subalgebra of* $B(\mathscr{H}_\varphi)$ *generated by* $\{\mathscr{N}_n | n=1,2,\ldots\}$.

Proof. Write $\varphi_n(x) = (x 1_{\varphi_n}, 1_{\varphi_n})$ $(x \in \mathfrak{B}_n)$ and set $\tilde{\varphi}_n(y) = (y 1_{\varphi_n}, 1_{\varphi_n})$ $(y \in B(\mathscr{H}_{\varphi_n}))$. Then $\tilde{\varphi}_n$ is a pure state on $B(\mathscr{H}_{\varphi_n})$. Let $\tilde{\varphi} = \bigotimes_{n=1}^{\infty} \tilde{\varphi}_n$ be the infinite product state on $\bigotimes_{n=1}^{\infty} B(\mathscr{H}_{\varphi_n})$. Then by 4.4.3, $\tilde{\varphi}$ is a pure state on $\bigotimes_{n=1}^{\infty} B(\mathscr{H}_{\varphi_n})$.

Since each 1_{φ_n} is a cyclic vector for $\pi_{\varphi_n}(\mathfrak{B}_n)$, we can consider $\mathscr{H}_\varphi = \mathscr{H}_{\tilde{\varphi}}$. Then $\pi_\varphi(\mathscr{A})'' \subset \pi_{\tilde{\varphi}}\left(\bigotimes_{n=1}^{\infty} B(\mathscr{H}_{\varphi_n})\right)'' = B(\mathscr{H}_{\tilde{\varphi}})$.

Clearly $R(\mathscr{N}_n | n=1,2,\ldots)' \supset \pi_\varphi(\mathscr{A})''$.

Consider the conditional expectation P_m considered in the proof of 4.4.2. It is a mapping of $\pi_\varphi\left(\bigotimes_{n=1}^{\infty} B(\mathscr{H}_{\varphi_n})\right)''$ onto $\pi_\varphi\left(\bigotimes_{n=1}^{m} B(\mathscr{H}_{\varphi_n}) \otimes 1 \otimes \ldots\right)''$.

For $x \in R(\mathscr{N}_n | n=1,2,\ldots)'$ and $y \in R(\mathscr{N}_n | n=1,2,\ldots,m)$, we have $P_m(xy) = P_m(x)y = P_m(yx) = yP_m(x)$. Hence

$$P_m(x) \in R(\mathscr{N}_n | n=1,2,\ldots,m)' \cap \pi_{\tilde{\varphi}}\left(\bigotimes_{n=1}^{m} B(\mathscr{H}_{\varphi_n}) \otimes 1 \otimes \cdots\right)''.$$

By the commutation theorem of tensor products (cf. 2.8.1),

$$P_m(x) \in \pi_\varphi\left(\bigotimes_{n=1}^{m} \mathfrak{B}_n \otimes 1 \otimes \cdots\right)''.$$

Since $P_m(x) \to x$ in the σ-weak topology, $x \in \pi_\varphi(\mathscr{A})''$. q.e.d.

Next we shall state results about a special class of factors (called hyperfinite factors) which is fairly manageable and an extensive study has been done for those factors. Those factors are also important, because they are appearing in mathematical physics and Ergodic theory. We will not give proofs to most of the results, because it requires too much space.

4.4.5. Definition. *A factor, with* $1_{\mathscr{H}}$, *on a separable Hilbert space* $\mathscr{H}$ *is said to be hyperfinite, if there exists an increasing sequence of type* I_{n_p}*-subfactors* $\{\mathscr{M}_p\}$ *with* $n_p < +\infty$ $(p=1,2,\ldots)$, *containing the identity of* $\mathscr{M}$, *such that the* σ*-closure of* $\bigcup_{p=1}^{\infty} \mathscr{M}_p$ *is* $\mathscr{M}$.

Let $\mathscr{A}$ be a separable uniformly hyperfinite C^*-algebra and let φ be a factorial state on $\mathscr{A}$. Then the W^*-algebra $\pi_\varphi(\mathscr{A})''$ is a hyperfinite factor.

Conversely let $\mathcal{M}$ be a hyperfinite factor on a separable Hilbert space $\mathcal{H}$ and let $\mathcal{M}$ be the σ-closure of $\bigcup_{p=1}^{\infty} \mathcal{M}_p$ with $\mathcal{M}_p \subset \mathcal{M}_{p+1}$ and $\mathcal{M}_p$, a type I_{n_p}-factor ($n_p < +\infty$). Let φ be a faithful normal state on $\mathcal{M}$, and let $\mathcal{A}$ be the uniform closure of $\bigcup_{p=1}^{\infty} \mathcal{M}_p$ in $\mathcal{M}$. Then $\mathcal{A}$ is a separable uniformly hyperfinite C^*-algebra and the restriction $\hat{\varphi}$ of φ to $\mathcal{A}$ is a factorial state and furthermore $\pi_{\hat{\varphi}}(\mathcal{A})''$ is *-isomorphic to $\mathcal{M}$.

4.4.6. Theorem. (cf. [121]). *Let $\mathcal{M}$ be a type II_1-factor on a separable Hilbert space $\mathcal{H}$. Then the following conditions are equivalent.*

1. *$\mathcal{M}$ is hyperfinite,*
2. *there exists an increasing sequence of finite-dimensional *-subalgebras $(\mathcal{N}_i)$ in $\mathcal{M}$ such that the σ-closure of $\bigcup_{i=1}^{\infty} \mathcal{N}_i$ is $\mathcal{M}$.*
3. *(α) $\mathcal{M}$ is the W^*-algebra generated by a family of countable elements,*
   *(β) for $\varepsilon > 0$ and $A_1, A_2, \ldots, A_m \in \mathcal{M}$, there exist a finite-dimensional *-subalgebra $\mathcal{N}$ of $\mathcal{M}$ and $B_1, B_2, \ldots, B_m \in \mathcal{N}$ such that $\|A_i - B_i\|_2 < \varepsilon$ for $i = 1, 2, \ldots, m$ ($\|\cdot\|_2$, the trace norm on $\mathcal{M}$).*
4. *there exists an increasing sequence of type I_{2^n}-subfactors $(\mathcal{M}_n)$ of $\mathcal{M}$ such that the σ-closure of $\bigcup_{n=1}^{\infty} \mathcal{M}_n$ is $\mathcal{M}$.*

This famous theorem of Murray-von Neumann implies that there exists one and only one hyperfinite type II_1-factor in a separable Hilbert space.

Now let $\mathcal{A} = \bigotimes_{n=1}^{\infty} \mathfrak{B}_n$ be the C^*-infinite tensor product of $\{\mathfrak{B}_n\}$ with $\mathfrak{B}_n$, a type I_{γ_n}-factor ($\gamma_n < +\infty$), and let φ_n be a state on $\mathfrak{B}_n$. Then $\varphi = \bigotimes_{n=1}^{\infty} \varphi_n$ is a factorial state on $\mathcal{A}$, because φ_n is factorial on $\mathfrak{B}_n$ for each n.

Suppose that $\mathfrak{B}_n = \mathfrak{B}$ with a type I_2-factor $\mathfrak{B}$ ($n = 1, 2, \ldots$), and let $(p_n | n = 1, 2, \ldots)$ be a sequence of positive numbers with $0 \leq p_n \leq \frac{1}{2}$.

For $\begin{pmatrix} \alpha & \beta \\ \gamma & \delta \end{pmatrix} \in \mathfrak{B}_n$ ($\alpha, \beta, \gamma, \delta$ complex numbers), define

$$\varphi_{p_n}\left(\begin{pmatrix} \alpha & \beta \\ \gamma & \delta \end{pmatrix}\right) = \alpha p_n + \delta(1 - p_n).$$

Then φ_{p_n} is a state on $\mathfrak{B}_n$. Set $\psi_{(p_n)} = \bigotimes_{n=1}^{\infty} \varphi_{p_n}$. Thus $\psi_{(p_n)}$ is a factorial state on $\mathcal{A}$.

If $p_n=0$ for each n, φ_{p_n} is a pure state on $\mathfrak{B}_n$. Therefore by 4.4.3, $\psi_{(p_n)}$ is a pure state on $\mathscr{A}$, and so $\pi_{\psi_{(p_n)}}(\mathscr{A})''$ is a type I_∞-factor.

If $p_n=\frac{1}{2}$ for each n, φ_{p_n} is a tracial state on $\mathfrak{B}_n$ and so $\psi_{(p_n)}$ is a tracial state on $\mathscr{A}$. Thus $\pi_{\psi_{(p_n)}}(\mathscr{A})''$ is a hyperfinite type II_1-factor.

Now let Ω_n $(n=1,2,\ldots)$ be the additive groups of integers, reduced mod 2; therefore Ω_n is a compact group composed of two elements $\{0,1\}$ as follows: $0+0=0$, $0+1=1+0=1$ and $1+1=0$. Let $\Omega=\prod_{n=1}^{\infty}\Omega_n$ be the weakly infinite direct product of $\{\Omega_n|n=1,2,\ldots\}$. Then Ω is a compact group. Ω is the set of all elements $\xi=(\xi_n|n=1,2,\ldots)$ with $\xi_n=0$ or 1. Let $\mathscr{G}$ be the set of those $a=(a_n|n=1,2,\ldots)$ in Ω for which $a_n\neq 0$ occurs for a finite number of n only. $\mathscr{G}$ is a countable group. For $a\in\mathscr{G}$ and $\xi\in\Omega$, define a homomorphism $\xi\to\xi a$ by $\xi a=\xi+a$.

Now let (p_n) be a sequence of positive numbers with $0<p_n\leq\frac{1}{2}$ for each n, and let μ_n be a Radon measure on Ω_n with $\mu_n\{(0)\}=p_n$ and $\mu\{(1)\}=1-p_n$. Let $\mu=\prod_{n=1}^{\infty}\mu_n$ be the infinite product measure on Ω. Then by 4.2.13, μ is quasi-invariant, and $\mathscr{G}$ is free and ergodic (cf. the first part of the proof of 4.2.16).

Now put $L^\infty(\Omega,\mu)=\mathscr{M}$ and we shall apply the construction in 4.2 to $\mathscr{M}$ and $\mathscr{G}$, and obtain a weakly closed subalgebra $\mathscr{N}$ of $B(\tilde{\mathscr{H}})$.

Now we shall use the notations in 4.2. Let $\rho(a_1,a_2,\ldots,a_k)(\xi)$ be the characteristic function of the set $\{(\xi_n)|\xi_n=a_n, n=1,2,\ldots,k\}$ with $a_n\in\Omega_n$. Let $\mathscr{N}_k$ be the *-subalgebra of $\mathscr{N}$ generated by $\{\tilde{U}_{\gamma_n}; n=1,2,\ldots,k\}$ and $\{\Phi(\rho(a_1,a_2,\ldots,a_k))|a_n\in\Omega_n\ (n=1,2,\ldots,k)\}$. Then one can easily see that $\mathscr{N}_k$ is a type I_{2^k}-factor. Therefore $\mathscr{N}_k$ is *-isomorphic to $\mathfrak{B}_1\otimes\mathfrak{B}_2\otimes\cdots\otimes\mathfrak{B}_k$.

In particular, let $\Psi(\tilde{U}_{\gamma_n})=1\otimes\cdots\otimes\begin{pmatrix}0&1\\1&0\end{pmatrix}\otimes 1\otimes\ldots$, where $\begin{pmatrix}0&1\\1&0\end{pmatrix}$ occurs in the n-th place and $\Psi(\rho(0,0,\ldots,a_n,0,\ldots))=1\otimes\cdots\otimes A_n\otimes 1\otimes\ldots$, where A_n occurs in the n-th place, and $A_n=\begin{pmatrix}1&0\\0&0\end{pmatrix}$ if $a_n=0$ and $A_n=\begin{pmatrix}0&0\\0&1\end{pmatrix}$ if $a_n=1$.

Then Ψ can be uniquely extended to a *-isomorphism (denoted by Ψ) of the C^*-subalgebra in $\mathscr{N}$, which is generated by $\Phi(C(\Omega))$ and $\{\tilde{U}_a|a\in\mathscr{G}\}$, onto $\mathscr{A}=\bigotimes_{n=1}^{\infty}\mathfrak{B}_n$. One can easily see that Ψ can be uniquely extended to a *-isomorphism of $\mathscr{N}$ onto $\pi_{\psi_{(p_n)}}(\mathscr{A})''$.

Therefore by 4.2.13, we have

4.4.7. Proposition. *Let* $\mathscr{A}=\bigotimes_{n=1}^{\infty}\mathfrak{B}_n$ *be the* C^**-infinite tensor product of* $\{\mathfrak{B}_n\}$ *with* $\mathfrak{B}_n$, *a type* I_2*-factor.*

If there exists a positive number δ with $\delta < p_n < \frac{1}{2} - \delta$ for each n, then $\pi_{\psi_{(p_n)}}(\mathscr{A})''$ is a type III*-factor.*

The reader is referred to [4], [19], [118], [226], about the types of more general $\pi_{\psi_{(p_n)}}(\mathscr{A})''$s.

Now we shall consider a special case where $p_n = \lambda$ for each n with $0 < \lambda < \frac{1}{2}$. In this case, we shall denote $\pi_{(\psi_{p_n})}(\mathscr{A})''$ by $\mathscr{M}_\lambda$ and call it Powers' factor.

4.4.8. Theorem. (cf. [137]). *Let $\mathscr{M}_{\lambda_1}$ and $\mathscr{M}_{\lambda_2}$ be two Powers' factors. If $\lambda_1 \neq \lambda_2$, $\mathscr{M}_{\lambda_1}$ is not *-isomorphic to $\mathscr{M}_{\lambda_2}$.*

Introducing the notion of asymptotic ratio, Araki and Woods [6] extended the results of Powers and gave a detailed classification of more general W^*-infinite tensor product factors of finite type I-factors. Moreover, Krieger [105], [106] extended the work of Araki-Woods, using the ergodic theory. In particular, he obtained examples of hyperfinite type III-factors $\mathscr{M}$ such that $\mathscr{M} \bar{\otimes} \mathscr{M}$ is not *-isomorphic to $\mathscr{M}$.

4.4.9. Problem. *Give reasonably simple conditions under which a hyperfinite factor is a W^*-infinite tensor product factor of finite type* I*-factors.*

4.4.10. Problem. *Is there a hyperfinite factor which is not a W^*-infinite tensor product factor of finite type* I*-factors?*

Let $\mathscr{U}$ be a hyperfinite II_1-factor. Then $\mathscr{U} \bar{\otimes} B(\mathscr{H})$ $(\dim \mathscr{H} = \infty)$ is a hyperfinite II_∞-factor.

4.4.11. Problem. *Is there a hyperfinite II_∞-factor which is not *-isomorphic to the above $\mathscr{U} \bar{\otimes} B(\mathscr{H})$?*

4.4.12. Problem. *Is there a type II_1-factor $\mathscr{N}$ which is not *-isomorphic to $\mathscr{N} \bar{\otimes} \mathscr{N}$?*

4.4.13. Theorem. (cf. [121], [46]). *In the construction of* 4.2, *suppose that $\mathscr{G}$ is free, ergodic and μ is $\mathscr{G}$-invariant. Furthermore suppose that $\mathscr{G}$ is commutative. Then the obtained II_1-factor $\mathscr{N}$ is always hyperfinite.*

4.4.14. Definition. *Let $\mathscr{M}$ be a W^*-algebra containing $1_{\mathscr{H}}$ on a Hilbert space $\mathscr{H}$ and let $\mathscr{M}'$ be the commutant of $\mathscr{M}$. For $x \in B(\mathscr{H})$, let $C(x)$ be the σ-closed convex subset of $B(\mathscr{H})$ generated by $\{u^* x u \mid u \in \mathscr{M}^u\}$, where $\mathscr{M}^u$ is the group of all unitary elements in $\mathscr{M}$. $\mathscr{M}$ is said to have the property P, if $C(x) \cap \mathscr{M}' \neq (\emptyset)$ for each $x \in B(\mathscr{H})$.*

4.4.15. Proposition. *Suppose that $\mathscr{M}$ has the property P. Then there exists a linear mapping P of $B(\mathscr{H})$ onto $\mathscr{M}'$ such that*

1. $P(x) \in C(x) \cap \mathscr{M}'$ $\quad (x \in B(\mathscr{H}))$;
2. $\|P(x)\| \leq \|x\|$ $\quad (x \in B(\mathscr{H}))$;
3. $P(h) \geq 0$ *for* $h \geq 0$ *and* $P(1_{\mathscr{H}}) = 1_{\mathscr{H}}$ $\quad (h \in B(\mathscr{H}))$;
4. $P(a' x b') = a' P(x) b'$ $\quad$ ($a', b' \in \mathscr{M}'$ *and* $x \in B(\mathscr{H})$).

Proof. Let $B(B(\mathscr{H}))$ be the algebra of all bounded linear operators on $B(\mathscr{H})$. Then $B(B(\mathscr{H}))=(B(\mathscr{H})\otimes_\gamma B(\mathscr{H})_*)^*$. We shall consider the topology $\sigma(B(B(\mathscr{H})), B(\mathscr{H})\otimes_\gamma B(\mathscr{H})_*)$ (called the σ-topology) on $B(B(\mathscr{H}))$. Then the unit sphere is σ-compact.

Let $\mathfrak{F}$ be the σ-closed convex subset of $B(B(\mathscr{H}))$ generated by $\{T^u: x\to u^*xu$ for $x\in B(\mathscr{H})$ with $u\in\mathscr{M}^u\}$. We introduce an order $\leq$ in $\mathfrak{F}$ as follows: for $T_1, T_2\in\mathfrak{F}$, define $T_1\leq T_2$ if

$$C(T_1(x))\supseteqq C(T_2(x)) \quad (\text{all } x\in B(\mathscr{H})).$$

Let $\{T_\alpha\}_{\alpha\in\mathbb{I}}$ be a linearly ordered subset of $\mathfrak{F}$ and put $\mathfrak{F}_\beta=$ the σ-closure of $\{T_\alpha | T_\alpha\geq T_\beta\}$. Take a $T\in\bigcap_{\beta\in\mathbb{I}}\mathfrak{F}_\beta$; then $T(x)\in$ the *σ-closure* of $\{T_\alpha(x) | T_\alpha\geq T_\beta\}$. Since $C(T_\alpha(x))\subseteqq C(T_\beta(x))$, $T(x)\subseteqq C(T_\beta(x))$ and so $T(x)\subseteqq\bigcap_{\beta\in\mathbb{I}} C(T_\beta(x))$. Hence $C(T(x))\subset C(T_\alpha(x))$ for $\alpha\in\mathbb{I}$, so that T is an upper bound of $\{T_\alpha\}_{\alpha\in\mathbb{I}}$. By Zorn's lemma there exists a maximal element T_0 in $\mathfrak{F}$. We shall show that $C(T_0(x))$ consists of a single element for all $x\in B(\mathscr{H})$. Suppose that $C(T_0(x_0))$ contains at least two points. Since $C(T_0(x))\cap\mathscr{M}'\neq(\emptyset)$, there exists a directed set of functions $(f_\gamma)_{\gamma\in\mathbb{I}}$ on $\mathscr{M}^u$, with $f_\gamma\geq 0$, such that $\sum_{u\in\mathscr{M}^u} f_\gamma(u)=1$ and

$$\sigma-\lim_\gamma \sum_{u\in\mathscr{M}^u} f_\gamma(u)u^* T_0(x_0)u=a'\in\mathscr{M}'.$$

Hence the σ-closed convex subset of $B(B(\mathscr{H}))$ generated by $\{T^u T_0\}$ contains an element T_1 with $T_1(x_0)=a'$, and so $C(T_1(x_0))=a'\subsetneqq C(T_0(x_0))$.

On the other hand, $C(T_1(x))\subset C(T_0(x))$. This contradicts the maximailty of T_0. Hence $C(T_0(x))$ consists of a single point. Since $C(T_0(x))\cap\mathscr{M}'\neq(\emptyset)$, we have $T_0(x)\in C(x)\cap\mathscr{M}'$. Put $P=T_0$. Then one can easily see that P satisfies the conditions of 4.4.15. q.e.d.

4.4.16. Proposition. *Let $\{\mathscr{M}_\alpha\}_{\alpha\in\mathbb{I}}$ be an increasing directed set of W^*-subalgebras, containing $1_{\mathscr{H}}$, of $B(\mathscr{H})$.*

If $\mathscr{M}_\alpha$ has the property P for each $\alpha\in\mathbb{I}$, then the σ-closure $\mathscr{M}$ of $\bigcup_{\alpha\in\mathbb{I}}\mathscr{M}_\alpha$ has again the property P.

Proof. Let $C_\alpha(x)$ be the σ-closed convex subset of $B(\mathscr{H})$ generated by $\{u^*xu | u\in\mathscr{M}_\alpha^u\}_{\alpha\in\mathbb{I}}$ $(x\in B(\mathscr{H}))$. Then $C_\alpha(x)\cap\mathscr{M}_\alpha'\neq(\emptyset)$.

Let $C(x)$ be the σ-closed convex subset of $B(\mathscr{H})$ generated by $\{u^*xu | u\in\mathscr{M}^u\}$. Then $C(x)\cap\mathscr{M}_\alpha'\neq(\emptyset)$ for all $\alpha\in\mathbb{I}$. Since $\{C(x)\cap\mathscr{M}_\alpha'\}$ is a decreasing directed set of σ-compact sets,

$$C(x)\cap\bigcap_{\alpha\in\mathbb{I}}\mathscr{M}_\alpha'=C(x)\cap\mathscr{M}'\neq(\emptyset). \qquad \text{q.e.d.}$$

4.4.17. Corollary. *A hyperfinite factor has the property P.*

Proof. Let $\{\mathcal{M}_\alpha\}_{\alpha\in\mathbb{I}}$ be an increasing directed set of finite type I-subfactors in $\mathcal{M}$ such that the σ-closure of $\bigcup_{\alpha\in\mathbb{I}} \mathcal{M}_\alpha$ is $\mathcal{M}$.

Since $\mathcal{M}_\alpha^u$ is a compact group, it has the Haar measure $d\mu_\alpha$ with $\mu_\alpha(\mathcal{M}_\alpha^u)=1$. For $x\in B(\mathscr{H})$, put $x^0=\int_{\mathcal{M}^u} u^* x u\, d\mu_\alpha(u)$; then $x^0\in\mathcal{M}_\alpha'$. q.e.d.

4.4.18. Proposition. *Let $\{\mathcal{N}_\alpha\}_{\alpha\in\mathbb{I}}$ be a mutually commuting family of W^*-subalgebras, containing $1_{\mathscr{H}}$, of $B(\mathscr{H})$, and let $\mathcal{N}$ be the W^*-algebra generated by $\{\mathcal{N}_\alpha\}_{\alpha\in\mathbb{I}}$. If $\mathcal{N}_\alpha$ has the property P for each $\alpha\in\mathbb{I}$, then $\mathcal{N}$ has again the property P.*

Proof. Let $C_\alpha(x)$ be the σ-closed convex subset of $B(\mathscr{H})$ generated by $\{u^*xu \mid u\in\mathcal{N}_\alpha^u\}$ $(x\in B(\mathscr{H}))$. Then $C_\alpha(x)\cap\mathcal{N}_\alpha'\neq(\emptyset)$. Let $\{\alpha_1,\dots,\alpha_n\}$ be a finite subfamily of $\mathbb{I}$. Take $y_1\in C_{\alpha_1}(x)\cap\mathcal{N}_{\alpha_1}'$; then $C_{\alpha_2}(y_1)\subset\mathcal{N}_{\alpha_1}'$ and $C_{\alpha_2}(y_1)\cap\mathcal{N}_{\alpha_2}'\neq(\emptyset)$. Hence $C_{\alpha_2}(y_1)\cap\mathcal{N}_{\alpha_1}'\cap\mathcal{N}_{\alpha_2}'\neq(\emptyset)$. Take $y_2\in C_{\alpha_2}(y_1)\cap\mathcal{N}_{\alpha_1}'\cap\mathcal{N}_{\alpha_2}'$. Then $C_{\alpha_3}(y_2)\cap\mathcal{N}_{\alpha_1}'\cap\mathcal{N}_{\alpha_2}'\cap\mathcal{N}_{\alpha_3}'\neq(\emptyset)$, and so on. Hence $C(x)\cap\bigcap_{i=1}^{n}\mathcal{N}_{\alpha_i}'\neq(\emptyset)$, where $C(x)$ is the σ-closed convex subset of $B(\mathscr{H})$ generated by $\{u^*xu \mid u\in R(\mathcal{N}_1,\mathcal{N}_2,\dots,\mathcal{N}_n)^u\}$.

Hence by 4.4.16, $\mathcal{N}$ has the property P. q.e.d.

4.4.19. Corollary. *Let $\mathcal{M}$ be a W^*-algebra containing $1_{\mathscr{H}}$ on $\mathscr{H}$. Suppose that there exists an increasing directed set of type I W^*-subalgebras $\{\mathcal{M}_\alpha\}$ $(\alpha\in\mathbb{I})$, containing $1_{\mathscr{H}}$, of $\mathcal{M}$ such that the σ-closure of $\bigcup_{\alpha\in\mathbb{I}}\mathcal{M}_\alpha$ is $\mathcal{M}$. Then $\mathcal{M}$ has the property P.*

Proof. By the structure theorem of type I W^*-algebras,

$$\mathcal{M}_\alpha=\sum_{\beta\in\mathbb{I}_\alpha}\oplus\,\mathcal{N}_{\alpha,\beta}\,\overline{\otimes}\,C_{\alpha,\beta},$$

where $\mathcal{N}_{\alpha,\beta}$ is a type I-factor and $C_{\alpha,\beta}$ is a commutative W^*-algebra.

Put $\mathscr{L}_{\alpha,\beta}=\mathcal{N}_{\alpha,\beta}\otimes 1_{\alpha,\beta}'\oplus\mathbb{C}(1_{\mathscr{H}}-z_{\alpha,\beta})$ and

$$\mathscr{R}_{\alpha,\beta}=1_{\alpha,\beta}\otimes C_{\alpha,\beta}\oplus\mathbb{C}(1_{\mathscr{H}}-z_{\alpha,\beta}),$$

where $z_{\alpha,\beta}$ is the identity of $\mathcal{N}_{\alpha,\beta}\,\overline{\otimes}\,C_{\alpha,\beta}$ and $1_{\alpha,\beta}$ (resp. $1_{\alpha,\beta}'$) is the identity of $\mathcal{N}_{\alpha,\beta}$ (resp. $C_{\alpha,\beta}$). Then $(\mathscr{L}_{\alpha,\beta},\mathscr{R}_{\alpha,\beta})_{\beta\in\mathbb{I}_\alpha}$ is a commuting family. By 4.4.17, $\mathscr{L}_{\alpha,\beta}$ has the property P and $\mathscr{R}_{\alpha,\beta}$ has the property P by Kakutani-Markoff theorem (cf. [16]). Hence by 4.4.18, $\mathcal{M}_\alpha$ has the property P and so by 4.4.16, $\mathcal{M}$ has the property P. q.e.d.

4.4.20. Definition. *Let G be a discrete group, and let $l^\infty(G)$ be the set of all bounded complex valued functions on G. Then $l^\infty(G)$ is a commutative C^*-algebra under the pointwise multiplication. Let φ be a state on $l^\infty(G)$ (equivalently, a finitely additive probability measure on G). G is said to be amenable if there exists a state φ on $l^\infty(G)$ which is invariant under the translations by G—i.e. $\varphi(f^a)=\varphi(f_a)=\varphi(f)$ $(f\in l^\infty(G))$ with $f_a(g)=f(ga^{-1})$ and $f^a(g)=f(ag)$ $(a,g\in G)$.*

Remark. It is known that if G has a left (or right) invariant state (i.e. $\varphi(f^a)=\varphi(f)$ (or $\varphi(f_a)=\varphi(f)$)), then it has an invariant state (cf. [25]).

4.4.21. Proposition. *$U(G)$ has the property P if and only if G is amenable.*

Proof. Suppose that G is amenable and let φ be an invariant state on $l^\infty(G)$ under G. We shall denote the corresponding finitely additive probability measure to φ by μ.

For $x \in B(l^2(G))$ and $\xi, \eta \in l^2(G)$, put

$$L(\xi,\eta) = \int_G (U(g)^* x U(g)\xi, \eta) d\mu(g).$$

Then $|L(\xi,\eta)| \leq \|x\| \, \|\xi\| \, \|\eta\|$. Hence L is a bounded conjugate bilinear functional on $\mathscr{H}$, and so there exists an element $T \in B(l^2(G))$ such that

$$(T\xi,\eta) = \int (U(g)^* x U(g)\xi, \eta) d\mu(g).$$

For $a \in G$,

$$\begin{aligned}(U(a)^* T U(a)\xi,\eta) &= \int (U(g)^* x U(g) U(a)\xi, U(a)\eta) d\mu(g) \\ &= \int (U(ga)^* x U(ga)\xi, \eta) d\mu(g) = (T\xi,\eta).\end{aligned}$$

Hence $U(a)^* T U(a) = T$ and so $T \in U(G)'$.

Since φ belongs to the $\sigma(l^\infty(G)^*, l^\infty(G))$-closure of $l^1(G)$ in $l^\infty(G)^*$, $T \in C(x)$ ($C(x)$, the σ-closed convex set of $B(l^2(G))$ generated by $\{u^* x u \mid u \in U(G)^u\}$, so that $U(G)$ has the property P.

Conversely suppose that $U(G)$ has the property P and let P be the linear mapping of $B(l^2(G))$ onto $U(G)'$ in 4.4.15.

For $f \in l^\infty(G)$ and $\xi \in l^2(G)$, define $(T_f\xi)(g) = f(g)\xi(g)$ $(g \in G)$. Then $T_f \in B(l^2(G))$.

Let τ be a tracial state on $U(G)' = \{V(a) \mid a \in G\}''$ and set $\varphi(f) = \tau(P(T_f))$ $(f \in l^\infty(G))$. Then

$$(V(a)^{-1} T_f V(a)\xi)(g) = V(a)^{-1} f(g)\xi(ga) = f(ga^{-1})\xi(g) \qquad (a, g \in G).$$

Hence $V(a)^{-1} T_f V(a) = T_{f_a}$. Thus

$$\varphi(f_a) = \varphi(P(V(a)^{-1} T_g V(a))) = \varphi(V(a)^{-1} P(T_f) V(a)) = \varphi(P(T_f)) = \varphi(f)$$

and so G has a right invariant finitely additive measure. q.e.d.

4.4.22. Corollary. *If G contains a subgroup which is the free group of two generators, then $U(G)$ does not have the property P.*

Proof. It is enough to show that G is not amenable, and so it suffices to assume that G is the free group of two generators.

Suppose that G has a right invariant finitely additive probability measure μ. Let g_1, g_2 be the generators of G and let E be the set of all elements in G ending with $\dots g_1^m$ $(m \neq 0)$, when written as a power of g_1, g_2 of minimum length. Then E, Eg_2, Eg_2^2 are mutually disjoint and so $\mu(E) \leq \frac{1}{3}$.

On the other hand $E \cup Eg_1 = G$ and so $\mu(E) \geq \frac{1}{2}$, a contradiction. q.e.d.

Remark. It is a conjecture of von Neumann that any non-amenable discrete group will contain a subgroup which is the free group of two generators.

4.4.23. Proposition. *Let $\mathcal{M}$ be a countably decomposable finite W^*-algebra and let $\mathcal{N}$ be a W^*-subalgebra, containing the identity of $\mathcal{M}$, of $\mathcal{M}$. Then there exists a σ-continuous linear mapping Q of $\mathcal{M}$ onto $\mathcal{N}$ satisfying the following conditions:*

1. $Q(h) \geqq 0$ *if* $h \geq 0$ $(h \in \mathcal{M})$, *and* $Q(h) = 0$ *if and only if* $h = 0$;
2. $\|Q(x)\| \leq \|x\|$ $(x \in \mathcal{M})$;
3. $Q(axb) = aQ(x)b$ $(a, b \in \mathcal{N}, x \in \mathcal{M})$;
4. $Q(x)^* Q(x) \leq Q(x^* x)$ $(x \in \mathcal{M})$.

Proof. Let τ be a faithful normal, tracial state on $\mathcal{M}$. For $h(\geq 0) \in \mathcal{M}$, put $\varphi_h(y) = \tau(hy)$ $(y \in \mathcal{N})$. Then $\varphi_h \leq \tau | \mathcal{N}$ and so there exists a positive element k in $\mathcal{N}$ such that $\varphi_h(y) = \tau(hy) = \tau(kyk) = \tau(k^2 y)$ $(y \in \mathcal{N})$ (cf. 1.24). Hence for each positive h in $\mathcal{M}$, there corresponds a positive element $Q(h)$ in $\mathcal{N}$ such that $\tau(hy) = \tau(Q(h)y)$. One can easily see that such a $Q(h)$ is unique—in fact if $\tau(Q(h)y) = \tau(Q(h)'y)$ for all $y \in \mathcal{N}$, then $Q(h) = Q(h)'$. By the linearity, we can extend uniquely the mapping Q to the linear mapping (denoted by Q again) of $\mathcal{M}$ into $\mathcal{N}$—in fact if

$$\sum_{i=1}^{n} \lambda_i h_i = 0 \qquad (\lambda_i \in \mathbb{C}, h_i \geq 0, h_i \in \mathcal{M}),$$

$$\tau\left(\sum_{i=1}^{n} \lambda_i Q(h_i) y\right) = \sum_{i=1}^{n} \lambda_i \tau(Q(h_i) y) = \tau\left(\left(\sum_{i=1}^{n} \lambda_i h_i\right) y\right) = 0$$

$(y \in \mathcal{N})$ and so $\sum_{i=1}^{n} \lambda_i Q(h_i) = 0$.

Hence we have a linear mapping Q of $\mathcal{M}$ into $\mathcal{N}$ satisfying $\tau(xy) = \tau(Q(x)y)$ $(x \in \mathcal{M}, y \in \mathcal{N})$.

One can easily see that this Q satisfies all the conditions in 4.4.23. q.e.d.

4.4.24. Proposition. *If G is a non-amenable discrete group, then the hyperfinite II_1-factor $\mathcal{U}$ can not contain a W^*-algebra $U(G)$ as a W^*-subalgebra.*

Proof. Suppose that the hyperfinite II_1-factor $\mathscr{U}$ contains $U(G)$ as a W^*-subalgebra.

Let τ be the unique tracial state on $\mathscr{U}$, and let $\mathscr{K}$ be an $\aleph_0$-dimensional Hilbert space.

Consider the amplification $\pi_\tau(\mathscr{U})\otimes 1_{\mathscr{K}}$ on $\mathscr{H}\otimes\mathscr{K}$. Then $(\pi_\tau(\mathscr{U})\otimes 1_{\mathscr{K}})'=\pi_\tau(\mathscr{U})'\,\bar{\otimes}\,B(\mathscr{K})$ as abstract W^*-algebras. Since $\pi_\tau(\mathscr{U})'$ is conjugate linear *-isomorphic to $\pi_\tau(\mathscr{U})$ (cf. 2.9.2), $\pi_\tau(\mathscr{U})'$ is hyperfinite. $B(\mathscr{H}_\tau)$ is also hyperfinite and so $\pi_\tau(\mathscr{U})'\,\bar{\otimes}\,B(\mathscr{K})$ is hyperfinite (cf. 4.4.18). Therefore there exists the mapping P in 4.4.15 of $B(\mathscr{H}\otimes\mathscr{K})$ onto $\pi_\tau(\mathscr{U})\otimes 1_{\mathscr{K}}$. Let e be the identity of $U(G)$. Then e is a projection in $\mathscr{U}$.

Let P_1 be a linear mapping of $B(\mathscr{H}\otimes\mathscr{K})$ onto $\pi_\tau(e\mathscr{U}e)\otimes 1_{\mathscr{K}}$ such that $P_1(x)=\pi_\tau(e)\otimes 1_{\mathscr{K}}P(x)\pi_\tau(e)\otimes 1_{\mathscr{K}}$ $(x\in B(\mathscr{H}\otimes\mathscr{K}))$.

On the other hand, let Q be the mapping of $\pi_\tau(e\mathscr{U}e)\otimes 1_{\mathscr{K}}$ onto $U(G)\otimes 1_{\mathscr{K}}$ in 4.4.23, and set

$$P_2(x)=Q(\pi_\tau(e)\otimes 1_{\mathscr{K}}P(x)\pi_\tau(e)\otimes 1_{\mathscr{K}})\qquad (x\in B(\mathscr{H}\otimes\mathscr{K})).$$

By 2.7.4 there exists a projection E' in $(\pi_\tau(U(G))\otimes 1_{\mathscr{K}})'$ such that $y\to(\pi_\tau(y)\otimes 1_{\mathscr{K}})E'$ $(y\in U(G))$ is equivalent to the standard representation of $U(G)$. (cf. 2.9.1).

Since we may consider $B(E'\mathscr{H}\otimes\mathscr{K})$ as a subalgebra of $B(\mathscr{H}\otimes\mathscr{K})$, there exists a conditional expectation P_3 of $B(l^2(G))$ onto $U(G)$.

Let τ be the trace of $U(G)$.

For $f\in l^\infty(G)$ and $\xi\in l^2(G)$, set $(T_f\xi)(g)=f(g)\xi(g)$ $(g\in G)$. Then $T_f\in B(l^2(G))$. Moreover

$$(U(a)^*T_fU(a)\xi)(g)=U(a)^*f(g)\xi(a^{-1}g)=f(ag)\xi(g)\qquad (a,g\in G).$$

Put $\varphi(f)=\tau(P_3(T_f))$. Then

$$\varphi(f^a)=\tau(P_3(U(a)T_fU(a)^*))=\tau(U(a)P_3(T_f)U(a)^*)=\tau(P_3(T_f))=\varphi(f).$$

Hence φ is a left invariant state on $l^\infty(G)$ under G, a contradiction. q.e.d.

4.4.25. Proposition. *Let G be a non-amenable discrete group and let $\mathscr{M}$ be a W^*-algebra. Then $U(G)\,\bar{\otimes}\,\mathscr{M}$ is not hyperfinite.*

Proof. Consider $U(G)$ (resp. $\mathscr{M}$) as a W^*-algebra on $l^2(G)$ (resp. Hilbert space $\mathscr{H}$).

If $U(G)\,\bar{\otimes}\,\mathscr{M}$ is hyperfinite, there exists a mapping P of $B(l^2(G)\otimes\mathscr{H})$ onto $R(U(G)\otimes 1_{\mathscr{H}},\ 1_{l^2(G)}\otimes\mathscr{M})'=R(U(G)'\otimes 1_{\mathscr{H}},\ 1_{l^2(G)}\otimes\mathscr{M}')$ in 4.4.15 (cf. 2.8). On the other hand, there exists a conditional expectation of $R(U(G)'\otimes 1_{\mathscr{H}},\ 1_{l^2(G)}\otimes\mathscr{M}')$ onto $U(G)'\otimes 1_{\mathscr{H}}$ (cf. the proof of 2.6.4). Hence we have a conditional expectation of $B(l^2(G))$ onto $U(G)'$.

Thus G must be amenable. q.e.d.

4.4.26. Corollary. *All factors in* 4.3 *are not hyperfinite.*

4.4.27. Problem. *Can we conclude that every subfactor of a hyperfinite* II_1*-factor is hyperfinite?*

Remark. Let $\mathscr{U}$ be a hyperfinite II_1-factor and let e be a non-zero projection in $\mathscr{U}$. Then the factor $e\mathscr{U}e$ is again hyperfinite (cf. [121]).

4.4.28. Problem. *Suppose that G is a discrete infinite conjugacy group. Can we conclude that the amenability of G is equivalent to that U(G) is hyperfinite?*

4.4.29. Problem. *Is there a* II_1*-factor* $\mathscr{M}$ *on a separable Hilbert space such that every* II_1*-factor on a separable Hilbert space is *-isomorphic to a subfactor of* $\mathscr{M}$*?*

4.4.30. Problem. *For each* II_1*-factor* $\mathscr{M}$ *on a separable Hilbert space, can we find a discrete group such that* $\mathscr{M}=U(G)$*?*

Remark. Let π be the group of all finite permutations of positive integers N onto itself. Then π is an infinite conjugacy group.

Let π_n be the subgroup of all permutations in π, fixing all positive integers greater than n. Then π_n is a finite group and $\pi=\bigcup_{n=1}^{\infty}\pi_n$.

Since $U(\pi_n)\subset U(\pi_{n+1})$, $U(\pi)$ is a hyperfinite II_1-factor by 4.4.6.

4.4.31. Problem. *Is every* W^**-algebra anti-*-isomorphic to itself?*

4.4.32. Problem. *Find a new construction of factors.*

4.4.33. Definition. *Let* $\mathscr{M}$ *be a* W^**-algebra and let* (x_n) *be a uniformly bounded sequence of elements in* $\mathscr{M}$. (x_n) *is called a central sequence if* $[x_n, y]\to 0$ *in the* $s(\mathscr{M},\mathscr{M}_*)$*-topology for* $y\in\mathscr{M}$ (cf. 4.3.21).

4.4.34. Definition. *A central sequence* (x_n) *in a* W^**-algebra* $\mathscr{M}$ *is said to be hypercentral if* $[x_n, y_n]\to 0$ *in the* $s(\mathscr{M},\mathscr{M}_*)$*-topology for every central sequence* (y_n) *in* $\mathscr{M}$.

4.4.35. Definition. *A central* (x_n) *in a* W^**-algebra is said to be trivial if there exists a sequence* (λ_n) *of complex numbers such that* $x_n-\lambda_n 1\to 0$ *in the* $s(\mathscr{M},\mathscr{M}_*)$*-topology.*

Now let $C_{\mathscr{M}}$ (resp. $H_{\mathscr{M}}$, $T_{\mathscr{M}}$) be the set of all central (resp. hypercentral, trivial) sequences in $\mathscr{M}$. Then clearly $T_{\mathscr{M}}\subset H_{\mathscr{M}}\subset C_{\mathscr{M}}$.

4.4.36. Proposition. (cf. [21], [39], [121]). I. *Let* F_2 *be the free group of two generators. Then* $T_{U(F_2)}=C_{U(F_2)}$.

II. $T_{U(L_1)}\subsetneqq H_{U(L_1)}=C_{U(L_1)}$, where L_1 is the discrete group appeared in 4.3.

4.4.37. Theorem. (cf. [113]). *Let $\mathcal{M}$ be a II_1-factor on a separable Hilbert space. If and only if $H_{\mathcal{M}} \neq C_{\mathcal{M}}$, $\mathcal{M}$ is *-isomorphic to $\mathcal{M} \bar{\otimes} \mathcal{U}$, where $\mathcal{U}$ is a hyperfinite II_1-factor.*

Remark 1. This theorem implies that if $H_{\mathcal{M}} \neq C_{\mathcal{M}}$, $\mathcal{M}$ is *-isomorphic to $e\mathcal{M}e$ for every non-zero projection e in $\mathcal{M}$.

In particular, $\mathcal{M}$ is *-isomorphic to $\mathcal{M} \otimes B(\mathcal{H})$ with $\dim(\mathcal{H}) < +\infty$.

Remark 2. All II_1-factors $\mathcal{M}$ in 4.3 except for $U(L_1)$ satisfy the condition "$H_{\mathcal{M}} \neq C_{\mathcal{M}}$".

Remark 3. If $\mathcal{M}$ is an infinite factor, it is *-isomorphic to $\mathcal{M} \bar{\otimes} B(\mathcal{H})$ with $\dim(\mathcal{H}) \leq \aleph_0$.

4.4.38. Problem. *Is there a II_1-factor $\mathcal{M}$ such that $\mathcal{M}$ is not *-isomorphic to $\mathcal{M} \otimes B(\mathcal{H})$ with $2 \leq \dim(\mathcal{H}) = n < +\infty$ for some n.*

Remark. We do not know whether or not $U(F_2)$ (resp. $U(L_1)$) is *-isomorphic to $U(F_2) \otimes B(\mathcal{H})$ (resp. $U(L_1) \otimes B(\mathcal{H})$) with $\dim(\mathcal{H}) < +\infty$.

4.4.39. Definition. *Let $\mathcal{M}$ be a W^*-algebra. $\mathcal{M}$ is said to be asymptotically abelian if there exists a sequence of *-automorphisms (Φ_n) in $\mathcal{M}$ such that $[\Phi_n(x), y] \to 0$ in the $s(\mathcal{M}, \mathcal{M}_*)$-topology $(x, y \in \mathcal{M})$.*

Now let $\mathcal{N}$ be a finite factor, and let φ be the normal tracial state on $\mathcal{N}$. Let $\mathcal{A} = \bigotimes_{n=1}^{\infty} \mathcal{N}_n$ with $\mathcal{N}_n = \mathcal{N}$ be the C^*-infinite tensor product of $\{\mathcal{N}_n\}$ and let $\tau = \bigotimes_{n=1}^{\infty} \varphi_n$ with $\varphi_n = \varphi$ be the infinite product tracial state on $\mathcal{A}$.

Let G be the group of finite permutations of positive integers N onto itself. Then g $(\in G)$ will define a *-automorphism, also denoted by g, of $\mathcal{A}$ by $(\sum \otimes a_n)^g = \sum \otimes a_{g(n)}$, where $a_n = 1$ for all but a finite number of indices.

For each integer n, we denote by g_n the permutations

$$g_n(k) = \begin{cases} 2^{n-1} + k & \text{if } 1 \leq k \leq 2^{n-1} \\ k - 2^{n-1} & \text{if } 2^{n-1} < k \leq 2^n \\ k & \text{if } 2^n < k. \end{cases}$$

Then one can easily see that $\lim_n \|[a^{g_n}, b]\| = 0$ $(a, b \in \mathcal{A})$. Clearly the tracial state τ on $\mathcal{A}$ is G-invariant—that is $\tau(a^g) = \tau(a)$ $(g \in G, a \in \mathcal{A})$.

$\pi_\tau(\mathcal{A})''$ is called the canonical W^*-infinite tensor product of finite factors $\{\mathcal{N}_n\}$ and is denoted by $\textcircled{\otimes}_{n=1}^{\infty} \mathcal{N}_n$.

Then there exists a unitary representation $g \to u_\tau(g)$ of G on $\mathscr{H}_\tau$ such that $u_\tau(g)1_\tau = 1_\tau$ $(g \in \mathscr{A})$ and $\pi_\tau(a^g) = u_\tau^*(g)\pi_\tau(g)u(g)$ $(a \in \mathscr{A})$.

The mapping $x \to u_\tau(g)^* x u_\tau(g)$ $(x \in \pi_\tau(\mathscr{A})'')$ will define a *-automorphism Φ_g on $\pi_\tau(\mathscr{A})''$.

4.4.40. Proposition (cf. [168]). *The finite factor* $\bigotimes_{n=1}^{\infty} \mathscr{N}_n$ *with* $\mathscr{N}_n = \mathscr{N}$, *a finite factor is asymptotically abelian under the above sequence* (Φ_{g_n}) *of *-automorphisms.*

Remark 1. $\left(\bigotimes_{n=1}^{\infty} \mathscr{N}_n\right) \bar{\otimes} \mathscr{U}$ is *-isomorphic to $\bigotimes_{n=1}^{\infty} \mathscr{N}_n$ and

$$\left(\bigotimes_{n=1}^{\infty} \mathscr{N}_n\right) \bar{\otimes} \left(\bigotimes_{n=1}^{\infty} \mathscr{N}_n\right)$$

is *-isomorphic to $\bigotimes_{n=1}^{\infty} \mathscr{N}_n$, where $\mathscr{U}$ is a hyperfinite II_1-factor.

Remark 2. $B(\mathscr{H})$ with $2 \leq \dim(\mathscr{H}) \leq \aleph_0$ is not asymptotically abelian (cf. [168], [217]); any type II_∞-factor is not asymptotically abelian (cf. [57]); $U(F_2)$ and $U(F_2) \bar{\otimes} \mathscr{U}$ is not asymptotically abelian (F_2, the free group of two generators; $\mathscr{U}$, the hyperfinite II_1-factor) (cf. [168]).

Remark 3. Zeller-Meier [224] introduced the notion of "inner asymptotic abelianness".

4.4.41. Problem. *Is there an asymptotically abelian* II_1*-factor which is not the canonical* W^**-infinite tensor product of* II_1*-factors* $\{\mathfrak{B}_n\}$ *with* $\mathfrak{B}_n = \mathfrak{B}$, *a* II_1*-factor* $(n = 1, 2, \ldots)$*?*

4.4.42. Problem. *Is there an asymptotically abelian infinite factor?*

4.4.43. Problem. *Is there a* II_1*-factor* $\mathscr{M}$ *such that* $\mathscr{M} \bar{\otimes} \mathscr{M}$ *is *-isomorphic to* $\mathscr{M}$, *but* $\mathscr{M}$ *is not asymptotically abelian?*

4.4.44. Problem. *Let* F_n $(n = \aleph_0, 2, 3, \ldots)$ *be the free group of n-generators. Is* $U(F_n)$ **-isomorphic to* $U(F_m)$ $(m \neq n)$*?*

4.4.45. Problem. *Is* $U(F_n) \bar{\otimes} U(F_n)$ **-isomorphic to* $U(F_n)$ $(n = \aleph_0, 2, 3, \ldots)$*?*

4.4.46. Problem. *Is* $U(F_n) \otimes B(\mathscr{H})$ $(\dim(\mathscr{H}) < \aleph_0)$ **-isomorphic to* $U(F_n)$ $(n = \aleph_0, 2, 3, \ldots)$*?*

4.4.47. Problem. *Let* $\mathscr{M}$ *be a* II_1*-factor on a separable Hilbert space. Is there an element x in* $\mathscr{M}$ *such that* $\mathscr{M}$ *is the* W^**-subalgebra generated by x.*

Remark. This is true for all infinite factors and many II_1-factors (cf. [42], [147], [199], [220]).

Concluding remarks on 4.4.

1. II_1-factors are very beautiful objects, because they do inherit many nice properties from finite-dimensional matrix algebras. But unfortunately, up to now, they have been used only for the representation theory of discrete groups (though a hyperfinite II_1-factor is used in mathematical physics). It would be a very important problem to find a fruitful application of II_1-factors in the field of functional analysis.

For example, can we apply the theory of II_1-factors to the theory of invariant measures on certain function spaces? (cf. [185], [186], [187]).

2. The property P was introduced by J. Schwartz [176]; Proposition 4.4.23 is due to Umegaki [215].

Additional references. [72], [167].

4.5. Global W^*-Algebras (Non Factors)

Let $\mathcal{M}$ be a W^*-algebra with a separable predual $\mathcal{M}_*$. Then by using the reduction theory in 3.2, $\mathcal{N}$ can be essentially uniquely expressed as an integral $\mathcal{M}=\int_\Omega \mathcal{M}(t)\,d\mu(t)$ of factors $\{\mathcal{M}(t)\}$.

If $\mathcal{M}(t)$ is *-isomorphic to a factor $\mathcal{N}$ for μ-almost all t in Ω, we can write $\mathcal{M}=\mathcal{N}\,\overline{\otimes}\,L^\infty(\Omega,\mu)$ (cf. [37], [205]). More generally, if there exist sequences of factors $\{\mathcal{N}_n\}$ and μ-measurable subsets $\{\Omega_n\}$ in Ω such that $\mathcal{M}(t)$ is *-isomorphic to $\mathcal{N}_n$ for each t in Ω_n, $\mu(\Omega_n)>0$ and $\mu\left(\Omega-\bigcup_{n=1}^{\infty}\Omega_n\right)=0$, then $\mathcal{M}=\sum_{n=1}^{\infty}\oplus\mathcal{N}_n\,\overline{\otimes}\,L^\infty(\Omega_n,\mu_n)$, where μ_n is the restriction of μ to Ω_n.

Therefore to assure that the reduction theory of W^*-algebras is non-trivial, the following question should be answered affirmatively: is there a global W^*-algebra $\mathcal{M}$, with a separable predual, whose reduction theory $\mathcal{M}=\int_\Omega \mathcal{M}(t)\,d\mu(t)$ satisfies the following condition: there exists a measurable subset Ω_0 in Ω such that $\mu(\Omega_0)=\mu(\Omega)$ and for arbitrary two distinct t_1,t_2 in Ω_0, $\mathcal{M}(t_1)$ is not *-isomorphic to $\mathcal{M}(t_2)$.

Obviously, to answer this question affirmatively, there must exist uncountably many examples of factors in a separable Hilbert space.

In this section, by using the examples in 4.3, we shall construct examples of global types II_1, II_∞ and III W^*-algebras satisfying the condition.

Now let Ω be the set of all infinite sequences (p_i) of positive integers with $p_i=1$ or i $(i=1,2,\ldots)$, and let $S=\{0,1\}$ be the compact group of order 2. Let $S_n=S$ for $n=2,3,\ldots$ and let $\Gamma=\prod_{n=2}^{\infty} S_n$ be the compact group obtained by the weakly infinite direct product of $\{S_n\}$. Let $\mathrm{II}_1=(p_i)$ be an element of Ω. We shall identify II_1 with the group $G[\mathrm{II}_1]$ in 4.3. Define $(\mathrm{II}_1)_n=1$ if $p_n=n$ and $(\mathrm{II}_1)_n=0$ if $p_n=1$ $(n=2,3,\ldots)$.

Then $((\mathrm{II}_1)_2,(\mathrm{II}_1)_3,\ldots)$ will define an element γ in Γ. Set $\rho(\mathrm{II}_1)=\gamma$. Then ρ is a one-to-one mapping of Ω onto Γ. By using the ρ, we shall identify Ω with Γ. Since Γ is a compact group, we have the corresponding compact group structure in Ω by this identification.

4.5.1. Lemma. *Let $\Delta_1=(G_1,G_2,\ldots;H_1,H_2,\ldots)$ and $\Delta_2=(G_1,G_2,\ldots;J_1,J_2,\ldots)$ (cf. the group construction of 4.3). Suppose that there exists a homomorphism ξ_i of H_i onto J_i for all i. Then we can define a homomorphism ξ of Δ_1 onto Δ_2 such that $\xi=$ the identity on G_i and $\xi=\xi_i$ on H_i for all i, where G_i and H_i are identified with the corresponding subgroups of $(G_1,G_2,\ldots;H_1,H_2,\ldots)$.*

This lemma will be almost trivial.

Now let F_∞ be the free group of denumberable generators and let $\Lambda=(G_1,G_2,\ldots;H_1,H_2,\ldots)$ with $G_i=Z$ and $H_i=F_\infty$. Let $(r_1,r_2,\ldots,r_n)$ be a finite sequence of positive integers such that $r_i=1$ or i $(i=1,2,\ldots,n)$. Consider the group $\sum_{i=1}^{n}\oplus L_{r_i}$. Then there exists a homomorphism ξ of F_∞ onto $\sum_{i=1}^{n}\oplus L_{r_i}$. We shall pick up one homomorphism ξ and fix it; we denote this ξ by $\xi(r_1,r_2,\ldots,r_n)$ and so ξ is a function of $(r_1,r_2,\ldots,r_n)$.

Let $\mathrm{II}_1\in\Omega$ with $\mathrm{II}_1=(p_1,p_2,\ldots)$. Then

$$G[\mathrm{II}_1]=(G_1,G_2,\ldots;M_1(\mathrm{II}_1),M_2(\mathrm{II}_1),\ldots)$$

with $G_i=Z$ $(i=1,2,\ldots)$. By 4.5.1, we can define a homomorphism $\xi(\mathrm{II}_1)$ of Λ onto $G[\mathrm{II}_1]$ such that $\xi(\mathrm{II}_1)=$ the identity on G_i and $\xi(\mathrm{II}_1)=\xi(p_1,p_2,\ldots,p_n)$ on H_n $(n=1,2,\ldots)$.

Let ν be the Haar measure on Λ such that $\nu(e)=1$ (e, the identity of Λ), and let $l^1(\Lambda)$ be the group algebra of Λ composed of all μ-integrable functions on Λ.

Let $\{\psi_\beta\}_{\beta\in\mathbb{J}}$ be the set of all positive definite functions on Λ with $\psi_\beta(e)=1$.

Define $\|x\|^2=\sup_{\beta\in\mathbb{J}}\int_\Lambda (x^* * x)(g)\,\psi_\beta(g)\,d\nu(g)$ for $x\in l^1(\Lambda)$ $(x^*(g)=x(g^{-1})$ and $x^* * x$ is the convolution of x^* and $x)$.

Then $\|\cdot\|$ will define a C^*-norm on $l^1(\Lambda)$. The completion of $l^1(\Lambda)$ under this C^*-norm $\|\cdot\|$ is called the group C^*-algebra of Λ and is denoted by $R(\Lambda)$.

For $\lambda\in\Gamma\equiv\Omega$, we shall define a trace τ_λ on $R(\Lambda)$ as follows: take the homomorphism $\xi(\lambda)$ of Λ onto $G[\lambda]$ and define $\tau_\lambda(g)=\delta_{\xi(\lambda)(g)}(e_\lambda)$, where $\delta_{\xi(\lambda)(g)}$ is the function on $G[\lambda]$ such that $\delta_{\xi(\lambda)(g)}(l)=1$ if $\xi(\lambda)(g)=l$ and $\delta_{\xi(\lambda)(g)}(l)=0$ if $\xi(\lambda)(g)\neq l$ $(l\in G[\lambda])$, and e_λ is the identity of $G[\lambda]$.

Then τ_λ is a central positive definite function on Λ with $\tau_\lambda(e)=1$; therefore it will define a unique tracial state (denoted by the same notation τ_λ) on $R(\Lambda)$.

Let $\{\pi_\lambda,\mathscr{H}_\lambda\}$ be the *-representation of $R(\Lambda)$ on a separable Hilbert space $\mathscr{H}_\lambda$ constructed via τ_λ. Then $\pi_\lambda(R(\Lambda))''$ is *-isomorphic to $U(G[\lambda])$.

Now suppose that a sequence (λ_n) of elements in Γ with $\lambda_n=(p_{n,1},p_{n,2},\ldots)$ converges to λ_0 in Γ with $\lambda_0=(p_{0,1},p_{0,2},\ldots)$ in the topology of Ω $(\equiv\Gamma)$.

Then for an arbitrary $g\in\Lambda$, there exists a positive integer n_0 such that g belongs to a subgroup of Λ generated by $G_1,G_2,\ldots,G_{n_0}$ and $H_1,H_2,\ldots,H_{n_0}$ $(\Lambda=(G_1,G_2,\ldots;H_1,H_2,\ldots)$ with $G_i=Z$ and $H_i=F_\infty$ for $i=1,2,\ldots)$.

Since $\lambda_n\to\lambda_0$, there exists an n_1 such that $p_{n,i}=p_{0,i}$ for $n\geq n_1$, $i=1,2,\ldots,n_0$. Hence $\lim_n \tau_{\lambda_n}(g)=\tau_{\lambda_0}(g)$ $(g\in\Lambda)$. This implies that the sequence (τ_{λ_n}) of traces on $R(\Lambda)$ converges to the trace τ_{λ_0} on $R(\Lambda)$ in the $\sigma(R(\Lambda)^*, R(\Lambda))$-topology ($R(\Lambda)^*$ the dual Banach space of $R(\Lambda)$).

Therefore the mapping $\eta\colon\lambda\to\tau_\lambda$ of Γ into the state space $\mathscr{S}$ of $R(\Lambda)$ is continuous; it is one-to-one, because $\pi_\lambda(R(\Lambda))''$ is not *-isomorphic to $\pi_{\lambda'}(R(\Lambda))''$ if $\lambda\neq\lambda'$.

The compactness of Γ implies that the η is homeomorphic. Let $d\lambda$ be the Haar measure on Γ $(\equiv\Omega)$ with the total mass 1; by using the η, we can introduce a Radon measure μ on $\mathscr{S}$ such that $d\mu(\tau_\lambda)=d\lambda$.

Define $\psi(x)=\int_{\mathscr{S}}\varphi(x)d\mu(\varphi)$ $(x\in R(\Lambda))$.

Now we shall show.

4.5.2. Theorem. *μ is the central measure of ψ.*

Proof. This is obtained immediately from 3.1.18.

4.5.3. Corollary. *There exists a type* II$_1$ *W^*-algebra $\mathscr{M}$ on a separable Hilbert space satisfying the following properties:*

1. *The center of $\mathscr{M}$ is isomorphic to $L^\infty(\Gamma,d\lambda)$, where Γ is an infinite compact group and $d\lambda$ is the Haar measure on Γ.*

2. *For its reduction theory $\mathscr{M}=\int_\Gamma\mathscr{M}(\lambda)d\lambda$, $\mathscr{M}(\lambda_1)$ is not *-isomorphic to $\mathscr{M}(\lambda_2)$ for arbitrary two distinct λ_1,λ_2 in Γ.*

4.5.4. Corollary. *There exist types* II$_\infty$ *and* III *W^*-algebras on a separable Hilbert space satisfying the similar properties to* 4.5.3.

Proof. Consider $\mathcal{M}\,\bar{\otimes}\,B(\mathcal{H})$ with $\dim(\mathcal{H})=\aleph_0$. Then

$$\mathcal{M}\,\bar{\otimes}\,B(\mathcal{H})=\int_\Gamma \mathcal{M}(\lambda)\,\bar{\otimes}\,B(\mathcal{H})\,d\lambda.$$

$\mathcal{M}(\lambda_1)\,\bar{\otimes}\,B(\mathcal{H})$ is not *-isomorphic to $\mathcal{M}(\lambda_2)\,\bar{\otimes}\,B(\mathcal{H})$ if $\lambda_1\neq\lambda_2$ (cf. 4.3).

Analogously consider $\mathcal{M}\,\bar{\otimes}\,\mathbb{P}$, where $\mathbb{P}$ is the III-factor of Pukanszky (cf. 4.3). q.e.d.

Remark. By using Powers' factors, we can construct another example of a non-trivial global type III-factor (cf. [169]).

Concluding remarks on 4.5.

Now the following problem would be very interesting. Let $\mathcal{M}$ be a global W^*-algebra on a separable Hilbert space. Find nice conditions under which $\mathcal{M}$ can be written as $\mathcal{N}\,\bar{\otimes}\,Z$ (Z, the center of $\mathcal{M}$; $\mathcal{N}$, a factor).

By using these conditions, can we find directly a global W^*-algebra $\mathcal{M}$ such that the reduction $\mathcal{M}=\int_{\mathcal{S}}\mathcal{M}(\varphi)\,d\mu(\varphi)$ satisfies that μ is a continuous measure and there exists a μ-measurable subset $\mathcal{S}_0$ in $\mathcal{S}$ with $\mu(\mathcal{S}-\mathcal{S}_0)=0$ such that for any two distinct φ_1, φ_2 in $\mathcal{S}_0$, $\mathcal{M}(\varphi_1)$ is not *-isomorphic to $\mathcal{M}(\varphi_2)$?

References. [169], [173].

4.6. Type I C^*-Algebras

In this section, we shall give some considerations to type I C^*-algebras which are important in the group representation theory, and in the theory of singular integral operators (cf. [24]).

Let G be a locally compact group, μ a left invariant Haar measure on G and let $L^1(G)$ be the Banach space of all complex valued μ-integrable functions on G.

For $f_1, f_2\in L^1(G)$, define a multiplication * and a *-operation as follows: $f_1 * f_2(x)=\int f_1(y)f_2(y^{-1}x)\,d\mu(y)$ and $f_1^*(x)=\rho(x)\overline{f(x^{-1})}$, where ρ is the modular function on G.

By a unitary representation $\{u,\mathcal{H}\}$ of G on a Hilbert space $\mathcal{H}$, we mean a mapping $x\to u(x)$ of G into the group of unitary operators in $\mathcal{H}$ satisfying the following conditions:

1. $u(x_1x_2)=u(x_1)u(x_2)$ $(x_1,x_2\in G)$;
2. a function $x\to(u(x)\xi,\eta)$ is continuous on G $(\xi,\eta\in\mathcal{H})$.

For $f \in L^1(G)$, put $\pi(f) = \int u(x) f(x) d\mu(x)$, where the integral is taken by using the strong operator topology. Then a mapping $f \to \pi(f)$ satisfies the following conditions:

1. $\pi(f_1 * f_2) = \pi(f_1)\pi(f_2)$ $(f_1, f_2 \in L^1(G))$;
2. $\pi(f^*) = \pi(f)^*$ $(f \in L^1(G))$;
3. $\|\pi(f)\| \leq \|f\|_1$ $(f \in L^1(G))$;
4. $[\pi(f)\mathscr{H} \mid f \in L^1(G)] = \mathscr{H}$—i.e. nowhere trivial, where $\|\cdot\|_1$ is the

norm on $L^1(G)$.

Such a mapping $\{\pi, \mathscr{H}\}$ of $L^1(G)$ into $B(\mathscr{H})$ is called a nowhere trivial *-representation of $L^1(G)$. Conversely we may show that given a nowhere trivial *-representation $\{\pi, \mathscr{H}\}$ of $L^1(G)$, we can construct a unitary representation $\{u, \mathscr{H}\}$ of G with $\pi(f) = \int u(x) f(x) d\mu(x)$ $(f \in L^1(G))$. Therefore the unitary representation theory of G can be reduced to the *-representation theory of $L^1(G)$.

Now let $\{\|\cdot\|_\alpha \mid \alpha \in \mathbb{I}\}$ be the set of all C^*-seminorms on $L^1(G)$, satisfying $\|f\|_\alpha \leq \|f\|_1$ $(f \in L^1(G))$, and define $\|f\| = \sup_{\alpha \in \mathbb{I}} \|f\|_\alpha$. One can easily see that $\|\cdot\|$ is a C^*-norm.

Let $R(G)$ be the completion of $L^1(G)$ under the C^*-norm $\|\cdot\|$. Then $R(G)$ is a C^*-algebra which is called the group C^*-algebra of G.

Then the unitary representation theory may be reduced to the *-representation theory of $R(G)$.

A basic principle in the representation theory is to pick up basic representations as building blocks, to determine all such basic representations and to construct an arbitrary representation by integrating basic ones. Therefore we have first to decide what kind of representations are suitable to basic ones. The representation theories of commutative groups and compact groups may suggest irreducible *-representations as the basic ones. But, from the reduction theory of W^*-algebras, we may guess that this choice is not reasonable unless a corresponding W^*-algebra $\overline{\pi(R(G))}$ is of type I.

For the general theory, we should choose factorial *-representations as the building blocks.

On the other hand, if $\overline{\pi(R(G))}$ is of type I for every *-representation $\{\pi, \mathscr{H}\}$, we may follow the classical line and expect a fairly nice representation theory.

Thus it is a very important problem "which groups have type I representations only?"

A locally compact group G is called a type I group if $u(G)''$ is of type I for every unitary representation $\{u, \mathscr{H}\}$. It is known that semi-simple Lie groups, nilpotent Lie groups are type I groups (cf. [34], [74], [102]). But there exists a solvable Lie group which is not of type I (cf. [114]).

Now we shall define.

4.6.1. Definition. *Let $\mathscr{A}$ be a C^*-algebra. $\mathscr{A}$ is said to be a type* I *C^*-algebra if for every *-representation $\{\pi, \mathscr{H}\}$ of $\mathscr{A}$, the weak closure $\overline{\pi(\mathscr{A})}$ of $\pi(\mathscr{A})$ on $\mathscr{H}$ is a type* I *W^*-algebra. (a *-representation $\{\pi, \mathscr{H}\}$ of a C^*-algebra $\mathscr{A}$ is said to be of type* I *(resp.* II, III*) if the weak closure $\overline{\pi(\mathscr{A})}$ of $\pi(\mathscr{A})$ on $\mathscr{H}$ is of type* I *(resp.* II, III*)).*

One can easily see that the study of type I C^*-algebras is important for the group representation theory.

On the other hand, we can not expect type I C^*-algebras in mathematical physics—in fact, the weak closures of the representations which appear in mathematical physics are of type III in many cases (cf. [3]).

In this section, we shall give characterizations of type I C^*-algebras.

4.6.2. Definition. *Let $\mathscr{A}$ be a C^*-algebra. $\mathscr{A}$ is said to satisfy the condition of Glimm if for every non-zero positive element h in $\mathscr{A}$, there exists an irreducible *-representation $\{\pi, \mathscr{H}\}$ of $\mathscr{A}$ such that the dimension of the range space of $\pi(h) \geq 2$.*

4.6.3. Definition. *Let $\mathscr{A}$ be a C^*-algebra. $\mathscr{A}$ is said to be smooth if for every non-zero irreducible *-representation $\{\pi, \mathscr{H}\}$ of $\mathscr{A}$, $\pi(\mathscr{A})$ contains a non-zero compact operator on $\mathscr{H}$.*

Then we shall show

4.6.4. Theorem. *Let $\mathscr{A}$ be a C^*-algebra. The following conditions are equivalent:*

1. *$\mathscr{A}$ is a type* I *C^*-algebra;*
2. *$\mathscr{A}$ is smooth;*
3. *$\mathscr{A}$ has no type* III *factorial *-representation;*
4. *any quotient C^*-algebra of $\mathscr{A}$ does not satisfy the condition of Glimm—i.e. for every non-zero quotient C^*-algebra $\mathfrak{B}$ of $\mathscr{A}$, there exists a positive element h (>0) in $\mathfrak{B}$ such that the rank of $\pi(h) \leq 1$ for all irreducible *-representations $\{\pi, \mathscr{H}\}$ of $\mathfrak{B}$.*

We shall prove first the easy part of the theorem.

4.6.5. Proof of 4.$\Rightarrow$1.: *Suppose every quotient C^*-algebra of $\mathscr{A}$ does not satisfy the condition of Glimm. If $\mathscr{A}$ is not of type* I, *then there exists a *-representation $\{\pi, \mathscr{H}\}$ of $\mathscr{A}$ such that $\overline{\pi(\mathscr{A})}$ does not have a type* I *W^*-direct summand.*

Let $\mathscr{I}$ = kernel of π. Then $\{\pi, \mathscr{H}\}$ can be considered as a *-representation of $\mathscr{A}/\mathscr{I}$. Hence it suffices to assume that $\mathscr{I}=(0)$. Let h_1 (>0) be an element in $\mathscr{A}$ with Rank of $\pi_\alpha(h_1) \leq 1$ for all irreducible *-representations $\{\pi_\alpha, \mathscr{H}_\alpha\}$ of $\mathscr{A}$. Hence $h_1 \mathscr{A} h_1$ is commutative and so $\pi(h_1 \mathscr{A} h_1) = \pi(h_1)\pi(\mathscr{A})\pi(h_1)$ is commutative.

Let $\pi(h_1)=\int_0^\infty \lambda dE_\lambda$ be the spectral decomposition of $\pi(h_1)$ and let $H=\int_\varepsilon^\infty \frac{1}{\lambda} dE_\lambda$ $(\varepsilon>0)$. Then $\pi(h_1)H=\int_\varepsilon^\infty dE_\lambda=1_{\mathscr{H}}-E_\varepsilon$ and

$$\pi(h_1)H\overline{\pi(\mathscr{A})}H\pi(h_1)=(1_{\mathscr{H}}-E_\varepsilon)\overline{\pi(\mathscr{A})}(1_{\mathscr{H}}-E_\varepsilon).$$

If ε is sufficiently small, $1_{\mathscr{H}}-E_\varepsilon\neq 0$. Hence $\overline{\pi(\mathscr{A})}$ contains an abelian projection and so $\overline{\pi(\mathscr{A})}$ must contain a type I W^*-direct summand, a contradiction. q.e.d.

Proof of 4.⇒2. Let $\{\pi,\mathscr{H}\}$ be a non-zero irreducible *-representation of $\mathscr{A}$, and let $\mathscr{I}=$kernel of π. Then $\{\pi,\mathscr{H}\}$ can be canonically considered as a *-representation of $\mathscr{A}/\mathscr{I}$. Let h (>0) be an element in $\mathscr{A}/\mathscr{I}$ with the rank of $\pi(h)\leq 1$. Then $\pi(h)$ is a non-zero compact operator. q.e.d.

The proof of 1.⇒3.⇒4. will be carried out in the following way. We shall show first that if a C^*-algebra $\mathscr{A}_1$ satisfies the condition of Glimm, it contains a separable C^*-subalgebra $\mathfrak{B}_1$ such that at least one quotient C^*-algebra of $\mathfrak{B}_1$ is an infinite-dimensional uniformly hyperfinite C^*-algebra. Next, we shall show that if a C^*-algebra $\mathscr{A}_2$ contains a separable C^*-subalgebra $\mathfrak{B}_2$ such that at least one quotient C^*-algebra of $\mathfrak{B}_2$ is an infinite dimensional uniformly hyperfinite C^*-algebra, then $\mathscr{A}_2$ has a type III factorial *-representation (this implies automatically that any C^*-subalgebra of a type I C^*-algebra is again a type I C^*-algebra). Thus we have that if $\mathscr{A}$ has a non-zero quotient C^*-algebra satisfying the condition of Glimm, it has a type III factorial *-representation.

Finally we shall give a proof of 2.⇒4. and conclude the proof of the theorem.

Now we shall prepare some lemmas to prove that if a C^*-algebra $\mathscr{A}_1$ satisfies the condition of Glimm, it contains a separable C^*-subalgebra $\mathfrak{B}_1$ such that at least one quotient C^*-subalgebra of $\mathfrak{B}_1$ is an infinite dimensional uniformly hyperfinite C^*-algebra.

Let B_n be the matrix algebra of all $2^n\times 2^n$ matrices over the complex field with a positive integer n. Then there exists a family of 2^n-mutually orthogonal minimal projections whose sum is the identity. We shall index such a family as follows: $\{e(a_1,a_2,\ldots,a_n)\mid a_1,a_2,\ldots,a_n\in\{0,1\}\}$.

Let $w(a_1,\ldots,a_n)$ be a partial isometry with

$$w(a_1,\ldots,a_n)^*w(a_1,\ldots,a_n)=e(0,\ldots,0),$$

$w(a_1,\ldots,a_n)w(a_1,\ldots,a_n)^*=e(a_1,\ldots,a_n)$ and $w(0,\ldots,0)=e(0,\ldots,0)$.

For simplicity, we shall denote $\underbrace{(0, \ldots, 0)}_{n}$ by (0_n). Then

$$\{w(a_1, \ldots, a_n)w(b_1, \ldots, b_n)^* \mid a_1, \ldots, a_n,\ b_1, \ldots, b_n \in \{0, 1\}\}.$$

is a system of matrix units in B_n, and if

$$(a_1, \ldots, a_n) \neq (b_1, \ldots, b_n), \qquad w(b_1, \ldots, b_n)^* w(a_1, \ldots, a_n) = 0.$$

Hence

$$\begin{aligned} &w(a_1, \ldots, a_n)w(b_1, \ldots, b_n)^* w(a'_1, \ldots, a'_n)w(b'_1, \ldots, b'_n)^* \\ &= \delta_{a'_1}^{b_1} \ldots \delta_{a'_n}^{b_n} w(a_1, \ldots, a_n)e(0_n)w(b'_1, \ldots, b'_n)^* \\ &= \delta_{a'_1}^{b_1} \ldots \delta_{a'_n}^{b_n} w(a_1, \ldots, a_n)w(b'_1, \ldots, b'_n)^*, \end{aligned}$$

where $\delta_{a'_i}^{b_i}$ $(i = 1, 2, \ldots, n)$ is the Kronecker symbol.

For $\varepsilon \in (0, 1]$, we shall denote by f_ε a continuous function on $(-\infty, \infty)$ such that $f(r) = 0$ on $(-\infty, 1-\varepsilon]$, $f(r) = 1$ on $[1-\varepsilon/2, +\infty)$ and $f(r)$ is linear on $[1-\varepsilon, 1-\varepsilon/2]$.

Then for $\varepsilon \in (0, \frac{1}{2}]$, we have $f_\varepsilon f_{2\varepsilon} = f_\varepsilon$.

4.6.6. Lemma. *Let $\mathscr{A}_1$ be a C^*-algebra, with an identity, satisfying the condition of Glimm and let d be a positive element of $\mathscr{A}_1$ with $\|d\| = 1$, and let $t \in (0, 1]$.*

Then there exist w, w', d' in $\mathscr{A}_1$ such that

I $\|w\| = \|w'\| = \|d'\| = 1$, $w \geq 0$, $d' \geq 0$, $w'^* w = 0$;

II $f_t(d)w = w$, $f_t(d)w' = w'$;

III $w^2 d' = d'$, $w'^* w' d' = d'$.

Proof. Put $s = t/8$. Choose $u, c \in \mathscr{A}_1$ with $\|u\| \leq 1$, $0 \leq c \leq 1$. Set $d_0 = f_{2s}(d)cf_{2s}(d)$ and $d_1 = f_{4s}(d) - d_0$. Then $0 \leq d_0 \leq 1$ and $-1 \leq d_1 \leq 1$.

Since $f_{4s}(d)d_0 = d_0 = d_0 f_{4s}(d)$, d_0 and $f_{4s}(d)$ generate a commutative C^*-algebra C of $\mathscr{A}_1$.

If χ is a character of C with $\chi(d_0) \neq 0$, then $\chi(f_{4s}(d)) = 1$ and so $\chi(d_1) = 1 - \chi(d_0)$.

Let g be a real valued continuous function on $(-\infty, \infty)$ with $g(r) = 0$ on $[0, \frac{1}{2}]$. Then $g(\chi(d_0))g(\chi(d_1)) = 0$ and so $\chi(g(d_0)g(d_1)) = 0$. Hence $g(d_0)g(d_1) = 0$.

Put $v = f_s(d_1)uf_s(d_0)$. Then $\|v\| \leq 1$ and $v^* v = f_s(d_0)u^* f_s(d_1)^2 u f_s(d_0)$. Hence $f_{2s}(d_0)v^* v = v^* v$, $v^* f_{2s}(d_1) = v^*$ and

$$v^*(v^* v) = v^* f_{2s}(d_1) f_{2s}(d_0) v^* v = 0.$$

Moreover $f_{8s}(d)d_0 = d_0$ and $f_{8s}(d)d_1 = d_1$. Hence $f_{8s}(d)p(d_0) = p(d_0)$ if p is a polynomial without a constant term, and so $f_{8s}(d)f(d_0) = f(d_0)$ if f is a continuous function vanishing at 0.

Analogously $f_{8s}(d)f(d_1) = f(d_1)$. Hence $f_{8s}(d)v = v$ and $f_{8s}(d)v^* = v^*$.

Put $d' = f_{\frac{1}{4}}(v^* v)$, $w = f_{\frac{1}{2}}(v^* v)^{\frac{1}{2}}$ and $w' = v k(v^* v)$ (k is a function on $(-\infty, \infty)$ such that $k(r) = (f_{\frac{1}{2}}(r) r^{-1})^{\frac{1}{2}}$ if $r \neq 0$; $= 0$ if $r = 0$). Then $0 \leq d' \leq 1$, $0 \leq w \leq 1$. Since $v^*(v^* v) = 0$, $w'^* w = 0$. Since $f_{8s}(d) v = v$ and $f_{8s}(d) v^* = v^*$, $f_{8s}(d) w = w$ and $f_{8s}(d) w' = w'$.

Moreover

$$w^2 = f_{\frac{1}{2}}(v^* v), \qquad w'^* w' = k^2(v^* v) v^* v = f_{\frac{1}{2}}(v^* v).$$

Hence $w^2 d' = f_{\frac{1}{2}}(v^* v) f_{\frac{1}{4}}(v^* v) = f_{\frac{1}{4}}(v^* v) = d'$, $w'^* w' d' = d'$ and

$$\|w'\| = \|w'^* w'\|^{\frac{1}{2}} \leq 1.$$

Now we shall show that if c and u are chosen suitably, we have $\|d'\| \geq 1$.

Then by the above equality, $\|w'\|$, $\|w\| \geq 1$ and so

$$\|w'\| = \|d'\| = \|w\| = 1.$$

Since $f_s(d) \neq 0$ and $\mathscr{A}_1$ satisfies the condition of Glimm, there exists an irreducible *-representation $\{\pi, \mathscr{H}\}$ of $\mathscr{A}_1$ such that the dimension of the range space of $\pi(f_s(d)) \geq 2$. Take two vectors in $\pi(f_s(d))\mathscr{H}$ with $\|\xi\| = \|\eta\| = 1$ and $(\xi, \eta) = 0$. Then there exists a self-adjoint element c in $\mathscr{A}_1$ such that $\pi(c)\xi = \xi$, $\pi(c)\eta = 0$ by 1.21.16. Let g be a continuous function on $(-\infty, \infty)$ as follows: $g(t) = 0$ if $t \leq 0$; $g(t) = t$ if $0 < t \leq 1$ and $g(t) = 1$ if $t > 1$. Then $0 \leq g(c) \leq 1$ and $\pi(g(c))\xi = \xi$ and $\pi(g(c))\eta = 0$. Hence we may assume that $0 \leq c \leq 1$.

Then there is a unitary element u in $\mathscr{A}_1$ with $\pi(u)\xi = \eta$ by 1.21.16.

Since $f_{2s}(d) f_s(d) = f_s(d)$, we have $\pi(f_{2s}(d))\xi = \xi$, $\pi(f_{2s}(d))\eta = \eta$ and $\pi(f_{4s}(d))\eta = \eta$.

Hence $\pi(d_0)\xi = \xi$, $\pi(d_0)\eta = 0$, $\pi(d_1)\eta = \eta$ and

$$\begin{aligned} \pi(v^* v)\xi &= f_s(\pi(d_0))\pi(u)^{-1} f_s(\pi(d_1))^2 \pi(u) f_s(\pi(d_0))\xi \\ &= f_s(\pi(d_0))\pi(u)^{-1} f_s(\pi(d_1))^2 \pi(u)\xi \\ &= f_s(\pi(d_0))\pi(u)^{-1} f_s(\pi(d_1))^2 \eta \\ &= f_s(\pi(d_0))\pi(u)^{-1} \eta = f_s(\pi(d_0))\xi = \xi. \end{aligned}$$

Hence $\pi(d')\xi = \xi$ and so $\|d'\| \geq 1$. q.e.d.

4.6.7. Lemma. *Let $\mathscr{A}_1$ be a C*-algebra, with an identity, satisfying the condition of Glimm. Then there exist non-zero elements $u(a_1, \ldots, a_n)$ in the unit sphere of $\mathscr{A}_1$ $(a_1, \ldots, a_n = 0$ or $1; n = 0, 1, 2, \ldots)$ satisfying the following conditions:*

I. *if $j \leq k$ and $(a_1, \ldots, a_j) \neq (b_1, \ldots, b_j)$, then $v(a_1, \ldots, a_j)^* v(b_1, \ldots, b_k) = 0$;*

II. *if $k \geq 1$, $v(a_1, \ldots, a_k) = v(a_1, \ldots, a_{k-1}) v(0_{k-1}, a_k)$, where*

$$(0_{k-1}, a_k) = (\underbrace{0, \ldots, 0}_{k-1}, a_k);$$

III. *if* $j<k$, $v(a_1,\ldots,a_j)^* v(a_1,\ldots,a_j) v(0_{k-1},a_k)=v(0_{k-1},a_k)$;
IV. $v(\emptyset)=1$, $v(0_k)\geq 0$;
V. $v(a_1,\ldots,a_n)^* v(a_1,\ldots,a_n) b(n)=b(n)$ *for some* $b(n)>0$ *with*

$$\|b(n)\|=1 \qquad (n=0,1,2,\ldots).$$

Proof. For $n=0$, put $v(\emptyset)=b(0)=1$. Suppose we construct $v(a_1,\ldots,a_j)$ and $b(j)$ for $j\leq n$. Then in the above lemma, let $d=b(n)$ and denote the elements w, w', d' by $v(0_{n+1})$, $v(0_n,1)$, $b(n+1)$.

Then

$$\begin{aligned} &v(a_1,\ldots,a_n)^* v(a_1,\ldots,a_n) v(0_n,a_{n+1}) \\ &= v(a_1,\ldots,a_n)^* v(a_1,\ldots,a_n) f_t(b(n)) v(0_n,a_{n+1}). \end{aligned}$$

Since $v(a_1,\ldots,a_n)^* v(a_1,\ldots,a_n) b(n)=b(n)$,

$$v(a_1,\ldots,a_n)^* v(a_1,\ldots,a_n) f_t(b(n))=f_t(b(n)).$$

Hence

$$v(a_1,\ldots,a_n)^* v(a_1,\ldots,a_n) v(0_n,a_{n+1})=f_t(b(n)) v(0_n,a_{n+1})=v(0_n,a_{n+1}).$$

On the other hand, $v(0_n,1)^* v(0_{n+1})=0$ and $v(0_{n+1})\geq 0$.

Now put $v(a_1,\ldots,a_{n+1})=v(a_1,\ldots,a_n) v(0_n,a_{n+1})$ for $(a_1,\ldots,a_n)\neq 0_n$. Then one can easily see that $v(a_1,\ldots,a_{n+1})$ satisfies the conditions I., II., III., and IV. Moreover if $(a_1,\ldots,a_n)\neq 0_n$.

$$\begin{aligned} &v(a_1,\ldots,a_{n+1})^* v(a_1,\ldots,a_{n+1}) b(n+1) \\ &= v(0_n,a_{n+1})^* v(a_1,\ldots,a_n)^* v(a_1,\ldots,a_n) v(0_n,a_{n+1}) b(n+1) \\ &= v(0_n,a_{n+1})^* v(0_n,a_{n+1}) b(n+1)=b(n+1). \end{aligned}$$

q.e.d.

4.6.8. Lemma. *Let $\mathscr{A}_1$ be a C^*-algebra satisfying the condition of Glimm. Then there exists a separable C^*-subalgebra $\mathfrak{B}_1$ in $\mathscr{A}_1$ such that at least one quotient C^*-algebra of $\mathfrak{B}_1$ is an infinite-dimensional uniformly hyperfinite C^*-algebra.*

Proof. Without loss of generality, we may assume that $\mathscr{A}_1$ has an identity. We realize the C^*-algebra $\mathscr{A}_1$ as an operator algebra on a Hilbert space $\mathscr{H}$.

Put $e(n)=\sum_{a_1,\ldots,a_n\in\{0,1\}} v(a_1,\ldots,a_n) v(a_1,\ldots,a_n)^*$ and let $\mathscr{H}(n)$ be the range space of $e(n)$ in $\mathscr{H}$—namely $\mathscr{H}(n)=[e(n)\mathscr{H}]$. Then $\mathscr{H}(n+1)\subset\mathscr{H}(n)$.

In fact if $\zeta\in\mathscr{H}$ with $e(n)\zeta=0$, $(v(a_1,\ldots,a_n) v^*(a_1,\ldots,a_n)\zeta,\zeta)=0$. Hence $v(a_1,\ldots,a_n)^*\zeta=0$ and so $v(a_1,\ldots,a_{n+1})^*\zeta=0$. Hence $e(n+1)\zeta=0$. By Property I in 4.6.7, $\{v(a_1,\ldots,a_n)\mathscr{H}\}$ are mutually orthogonal for a fixed n. Let $P(n)$ be the orthogonal projection of $\mathscr{H}$ onto $\mathscr{H}(n)$. Then

$P(n)=\sum_{a_1,\dots,a_n\in\{0,1\}} s(v(a_1,\dots,a_n)v(a_1,\dots,a_n{}^*))$, where the support projection $s(\cdot)$ is taken in $B(\mathscr{H})$. For $\eta\in\mathscr{H}$ and $\xi=v(c_1,\dots,c_{n+1})\eta$,

$$\begin{aligned}v(a_1,\dots,a_n)v(b_1,\dots,b_n)^*\,v(c_1,\dots,c_n)v(0_n,c_{n+1})\eta\\ =\delta_{b_1}^{c_1}\dots\delta_{b_n}^{c_n}v(a_1,\dots,a_n,c_{n+1})\eta.\end{aligned}$$

Therefore $\mathscr{H}(n+1)$ is invariant under $v(a_1,\dots,a_n)v(b_1,\dots,b_n)^*$.

Moreover,

$$\begin{aligned}&v(a_1,\dots,a_n)v(b_1,\dots,b_n)^*\,v(a_1',\dots,a_n')v(b_1',\dots,b_n')^*\,\xi\\ &\quad=\delta_{b_1'}^{c_1}\dots\delta_{b_n'}^{c_n}v(a_1,\dots,a_n)v(b_1,\dots,b_n)^*\,v(a_1',\dots,a_n')v(0_n,c_{n+1})\eta\\ &\quad=\delta_{b_1'}^{c_1}\dots\delta_{b_n'}^{c_n}\delta_{a_1'}^{b_1}\dots\delta_{a_n'}^{b_n}v(a_1,\dots,a_n)v(0_n,c_{n+1})\eta\\ &\quad=\delta_{a_1'}^{b_1}\dots\delta_{a_n'}^{b_n}v(a_1,\dots,a_n)v(b_1',\dots,b_n')^*\,\xi.\end{aligned}$$

Hence

$$\begin{aligned}v(a_1,\dots,a_n)v(b_1,\dots,b_n)^*\,P(n+1)\,v(a_1',\dots,a_n')v(b_1',\dots,b_n')^*\,P(n+1)\\ =\delta_{a_1'}^{b_1}\dots\delta_{a_n'}^{b_n}v(a_1,\dots,a_n)v^*(b_1',\dots,b_n')^*\,P(n+1)\\ (a_1',\dots,a_n';\ b_1',\dots,b_n'\in\{0,1\}).\end{aligned}$$

Since $\mathscr{H}(n+1)$ is invariant under $v(a_1,\dots,a_n)v(b_1,\dots,b_n)^*$,

$$\begin{aligned}P(n+1)v(a_1,\dots,a_n)v(b_1,\dots,b_n)^*\,P(n+1)\\ =v(a_1,\dots,a_n)v(b_1,\dots,b_n)^*\,P(n+1).\end{aligned}$$

Hence

$$\begin{aligned}P(n+1)v(b_1,\dots,b_n)v(a_1,\dots,a_n)^*\\ =P(n+1)v(b_1,\dots,b_n)v(a_1,\dots,a_n)^*\,P(n+1).\end{aligned}$$

Hence

$$P(n+1)v(a_1,\dots,a_n)v(b_1,\dots,b_n)^*=v(a_1,\dots,a_n)v(b_1,\dots,b_n)^*\,P(n+1).$$

Therefore the set $\mathscr{D}_n$ of all linear combinations of

$$\{v(a_1,\dots,a_n)v(b_1,\dots,b_n)^*\,P(n+1)\}$$

is a C^*-algebra for a fixed n. Let Φ_n be a linear mapping of the C^*-algebra B_n onto $\mathscr{D}_n$ as follows:

$$\Phi_n(w(a_1,\dots,a_n)w(a_1,\dots,a_n)^*)=v(a_1,\dots,a_n)v(b_1,\dots,b_n)^*\,P(n+1).$$

Then Φ_n is a $*$-homomorphism; since B_n is simple, Φ_n is an isomorphism if $\Phi_n(B_n)\neq(0)$.

Since

$$v(a_1,\dots,a_n)v(a_1,\dots,a_n)^*v(a_1,\dots,a_{n+1})v(a_1,\dots,a_{n+1})^* \\ = v(a_1,\dots,a_n)v(0_n,a_{n+1})v(a_1,\dots,a_{n+1})^* \\ = v(a_1,\dots,a_{n+1})v(a_1,\dots,a_{n+1})^*,$$

$\Phi_n(B_n)\neq(0)$.

Moreover,

$$v(a_1,\dots,a_{n-1},0)v(b_1,\dots,b_{n-1},0)^*\xi+v(a_1,\dots,a_{n-1},1)v(b_1,\dots,b_{n-1},1)^*\xi \\ =\delta_{b_1}^{c_1}\dots\delta_{b_{n-1}}^{c_{n-1}}\delta_0^{c_n}v(a_1,\dots,a_{n-1},0,c_{n+1})\eta \\ +\delta_{b_1}^{c_1}\dots\delta_{b_{n-1}}^{c_{n-1}}\delta_1^{c_n}v(a_1,\dots,a_{n-1},1,c_{n+1})\eta \\ =\delta_{b_1}^{c_1}\dots\delta_{b_{n-1}}^{c_{n-1}}v(a_1,\dots,a_{n-1},c_n,c_{n+1})\eta$$

and,

$$v(a_1,\dots,a_{n-1})v(b_1,\dots,b_{n-1})^*\xi \\ =v(a_1,\dots,a_{n-1})v(b_1,\dots,b_{n-1})^*v(c_1,\dots,c_{n-1})v(0_{n-1},c_n)v(0_n,c_{n+1})\eta \\ =\delta_{b_1}^{c_1}\dots\delta_{b_{n-1}}^{c_{n-1}}v(a_1,\dots,a_{n-1})v(0_{n-1},c_n)v(0_n,c_{n+1})\eta \\ =\delta_{b_1}^{c_1}\dots\delta_{b_{n-1}}^{c_{n-1}}v(a_1,\dots,a_{n-1},c_n,c_{n+1})\eta.$$

Hence

$$v(a_1,\dots,a_{n-1})v(b_1,\dots,b_{n-1})^*P(n+1) \\ =\{v(a_1,\dots,a_{n-1},0)v(b_1,\dots,b_{n-1},0)^* \\ +v(a_1,\dots,a_{n-1},1)v(b_1,\dots,b_{n-1},1)^*\}P(n+1),$$

and so $\mathscr{H}(n+1)$ is invariant under $\{v(a_1,\dots,a_{n-1})v(b_1,\dots,b_{n-1})^*\}$.

Since $\{v(a_1,\dots,a_{n-1})v(b_1,\dots,b_{n-1})^*\}$ is a self-adjoint family, $P(n+1)$ commutes with its elements.

Now we shall show that $v(a_1,\dots,a_n)v(b_1,\dots,b_n)^*$ commutes with $P(r)$ $(r>n)$.

Put $r-n=s\geq 1$. We proved this is true if $s=1$. Suppose that it is true for $s\leq s_0$. Let $r=n+s_0+1$ $(\geq n+2)$. By the assumption, $\mathscr{H}(r)$ is invariant under $\{v(a_1,\dots,a_{n+1})v(b_1,\dots,b_{n+1})^*\}$. By the previous result, $v(a_1,\dots,a_n)v(b_1,\dots,b_n)^*$ operates on $\mathscr{H}(n+2)$ as a linear combination of $\{v(a_1,\dots,a_{n+1})v(b_1,\dots,b_{n+1})^*\}$.

Since $\mathscr{H}(r)\subset\mathscr{H}(n+2)$, $\mathscr{H}(r)$ is invariant under

$$\{v(a_1,\dots,a_n)v(b_1,\dots,b_n)^*\}.$$

Now let $\mathfrak{B}(n)$ be the C^*-subalgebra of $\mathscr{A}_1$ generated by

$$\{v(a_1,\dots,a_n)v(b_1,\dots,b_n)^*\}$$

and the identity. Then $P(r)$ $(r>n)$ commutes with $\mathfrak{B}(n)$. Since $\mathfrak{B}(n)P(n+1)$ is the matrix algebra of all $2^n\times 2^n$ matrices over the complex field, $\mathfrak{B}(n)P(r)$ $(r>n+1)$ is again the matrix algebra of all $2^n \times 2^n$ matrices. Let $\mathscr{A}(n)$ be the C^*-subalgebra of $\mathscr{A}_1$ generated by $\{\mathfrak{B}(i) \mid i \leq n\}$. Since $\mathfrak{B}(n-1)$ acts as linear combinations of elements in $\mathfrak{B}(n)$ on $P(n+1)$, $\mathfrak{B}(n-1)P(n+1) \subset \mathfrak{B}(n)P(n+1)$, and since $\mathfrak{B}(n-2)$ acts as linear combinations of $\mathfrak{B}(n-1)$ on $\mathscr{H}(n)$ $(\supset \mathscr{H}(n+1))$,

$$\mathfrak{B}(n-2)P(n+1) \subset \mathfrak{B}(n)P(n+1)$$

and so on.

Hence we have $\mathscr{A}(n)P(n+1) = \mathfrak{B}(n)P(n+1)$.

Let $\mathscr{I}(n) = \{x \mid xP(n+1) = 0, x \in \mathscr{A}(n)\}$. Then $\mathscr{I}(n)$ is a closed two-sided ideal of $\mathscr{A}(n)$ and $\mathscr{A}(n)/\mathscr{I}(n)$ is the matrix algebra of all $2^n \times 2^n$ matrices over the complex field.

Clearly $\mathscr{A}(n) \subset \mathscr{A}(n+1)$ and $\mathscr{I}(n) \subset \mathscr{I}(n+1)$. But $\mathscr{A}(n)/\mathscr{I}(n)$ is a simple C^*-algebra, so that $\mathscr{A}(n) \cap \mathscr{I}(n+1) = \mathscr{I}(n)$. Now let $\mathfrak{B}_1$ be the C^*-subalgebra of $\mathscr{A}_1$ generated by $\bigcup_{n=1}^{\infty} \mathscr{A}(n)$ and let $\mathscr{I}$ be the uniform closure of $\bigcup_{n=1}^{\infty} \mathscr{I}(n)$ in $\mathfrak{B}_1$. Then $\mathscr{I}$ is a closed two-sided ideal of $\mathfrak{B}_1$—in fact, $\mathscr{A}(n)\mathscr{I}\mathscr{A}(n) \subset \mathscr{I}$ and so $\mathfrak{B}_1 \mathscr{I} \mathfrak{B}_1 \subset \mathscr{I}$.

Since $\mathscr{A}(n)/\mathscr{I}(n)$ is simple, $\mathscr{A}(n) \cap \mathscr{I} = \mathscr{I}(n)$. Consider a quotient algebra $\mathfrak{B}/\mathscr{I}$. $\mathscr{A}(n) + \mathscr{I}/\mathscr{I} = \mathscr{A}(n)/\mathscr{I}(n)$ is a type I_{2^n}-factor. Hence $\mathfrak{B}/\mathscr{I}$ is a uniformly hyperfinite C^*-algebra such that there exists an increasing sequence of type I_{2^n}-factor $\{\mathscr{M}_n\}$ with the uniform closure of $\bigcup_{n=1}^{\infty} \mathscr{M}_n = \mathfrak{B}/\mathscr{I}$ (such a uniformly hyperfinite algebra is said to be of type $(2, 2^2, \ldots, 2^n, \ldots)$). q.e.d.

4.6.9. Lemma. *Let $\mathscr{A}_2$ be a C^*-algebra with identity, $\mathfrak{B}_2$ a C^*-subalgebra, containing* 1, *of $\mathscr{A}_2$ and let $\mathscr{M}$ be a type* III*-factor on a separable Hilbert space $\mathscr{H}$.*

Suppose that there exists a linear mapping P of $\mathscr{A}_2$ into $\mathscr{M}$ satisfying the following conditions:

I. $P(x^*) = P(x)^*$ $(x \in \mathscr{A}_2)$;

II. $P(h) \geq 0$ *($h \in \mathscr{A}_2$ with $h \geq 0$)*;

III. $P(axb) = P(a)P(x)P(b)$ $(a, b \in \mathfrak{B}_2;\ x \in \mathscr{A}_2)$;

IV. $P(\mathfrak{B}_2)$ *is σ-dense in $\mathscr{M}$.*

Then $\mathscr{A}_2$ has a type III *factorial *-representation.*

To prove this lemma, we shall provide some sublemmas.

Let Ω be the set of all linear mappings Q of $\mathscr{A}_2$ into $\mathscr{M}$ satisfying the conditions I., II., III. and $Q(a) = P(a)$ $(a \in \mathfrak{B}_2)$. Let $B(\mathscr{A}_2, \mathscr{M})$ be the Banach space of all bounded linear mappings of $\mathscr{A}_2$ into $\mathscr{M}$. Then it is the dual

Banach space of a Banach $\mathscr{A}_2 \otimes_\gamma \mathscr{M}_*$, where γ is the greatest cross norm.

4.6.10. Sublemma. *Ω is a $\sigma(B(\mathscr{A}_2, \mathscr{M}), \mathscr{A}_2 \otimes_\gamma \mathscr{M}_*)$-compact convex subset in $B(\mathscr{A}_2, \mathscr{M})$, and each $Q \in \Omega$ satisfies $Q(x^* x) \geq Q(x)^* Q(x)$ for $x \in \mathscr{A}_2$.*

Proof. The first part is clear. We shall prove the second part. By the assumptions I., II., IV. and the density theorem of Kaplansky, there exists a direct set (a_α) in $\mathfrak{B}_2$ with $\|a_\alpha\| \leq \|Q(x)\|$ and $Q(a_\alpha) \to Q(x)$ (in the s-topology) in $\mathscr{M}$.

Thus for $\varphi\ (\geq 0) \in \mathscr{M}_*$,

$$\begin{aligned}\langle Q(x)^* Q(x), \varphi\rangle &= \lim_\alpha \langle Q(x)^* Q(a_\alpha), \varphi\rangle \\ &= \lim_\alpha \langle Q(x^* a_\alpha), \varphi\rangle = \lim_\alpha \langle x^* a_\alpha, Q^*(\varphi)\rangle \\ &\leq \lim_\alpha \sup \langle x^* x, Q^*(\varphi)\rangle^{\frac{1}{2}} \langle a_\alpha^* a_\alpha, Q^*(\varphi)\rangle^{\frac{1}{2}} \\ &\quad \text{(because } Q^*(\varphi) \geq 0 \text{ by II.)} \\ &= \lim_\alpha \sup \langle Q(x^* x), \varphi\rangle^{\frac{1}{2}} \langle Q(a_\alpha)^* Q(a_\alpha), \varphi\rangle^{\frac{1}{2}}.\end{aligned}$$

Hence $Q(x^* x) \leq Q(x^* x)$ $(x \in \mathscr{A}_2)$. q.e.d.

Let φ be a normal faithful state on $\mathscr{M}$. For $Q \in \Omega$, define a state φ_Q on $\mathscr{A}_2$ by $\varphi_Q(x) = \varphi(Q(x))$ for $x \in \mathscr{A}_2$. Let $\mathscr{E} = \{\varphi_Q \mid Q \in \Omega\}$. Then by the above sublemma, one can easily see that $\mathscr{E}$ is a compact convex set in the state space of $\mathscr{A}_2$. Let φ_Q $(Q \in \Omega)$ be an extreme point of $\mathscr{E}$ and let $\{\pi_Q, \mathscr{H}_Q\}$ be the *-representation of $\mathscr{A}_2$ on a Hilbert space $\mathscr{H}_Q$ constructed via φ_Q. Let $\mathscr{N}$ be the weak closure of $\pi_Q(\mathscr{A}_2)$ on $\mathscr{H}_Q$. For $f \in \mathscr{M}_*$, define $F(\pi_Q(x)) = f(Q(x))$ for $x \in \mathscr{A}_2$. This is well-defined, since $\pi_Q(x) = 0$ implies $\varphi_Q(x^* x) = \varphi(Q(x^* x)) = 0$ and so $Q(x^* x) \geq Q(x)^* Q(x) = 0$, so that $Q(x) = 0$.

If $\{\pi_Q(x_\alpha) \mid x_\alpha \in \mathscr{A}_2, \|x_\alpha\| \leq 1\}$ is a directed set with $\pi_Q(x_\alpha) \to 0$ (in the s-topology) in $\mathscr{N}$, then $\varphi_Q(x_\alpha^* x_\alpha) = \varphi(Q(x_\alpha^* x_\alpha)) \geq \varphi(Q(x_\alpha)^* Q(x_\alpha)) \to 0$. Hence $\{Q(x_\alpha)\}$ converges to 0 in the s-topology of $\mathscr{M}$, since $\{Q(x_\alpha)\}$ is bounded and φ is faithful and normal. Hence $F(\pi_Q(x_\alpha)) \to 0$.

Therefore F can be extended uniquely to a σ-continuous linear functional $\overline{F}$ on $\mathscr{N}$ with $\|\overline{F}\| = \|F\|$.

In fact, let $\pi_Q(\mathscr{A}_2)^{**}$ be the second dual of $\pi_Q(\mathscr{A}_2)$ and let $s(\pi_Q)$ be the central support of $\{\pi_Q, \mathscr{H}_Q\}$ in $\pi_Q(\mathscr{A}_2)^{**}$. Then $\mathscr{N} = \pi_Q(\mathscr{A}_2)^{**} s(\pi_Q)$. Suppose $F \notin R_{s(\pi_Q)} \pi_Q(\mathscr{A}_2)^*$; then there exists a directed set (h_α) of positive elements in $\pi_Q(\mathscr{A}_2)$ with $\|h_\alpha\| \leq 1$ and $h_\alpha s(\pi_Q) \to 0$ (in the $s(\pi_Q(\mathscr{A}_2)^{**}, \pi_Q(\mathscr{A}_2)^*)$-topology), but $F(h_\alpha) \not\to 0$.

Now put $\Phi(f) = \overline{F}$ for $f \in \mathscr{M}_*$. Then Φ is a bounded linear mapping of $\mathscr{M}_*$ into $\mathscr{N}_*$. Let Φ^* be the dual of Φ. Then Φ^* is a σ-continuous linear mapping of $\mathscr{N}$ into $\mathscr{M}$.

4.6.11. Sublemma. *Φ^* satisfies the following conditions:*

I. $\Phi^*(\pi_Q(x))=Q(x)$ $(x\in\mathscr{A}_2)$;
II. $\Phi^*(y^*)=\Phi^*(y)^*$ $(y\in\mathscr{N})$;
III. $\Phi^*(h)\geq 0$ $(h\in\mathscr{N}$ *with* $h\geq 0)$;
IV. $\Phi^*(uyv)=\Phi^*(u)\Phi^*(y)\Phi^*(v)$ *($u,v\in$ the σ-closure of $\pi_Q(\mathfrak{B}_2)$ in $\mathscr{N}$ and $y\in\mathscr{N}$);*
V. $\Phi^*(y^*y)\geq\Phi^*(y)^*\Phi(y)$ $(y\in\mathscr{N})$.

Proof. I. For $f\in\mathscr{M}_*$, $\langle\Phi^*(\pi_Q(x)),f\rangle=\langle\pi_Q(x),\Phi(f)\rangle=\langle Q(x),f\rangle$. Hence $\Phi^*(\pi_Q(x))=Q(x)$.

II. If (x_α) is a bounded directed set of elements in $\mathscr{A}_2$ such that $\pi_Q(x_\alpha)\to y$ (in the σ-topology) in $\mathscr{N}$, then

$$\begin{aligned}\Phi^*(\pi_Q(x_\alpha)^*)&=\Phi^*(\pi_Q(x_\alpha^*))=Q(x_\alpha^*)=Q(x_\alpha)^*=\Phi^*(\pi_Q(x_\alpha))^*\\&\to\Phi^*(y)^*\quad\text{(in the }\sigma\text{-topology) in }\mathscr{M}.\end{aligned}$$

Hence $\Phi^*(y^*)=\Phi^*(y)^*$.

III. The proof is similar to the proof of II.

IV. $\Phi^*(\pi_Q(a)\pi_Q(x)\pi_Q(b))=\Phi^*(\pi_Q(axb))=Q(axb)=Q(a)Q(x)Q(b)$ $(a,b\in\mathfrak{B}_2;\ x\in\mathscr{A}_2)$. Therefore by the σ-continuity of Φ^* and the density of $\pi_Q(\mathscr{A}_2)$ in $\mathscr{N}$, one can easily prove the equality IV.

V. Let $\{Q(x_\alpha)\}$ $(x_\alpha\in\mathfrak{B}_2)$ be a directed set such that $Q(x_\alpha)\to\Phi^*(y)$ in the s-topology.

Then for $\psi\ (\geq 0)\in\mathscr{M}_*$,

$$\begin{aligned}\langle\Phi^*(y)^*\Phi^*(y),\psi\rangle&=\lim_\alpha\langle\Phi^*(y)^*Q(x_\alpha),\psi\rangle\\&=\lim_\alpha\langle\Phi^*(y)^*\Phi^*(\pi_Q(x_\alpha)),\psi\rangle=\lim_\alpha\langle\Phi^*(y^*\pi_Q(x_\alpha)),\psi\rangle\\&=\lim_\alpha\langle y^*\pi_Q(x_\alpha),\Phi(\psi)\rangle\\&\leq\lim_\alpha\sup\langle y^*y,\Phi(\psi)\rangle^{\frac{1}{2}}\langle\pi_Q(x_\alpha)^*\pi_Q(x_\alpha),\Phi(\psi)\rangle^{\frac{1}{2}}\\&=\lim_\alpha\sup\langle\Phi^*(y^*y),\psi\rangle^{\frac{1}{2}}\langle Q(x_\alpha)^*Q(x_\alpha),\psi\rangle^{\frac{1}{2}}\\&=\langle\Phi^*(y^*y),\psi\rangle^{\frac{1}{2}}\langle\Phi^*(y)^*\Phi^*(y),\psi\rangle^{\frac{1}{2}}\end{aligned}$$

Hence $\langle\Phi^*(y)^*\Phi^*(y),\psi\rangle\leq\langle\Phi^*(y^*y),\psi\rangle$. q.e.d.

4.6.12. Sublemma. *$\mathscr{N}$ is a factor.*

Proof. Let Z_p be the set of all central projections in $\mathscr{N}$. If $z\in Z_p$, $\Phi^*(z)$ is a central element in $\mathscr{M}$, because

$$\Phi^*(z)Q(b)=\Phi^*(z\pi_Q(b))=\Phi^*(\pi_Q(b)z)=Q(b)\Phi^*(z)\quad\text{for } b\in\mathfrak{B}_2.$$

Since $\mathscr{M}$ is a factor, $\Phi^*(z)=\lambda(z)1$ with $0\leq\lambda(z)\leq 1$.

If $0<\lambda(z)<1$, then we shall define two linear mappings Q_1 and Q_2 of $\mathscr{A}_2$ into $\mathscr{M}$ as follows: $Q_1(x)=\left(\frac{1}{\lambda(z)}\right)\Phi^*(\pi_Q(x)z)$ and

$$Q_2(x)=\left(\frac{1}{1-\lambda(z)}\right)\Phi^*(\pi_Q(x)(1-z)) \qquad (x\in\mathscr{A}_2).$$

Then Q_1, Q_2 belong to Ω, and $\varphi_Q=\lambda(z)\varphi_{Q_1}+(1-\lambda(z))\varphi_{Q_2}$. By the extremity of φ_Q, $\varphi_Q=\varphi_{Q_1}=\varphi_{Q_2}$ on $\mathscr{A}_2$ and so

$$\frac{1}{\lambda(z)}\langle\pi_Q(x)z,\Phi(\varphi)\rangle = \frac{1}{1-\lambda(z)}\langle\pi_Q(x)(1-z),\Phi(\varphi)\rangle$$

for $x\in\mathscr{A}_2$.

Take a directed set $\{\pi_Q(x_\alpha)\}$ with $x_\alpha\in\mathscr{A}_2$, $\|x_\alpha\|\leq 1$ and $\pi_Q(x_\alpha)\to z$ in the s-topology of $\mathscr{N}$; then

$$\langle\pi_Q(x_\alpha)z,\Phi(\varphi)\rangle\to\langle z,\Phi(\varphi)\rangle = \langle\Phi^*(z),\varphi\rangle = \lambda(z)\neq 0.$$

On the other hand,

$$\langle\pi_Q(x_\alpha)(1-z),\Phi(\varphi)\rangle\to\langle z(1-z),\Phi(\varphi)\rangle=0,$$

a contradiction.

Hence $\lambda(z)=0$ or 1. If $\lambda(z)=0$, $\langle z,\Phi(\varphi)\rangle=\langle\Phi^*(z),\varphi\rangle=\lambda(z)=0$. But $\langle z,\Phi(\varphi)\rangle=(z1_Q,1_Q)$, where (,) is the inner product of $\mathscr{H}_Q$.

Hence $z1_Q=0$ and so $\pi_Q(\mathscr{A}_2)z1_Q=z\pi_Q(\mathscr{A}_2)1_Q=0$, so that $z=0$. Similarly if $\lambda(z)=1$, then z must be 1. q.e.d.

Proof of 4.6.9. Let E' be the orthogonal projection of $\mathscr{H}_Q$ onto $[\pi_Q(\mathfrak{B}_2)1_Q]$; then $E'\in\pi_Q(\mathfrak{B}_2)'$.

Since $\varphi_Q(a^*bc)=\varphi(Q(a^*bc))=\varphi(Q(a)^*Q(b)Q(c))$ for $a,b,c\in\mathfrak{B}_2$ and $Q(\mathfrak{B}_2)$ is σ-dense in $\mathscr{M}$, a *-isomorphism $\pi_Q(b)E'\to Q(b)$ of $\pi_Q(\mathfrak{B}_2)E'$ into $\mathscr{M}$ can be uniquely extended to a *-isomorphism Γ of a W^*-algebra $\pi_Q(\mathfrak{B}_2)''E'$ onto $\mathscr{M}$. Hence $\pi_Q(\mathfrak{B}_2)''E'$ and $E'\pi_Q(\mathfrak{B}_2)'E'$ are type III-factors.

Let F' be the central envelope of E' in $\pi_Q(\mathfrak{B}_2)'$. Then $F'\in\pi_Q(\mathfrak{B}_2)''$. A mapping $\Psi: xF'\to xE'$ $(x\in\pi_Q(\mathfrak{B}_2)'')$ of $\pi_Q(\mathfrak{B}_2)''F'$ onto $\pi_Q(\mathfrak{B}_2)''E'$ is a *-isomorphism. Hence the mapping $\Gamma\cdot\Psi$ of $\pi_Q(\mathfrak{B}_2)''F'$ onto $\mathscr{M}$ is a *-isomorphism.

Now suppose that $\mathscr{N}$ is semi-finite.

Since $\langle\Phi^*(F'),\varphi\rangle=\langle F',\Phi(\varphi)\rangle=(F'1_Q,1_Q)$ and $1_Q\in E'\mathscr{H}_Q$, $\Phi^*(F')\neq 0$.

Since $F'\in\pi_Q(\mathfrak{B}_2)''\subset\pi_Q(\mathscr{A}_2)''$ and Φ^* is σ-continuous, there exists a non-zero finite projection e in $\mathscr{N}$ with $e\leq F'$ and $\Phi^*(e)\neq 0$, and so there is a non-zero projection p with $\lambda p\leq\Phi^*(e)$ for some positive number λ.

Suppose that a directed set (a_α) $(\|a_\alpha\| \leq 1, a_\alpha \in p\mathcal{M}p)$ converges to 0 in the s-topology of $\mathcal{M}$. Then $\{\Psi^{-1}\Gamma^{-1}(a_\alpha)\}$ converges to 0 in the s-topology of $\pi_Q(\mathfrak{B}_2)'' F'$. Since $\pi_Q(\mathfrak{B}_2)'' F' \subset \pi_Q(\mathcal{A}_2)'' = \mathcal{N}$ and the s-topology of $\pi_Q(\mathfrak{B}_2)'' F'$ is the restriction of the s-topology of $\mathcal{N}$ to $\pi_Q(\mathfrak{B}_2)'' F'$, $\{\Psi^{-1}\Gamma^{-1}(a_\alpha)\}$ converges to 0 in the s-topology of $\mathcal{N}$, and so $\{\Psi^{-1}\Gamma^{-1}(a_\alpha)e\}$ converges again to 0 in the s-topology. Since e is finite in $\mathcal{N}$, $\{e\Psi^{-1}\Gamma^{-1}(a_\alpha)^*\}$ converges to 0 in the s-topology by 2.5.6.

Thus

$$\Phi^*\left((e(\Psi^{-1}\Gamma^{-1}(a_\alpha)^*)^*(e(\Psi^{-1}\Gamma^{-1}(a_\alpha))^*)\right)$$

$$\geq \Phi^*\left(e(\Psi^{-1}\Gamma^{-1}(a_\alpha)^*)^*\right)\Phi^*\left(e(\Psi^{-1}\Gamma^{-1}(a_\alpha))^*\right) \to 0 \quad (\sigma\text{-weakly) in } \mathcal{M}.$$

Hence $\{\Phi^*(e(\Psi^{-1}\Gamma^{-1}(a_\alpha))^*)\}$ converges to 0 in the s-topology of $\mathcal{M}$.

For $a \in \mathcal{M}$, choose a bounded directed set $\{Q(b_\beta)\}$ with $b_\beta \in \mathfrak{B}_2$ and $Q(b_\beta) \to a$ (σ-weakly) in $\mathcal{M}$; then $\Psi^{-1}\Gamma^{-1}(Q(b_\beta)) = \pi_Q(b_\beta)F' \to \Psi^{-1}\Gamma^{-1}(a)$ (σ-weakly) in $\pi_Q(\mathfrak{B}_2)''$.

Moreover

$$\Phi^*\left(\Psi^{-1}\Gamma^{-1}(Q(b_\beta))\right) = \Phi^*(\pi_Q(b_\beta)F') = \Phi^*(\pi_Q(b_\beta))\,\Phi^*(F')$$

$$= Q(b_\beta)\,\Phi^*(F') \to a\,\Phi^*(F') \quad (\sigma\text{-weakly) in } \mathcal{M}.$$

Hence $a\Phi^*(F') = \Phi^*(\Psi^{-1}\Gamma^{-1}(a))$ for $a \in \mathcal{M}$. Therefore

$$\begin{aligned}
&\{p\Phi^*(e)p + (1-p)\}^{-1} p\Phi^*\left(e(\Psi^{-1}\Gamma^{-1}(a_\alpha))^*\right)\\
&\quad = \{p\Phi^*(e)p + (1-p)\}^{-1} p\Phi^*(e)\Phi^*(\Psi^{-1}\Gamma^{-1}(a_\alpha))^*\\
&\qquad (\text{because } \Psi^{-1}\Gamma^{-1}(a_\alpha) \in \pi_Q(\mathfrak{B}_2)'' \text{ and } \Phi^*(e)\Phi^*(F')\\
&\quad = \Phi^*(eF) = \Phi^*(e))\\
&\quad = \{p\Phi^*(e)p + (1-p)\}^{-1} p\Phi^*(e)a_\alpha^* = a_\alpha^* \to 0\\
&\qquad \text{in the } s\text{-topology of } \mathcal{M}.
\end{aligned}$$

Hence the *-operation is s-continuous on bounded spheres of $p\mathcal{M}p$, but $p\mathcal{M}p$ is of type III. This is a contradiction by 2.5.6. q.e.d.

4.6.13. Lemma. *Let $\mathcal{A}_1$ be a C^*-algebra satisfying the condition of Glimm. Then $\mathcal{A}_1$ has a type* III *factorial *-representation.*

Proof. Without loss of generality, we may assume that $\mathcal{A}_1$ has an identity. By 4.6.8, there exists a separable C^*-subalgebra $\mathfrak{B}_1$ of $\mathcal{A}_1$ such that at least one quotient C^*-subalgebra of $\mathfrak{B}_1$ is a uniformly hyperfinite C^*-algebra of type $(2, 2^2, \ldots, 2^n, \ldots)$ (cf. the proof of 4.6.8.).

Now let $\mathcal{D}$ be a uniformly hyperfinite C^*-algebra of type $(2, 2^2, \ldots, 2^n, \ldots)$.

Then $\mathcal{D} = \bigotimes_{n=1}^{\infty} \mathcal{D}_n$ with $\mathcal{D}_n = B$ for each n (B, a type I_2-factor).

Consider the infinite product state $\psi_{(p_n)} = \bigotimes_{n=1}^{\infty} \varphi_{p_n}$ on $\mathscr{D}$ with $p_n = \lambda$ $(0 < \lambda < \frac{1}{2})$ for each n (cf. 4.4). We shall denote $\psi_{(p_n)}$ by ψ_λ. Then $\pi_{\psi_\lambda}(\mathscr{D})''$ is a type III-factor; moreover by 4.4.4, $\pi_{\psi_\lambda}(\mathscr{D})'$ is a hyperfinite factor. Hence by 4.4.15 and 4.4.17 there exists a mapping P of $B(\mathscr{H}_{\psi_\lambda})$ onto $\pi_{\psi_\lambda}(\mathscr{D})''$ in 4.4.15.

Now let Φ be a *-homomorphism of $\mathfrak{B}_1$ onto $\mathscr{D}$ and set

$$\omega_\lambda(a) = \psi_\lambda(\Phi(a)) \qquad (a \in \mathfrak{B}_1).$$

Then ω_λ is a state on $\mathfrak{B}_1$. Let $\bar{\omega}_\lambda$ be a state on $\mathscr{A}_1$ with $\bar{\omega}_\lambda = \omega_\lambda$ on $\mathfrak{B}_1$.

Consider the *-representation $\{\pi_{\bar{\omega}_\lambda}, \mathscr{H}_{\bar{\omega}_\lambda}\}$ of $\mathscr{A}_1$, and let E' be the orthogonal projection of $\mathscr{H}_{\bar{\omega}_\lambda}$ onto $[\pi_{\bar{\omega}_\lambda}(\mathfrak{B}_1) 1_{\bar{\omega}_\lambda}]$. Then the *-representation $b \to \pi_{\bar{\omega}_\lambda}(b) E'$ of $\mathfrak{B}_1$ is equivalent to the *-representation $\{\pi_{\omega_\lambda}, \mathscr{H}_{\omega_\lambda}\}$ of $\mathfrak{B}_1$.

Hence the mapping P can be considered as the mapping of $B(E' \mathscr{H}_{\bar{\omega}_\lambda})$ onto $\pi_{\bar{\omega}_\lambda}(\mathfrak{B}_1)'' E'$.

Now, define a mapping P_1 of $\pi_{\bar{\omega}_\lambda}(\mathscr{A}_1)$ into $\pi_{\bar{\omega}_\lambda}(\mathfrak{B}_1)'' E'$ as follows:

$$P_1(\pi_{\bar{\omega}_\lambda}(a)) = P_1(E' \pi_{\bar{\omega}_\lambda}(a) E') \qquad (a \in \mathscr{A}_1).$$

Set $\mathscr{A}_2 = \pi_{\bar{\omega}_\lambda}(\mathscr{A}_1)$, $\mathfrak{B}_2 = \pi_{\bar{\omega}_\lambda}(\mathfrak{B}_1)$ and $\mathscr{M} = \pi_{\bar{\omega}_\lambda}(\mathfrak{B}_1)'' E'$ and take P_1 as the P in 4.6.9.

Then one can easily see that this system satisfies all the conditions of 4.6.9. Hence $\pi_{\bar{\omega}_\lambda}(\mathscr{A}_1)$ and so $\mathscr{A}_1$ has a type III factorial *-representation. q.e.d.

4.6.14. Lemma. *If a C^*-algebra $\mathscr{A}$ is smooth, $\mathscr{A}$ is a type* I *C^*-algebra.*

Proof. It suffices to assume that $\mathscr{A}$ has an identity. Suppose that $\mathscr{A}$ is not of type I. Then there exists a quotient C^*-algebra of $\mathscr{A}$ satisfying the condition of Glimm. Therefore there exists a separable C^*-subalgebra $\mathfrak{B}$ of $\mathscr{A}$ such that at least one quotient C^*-algebra of $\mathfrak{B}$ is a uniformly hyperfinite C^*-algebra of type $(2, 2^2, \ldots, 2^n, \ldots)$.

Let $\mathscr{D} = \bigotimes_{n=1}^{\infty} \mathscr{D}_n$ with $\mathscr{D}_n = B$, a type I_2-factor. Let φ be a pure state on B, and let $\varphi_n = \varphi$ and $\psi = \bigotimes_{n=1}^{\infty} \varphi_n$. Then $\pi_\psi(\mathscr{D})''$ is $B(\mathscr{H}_\psi)$.

Since $\mathscr{D}$ is simple, $\pi_\psi(\mathscr{D})$ is simple. Since $1_{\mathscr{H}_\psi} \in \pi_\psi(\mathscr{D})$, clearly $\pi_\psi(\mathscr{D})$ does not contain any non-zero compact operator in $\mathscr{H}_\psi$.

Let Φ be a *-homomorphism of $\mathfrak{B}$ onto $\mathscr{D}$ and put $\omega(a) = \psi(\Phi(a))$ $(a \in \mathfrak{B})$. Then ω is a pure state on $\mathfrak{B}$. Let $\bar{\omega}$ be a pure state on $\mathscr{A}$ with $\bar{\omega} = \omega$ on $\mathfrak{B}$.

Consider the *-representation $\{\pi_{\bar{\omega}}, \mathscr{H}_{\bar{\omega}}\}$ of $\mathscr{A}$.

Let E' be the projection of $\mathscr{H}_{\bar{\omega}}$ onto $[\pi_{\bar{\omega}}(\mathfrak{B}) 1_{\bar{\omega}}]$. Then $\pi_{\bar{\omega}}(\mathfrak{B}) E'$ does not contain any non-zero compact operator. Now let $\mathscr{E}_0$ be the set of all pure states ζ on $\mathscr{A}$ with $\bar{\omega}=\zeta$ on $\mathfrak{B}$. We shall define a partial ordering $\prec$ on $\mathscr{E}_0$ in the following way. Take $\zeta\in\mathscr{E}_0$; then $\pi_\zeta(\mathscr{A})$ contains a non-zero compact operator; hence $\pi_\zeta(\mathscr{A})$ contains all compact linear operators on $\mathscr{H}_\zeta$. In fact, $\pi_\zeta(\mathscr{A})\cap C(\mathscr{H}_\zeta)$ is a closed two-sided ideal of $\pi_\zeta(\mathscr{A})$ and so $\pi_\zeta(\mathscr{A})\cap C(\mathscr{H}_\zeta)$ is an irreducible family. If

$$\pi_\zeta(\mathscr{A})\cap C(\mathscr{H}_\zeta)\subsetneqq C(\mathscr{H}_\zeta),\qquad (\pi_\zeta(\mathscr{A})\cap C(\mathscr{H}_\zeta))^{00}\subsetneqq C(\mathscr{H}_\zeta)^{**}=B(\mathscr{H}_\zeta).$$

This contradicts the fact that $\pi_\zeta(\mathscr{A})\cap C(\mathscr{H}_\zeta)$ is irreducible.

Set $\mathscr{I}(\zeta)=\pi_\zeta^{-1}(C(\mathscr{H}_\zeta))$. Then $\mathscr{I}(\zeta)$ is a closed two-sided ideal of $\mathscr{A}$. For $\zeta_1,\zeta_2\in\mathscr{E}_0$, we shall define the order as follows: $\zeta_1\prec\zeta_2$ if $\mathscr{I}(\zeta_1)\subset\mathscr{I}(\zeta_2)$.

Let $\{\zeta_\alpha|\alpha\in\mathbb{I}\}$ be a linearly ordered subset of $\mathscr{E}_0$ and let $\mathscr{I}$ be the uniform closure of $\bigcup_{\alpha\in\mathbb{I}}\mathscr{I}(\zeta_\alpha)$. Then $\mathscr{I}$ is a two-sided ideal. Let $\mathscr{L}$ be the kernel of the representation $\{\pi_\omega,\mathscr{H}_\omega\}$ of $\mathfrak{B}$.

Then first of all we shall show that $\mathfrak{B}\cap\mathscr{I}\subset\mathscr{L}$. Suppose that $\mathfrak{B}\cap\mathscr{I}\not\subset\mathscr{L}$; then there is an element $b\in(\mathfrak{B}\cap\mathscr{I})\cap\mathscr{L}^c$ and $b_n\in\mathscr{I}(\zeta_{\alpha_n})$ $(n=1,2,\ldots)$ such that $\|\pi_\omega(b)\|=1$ and $\|b-b_n\|<1/n$ $(n=1,2,\ldots)$, where $\mathscr{L}^c$ is the complement of $\mathscr{L}$ in $\mathfrak{B}$.

Consider the *-representation $\{\pi_{\zeta_{\alpha_n}},\mathscr{H}_{\zeta_{\alpha_n}}\}$ of $\mathscr{A}$. Then

$$\|\pi_{\zeta_{\alpha_n}}(b)-\pi_{\zeta_{\alpha_n}}(b_n)\|<1/n.$$

Let E_n' be the orthogonal projection of $\mathscr{H}_{\zeta_{\alpha_n}}$ on $[\pi_{\zeta_{\alpha_n}}(\mathfrak{B})1_{\zeta_{\alpha_n}}]$. Then the representation $y\to\pi_{\zeta_{\alpha_n}}(y)E_n'$ $(y\in\mathfrak{B})$ of $\mathfrak{B}$ is equivalent to $\{\pi_\omega,\mathscr{H}_\omega\}$.

On the other hand

$$\|E_n'\pi_{\zeta_{\alpha_n}}(b)E_n'-E_n'\pi_{\zeta_{\alpha_n}}(b_n)E_n'\|<1/n,$$

and $E_n'\pi_{\zeta_{\alpha_n}}(b_n)E_n'$ is a compact operator on $E_n'\mathscr{H}_{\zeta_{\alpha_n}}$. Hence there exists a compact operator T_n on $\mathscr{H}_\omega$ such that $\|\pi_\omega(b)-T_n\|<1/n$, since $E_n'\pi_{\zeta_{\alpha_n}}(b)E_n'=\pi_{\zeta_{\alpha_n}}(b)E_n'$. This is a contradiction, because $\pi_\omega(b)$ is not compact. Hence $\mathfrak{B}\cap\mathscr{I}\subset\mathscr{L}$.

Next let us consider a C^*-algebra $\mathscr{A}/\mathscr{I}$. Then $\mathfrak{B}+\mathscr{I}/\mathscr{I}$ is a C^*-subalgebra of $\mathscr{A}/\mathscr{I}$.

The state ω on $\mathfrak{B}$ can be canonically considered as a pure state on $\mathfrak{B}+\mathscr{I}/\mathscr{I}$, because $\mathfrak{B}\cap\mathscr{I}\subset\mathscr{L}$, and the C^*-algebra $\mathfrak{B}+\mathscr{I}/\mathscr{I}$ is isomorphic to $\mathfrak{B}/\mathfrak{B}\cap\mathscr{I}$.

Take a pure state extension $\tilde{\omega}$ of ω to $\mathscr{A}/\mathscr{I}$; then we can define a pure state ζ_0 of $\mathscr{A}$ by $\zeta_0(y)=\tilde{\omega}(y+\mathscr{I})$ $(y\in\mathscr{A})$. Then $\zeta_0=\omega$ on $\mathfrak{B}$ and so $\zeta_0\in\mathscr{E}_0$.

Clearly $\mathscr{I}(\zeta_\alpha)\subset\mathscr{I}(\zeta_0)$ for all $\alpha\in\mathbb{I}$. Hence $\zeta_\alpha\prec\zeta_0$. Hence by Zorn's lemma there exists a maximal element ζ_1 in $\mathscr{E}_0$.

Now we shall show $\mathscr{I}(\zeta_1)\cap\mathfrak{B}\not\subset\mathscr{L}$. Assume that $\mathscr{I}(\zeta_1)\cap\mathfrak{B}\subset\mathscr{L}$. Then by the similar reasoning to the above one, ζ_1 can be considered canonically as a pure state on the C^*-subalgebra $\mathfrak{B}+\mathcal{I}(\zeta_1)/\mathcal{I}(\zeta_1)$ of $\mathcal{A}/\mathcal{I}(\zeta_1)$. Therefore we can have a pure state ζ_2 on $\mathcal{A}$ such that $\zeta_2(\mathcal{I}(\zeta_1))=0$ and $\zeta_2=\omega$ on $\mathfrak{B}$. Hence $\mathcal{I}(\zeta_2)\supsetneqq\mathcal{I}(\zeta_1)$, a contradiction.

On the other hand, $\mathcal{I}(\zeta_1)\cap\mathfrak{B}\not\subset\mathcal{L}$ again implies a contradiction, because $\pi_{\zeta_1}(b)$ is a compact operator on $\mathcal{H}_{\zeta_1}$ for some $b\in(\mathcal{I}(\zeta_1)\cap\mathfrak{B})\cap\mathcal{L}^c$.

In fact, $\pi_{\zeta_1}(b)E'$ is compact, where E' is the orthogonal projection of $\mathcal{H}_{\zeta_1}$ onto $[\pi_{\zeta_1}(\mathfrak{B})1_{\zeta_1}]$ and so $\pi_\omega(b)=0$, so that $b\in\mathcal{L}$, a contradiction. q.e.d.

4.6.15. Definition. *Let $\mathcal{A}$ be a C^*-algebra. An increasing family of closed two sided ideals $(\mathcal{I}_\rho)_{0\leq\rho\leq\alpha}$ of $\mathcal{A}$, indexed by the ordinals ρ between 0 and a certain ordinal α is called a composition series of $\mathcal{A}$ if it satisfies the following conditions:*

1. $\mathcal{I}_0=0$, $\mathcal{I}_\alpha=\mathcal{A}$;
2. *if $\rho\leq\alpha$ is a limit ordinal, $\mathcal{I}_\rho=$ the uniform closure of* $\bigcup_{\rho'<\rho}\mathcal{I}_{\rho'}$.

4.6.16. Corollary. *Let $\mathcal{A}$ be a C^*-algebra. The following conditions are equivalent:*

1. *$\mathcal{A}$ is a type* I *C^*-algebra;*
2. *$\mathcal{A}$ has a composition series $(\mathcal{I}_\rho)_{0\leq\rho\leq\alpha}$ such that $\mathcal{I}_{\rho+1}/\mathcal{I}_\rho$ contains a positive element h such that for every non-zero irreducible *-representation $\{\pi,\mathcal{H}\}$ of $\mathcal{I}_{\rho+1}/\mathcal{I}_\rho$, the rank of $\pi(h)$ is one-dimensional.*

Proof. We proved already that $2.\Rightarrow 1$. Conversely suppose that $\mathcal{A}$ is of type I. By the transfinite induction, we shall construct a composition series satisfying the condition.

If ρ is a limit ordinal, $\mathcal{I}_\rho$ is the uniform closure of $\bigcup_{\rho'<\rho}\mathcal{I}_{\rho'}$. If ρ is not a limit ordinal, there exists an ordinal ρ' with $\rho=\rho'+1$.

Consider the quotient algebra $\mathcal{A}/\mathcal{I}_{\rho'}$. By 4.6.4, there exists a positive element h in $\mathcal{A}/\mathcal{I}_{\rho'}$ such that for each irreducible *-representation $\{\pi,\mathcal{H}\}$ of $\mathcal{A}/\mathcal{I}_{\rho'}$, the rank of $\pi(h)\leq 1$. Now let $\mathcal{L}$ be the closed two sided ideal of $\mathcal{A}/\mathcal{I}_{\rho'}$ generated by h.

Let $\mathcal{I}_\rho$ be the inverse image of $\mathcal{L}$ in $\mathcal{A}$. Then a closed two-sided ideal of $\mathcal{I}_\rho/\mathcal{I}_{\rho'}$ containing h equals $\mathcal{I}_\rho/\mathcal{I}_{\rho'}$.

Let $\{\pi,\mathcal{H}\}$ be an arbitrary non-zero irreducible *-representation of $\mathcal{I}_\rho/\mathcal{I}_{\rho'}$, then $\pi(h)\neq 0$—in fact if $\pi(h)=0$ for some π, the kernel of π must contain h and so $\pi\equiv 0$.

Hence the rank of $\pi(h)=1$. q.e.d.

4.6.17. Proposition. *Let $\mathcal{A}$ be a type* I *C^*-algebra, and let $\{\pi_1,\mathcal{H}_1\}$ and $\{\pi_2,\mathcal{H}_2\}$ be two irreducible *-representations of $\mathcal{A}$. Put*

$$K_i=\{x\,|\,\pi_i(x)=0,\,x\in\mathcal{A}\}.$$

If $K_1=K_2$, $\{\pi_1,\mathcal{H}_1\}$ is equivalent to $\{\pi_2,\mathcal{H}_2\}$.

Proof. It suffices to assume that $K_1 = K_2 = (0)$. Put $\mathscr{A}_i = \pi_i^{-1}(C(\mathscr{H}_i))$. Since $C(\mathscr{H}_i)^* = T(\mathscr{H}_i)$ (cf. 1.19), $C(\mathscr{H}_i)$ (and so $\mathscr{A}_i$) has only one non-zero irreducible *-representation and so $\mathscr{A}_1 = \mathscr{A}_2$. Let $\mathscr{A}_1^{00}$ be the bipolar of $\mathscr{A}_1$ in $\mathscr{A}^{**}$. Then there exists a central projection z in $\mathscr{A}^{**}$ such that $\mathscr{A}^{**}z = \mathscr{A}_1^{00}$, because $\mathscr{A}_1^{00}$ is a two-sided ideal in $\mathscr{A}^{**}$. Since $\mathscr{A}_1^{00} = \mathscr{A}_1^{**} = B(\mathscr{H}_1)$ as abstract W^*-algebras (cf. 1.19), z is a minimal central projection in $\mathscr{A}^{**}$. Let $\{\pi_i^W, \mathscr{H}_i\}$ be the corresponding W^*-representation of $\mathscr{A}^{**}$ to $\{\pi_i, \mathscr{H}_i\}$. Then $1 - k(\pi_i^W)$ is a minimal central projection in $\mathscr{A}^{**}$. Clearly $1 - z \leq k(\pi_i^W)$. Hence $1 - k(\pi_i^W) \leq z$ and so $1 - k(\pi_i^W) = z$. Therefore $\{\pi_1^W, \mathscr{H}_1\}$ and $\{\pi_2^W, \mathscr{H}_2\}$ are quasi-equivalent. Since both are irreducible, this implies that they are equivalent. q.e.d.

Remark 1. If $\mathscr{A}$ is separable, the converse is again true (cf. [33], [60]). It is not known whether or not the converse is still true without the assumption of the separability.

In particular, if $\mathscr{A}$ is a separable C^*-algebra which has only one irreducible *-representation, then $\mathscr{A} = C(\mathscr{H})$ (cf. [143]).

The following problem (Naimark [123]) is open: Suppose $\mathscr{A}$ is a C^*-algebra which has only one irreducible *-representation. Then can we conclude that $\mathscr{A} = C(\mathscr{H})$?

Remark 2. If $\mathscr{A}$ is a separable non-type I C^*-algebra, $\mathscr{A}$ has a type II factorial *-representation (cf. [60]). It is an open question whether or not any non-separable, non-type I C^*-algebra has a type II factorial *-representation (even a global type II *-representation) (cf. [161]).

Remark 3. Let $\mathscr{H}$ be an $\aleph_0$-dimensional Hilbert space. The C^*-algebra $B(\mathscr{H})/C(\mathscr{H})$ is called the Calkin algebra (cf. [20]).

Clearly the Calkin algebra is not of type I, so that it has a type III factorial *-representation.

Does the Calkin algebra have a type II *-representation?

Concluding Remark on 4.6.

Theorem 4.6.4 is due to Glimm [60] in case where $\mathscr{A}$ is separable.

References. [33], [36], [51], [60], [96], [161], [162], [163].

4.7. On a Stone-Weierstrass Theorem for C^*-Algebras

Let C be the C^*-algebra of all complex valued continuous functions, vanishing at infinity, on a locally compact Hausdorff space. The classical Stone-Weierstrass theorem gives the conditions under which a C^*-subalgebra D of C coincides with C.

A plausible non-commutative extension of the Stone-Weierstrass theorem is

4.7.1. Conjecture. *Let $\mathscr{A}$ be a C^*-algebra, $\mathfrak{B}$ a C^*-subalgebra of $\mathscr{A}$, $\mathscr{E}_{\mathscr{A}}$ the set of all pure states on $\mathscr{A}$ and let 0 be the identically zero function on $\mathscr{A}$. Suppose that $\mathfrak{B}$ separates $\mathscr{E}_{\mathscr{A}} \cup (0)$—i.e. for any two distinct $\varphi_1, \varphi_2 \in \mathscr{E}_{\mathscr{A}} \cup (0)$, there exists an element b in $\mathfrak{B}$ such that $\varphi_1(b) \neq \varphi_2(b)$. Then $\mathscr{A} = \mathfrak{B}$.*

In this section, we shall state some partial results and related problems concerning this conjecture.

If $\mathscr{A}$ does not have an identity, we shall consider the C^*-algebras $\mathscr{A}_1 = \mathscr{A} + \mathbb{C}1$ and $\mathfrak{B}_1 = \mathfrak{B} + \mathbb{C}1$. Since any pure state φ on $\mathscr{A}$ can be uniquely extended to a pure state $\tilde{\varphi}$ on $\mathscr{A}_1$, $\mathscr{E}_{\mathscr{A}_1} = \tilde{\mathscr{E}}_{\mathscr{A}} \cup (\varphi_0)$ (φ_0, the pure state on $\mathscr{A}_1$ with $\varphi_0(\mathscr{A}) = 0$).

If $\mathfrak{B}$ separates $\mathscr{E}_{\mathscr{A}} \cup (0)$, $\mathfrak{B}_1$ separates $\mathscr{E}_{\mathscr{A}_1} \cup (0)$. Therefore it suffices to assume that $\mathscr{A}$ has an identity.

4.7.2. Lemma. *If $\mathfrak{B}$ separates $\mathscr{E}_{\mathscr{A}} \cup (0)$, then $\mathfrak{B}$ contains the identity of $\mathscr{A}$.*

Proof. Suppose that $1 \notin \mathfrak{B}$. Then $\|b+1\| \geq 1$ for $b \in \mathfrak{B}$—in fact, if $\|b+1\| < 1$ for some b in $\mathfrak{B}$, b is invertible and $b^{-1} \in \mathfrak{B}$, so that $1 \in \mathfrak{B}$. Take a bounded linear functional f on $\mathscr{A}$ with $f(\mathfrak{B}) = 0$ and $\|f\| = f(1) = 1$. Then f is a state. Let $\mathscr{L} = \{x \mid f(x^* x) = 0, x \in \mathscr{A}\}$. Then $\mathscr{L}$ is a closed left ideal with $\mathfrak{B} \subset \mathscr{L}$. Let $\mathscr{L}_1$ be a maximal left ideal of $\mathscr{A}$ with $\mathscr{L} \subset \mathscr{L}_1$. Then there exists a pure state φ on $\mathscr{A}$ such that $\varphi(\mathscr{L}_1) = 0$ (cf. 1.21.19).

Thus $\mathfrak{B}$ cannot separate φ and 0, a contradiction. q.e.d.

Now it suffices to assume that $\mathscr{A}$ has an identity and $\mathfrak{B}$ contains it. In this case, the separation of $\mathscr{E}_{\mathscr{A}} \cup (0)$ by $\mathfrak{B}$ is equivalent to the separation of $\mathscr{E}_{\mathscr{A}}$ by $\mathfrak{B}$.

4.7.3. Definition. *Let φ be a state on a C^*-algebra $\mathscr{D}$. φ is said to be atomic if $\pi_\varphi(\mathscr{D})''$ is a direct sum of type I-factors.*

4.7.4. Lemma. *Let φ_1, φ_2 be two states on $\mathscr{A}$ such that their restrictions $\varphi_1|\mathfrak{B}, \varphi_2|\mathfrak{B}$ to $\mathfrak{B}$ are atomic. If $\mathfrak{B}$ separates $\mathscr{E}_{\mathscr{A}}$ and $\varphi_1 = \varphi_2$ on $\mathfrak{B}$, then $\varphi_1 = \varphi_2$ on $\mathscr{A}$.*

Proof. Put $\varphi = \dfrac{\varphi_1 + \varphi_2}{2}$. Then $\varphi(a) = (\pi_\varphi(a) 1_\varphi, 1_\varphi)$ $(a \in \mathscr{A})$. Let e' be the projection of $\mathscr{H}_\varphi$ onto the closed subspace $[\pi_\varphi(\mathfrak{B}) 1_\varphi]$. Then $\pi_\varphi(\mathfrak{B})'' e'$ is a direct sum of type I-factors. Let $c(e')$ be the central support of e' in $\pi_\varphi(\mathfrak{B})'$. Then $\pi_\varphi(\mathfrak{B})'' c(e')$ is *-isomorphic to $\pi_\varphi(\mathfrak{B})'' e'$. Hence $\pi_\varphi(\mathfrak{B})''$ contains a type I-factor as a direct summand. Let p' be a minimal projection in $\pi_\varphi(\mathfrak{B})'$. Then $b \to \pi_\varphi(b) p'$ $(b \in \mathfrak{B})$ is irreducible. For $\eta \in p' \mathscr{H}_\varphi$ with $\|\eta\| = 1$, put $\psi_0(a) = (\pi_\varphi(a)\eta, \eta)$ $(a \in \mathscr{A})$. Then $\psi_0 | \mathfrak{B}$ is pure.

Let $\Gamma=\{\psi|\psi=\psi_0 \text{ on } \mathfrak{B}, \psi\in\mathscr{S}\}$ ($\mathscr{S}$, the state space of $\mathscr{A}$). Then Γ is a compact convex set. One can easily see that an extreme point in Γ is again extreme in $\mathfrak{S}$. Hence it is pure, and so Γ consists of only one point and it is pure. If $p'\mathscr{H}_\varphi \subsetneqq [\pi_\varphi(\mathscr{A})\eta]$, take a non-zero ξ_1 (resp. ξ_2)$\in p'\mathscr{H}_\varphi$ (resp. $(p'\mathscr{H}_\varphi)^\perp \cap [\pi_\varphi(\mathscr{A})\eta]$) with $\|\xi_1+\xi_2\|=1$, and put

$$g_1(a)=(\pi_\varphi(a)(\xi_1+\xi_2),(\xi_1+\xi_2))$$

and $g_2(a)=(\pi_\varphi(a)(\xi_1-\xi_2),(\xi_1-\xi_2))$ $(a\in\mathscr{A})$. Then g_1, g_2 are pure on $\mathscr{A}$ and $g_1=g_2$ on $\mathfrak{B}$, so that $g_1=g_2$ on $\mathscr{A}$. Hence $\xi_1+\xi_2=\lambda(\xi_1-\xi_2)$ for some complex number λ ($|\lambda|=1$). This is impossible and so $[\pi_\varphi(\mathscr{A})\eta]=[\pi_\varphi(\mathfrak{B})\eta]$. Hence $p'\in\pi_\varphi(\mathscr{A})'$. Let c be the greatest central projection in $\pi_\varphi(\mathfrak{B})'$ such that $\pi_\varphi(\mathfrak{B})'c$ is a direct sum of type I-factors. Since any non-zero projection in $\pi_\varphi(\mathfrak{B})'c$ is a sum of mutually orthogonal minimal projections, $c\in\pi_\varphi(\mathscr{A})'$. Since $1_\varphi\in c\mathscr{H}_\varphi$, $[\pi_\varphi(\mathscr{A})1_\varphi]\subset c\mathscr{H}_\varphi$ and so $c=1_{\mathscr{H}}$. Hence $\pi_\varphi(\mathfrak{B})'\subset\pi_\varphi(\mathscr{A})'$ and so $\pi_\varphi(\mathfrak{B})''=\pi_\varphi(\mathscr{A})''$.

Since $\varphi_1,\varphi_2\leq 2\varphi$, there exist elements η_1, η_2 in $\mathscr{H}_\varphi$ such that $\varphi_i(a)=(\pi_\varphi(x)\eta_i,\eta_i)$ $(a\in\mathscr{A})$ $(i=1,2)$.

For $a\in\mathscr{A}$, take a directed set $\{\pi_\varphi(b_\alpha)\}$ $(b_\alpha\in\mathfrak{B})$ with $\pi_\varphi(b_\alpha)\to\pi_\varphi(a)$ in the σ-topology. Then $\varphi_1(b_\alpha)=\varphi_2(b_\alpha)$ implies $\varphi_1(a)=\varphi_2(a)$. q.e.d.

4.7.5. Lemma. *Let φ_1, φ_2 be two states on $\mathscr{A}$ and suppose that one of them is atomic and $\varphi_1=\varphi_2$ on $\mathfrak{B}$. Then $\varphi_1=\varphi_2$ on $\mathscr{A}$.*

Proof. Suppose that φ_1 is atomic. Then $\pi_{\varphi_1}(\mathscr{A})'$ is a direct sum of type I-factors. Hence there exists a family of mutually orthogonal minimal projections $(e'_\alpha|a\in\mathbb{I})$ in $\pi_\varphi(\mathscr{A})'$ such that $\sum_{\alpha\in\mathbb{I}} e'_\alpha=1_{\mathscr{H}_{\varphi_1}}$. Thus

$$\varphi_1(a)=\sum_{\alpha\in\mathbb{I}}(\pi_{\varphi_1}(a)e'_\alpha 1_{\varphi_1}, e'_\alpha 1_{\varphi_1})$$

$$=\sum_{\alpha\in\mathbb{I}}\frac{1}{\|e'_\alpha 1_{\varphi_1}\|^2}\left(\pi_{\varphi_1}(a)\frac{e'_\alpha 1_{\varphi_1}}{\|e'_\alpha 1_{\varphi_1}\|}, \frac{e'_\alpha 1_{\varphi_1}}{\|e'_\alpha 1_{\varphi_1}\|}\right)$$

Since $\left(\pi_{\varphi_1}(a)\frac{e'_\alpha 1_{\varphi_1}}{\|e'_\alpha 1_{\varphi_1}\|}, \frac{e'_\alpha 1_{\varphi_1}}{\|e'_\alpha 1_{\varphi_1}\|}\right)$ is pure, its restriction to $\mathfrak{B}$ is again pure (cf. the proof of 4.7.4). Hence $\varphi_1|\mathfrak{B}$ is atomic and so by 4.7.4, $\varphi_1=\varphi_2$ on $\mathscr{A}$. q.e.d.

Now we shall assume that $\mathscr{A}$ is uniformly separable. Let $\{\pi,\mathscr{H}\}$ be a *-representation of $\mathscr{A}$ on a separable Hilbert space $\mathscr{H}$. Put $\mathscr{A}_0=\pi(\mathscr{A})$ and $\mathfrak{B}_0=\pi(\mathfrak{B})$, and let C be a maximal commutative *-subalgebra of $\mathscr{A}'_0$. Then the W^*-subalgebra $R(\mathscr{A}_0,C)$ generated by $\mathscr{A}_0$ and C is of type I, because $R(\mathscr{A}_0,C)'=\mathscr{A}'_0\cap C=C$.

4.7.6. Theorem. *Suppose that* $\mathfrak{B}$ *separates* $\mathscr{E}_{\mathscr{A}} \cup (0)$, *and let* Φ *be a linear mapping of* $\mathscr{A}_0$ *into* $R(\mathscr{A}_0, C)$ *satisfying*

1. $\|\Phi(a)\| \leq \|a\|$ $(a \in \mathscr{A}_0)$;
2. $\Phi(b) = b$ $(b \in \mathfrak{B}_0)$.

Then $\Phi(a) = a$ *for* $a \in \mathscr{A}_0$.

Proof. Suppose that $\Phi(a_0) \neq a_0$ for some $a_0 \in \mathscr{A}_0$. Then there exists a normal state ψ of $R(\mathscr{A}_0, C)$ such that $\psi(\Phi(a_0)) \neq \psi(a_0)$.

By the reduction theory (3.2), $R(\mathscr{A}_0, C) = \sum_{i=1}^{\infty} \oplus L^\infty(\Omega_i, \mu_i, B_i)$, where B_i is a type I_{n_i}-factor $(n_i \leq \aleph_0;\ i = 1, 2, \ldots)$.

Let $\mathscr{D}$ be the C^*-subalgebra of $R(\mathscr{A}_0, C)$ generated by $\mathscr{A}_0$ and $\Phi(a_0)$. Then $\mathscr{D}$ is separable.

Let $\mu = \sum_{i=1}^{\infty} \oplus \mu_i$ be the direct sum of measures $\{\mu_i\}$ on $\sum_{i=1}^{\infty} \Omega_i$ (say Ω). Let $\{x_n\}$ be a sequence of elements in $\mathscr{D}$ which is uniformly dense in $\mathscr{D}$, and let $\mathscr{D}_1$ be a *-subalgebra of $\mathscr{D}$ over the field of complex rational numbers λ (i.e. the real and imaginary parts of λ are rationals) generated by $\{x_n\}$. Then $\mathscr{D}_1$ is countable. Then for $x, y \in \mathscr{D}_1$ with $x = \int_\Omega x(t)$ and $y = \int_\Omega y(t)$, we have: $(x+y)(t) = x(t) + y(t)$, $(xy)(t) = x(t)y(t)$, $x^*(t) = x(t)^*$, $(\lambda x)(t) = \lambda x(t)$ for all complex rationals λ and $\|x(t)\| \leq \|x\|$ for μ-almost all $t \in \Omega$.

Therefore there is a μ-null set Q in Ω such that the above equalities and inequalities are held for all $t \in \Omega - Q$.

For $y \in \mathscr{D}$, there exists a sequence $\{y_n\}$ in $\mathscr{D}_1$ with $\|y - y_n\| \to 0$ $(n \to \infty)$.

Hence $\|y(t) - y_n(t)\| \leq \|y - y_n\|$ $(n \to \infty)$ μ—a.e.

For $t \in \Omega - Q$, $\|y_m(t) - y_n(t)\| \leq \|y_m - y_n\|$. Hence $\{y_n(t)\}$ converges uniformly to an element $y^0(t)$ and so $y = \int y^0(t)$.

One can easily see that $\{y^0(t) \mid t \in \Omega - Q\}$ does not depend on the choice of a sequence $\{y_n\}$ in $\mathscr{D}_1$. Hence we can define uniquely a mapping $y \to y(t)$ of $\mathscr{D}$ into B_i $(t \in \Omega_i - Q\ (i = 1, 2, \ldots))$. One can easily see that the mapping $y \to y(t)$ $(y \in \mathscr{D})$ is a *-homomorphism for all t in $\Omega - Q$.

Since $t \to \mathscr{A}_0(t)''$ is a measurable family, $\overline{\mathscr{A}_0(t)} = R(\mathscr{A}_0, C)(t) = B_i$ for μ-almost all $t \in \Omega$ $(i = 1, 2, \ldots)$ ($\overline{\mathscr{A}_0(t)}$ is the σ-closure of $\mathscr{A}_0(t)$ in B_i).

Hence without loss of generality, we may assume that $\overline{\mathscr{A}_0(t)} = B_i$ for all $t \in \Omega_i - Q$ $(i = 1, 2, \ldots)$.

Let $\psi = \int \psi(t)$. Then $\psi(t)$ is a normal positive linear functional on B_i for μ-almost all t in $\Omega_i - Q$ $(i = 1, 2, \ldots)$. Hence we may assume that ψ is a normal positive linear functional on B_i for all $t \in \Omega_i - Q$ $(i = 1, 2, \ldots)$. Since $\psi(a_0) = \int \psi(t)(a_0(t))\, d\mu(t)$ and $\psi(\Phi(a_0)) = \int \psi(t)(\Phi(a_0)(t))\, d\mu(t)$, $\psi(a_0) \neq \psi(\Phi(a_0))$ implies that there exists a μ-measurable set G, with $\mu(G) > 0$, in Ω such that $\psi(t)(a_0(t)) \neq \psi(t)(\Phi(a_0)(t))$ for all $t \in G$.

Clearly $G \cap (\Omega - Q) \neq (\emptyset)$. Take a $t_0 \in G \cap (\Omega - Q)$ and define a linear functional ψ_1 on $\mathscr{A}$ as follows: $\psi_1(c) = \psi(t_0)(\pi(c)(t_0))$ $(c \in \mathscr{A})$. Then ψ_1 is an atomic state on $\mathscr{A}$, since $\overline{\pi(\mathscr{A})(t_0)} = B_i$ for some i.

Put $a_0 = \pi(c_0)$ for some c_0 in $\mathscr{A}$, and define a linear functional ψ'_2 on $\mathfrak{B} + \mathbb{C} c_0$ as follows: $\psi'_2(d + \lambda c_0) = \psi(t_0)(\pi(d)(t_0) + \lambda \Phi(a_0)(t_0))$ for $d \in \mathfrak{B}$ and a complex number λ.

Then

$$\begin{aligned}|\psi'_2(d+\lambda c_0)| &\leq \|\psi(t_0)\| \, \|\pi(d)(t_0) + \lambda \Phi(a_0)(t_0)\| \\ &\leq \|\psi(t_0)\| \, \|\Phi(\pi(d) + \lambda \pi(c_0))(t_0)\| \\ &\leq \|\psi(t_0)\| \, \|\Phi(\pi(d) + \lambda \pi(c_0))\| \leq \|\psi(t_0)\| \, \|\pi(d) + \lambda \pi(c_0)\| \\ &\leq \|\psi(t_0)\| \, \|d + \lambda c_0\| \,.\end{aligned}$$

Hence ψ'_2 is well defined and is bounded.

Let ψ_2 be a linear functional on $\mathscr{A}$ with $\|\psi_2\| = \|\psi'_2\|$, and $\psi_2 = \psi'_2$ on $\mathfrak{B} + \mathbb{C} c_0$.

Since $\psi_2(1) = \psi'_2(1) = \|\psi(t_0)\|$, ψ_2 is positive, and clearly $\psi_1 = \psi_2$ on $\mathfrak{B}$. Hence by 4.7.5, $\psi_1 = \psi_2$ on $\mathscr{A}$, so that

$$\psi_1(c_0) = \psi(t_0)(\pi(c_0)(t_0)) = \psi(t_0)(a_0(t_0)) = \psi_2(c_0) = \psi(t_0)(\Phi(c_0)(t_0)),$$

a contradiction. q.e.d.

4.7.7. Definition. *Let $\mathscr{A}$ be a C^*-algebra. $\mathscr{A}$ is said to be amenable if there exists an increasing directed set $\{\mathscr{A}_\alpha\}_{\alpha \in \mathbb{I}}$ of type* I *C^*-subalgebras in $\mathscr{A}$ such that the uniform closure of $\bigcup_{\alpha \in \mathbb{I}} \mathscr{A}_\alpha$ is $\mathscr{A}$.*

Remark. Every type I C^*-algebra is obviously amenable, and every uniformly hyperfinite C^*-algebra is also amenable.

4.7.8. Corollary. *Let $\mathscr{A}$ be a separable C^*-algebra, $\mathfrak{B}$ an amenable C^*-subalgebra of $\mathscr{A}$. If $\mathfrak{B}$ separates $\mathscr{E}_{\mathscr{A}} \cup (0)$, then $\mathfrak{B} = \mathscr{A}$.*

Proof. It suffices to assume that $\mathscr{A}$ has an identity and $\mathfrak{B}$ contains it.

Suppose that $\mathfrak{B} \subsetneqq \mathscr{A}$. Take a self-adjoint element f, with $\|f\| = 1$, in the polar $\mathfrak{B}^0$ of $\mathfrak{B}$ in $\mathscr{A}^*$, and let $f = f^+ = f^-$ be the orthogonal decomposition of f. Put $\varphi = f^+ + f^-$. Then clearly $\pi_\varphi(\mathfrak{B})'' \subsetneqq \pi_\varphi(\mathscr{A})''$.

Put $\mathscr{A}_0 = \pi_\varphi(\mathscr{A})$ and $\mathfrak{B}_0 = \pi_\varphi(\mathfrak{B})$.

Since $\mathfrak{B}$ is amenable, there exists an increasing directed set $\{\mathfrak{B}_\alpha\}_{\alpha \in \mathbb{I}}$ of type I C^*-algebras in $\mathfrak{B}$ such that the uniform closure of $\bigcup_{\alpha \in \mathbb{I}} \mathfrak{B}_\alpha$ is $\mathfrak{B}$.

Without loss of generality, we may assume that $1 \in \mathfrak{B}_\alpha$ for all $\alpha \in \mathbb{I}$.

Since $\pi_\varphi(\mathfrak{B}_\alpha)''$ is a type I W^*-algebra, $\pi_\varphi(\mathfrak{B}_\alpha)'$ is a type I W^*-algebra. Hence by 4.4.19, there exists a conditional expectation P_α of $B(\mathscr{H}_\varphi)$ onto $\pi_\varphi(\mathfrak{B}_\alpha)''$.

Let $B(B(\mathscr{H}_\varphi))$ be the algebra of all bounded linear operators on $B(\mathscr{H}_\varphi)$. Then $B(B(\mathscr{H}_\varphi))=(B(\mathscr{H}_\varphi)\otimes_\gamma B(\mathscr{H}_\varphi)_*)^*$.

Take an accumulation point Φ of $\{P_\alpha\}_{\alpha\in\mathbb{I}}$ in the σ-topology of $B(B(\mathscr{H}_\varphi))$.

Then $P_\alpha(b)=b$ $(b\in\pi_\varphi(\mathfrak{B}_\beta))$ for $\alpha\geq\beta$, and so $\Phi(b)=b$ $(b\in\pi_\varphi(\mathfrak{B}_\beta))$ for $\beta\in\mathbb{I}$. Therefore Φ is the identity operator on $\bigcup_{\beta\in\mathbb{I}}\pi_\varphi(\mathfrak{B}_\beta)$.

Since Φ is uniformly continuous, Φ is again the identity operator on the uniform closure of $\bigcup_{\beta\in\mathbb{I}}\pi_\varphi(\mathfrak{B}_\beta)$ $(=\pi_\varphi(\mathfrak{B}))$.

On the other hand, $\Phi(B(\mathscr{H}_\varphi))\subseteq$ the σ-closure of $\bigcup_{\alpha\in\mathbb{I}}P_\alpha(B(\mathscr{H}_\varphi))$ and so $\Phi(B(\mathscr{H}_\varphi))\subset$ the σ-closure of $\bigcup_{\alpha\in\mathbb{I}}\pi_\varphi(\mathfrak{B}_\alpha)''$ (= the σ-closure of $\pi_\varphi(\mathfrak{B})$). Hence by 4.7.6, $\Phi(a)=a$ for $a\in\pi_\varphi(\mathscr{A})$, and so $\pi_\varphi(\mathscr{A})\subset\pi_\varphi(\mathfrak{B})''$, a contradiction. q.e.d.

Remark 1. If $\mathscr{A}$ is a type I C^*-algebra, then any C^*-subalgebra $\mathfrak{B}$ is again a type I C^*-subalgebra (cf. 4.6). Therefore this theorem implies that the Stone-Weierstrass theorem is true for all separable type I C^*-algebras. Kaplansky [96] proved this theorem without the assumption of separability, using a different method.

Remark 2. Let $\mathscr{A}_1$ (resp. $\mathscr{A}_2$) be a separable uniformly hyperfinite C^*-algebra with type $(p_1,p_2,\ldots)$ (resp. $(q_1,q_2,\ldots)$)—i.e. there exists an increasing sequence $\{\mathscr{A}_{1,n}\}$ (resp. $\mathscr{A}_{2,n}$) of type I_{p_n} (resp. I_{q_n})-factors with $p_n<+\infty$ (resp. $q_n<+\infty$) in $\mathscr{A}_1$ (resp. $\mathscr{A}_2$), containing the identity of $\mathscr{A}_1$ (resp. $\mathscr{A}_2$), such that the uniform closure of $\bigcup_{n=1}^{\infty}\mathscr{A}_{1,n}$ (resp. $\bigcup_{n=1}^{\infty}\mathscr{A}_{2,n}$) is $\mathscr{A}_1$ (resp. $\mathscr{A}_2$).

Then Glimm [58] proved that $\mathscr{A}_1$ and $\mathscr{A}_2$ are mutually *-isomorphic if and only if for each positive integer n, there is an integer $m\geq n$ such that p_n is a divisor of q_m and q_n is a divisor of p_m.

Moreover he proved that if $\mathscr{A}_2$ is a C^*-subalgebra, containing the identity of $\mathscr{A}_1$, of $\mathscr{A}_1$, then for each positive integer n there exists an integer $m\geq n$ such that q_n is a divisor of p_m.

One can easily see that conversely if the sequence (q_n) satisfies the above condition, we can find a uniformly hyperfinite C^*-subalgebra $\mathfrak{B}$ of $\mathscr{A}_1$ with type $(q_1,q_2,\ldots)$ such that $\mathfrak{B}_n\subset\mathscr{A}_{1,m}$ $(n=1,2,\ldots)$, where $\{\mathfrak{B}_n\}$ is an increasing sequence of type I_{q_n}-factors in $\mathfrak{B}$ with the uniform closure of $\bigcup_{n=1}^{\infty}\mathfrak{B}_n=\mathfrak{B}$.

The following problem would be interesting: let $\mathfrak{B}$ be a uniformly hyperfinite C^*-subalgebra, with type $(q_1,q_2,\ldots)$, of the C^*-algebra $\mathscr{A}_1$.

Then can we find an increasing sequence $\{\mathscr{D}_n\}$ of type I_{p_n}-factors in $\mathscr{A}_1$ such that $\mathfrak{B}_n \subset \mathscr{D}_n$ and the uniform closure of $\bigcup_{n=1}^{\infty} \mathscr{D}_n$ is $\mathscr{A}_1$.

The above Stone-Weierstrass theorem might be useful to attack this problem.

4.7.9. Corollary. *Let $\mathscr{A}$ be a separable C^*-algebra, $\mathfrak{B}$ a C^*-subalgebra. Suppose that there exists a *-representation $\{\pi, \mathscr{H}\}$ of $\mathscr{A}$ such that $\pi(\mathscr{A})''$ is a finite W^*-algebra and $\pi(\mathfrak{B})'' \subsetneqq \pi(\mathscr{A})''$.*

Then $\mathfrak{B}$ can not separate $\mathscr{E}_{\mathscr{A}} \cup (0)$.

Proof. It suffices to assume that $\dim(\mathscr{H}) \leq \aleph_0$. Since $\pi(\mathscr{A})''$ is finite, by 4.4.23, there exists a conditional expectation Q of $\pi(\mathscr{A})''$ onto $\pi(\mathfrak{B})''$. Then if $\mathfrak{B}$ separates $\mathscr{E}_{\mathscr{A}} \cup (0)$, by 4.7.6, $Q(\pi(\mathscr{A})) = \pi(\mathscr{A}) \subseteqq \pi(\mathfrak{B})''$, a contradiction. q.e.d.

4.7.10. Corollary. *Let G be a countable discrete group, H a proper subgroup of G. Then there exist two elementary positive definite functions φ_1, φ_2 on G such that $\varphi_1 = \varphi_2$ on H, but $\varphi_1 \neq \varphi_2$ on G.*

Proof. Consider the W^*-algebras $U(G)$ and $U(H)$. Then $U(H) \subsetneqq U(G)$. q.e.d.

Remark 1. The following problem would be interesting. Let G be a locally compact group, H a proper closed subgroup of G. Then do there exist two elementary positive definite continuous functions on G such that $\varphi_1 = \varphi_2$ on H, but $\varphi_1 \neq \varphi_2$ on G? (cf. [62]).

Remark 2. It would be worthwhile to attach a weaker form of Stone-Weierstrass theorem as follows.

Let $\mathfrak{F}_{\mathscr{A}}$ be the set of all factorial states on a C^*-algebra $\mathscr{A}$. Suppose that a C^*-subalgebra $\mathfrak{B}$ of $\mathscr{A}$ separates $\mathfrak{F}_{\mathscr{A}} \cup (0)$. Then can we conclude that $\mathscr{A} = \mathfrak{B}$?

As a related problem to this, the following problem would also be interesting.

Let $\mathfrak{B}$ be a C^*-subalgebra of a C^*-algebra $\mathscr{A}$, and let φ be a factorial state on $\mathfrak{B}$. Then can we extend the φ to a factorial state on $\mathscr{A}$?

Concluding remarks on 4.7.

There are other results about the non-commutative extension of the Stone-Weierstrass theorem (cf. [2], [59], [142]).

Generally speaking, the theory of C^*-algebras is very far from the completion, comparing with the theory of W^*-algebras.

The following problem is one of outstanding problems in the theory of C^*-algebras (cf. [99]).

Is there an infinite-dimensional simple C^*-algebra which does not contain any non-trivial projection?

References. [172].

Bibliography

1. Akemann, C. A.: The dual space of an operator algebra. Trans. Amer. Math. Soc. **126,** 286—302 (1967).
2. — The Stone-Weierstrass problem of C^*-algebras. J. Functional Analysis **4,** 277—294 (1969).
3. Araki, H.: von Neumann algebras of local observables for free scalar field, J. Math. phys. **5,** 1—13 (1964).
4. — Type of von Neumann algebra associated with free field. Progr. Theoret. Phys. **32,** 956—965 (1964).
5. — Einführung in die axiomatische Quantenfeldtheorie. E. T. H. Lecture notes, Zürich 1961/62.
6. — Woods, E. J.: A classification of factors. Publ. Res. Instit. Math. Sci. **4,** 51—130 (1968).
7. Arens, R.: The adjoint of a bilinear operation. Proc. Amer. Math. Soc. **2,** 839—848 (1951).
8. Banach, S.: Théorie des Opérations Linéaires. Warsaw 1932.
9. Pallu de la Barrière, R.: Sur les algèbres d'opérateurs dans les espaces hilbertiens, Bull. Soc. Math. France, **82,** 1—51 (1954).
10. Bishop, E., de Leeuw, K.: The representation of linear functionals by means on the set of extreme points. Ann. Inst. Fourier (Grenoble) **9,** 305—331 (1959).
11. Blattner, R. J.: Automorphic group representations. Pacific J. Math. **8,** 665—677 (1958).
12. Bohnenblust, H. F., Karlin, S.: Geometrical properties of the unit sphere of Banach algebras. Ann. of Math. **62,** 217—229 (1955).
13. Borchers, H. J.: Energy and momentum as obserbables in quantum field theory. Comm. Math. Phys. **2,** 49—54 (1966).
14. Bourbaki, N.: Topologie générale, Chap. IX, 2ᵉ éd., Act. Sc. Ind., n° 1045. Paris: Hermann 1958.
15. — Topologie générale. Chap. X, 2ᵉ éd., Act. Sc. Ind., n° 1084. Paris: Hermann 1961.
16. — Espaces vectoriels topologiques. Chap. I—II, Act. Sc. Ind., n° 1189. Paris: Hermann 1953.
17. — Espaces vectoriels topologiques. Chap. III—IV—V, Act. Sc. Ind., n° 1229. Paris: Hermann 1955.
18. — Intégration. Chap. VII—VIII, Act. Sc. Ind., n° 1306. Paris: Hermann 1963.
19. Bures, D. J. C.: Certain factors constructed as infinite tensor products. Compositio Math. **15,** 169—191 (1963).
20. Calkin, J. W.: Two sided ideals and congruences in the ring of bounded operators in Hilbert space. Ann. of Math. **42,** 839—873 (1941).
21. Ching, Wai-Mee: Non-isomorphic non-hyperfinite factors. Can. J. Math. **21,** 1293—1308 (1969).

22. Choquet, G.: Existence de représentations intégrales au moyen des points extrémaux dans les cônes convexes. C. R. Acad. Sci., Paris **243,** 699—702 (1956).
23. — Meyer, P. A.: Existence et unicité des représentations intégrales dans les convexes compacts quelconques. Ann. Inst. Fourier (Grenoble) **13,** 139—154 (1963).
24. Cordes, H.: To appear in Ergebnisse der Mathematik.
25. Day, M. M.: Amenable semigroup. Illinois J. Math. **1,** 509—544 (1957).
26. — Normed linear Spaces. New York: Academic Press, Inc., 1962.
27. Dell'Antonio, G. F.: On the limits of sequences of normal states. Comm. Pure Appl. Math. **20,** 413—429 (1967).
28. Diedonne, J.: Recent developments in the theory of locally convex spaces. Bull. Amer. Math. Soc. **59,** 495—512 (1953).
29. Dixmier, J.: Les anneaux d'opérateurs de classe finie. Ann. Sci. École Norm. Sup. **66,** 209—261 (1949).
30. — Les fonctionelles linéaires sur l'ensemble des opérateurs bornés d'un espace de Hilbert. Ann. of Math. **51,** 387—408 (1950).
31. — Sur certains espaces considérés par M. H. Stone. Summa Brasil. Math. **2,** fasc. 11, 151—182 (1951).
32. — Formes linéaires sur un anneau d'opérateurs. Bull. Soc. Math. France **81,** 9—39 (1953).
33. — Sur les C^*-algèbres, Bull. Soc. Math. France **88,** 95—112 (1960).
34. — Sur les representations unitaires des groupes de Lie nilpotent, I. Amer. J. Math. **81,** 160—170 (1959).
35. — Dual et quasi-dual d'une algèbre de Banach involutive. Trans. Amer. Math. Soc. **104,** 278—283 (1962).
36. — Les C^*-algèbres et leurs représentations. Paris: Gauthier-Villars 1964.
37. — Les algèbres d'operateurs l'espace hilbertien, 2e edition. Paris: Gauthier-Villars 1969.
38. — Sur les groupes d'automorphisms normiquement continus des C^*-algèbres. C. R. Acad. Sci. Paris Ser. A. **269,** 643—644 (1969).
39. — Lance, E. C.: Deux nouveaux facteurs de type II_1. Invent. Math. **17,** 226—234 (1969).
40. Doplicher, S., Kadison, R., Kastler, D., Robinson, D.: Asymptotically abelian system. Comm. Math. Phys. **6,** 101—120 (1967).
41. — Kastler, D., Størmer, E.: Invariant states and asymptotic abelianness. J. Functional Analysis **3,** 419—434 (1969).
42. Douglas, R. G., Pearcy, C.: Von Neumann algebras with a single generator. Michigan Math. J. **16,** 21—26 (1969).
43. Dunford, Pettis: Linear operations on summable functions. Trans. Amer. Math. Soc. **47,** 323—392 (1940).
44. Dye, H. A.: The Radon-Nikodym theorem for finite rings of operators. Trans. Amer. Math. Soc. **72,** 243—280 (1952).
45. — On groups of measure preserving transformations, I. Amer. J. Math. **81,** 119—159 (1959).
46. — On groups of measure preserving transformations, II. Amer. J. Math. **85,** 551—576 (1963).
47. Effros, E. G.: The Borel space of von Neumann algebras on a separable Hilbert space. Pacific. J. Math. **15,** 1153—1164 (1965).
48. Elliot, G.: Derivations of Matroid C^*-algebras. Invent. Math. **9,** 253—269 (1970).
49. Automorphisms of post liminal C^*-algebras. To appear in proceeding of Amer. Math. Soc.

50. Feldman, J.: Borel sets of states and of representations. Michigan Math. J. **12,** 363—365 (1965).
51. Fell, J. M. G.: C^*-algebras with smooth dual. Illinois J. Math. **4,** 221—230 (1960).
52. Fukamiya, M.: On a theorem of Gelfand and Neumark and the B^*-algebra. Kumamoto J. Sci. **1,** 17—22 (1952).
53. Gardner, T.: On isomorphisms of C^*-algebras. Amer. J. Math. **87,** 384—396. (1965).
54. Gelfand, I.: On normed rings. Dokl. Akad. Nauk SSSR. **23,** 430—432 (1939).
55. — Neumark, M.: On the imbedding of normed rings into the ring of operators in Hilbert space, Mat. USSR-Sb. **12,** 197—213 (1943).
56. — Raikov, D.: Continuous unitary representations of locally bicompact groups. Math. USSR-Sb. **13,** 301—316 (1943).
57. Glaser, M.: Asymptotic abelianness of infinite factors. To appear.
58. Glimm, J.: On a certain class of operator algebras. Trans. Amer. Math. Soc. **95,** 216—244 (1960).
59. — A Stone-Weierstrass theorem for C^*-algebras. Ann. Math. **72,** 216—244. (1960).
60. — Type I C^*-algebras. Ann. of Math. **73,** 572—612 (1961).
61. — Kadison, R. V.: Unitary operators in C^*-algebras. Pacific J. Math. **10,** 547—556 (1960).
62. Godement, R.: Les fonctions de type positif et la théorie des groupes. Trans. Amer. Math. Soc. **63,** 1—84 (1948).
63. Griffin, E. L.: Some contributions to the theory of rings of operators. Trans. Amer. Math. Soc. **75,** 471—504 (1953).
64. — Some contributions to the theory of rings of operators, II. Trans. Amer. Math. Soc. **79,** 389—400 (1955).
65. Grothendieck, A.: Produits tensoriels topologiques et espaces nucléaires. Mem. Amer. Math. Soc. **16,** (1955).
66. — Un résultat sur le dual d'une C^*-algèbre. J. Math. Pures Appl. **36,** 97—108 (1957).
67. Guichardet, A.: Sur un probleme posé par G. W. Mackey. C. R. Acad. Sci. **250,** 962—963 (1960).
68. — Tensor products of C^*-algebras. Soviet Math. Dokl. **6,** 210—213 (1965).
69. — Produits tensoriels infinies et représentations des relations d'anticommutation. Ann. Sci. École Norm. Sup. **83,** 1—52 (1966).
70. Guichardet, A., Kastler, D.: Des integration des etats Quasi-invariants des C^*-algebres. J. Math. Pures et Appl. **49,** 349—380 (1970).
71. Haag, R., Kastler, D., Michel, L.: Central decomposition of Ergodic states, mimeographed notes.
72. Hakeda, J., Tomiyama, J.: On some extension properties of von Neumann algebras. Tôhoku Math. J. (2) **19,** 315—323 (1967).
73. Halmos, P. R.: Introduction to Hilbert space and the theory of multiplicity. New York: Chelsea 1951.
74. Harish-Chandra: Trans. Amer. Math. Soc. **75,** 185—243 (1953).
75. Hille, E., Phillips, R. S.: Functional analysis and semi-groups. Amer. Math. Soc. Coll. Publ. **31,** Providence 1957.
76. Hugenholtz, N. M.: On the factor type of equilibrium states in quantum statistical mechanics. Comm. Math. Phys. **6,** 189—193 (1967).
77. Johnson, B. E.: The uniqueness of the (complete) norm topology. Bull. Amer. Math. Soc. **73,** 537—539 (1967).
78. — Continuity of Derivations on commutative Algebras. Amer. J. Math. **91,** 1—10 (1969).

79. — Sinclair, A. M.: Continuity of derivations and a problem of Kaplansky. Amer. J. Math. **90,** 1067—1073 (1968).
80. — Ringrose, J. R.: Derivations of operator algebras and discrete group algebras. Bull. London Math. Soc. **1,** 70—74 (1969).
81. Kadison, R. V.: Isometries of operator algebras. Ann. of Math. **54,** 325—338 (1951).
82. — On the additivity of the trace in finite factors. Proc. Nat. Acad. Sci. USA **41,** 385—387 (1955).
83. — Isomorphisms of factors of infinite type. Canad. J. Math. **7,** 322—327 (1955).
84. — Operator algebras with a faithful weakly-closed representation. Ann. of Math. **64,** 175—181 (1956).
85. — Irreducible operator algebras. Proc. Nat. Acad. Sci. USA 273—276 (1957).
86. — Derivation of operator algebras. Ann. of Math. **83,** 280—293 (1966).
87. — Ringrose, J. R.: Derivations of operator group algebras. Amer. J. Math. **88,** 562—576 (1966).
88. — Derivations and automorphisms of operator algebras. Comm. Math. Phys. **4,** 32—63 (1967).
89. — Lance, E. C., Ringrose, J. R.: Derivations and automorphisms of operator algebras, II. J. Functional Analysis **1,** 204—221 (1967).
90. — Ringrose, J. R.: Cohomology of operator algebras I, Type I von Neumann algebras. To appear.
91. Kallman, R. R.: Unitary groups and automorphisms of operator algebras. Amer. J. Math. **91,** 785—806 (1969).
92. Kaplansky, I.: Normed algebras. Duke Math. J. **16,** 399—418 (1949).
93. — Quelques résultats sur les anneaux d'opérateurs. C. R. Acad. Sci. Paris **231,** 485—486 (1950).
94. — Projections in Banach algebras. Ann. of Math. **53,** 235—249 (1951).
95. — A theorem on rings of operators. Pacific J. Math. **1,** 227—232 (1951).
96. — The structure of certain operator algebras. Trans. Amer. Math. Soc. **70,** 219—255 (1951).
97. — Algebras of type I. Ann. of Math. **56,** 460—472 (1952).
98. — Modules over operator algebras. Amer. J. Math. **75,** 839—853 (1953).
99. — Functional analysis, some aspects of analysis and probability, 1—34. New York: John Wiley and Sons 1958.
100. — Rings of operators. New York—Amsterdam: W. A. Benjamin, Inc., 1968.
101. Kelley, J. L., Vaught, R. L.: The positive cone in Banach algebras. Trans. Amer. Math. Soc. **74,** 44—55 (1953).
102. Kirillov, A.: Unitary representations of nilpotent Lie groups. Uspehi. Mat. Nauk **17,** 57—110 (1962).
103. Kovacs, I., Szücs, J.: Ergodic type theorems in von Neumann algebras. Acta. Sci. Math. (Szeged) **27,** 233—246 (1966).
104. Krieger, W.: On constructing non *-isomorphic hyperfinite factors of type III. To appear in J. of Functional Analysis.
105. — On hyperfinite factors and non-singular transformations of a measure space. To appear.
106. —On a class of hyperfinite factors that arise from null-Recurrent Markov chains. To appear.
107. Lance, E. C.: Automorphisms of certain operator algebras. Amer. J. Math. **91,** 160—174 (1967).
108. — Inner automorphisms of UHF algebras. J. London Math. Soc. **43,** 681—688 (1968).

109. — Automorphisms of postliminal C^*-algebras. Pacific J. Math. **23,** 547—555 (1967).
110. Lanford, O., Ruelle, D.: Integral representations of invariant states on a B^*-algebra. J. Mathematical Phys. **8,** 1460—1463 (1967).
111. McDuff, D.: A countable infinity of II_1-factors. Ann. of Math. **90,** 361—371 (1969).
112. — Uncountably many II_1-factors. Ann. of Math. **90,** 372—377 (1969).
113. — Central sequences and the hyperfinite factor. Proc. London Math. Soc. XXI, 443—461 (1970).
114. Mautner, F.: Unitary representations of locally compact groups II. Ann. of Math. **52,** 528—556 (1950).
115. — The completeness of the irreducible unitary representations of a locally compact group. Proc. Nat. Acad. Sci. USA **34,** 52—54 (1948).
116. Misonou, Y.: Unitary equivalence of factors of type III. Proc. Japan Acad. **29,** 482—485 (1953).
117. — On the direct product of W^*-algebras. Tôhoku Math. J. **6,** 189—204 (1954).
118. Moore, C. C.: Invariant measures on product spaces. Fifth Berkeley symp. Math. Stat. Proba. vol. 2, part 2, 447—459.
119. Murray, F. J., von Neumann, J.: On rings of operators. Ann. of Math. **37,** 116—229 (1936).
120. — — On rings of operators II. Trans. Amer. Math. Soc. **41,** 208—248 (1937).
121. — — On rings of operators IV. Ann. of Math. **44,** 716—808 (1943).
122. Nagy, B. Sz.: Spektraldarstellung linearerTransformationen des Hilbertschen Raumes. Egr. der Math. Berlin: Springer 1942.
123. Naimark, M. A.: On a problem of the theory of rings with involution. Uspehi Mat. Nauk **6,** 160—164 (1951).
124. — Normed rings. Moscow 1956.
125. Nakamura, M.: On the direct product of finite factors. Tôhoku Math. J. **6,** 205—207 (1954).
126. von Neumann, J.: Zur Algebra der Functionaloperationen und Theorie der normalen Operatoren. Math. Ann. **102,** 370—427 (1929).
127. — On a certain topology for rings of operators. Ann. of Math. **37,** 111—115 (1936).
128. — On an algebraic generalization of the quantum mechanical formalism 1. Mat. Sb. **1,** 415—484 (1936).
129. — On infiinite direct products. Compisitio Math. **6,** 1—77 (1938).
130. — On rings of Operators III. Ann. of Math. **41,** 94—161 (1940).
131. — On some algebraical properties of operator rings. Ann. of Math. **44,** 709—715 (1943).
132. — On rings of operators, Reduction theory. Ann. of Math. **50,** 401—485 (1949).
133. Niiro, F.: Sur l'unicite de la décomposition d'une trace. Sci. Papers College Gen. Ed. Univ. Tokyo **13,** 159—162 (1963).
134. Okayasu, T.: On the tensor products of C^*-algebras. Tôhoku Math. J. **18,** 325—331 (1966).
135. — A Structure theorem of automorphisms of von Neumann algebras. Tôhoku Math. J. **20,** 199—206 (1968).
136. Ono, T.: Note on a B^*-algebra. J. Math. Soc. Japan **11,** 140—158 (1959).
137. Powers, R.T.: Representations of uniformly hyperfinite algebras and the associated von Neumann rings. Ann. of Math. **86,** 138—171 (1967).
138. Prosser, R.T.: On the ideal structure of operator algebras. Mem. Amer. Math. Soc. **45,** 1963.

139. Pukanszky, L.: Some examples of factors. Publ. Math. Debrecen **4,** 135—156 (1958).
140. Rickart, C. E.: The uniqueness of norm problem in Banach algebras. Ann. of Math. **51,** 615—628 (1950).
141. — General theory of Banach algebras. New York: D. von Nostrand 1960.
142. Ringrose, J.: On subalgebras of a C^*-algebra. Pacific J. Math. **15,** 1377—1382 (1965).
143. Rosenberg, A.: The number of irreducible representations of simple rings with no minimal ideals. Amer. J. Math. **75,** 523—530 (1953).
144. Ruelle, D.: States of physical systems. Comm. Math. Phys. **3,** 133–150 (1966).
145. — Statistical mechanics, Rigorous results. New York: Benjamin 1969.
146. — Integral representations of states on a C^*-algebra. J. Functional Analysis **6,** 116—151 (1970).
147. Saito, T.: On generators of von Neumann algebras. Michigan Math. J. **15,** 1—4 (1968).
148. — Generators of certain von Neumann algebras. Tôhoku Math. J. (2) **20,** 101—105 (1968).
149. Sakai, S.: A characterization of W^*-algebras. Pacific J. Math. **6,** 763—773 (1956).
150. — On the σ-weak topology of W^*-algebras. Proc. Japan Acad. **32,** 329—332 (1956).
151. — On topological properties of W^*-algebras. Proc. Japan Acad. **33,** 439—444 (1957).
152. — On linear functionals of W^*-algebras. Proc. Japan Acad. **34,** 571—574 (1958).
153. — On a conjecture of Kaplansky. Tôhoku Math. J. **12,** 31—33 (1960).
154. — The theory of W^*-algebras. Lecture notes, Yale University 1962.
155. — Weakly compact operators on operator algebras. Pacific J. Math. **14,** 659—664 (1964).
156. — On the reduction theory of von Neumann. Bull. Amer. Math. Soc. **70,** 393—398 (1964).
157. — On topologies of finite W^*-algebras. Illinois J. Math. **9,** 236—241 (1965).
158. — A Radon-Nikodym theorem in W^*-algebras. Bull. Amer. Math. Soc. **71,** 149—151 (1965).
159. — On the central decomposition for positive functionals on C^*-algebras. Trans. Amer. Math. Soc. **118,** 406—419 (1965).
160. — Derivations of W^*-algebras. Ann. of Math. **83,** 273—279 (1966).
161. — A problem of Calkin. Amer. J. Math. **88,** 935—941 (1966).
162. — A characterization of type I C^*-algebras. Bull. Amer. Math. Soc. **72,** 508—512 (1966).
163. — On type I C^*-algebras. Proc. Amer. Math. Soc. **18,** 861—863 (1967).
164. — Derivations of uniformly hyperfinite C^*-algebras. Publ. Res. Inst. Math. Sci. **3,** 167—175 (1967).
165. — Derivations of simple C^*-algebras. J. Functional Analysis **2,** 202—206 (1968).
166. — On the tensor product of W^*-algebras. Amer. J. Math. **90,** 935—941 (1968).
167. — On the hyperfinite II_1-factor. Proc. Amer. Math. Soc. **19,** 589—591 (1968).
168. — Asymptotically abelian II_1-factors. Publ. Res. Inst. Sci. **4,** 299—307 (1968).
169. — On global W^*-algebras. J. Functional Analysis, **3,** 79—84 (1969).

170. — An uncountable number of II_1 and II_∞-factors. J. Functional Analysis **5,** 236—246 (1970).
171. — An uncountable family of non-hyperfinite type III-factors. To appear in the Monterey Symposium of Functional Analysis.
172. — On a Stone-Weierstrass theorem for C^*-algebras. To appear in Tôhoku Math. J.
173. — On global type II_1 W^*-algebras. To appear in J. Functional Analysis.
174. Schatten, R.: A theory of cross spaces. Ann. Math. Studies, n° 26. Princeton: Princeton University Press 1950.
175. Schatz, J.A.: Math. Reviews **14,** 884 (1953).
176. Schwartz, J.: Two finite, non-hyperfinite, non-isomorphic factors. Comm. Pure Appl. Math. **16,** 19—26 (1963).
177. — Non-isomorphism of a pair of factors of type III. Comm. Pure Appl. Math. **16,** 111—120 (1963).
178. — Type II-factors in a central decomposition. Comm. Pure Appl. Math. **16,** 247—252 (1963).
179. — W^*-algebras. New York: Gordon and Breach 1967.
180. Segal, I.E.: Irreducible representations of operator algebras. Bull. Amer. Math. Soc. **53,** 73—88 (1947).
181. — The two-sided ideals in operator algebras. Ann. of Math. **50,** 856—865 (1949).
182. — Equivalence of measure spaces. Amer. J. Math. **73,** 275—313 (1951).
183. — Decomposition of operator algebras, I and II. Mem. Amer. Math. Soc. **9,** 1—67 (1951).
184. — A non-commutative extension of abstract integration. Ann. of Math. **57,** 401—457 (1953).
185. Tensor algebras over Hilbert spaces. Trans. Amer. Math. Soc. **81,** 106—134 (1956).
186. — Tensor algebras over Hilbert spaces, II. Ann. of Math. **63,** 160—175 (1956).
187. — Distributions in Hilbert space and canonical system of operators. Trans. Amer. Math. Soc. **88,** 12—41 (1958).
188. Sherman, S.: The second adjoint of a C^*-algebra. Proc. Intern. Congr. Math., Cambridge **1,** 470 (1950).
189. Singer, I.M.: Automorphisms of finite factors. Amer. J. Math. **77,** 117—133 (1955).
190. Stampfli, J.: The norms of derivations. To appear in Pacific J. Math.
191. Stone, M.H.: Linear transformations in Hilbert space. Amer. Math. Soc. Coll. Publ. **15,** New York 1950.
192. — Canad. J. Math. **1,** 176—186 (1949).
193. Streater, R.F., Wightman, A.S.: PCT, Spin and Statistics. New York: W.A. Benjamin 1964.
194. Størmer, E: Large groups of automorphisms of C^*-algebras. Comm. Math. Phys. **5,** 1—22 (1967).
195. — Types of von Neumann algebras associated with extremal invariant states. Comm. Math. Phys. **6,** 194—204 (1967).
196. — Symmetric states of infinite tensor products of C^*-algebras. J. Functional Analysis **3,** 48—68 (1969).
197. — States and invariant maps of operator algebras. J. Functional Analysis **5,** 44—65 (1970).
198. Suzuki, N.: Crossed products of rings of finite factors. Tôhoku Math. J. **11,** 113—124 (1959).
199. — Saito, T.: On the operators which generate continuous von Neumann algebras. Tôhoku Math. J. **15,** 277—280 (1963).

200. Takeda, Z.: Conjugate spces of operator algebras. Proc. Japan Acad. **28,** 90—95 (1954).
201. — Inductive limit and infinite direct product of operator algebras. Tôhoku Math. J. **7,** 67—86 (1955).
202. Takesaki, M.: On the conjugate space of an operator algebra. Tôhoku Math. J. **10,** 194—203 (1958).
203. — On the singularity of a positive linear functional on operator algebra. Proc. Japan. Acad. **35,** 365—366 (1959).
204. — On the cross-norm of the direct product of C^*-algebras. Tôhoku Math. J. **16,** 111—122 (1964).
205. — Remarks on the reduction theory of von Neumann algebras. Proc. Amer. Math. Soc. **20,** 434—438 (1969).
206. — Tomita's theory of modular Hilbert algebras and its applications. Lecture notes in mathematics **128.** Berlin: Springer.
207. Thoma, E.: Über unitäre Darstellungen abzählbar, diskreter Gruppen. Math. Ann. **153,** 111—138 (1964).
208. Tomita, M.: On rings of operators in non-separable Hilbert spaces. Mem. Fac. Sci. Kyushu Univ. **7,** 129—168 (1953).
209. — Spectral theory of operator algebras II. Math. J. Okayama Univ. **10,** 19—60 (1960).
210. — Quasi-standard von Neumann algebras. To appear.
211. Tomiyama, J.: On the projection of norm one in W^*-algebras. Proc. Japan Acad. **33,** 608—612 (1957).
212. Turumaru, T.: On the direct product of Operator algebras. Tôhoku Math. J. **4,** 242–251 (1952).
213. — On the direct product of operator algebras, II. Tôhoku Math. J. **5,** 1—7 (1953).
214. — On the direct product of operator algebras, IV. Tôhoku Math. J. **8,** 281—285 (1956).
215. Umegaki, H.: Conditional expectations in an operator algebra, I. Tôhoku Math. J. **6,** 177—181 (1954).
216. Vowden, B.J.: A new proof in the spatial theory of von Neumann algebras. J. London Math. Soc. **44,** 429—432 (1969).
217. Wilig, P.: $B(H)$ is very noncommutative. Proc. Amer. Math. Soc. **24,** 204—205 (1970).
218. Wils, W.: Dés intégration central des formes positives sur les C^*-algèbres. C. R. Acad. Sci. Paris. **267,** 810—812 (1968).
219. — Des intégration centrale dans une partie convexe compacte d'un espace localment convexe, C. R. Acad. Sci. Paris **269,** 702—704 (1969).
220. Wogen, W.: On generators for von Neumann algebras. Bull. Amer. Math. Soc. **75,** 95—99 (1969).
221. Wolfsohn, A.: Le produit tensoriel de C^*-algèbres. Bull. Sci. Math. **87,** 13—27 (1963).
222. Ogasawara, T., Yoshinaga, K.: A non-commutative theory of integration for operators. J. Sci. Hiroshima Univ. **18,** 273—299 (1955).
223. Zeller-Meier, G.: Sur les automorphisms des algèbres de Banach. C. R. Acad. Sci. Paris **264,** 1131—1132 (1967).
224. — Deux autres facteurs de type II_1. Invent. Math. **17,** 235—242 (1969).
225. Alfsen, E. M.: Compact convex sets and boundary integrals, to appear in Ergebnisse der Mathematik.
226. Takenouchi, O.: On type classification of factors constructed as infinite tensor products. Publ. Res. Inst. Math. Sci., **4,** 467—482 (1968).
227. Vesterström, J.: Quotient algebras of finite W^*-algebras, to appear.

Subject Index

Abelian projection 85
Absolute value 8, 27
— of linear functional 32
Adjoint 9, 33
— operation 33
Amenable C^*-algebra 240
— group 150, 209
Amplification 102
Anti-isomorphic 113
Associated closed operator to a W^*-algebra 106
Asymptotically abelian C^*-algebra 128
— — W^*-algebra 214
Atomic 237
Automorphism 153
*-automorphism 119

Bicommutant 48
B T theorem 106

C-isomorphism 123
C-measure 122
C^*-algebra 1
C^*-algebra of all compact linear operators 46
C^*-norm 60
C^*-seminorm 220
C^*-subalgebra 2
C^*-tensor product 66
Canonical W^*-infinite tensor product 214
Central decomposition 125
— — of states 125, 146
— homomorphism 147
— measure 125, 147
— projection 25
— sequence 183, 197
— support (envelope) 25
Centrally orthogonal 148

Classification of W^*-algebras 83
Commutant 48
Commutation theorem of von Neumann 48
— — of tensor products 108
Commutative C^*-algebra 3
— W^*-algebra 45
Comparability theorem 79, 80
Complete additivity 30
Composition series 235
Concrete C^*-algebra 33
— W^*-algebra 33
Conditional expectation 101
Condition of Glimm 221
Continuous W^*-algebra 86
Countably decomposable 80
Coupling function 115, 116, 117
Criterion of Types 95
Cross norm 58
Cyclic 50
— vector 50
— projection 116

Decomposition of states 121, 140
Derivation 153
Diagonalizable operators 134
Direct integral of Hilbert spaces 137
— sum 2
Discrete W^*-algebra 86
Disjoint 148

ε-central sequence 184
Equivalence of central sequences 183
— of projections 79
— of representations 51
Ergodic 126, 175
— decomposition 127, 175
— state 120

Extremal decomposition 123
Extreme points 10

Factor 80
Factorial representation 142
— states 125
Faithful family of representations 53
— — of states 31
— representation 41
— state 31
— trace 95
Final projection 3
Finite projection 83
— W^*-algebra 83, 89
— trace 95
Free 175

G-abelian system 126
Glimm's condition 221
Global W^*-algebras 216
Greatest cross norm 58
Group C^*-algebra 217, 220

Hilbert-Schmidt class 35
Homogeneous projection 88
Homomorphism 3
*-homomorphism 4, 5
Hypercentral sequence 213
Hyperfinite factor 204
Hyper-Stonean 46

Ideal 24
Increasing directed set 6, 15
Induction of representation 102
Inductive limit of C^*-algebras 70, 71
— — of states 74
Infinite conjugacy group 182
— product state 75
— projection 83
— tensor product of C^*-algebras 70, 75
Initial projection 3
Inner asymptotic abelianness 215
— derivation 153
Invariant 19
Involution 1, 111
Irreducible *-representation 52
Isomorphism 4
*-isomorphism 4, 5
*-isomorphic 4

Kaplansky's Density Theorem 22
Kernel of *-representation 54

Least cross norm 58
Left support 25
Localizable 45
Locally countably decomposable 117

Mackey topology 19
Measurable 175
— family of Hilbert spaces 138
— — of W^*-algebras 136
Minimal projection 132
$\natural$-operation 93

Non-measurable 175
Norm 1
Normal 3
— linear functional 30
— positive linear functional 28
— trace 95
Nowhere trivial *-representation 51

Orthogonal 31
— decomposition 8, 31
Outer automorphism 166
— derivation 153

Partial isometry 3
Polar decomposition of operators 27, 109
— — of linear functionals 31, 32
Polish space 141
Positive element 7
— linear functional 9
Powers' factors 207
Predual 1
Projection 3
Properly infinite 84
Property P 207
Pure state 53
Purely infinite 83

Quasi-equivalent 54
Quasi-invariant measure 175

Radon-Nikodym theorems in W^*-algebras 75
Reduction theory 131
Regular maximal left ideal 58
*-representation 40, 50
*-representation constructed via a state 40

Representation theorems for C^*-algebras and W^*-algebras 40
Residual 184
Right support 25

Second dual of C^*-algebras 42
Self-adjoint 2, 3, 9, 19
Semi-central measure 148
Semi-finite 84
— trace 95
Separating vector 50
Simple 74
Smooth 221
Spatial isomorphism 111, 118
Spectral resolution 26
Spectrum 3
— space 4
Standard representation 111
State 10
— space 41
Stonean space 6
Stone-Weierstrass theorem for C^*-algebras 236
s-topology 20
s^*-topology 20
Strong operator topology 33
— topology 20
— *-topology 20
Strongest operator topology 33
Strongly residual 184
— residual sequence 185
σ-topology 2, 14
σ-weak operator topology 34
Subrepresentation 51
Sum of representations 51
Support of normal state 31
— of normal trace 97
— of operators 25
— of W^*-representation 54

Tensor product of C^*-algebras 58
— product of W^*-algebra 58
Trace class 36
Traces 95
Tracial decomposition 130
— state 90
Trivial central sequence 213
Type I C^*-algebra 219, 221
— I group 220
— I *-representation 221
— II *-representation 221
— III *-representation 221
— I W^*-algebra 86, 87
— I_n W^*-algebra 88
— II W^*-algebra 86
— II_1 W^*-algebra 86
— II_∞ W^*-algebra 86
— III W^*-algebra 86
— $P\gamma$ 117
— $S\gamma$ 115
— of Tensor products of W^*-algebras 98
— $(p_1, p_2, \ldots)$, 24
τ-topology 21

Uncountable families of factors of type II_1, II_∞, III 183
Uniform topology 2, 33
Uniformly hyperfinite 73
Unitary 3
Universal *-representation 41

Vector state 103

Weak operator topology 34
Weak topology 2, 14
Weakly closed self-adjoint algebra 35
W^*-algebra 1
W^*-algebra generated by a set 18
W^*-algebra on a Hilbert space 103
W^*-homomorphism 140
W^*-infinite tensor product 202
W^*-representation 40
W^*-subalgebra 2
W^*-tensor product 67

List of Symbols

$\mathscr{M}_*$, 1
$(\mathscr{M}_*)^*$, 1
$\sigma(\mathscr{M},\mathscr{M}_*)$, 2
$\mathscr{N}^0$, 2
$\sum_{\alpha\in\mathbb{I}} \oplus \mathscr{A}_\alpha$, 2
$\mathrm{Sp}(a)$, 3
$C(K)$, 4
$C_0(\Omega)$, 4
$\bar{K}_1$, 5
Λ, 6
$f|\mathscr{A}$, 7
$\mathscr{A}^s$, 8
a^+, 8
a^-, 8
$|a|$, 8
f^*, 9
T, 15
$\bar{\mathscr{A}}$, 18
$\langle x,f\rangle$, 19
$L_a f$, 19
$R_a f$, 19
$\tau(\mathscr{M},\mathscr{M}_*)$, 19
α_φ, 20
$s(\mathscr{M},\mathscr{M}_*)$, 20
s, 20
α^*_φ, 20
$s^*(\mathscr{M},\mathscr{M}_*)$, 20
s^*, 20
$\mathbb{C}$, 23
$\mathscr{M}^p$, 24
$l(a)$, 25
$r(a)$, 25
$s(h)$, 25
$c(p)$, 25
$\sum_{\alpha\in\mathbb{I}} e_\alpha$, 30
$s(\varphi)$, 31
f^+, 31
f^-, 31
$|g|$, 32
$B(\mathscr{H})$, 33
H, 35
$T(\mathscr{H})$, 36
φ_a, 37
$Tr(h)$, 36
$\dim(\mathscr{H})$, 38
$\{\pi,\mathscr{H}\}$, 40
$\{\pi_\varphi,\mathscr{H}_\varphi\}$, 40
$\{\pi^w,\mathscr{H}\}$, 40
x_φ, 40
$\mathscr{S}$, 41
$\{U,K\}$, 41
$\mathscr{A}^{**}$, 42
$L^1(\Omega,\mu)$, 45
$L^\infty(\Omega,\mu)$, 45
$C(\mathscr{H})$, 46
$\mathscr{H}_1\otimes\mathscr{H}_2$, 48
$\mathscr{L}'$, 48
$\mathscr{L}''$, 48
$1_{\mathscr{H}_1}$, 48
$[\pi(\mathscr{A})\xi]$, 50
$R(\mathscr{M},\mathscr{N})$, 50
$\sum_{\alpha\in\mathbb{I}} \{\pi_\alpha,\mathscr{H}_\alpha\}$, 51
$\{\pi E', E'\mathscr{H}\}$, 51
Φ^*, 53
Φ^{**}, 53
$s(\pi)$, 54
$E\odot F$, 58
$E\otimes_\beta F$, 58
$C_0(\Omega,F)$, 59
$C_0(\Omega)\otimes_\lambda F$, 59
$L^1(\Omega,\mu,F)$, 59
$L^1(\Omega,\mu)\otimes_\gamma F$, 59
$\varphi\otimes\psi$, 60
$\mathscr{A}\otimes_\alpha\mathfrak{B}$, 60
$\mathscr{A}\otimes_{\alpha_0}\mathfrak{B}$, 66
$\mathscr{A}\otimes\mathfrak{B}$, 66
$\mathscr{S}_{\mathscr{A}}$, 66
α_0^*, 66
$\mathscr{M}\bar{\otimes}\mathscr{N}$, 67
$\mathrm{Card}(\mathbb{I})$, 69
$\bigodot_{\alpha\in\mathbb{I}} \mathscr{A}_\alpha$, 75
$\bigotimes_{\alpha\in\mathbb{I}} A_\alpha$, 75
$\bigotimes_{\alpha\in\mathbb{I}} \varphi_\alpha$, 75
$p\sim q$, 79
$p\precsim q$, 79
$p\prec q$, 79
$\mathscr{M}^u$, 82
$\natural$, 93
$\mathscr{A}^+$, 95
V, 97
$k(\pi_1)$, 102
$\pi\otimes 1_{\mathscr{K}}$, 102
$(e\mathscr{N}e)'$, 103
$(\mathscr{N}e)'$, 103
φ_ξ, 103
$P_{[\mathscr{N}'\xi]}$, 104
$t\hat{\in}\mathscr{N}$, 106
J, 111
$L^2(\Omega,\mu)$, 112
$\hat{a}(\varphi)$, 121
$\Omega(\psi)$, 121
$\mu_1\prec\mu_2$, 123
$\mathscr{E}$, 123
$\mathfrak{F}$, 125
$\{\mathscr{A},G\}$, 125
a^g, 125
$u_\varphi(g)$, 125
$\mathscr{S}_G$, 126
$\mathscr{E}_G$, 127
$\mathfrak{J}$, 130

$\int a(t)\,d\mu(t)$, 134
$\int \mathscr{A}(t)\,d\mu(t)$, 136
$\int \mathscr{H}(t)\,d\mu(t)$, 140
∂C, 140
$\mathbb{R}$, 141
$U(G)$, 182
$\mathcal{N} \overset{\delta}{\subset} \mathcal{M}$, 183
$(\mathcal{N})_1$, 183
$M_n(\mathbb{I}_1)$, 185
$G[\mathbb{I}_1]$, 185
$Q(G,n)$, 186
$Q(G,m,n)$, 186
$Q_i(h,k)$, 188
$Q_i^2[(h,k),n]$, 188
$RQ_i^2[(h,k)(r,s)]$, 189
$\mathbb{P}$, 193
$\psi_{(p_n)}$, 205
$\mathcal{M}_\lambda$, 207
f^a, 209
f_a, 209
$R(G)$, 220

$\lim_{\alpha} \{\mathscr{A}_\alpha; \Phi_{\beta,\alpha} | (\beta,\alpha) \in \mathbb{I} \times \mathbb{I} \text{ and } \beta \geqq \alpha\}$, 71
$\lim_{\alpha} \{\varphi_\alpha; \Phi_{\beta,\alpha} | (\beta,\alpha) \in \mathbb{I} \times \mathbb{I}, \beta \geqq \alpha\}$, 74

Ergebnisse der Mathematik und ihrer Grenzgebiete

---

1. Bachmann: Transfinite Zahlen. DM 42,—; US $ 11.60
2. Miranda: Partial Differential Equations of Elliptic Type. DM 58,—; US $ 16.00
4. Samuel: Méthodes d'Algèbre Abstraite en Géométrie Algébrique. DM 29,—; US $ 8.00
5. Dieudonné: La Géométrie des Groupes Classiques. DM 38,—; US $ 10.50
6. Roth: Algebraic Threefolds with Special Regard to Problems of Rationality. DM 22,—; US $ 6.10
7. Ostmann: Additive Zahlentheorie. 1. Teil: Allgemeine Untersuchungen. DM 42,—; US $ 11.60
8. Wittich: Neuere Untersuchungen über eindeutige analytische Funktionen. DM 31,—; US $ 8.60
11. Ostmann: Additive Zahlentheorie. 2. Teil: Spezielle Zahlenmengen. DM 28,—; US $ 7.70
13. Segre: Some Properties of Differentiable Varieties and Transformations. DM 46,—; US $ 12.70
14. Coxeter/Moser: Generators and Relations for Discrete Groups. DM 35,—; US $ 9.70
15. Zeller/Beckmann: Theorie der Limitierungsverfahren. DM 64,—; US $ 17.60
16. Cesari: Asymptotic Behavior and Stability Problems in Ordinary Differential Equations. 3rd edition in preparation
17. Severi: Il teorema di Riemann-Roch per curve-superficie e varietà questioni collegate. DM 26,—; US $ 7.20
18. Jenkins: Univalent Functions and Conformal Mapping. DM 37,—; US $ 10.20
19. Boas/Buck: Polynomial Expansions of Analytic Functions. DM 18,—; US $ 5.00
20. Bruck: A Survey of Binary Systems. DM 46,—; US $ 12.70
21. Day: Normed Linear Spaces. DM 17,80; US $ 4.90
23. Bergmann: Integral Operators in the Theory of Linear Partial Differential Equations. DM 40,—; US $ 11.00
25. Sikorski: Boolean Algebras. DM 42,—; US $ 11.60
26. Künzi: Quasikonforme Abbildungen. DM 43,—; US $ 11.90
27. Schatten: Norm Ideals of Completely Continuous Operators. DM 26,—; US $ 7.20
28. Noshiro: Cluster Sets. DM 40,—; US $ 11.00
29. Jacobs: Neuere Methoden und Ergebnisse der Ergodentheorie. Vergriffen
30. Beckenbach/Bellmann: Inequalities. DM 38,—; US $ 10.50
31. Wolfowitz: Coding Theorems of Information Theory. DM 30,—; US $ 8.30
32. Constantinescu/Cornea: Ideale Ränder Riemannscher Flächen. DM 75,—; US $ 20.70
33. Conner/Floyd: Differentiable Periodic Maps. DM 29,—; US $ 8.00
34. Mumford: Geometric Invariant Theory. DM 24,—; US $ 6.60
35. Gabriel/Zisman: Calculus of Fractions and Homotopy Theory. DM 42,—; US $ 11.60
36. Putman: Commutation Properties of Hilbert Space Operators and Related Topics. DM 31,—; US $ 8.60
37. Neumann: Varities of Groups. DM 51,—; US $ 14.10
38. Boas: Integrability Theorems for Trigonometric Transforms. DM 20,—; US $ 5.50
39. Sz.-Nagy: Spektraldarstellung linearer Transformationen des Hilbertschen Raumes. DM 20,—; US $ 5.50
40. Seligman: Modular Lie Algebras. DM 43,—; US $ 11.90
41. Deuring: Algebren. DM 26,—; US $ 7.20
42. Schütte: Vollständige Systeme modaler und intuitonistischer Logik. DM 26,—; US $ 7.20
43. Smullyan: First-Order Logic. DM 36,—; US $ 9.90

44. Dembowski: Finite Geometries. DM 68,—; US $ 17.00
45. Linnik: Ergodic Properties of Algebraic Fields. DM 44,—; US $ 12.10
46. Krull: Idealtheorie. DM 28,—; US $ 7.70
47. Nachbin: Topology on Spaces of Holomorphic Mappings. DM 18,—; US $ 5.00
48. A. Ionescu Tulcea/C. Ionescu Tulcea: Topics in the Theory of Lifting. DM 36,—; US $ 9.90
49. Hayes/Pauc: Derivation and Martingales. DM,—; US $ 13.20
50. Kahane: Séries de Fourier Absolument Convergents. DM 44,—; US $ 12.10
51. Behnke/Thullen: Theorie der Funktionen mehrerer komplexer Veränderlichen. DM 48,—; US $ 13.20
52. Wilf: Finite Sections of Some Classical Inequalities. DM 28,—; US $ 7.70
53. Ramis: Sous-ensembles analytiques d'une variété banachique complexe. DM 36,—; US $ 9.90
54. Busemann: Recent Synthetic Differential Geometry. DM 32,—; US $ 8.80
55. Walter: Differential and Integral Inequalities. DM 74,—; US $ 20.40
56. Monna: Analyse non-archimédienne. DM 38,—; US $ 10.50
57. Alfsen: Compact Convex Sets and Boundary Integrals. DM 46,—; US $ 12.70
58. Greco/Salmon: Topics in $\mathfrak{m}$-Adic Topologies. DM 24,—; US $ 6.60
59. López de Medrano: Involutions on Manifolds. DM 36,—; US $ 9.90
60. Sakai: C*-Algebras and W*-Algebras. DM 66,—; US $ 18.20
61. Zariski: Algebraic Surfaces. DM 54,—; US $ 14.90